Springer-Verlag Berlin Heidelberg GmbH

Progress in Colloid and Polymer Science

Editors: F. Kremer, Leipzig and G. Lagaly, Kiel

Volume 123 · 2004

Trends in Colloid and Interface Science XVI

Volume Editors:

M. Miguel

H. D. Burrows

Springer

The series Progress in Colloid and Polymer Science is also available electronically (ISSN 1437-8027)

- Access to tables of contents and abstracts is *free* for everybody.
- Scientists affiliated with departments/institutes subscribing to Progress in Colloid and Polymer Science as a whole also have full access to all papers in PDF form. Point your librarian to the LINK access registration form at http://springeronline.com

ISSN 0340-255X
ISBN 978-3-662-14580-7
ISBN 978-3-540-36462-7 (eBook)
DOI 10.1007/978-3-540-36462-7

http://www.springeronline.com

Originally published Springer-Verlag Berlin Heidelberg New York in 2004
Softcover reprint of the hardcover 1st edition 2004

Typesetting: SPS, Chennai, India

Cover: Estudio Calamar,
F. Steinen-Broo, Pau/Girona, Spain

Printed on acid-free paper

Progr Colloid Polym Sci (2004) 123: V

PREFACE

The fifteenth meeting of the European Colloid and Interface Society (ECIS) was held in the historic university city of Coimbra from 16th to 21st September 2001. This follows in the tradition of these annual meetings, which started in Como, Italy, in 1987. From the beginning these were intended as interdisciplinary meetings, with participation from chemists, physicists, life and materials scientists, both from academia and industry. The 15th meeting followed this tradition. There was a broad scientific programme, with sessions on *Self Assembly in Mixed Systems*, *Surface Modification*, *Biological and Biomimetic Systems*, *Theory and Modelling*, *New Techniques and Developments*, *Food and Pharmaceuticals*, *Dynamics at Interfaces* and *Mesoscopic and Mesoporous Systems*. In spite of the shadow of the tragic events of September 11th, the meeting attracted 340 participants from 34 countries. It was especially gratifying that our aim to encourage participation of younger scientists succeeded.

The meeting had a very strong scientific programme. We were particularly pleased to be host to the first *Rhodia* Colloid Prize Lecture, which was presented by Professor Kåre Larsson of the University of Lund, Sweden. In addition there were 16 invited lectures, 62 oral presentations and 184 posters.

This special issue of *Progress in Colloid and Polymer Science* contains a selection of the contributions, all of which have been peer reviewed. We take this opportunity to thank all the colleagues who accepted to review these manuscripts. We also thank all the members of the scientific committee, the local organising committee and the sponsors who helped to make ECIS 2001 such a memorable meeting. Finally, we hope that you will enjoy reading the contributions to this special issue, which we feel highlights some of the important contemporary advances in the area of colloid and interface science.

Maria da Graça Miguel
Hugh D. Burrows

Progr Colloid Polym Sci (2004) 123: VI–IX

CONTENTS

Food and Pharmaceuticals

Progr Colloid Polym Sci (2004) 123 : 1–4
DOI 10.1007/b11608

G. Hungerford
M.E.C.D. Real Oliveira
E.M.S. Castanheira
H.D. Burrows
M. da G. Miguel

Transitions in ternary surfactant/alkane/water microemulsions as viewed by fluorescence

G. Hungerford (✉)
M.E.C.D. Real Oliveira
E.M.S. Castanheira
Departamento de Física,
Universidade do Minho,
4710-057 Braga, Portugal
e-mail: graham@fisica.uminho.pt
Fax: +351-253-678981

H.D. Burrows · M. da G. Miguel
Departamento de Química,
Universidade de Coimbra,
3004-535 Coimbra, Portugal

Abstract Fluorescent probes incorporated in ternary surfactant systems have proved valuable in elucidating structure, dynamics and phase behaviour. The intriguing case of microemulsions using non-ionic surfactants, such as alkyloligoethylene oxides, can form phases simultaneously bicontinuous in oil and water. Fluorescence analysis provides both sensitivity and selectivity to monitor these systems. We have used pyrene and rhodamine 6G as probes to enrich our knowledge of the $C_{12}E_5$/alkane/water system, with particular relevance to the bicontinuous phase. Pyrene has been used as a dynamic probe, studying both excimer formation and quenching by molecular oxygen. This provides a useful tool to monitor transitions between microemulsion phases. Both steady state and time-resolved fluorescence measurements indicate a change in localisation on passing from one phase to another. Information was also obtained on microviscosities in these systems using fluorescence depolarisation making use of the well-known laser dye rhodamine 6G. Comparison of the fluorescence characteristics of these systems provides a means to monitor at the microscopic level changes in phase behaviour.

Keywords Anisotropy · $C_{12}E_5$ · Fluorescence · Pyrene · Rhodamine 6G

Introduction

The physical structures and phases created in microemulsions containing the non-ionic poly(oxyethylene) surfactant $C_{12}E_5$ [$C_{12}H_{25}(OCH_2CH_2)_5OH$] have generated considerable interest. The phase diagrams for $C_{12}E_5$/water/alkane systems have been extensively studied [1–3]. Particularly valuable information has come from ^{1}H-NMR self-diffusion measurements using pulsed-gradient spin echo techniques [1], where the observation of high diffusion coefficients for both water and oil indicates that neither can be in a confined environment, such as a droplet. An important feature of these systems is their ability, under certain conditions of composition and temperature, to form phases that are bicontinuous in oil and water [4, 5]. This is of both practical and theoretical importance and although experimental evidence exists for these phases [1, 2], ideas on their exact structures are still speculative. We have previously used solvatochromic fluorescence probes to investigate this region [6]. We report an extension of this study using other fluorescent probes.

Fluorescence techniques have proved useful in elucidating the shape of aggregates and the rate and dimensionality of diffusion in microheterogeneous systems [7, 8]. A molecule of particular note for use to probe this type of system is pyrene, as the ratio of the intensity of the first and third bands of its emission spectrum (I_1/I_3) provides a measure of the polarity of the local environment [9, 10]. Its ability to form excimers has also been put to use to follow the transition through the bicontinuous region in a $C_{12}E_5$/water/tetradecane system [11]. The fact that its fluorescence lifetime and quantum yield are sensitive to the presence of oxygen also provides an interesting property for studying dynamic behaviour. We also consider the use of fluorescence depolarisation for studying fluidity in the three microemulsion regions of this system.

Experimental

Samples of the ternary surfactant systems of $C_{12}E_5$/water/tetradecane containing ca. 10^{-5} M pyrene or rhodamine 6G were prepared in the manner described previously [11] to provide samples that were (i) rich in water (*o/w*) present at 34 °C, (ii) bicontinuous in both oil and water (*bic*) present at 45 °C, and (iii) rich in oil (*w/o*), at 57 °C. These correspond to weight fractions [$C_{14}H_{30}$/($H_2O+C_{14}H_{30}$)] of tetradecane of 0.1, 0.45, and 0.9, respectively.

Steady state fluorescence and absorption measurements were performed using Spex Fluorolog and Shimadzu UV-3101PC spectrometers, respectively. The time-resolved fluorescence measurements were performed using a single-photon counting spectrometer equipped with a nanosecond coaxial flashlamp filled with a nitrogen/hydrogen gas mixture for the pyrene measurements and hydrogen for the rhodamine 6G measurements. The detection of the fluorescence, monitored at a right angle to the excitation, was made using a Philips XP2020 photomultiplier. The decays and anisotropy were analysed using software provided by IBH Consultants Ltd. The pre-exponential factors (α_i) are shown normalised to 1 and the errors are taken as 3 standard deviations. The goodness of fit was judged both in terms of a chi-squared (χ^2) value and weighted residuals.

Results and discussion

Pyrene has proved to be a valuable probe molecule as the intensity ratio between the first and third emission peaks can be used to ascertain the polarity of the probes environment and confinement effects can be observed via its ability to form excimers. This coupled with the fact that the excited state lifetime is drastically affected by oxygen quenching makes pyrene a versatile probe molecule. Pyrene's excimer forming properties have been used to probe phase transitions in the ternary system of $C_{12}E_5$/water/tetradecane [11]. The transition through the bicontinuous phase (oil fraction 0.45) is clearly seen by a decrease in the excimer/monomer ratio (I_E/I_M).

In order to take advantage of the other properties (I_1/I_3 ratio and sensitivity to oxygen) preliminary measurements were performed using (ca. 10^{-5} M) pyrene in pure constituent solvents. The outcome is summarised in Table 1. This table shows that the fluorescence decay time for pyrene in pure tetradecane is significantly shorter than that obtained using $C_{12}E_5$ and that the effect of degassing is more pronounced when using tetradecane as the solvent. This can relate to high oxygen solubility in tetradecane [12]. The values obtained for the I_1/I_3 ratio confirm the less polar environment of the oil and a more polar one in $C_{12}E_5$, although this is still much lower than the value of 1.87 found for water [10].

Table 1 Fluorescence decay times for pyrene in pure $C_{12}E_5$ and tetradecane. The excitation wavelength was 340 nm and the emission was 393 nm. The effect of degassing (DG) the sample is also shown along with the I_1/I_3 ratio (range for all temperatures) from the steady state spectrum

Temp [°C]	$C_{12}E_5$			Tetradecane		
	τ/ [ns]	χ^2	I_1/I_3	τ [ns]	χ^2	I_1/I_3
34 (DG)	306.0 ± 3.0	1.06		194.9 ± 0.9	1.06	
34	134.6 ± 0.6	1.07	1.2–1.3	27.5 ± 0.12	1.19	0.4–0.6
45	115.0 ± 0.6	1.07		22.4 ± 0.13	1.26	
57	102.4 ± 0.6	1.13		18.6 ± 0.12	1.18	

Table 2 Fluorescence decay times for pyrene in $C_{12}E_5$/water/tetradecane. The effect of degassing the sample (DG) is also shown along with the I_1/I_3 ratio from the steady state spectrum

Region	Temp [°C]	τ_1 [ns]	α_1	τ_2 [ns]	α_2	χ^2	I_1/I_3
o/w	34	10.6 ± 0.12	0.10	92.0 ± 0.45	0.90	1.18	0.96
DG	34	264.2 ± 0.8	1			1.08	
bic	49	8.6 ± 6.0	0.08	37.0 ± 0.39	0.92	1.12	0.80
DG	49	302.4 ± 1.8	1			1.03	
w/o	57	27.4 ± 21.0	0.72	37.7 ± 4.2	0.28	1.11	0.71
DG	57	231.2 ± 1.2	1			1.09	

A comparative study using the ternary system is presented in Table 2. This shows that the effect of degassing is not only limited to increasing the decay time, but also affects the number of fluorescence components required to give an adequate fit to the decay. In all cases the degassed lifetimes could be fitted to a monoexponential decay model. The recovered decay times are generally less than those obtained for pyrene in pure $C_{12}E_5$, but greater than in pure tetradecane. Given the hydrophobic nature of pyrene, this can relate to its location in the tail region of the surfactant close to the oil phase. Also the values obtained for the I_1/I_3 ratio tend to confirm this fact, although in the bicontinuous region the decay time is about the same as that obtained for pure $C_{12}E_5$.

Further information can be ascertained from the results from the corresponding aerated samples. In all three cases a sum of two exponential components was required to fit the data. A possible explanation involves the pyrene occupying environments with different oxygen concentrations (solubility) and/or quenching dimensionalities [8]. From the overall trends of the lifetimes obtained coupled with the I_1/I_3 ratio it is apparent that the pyrene has a preference for the surfactant tail region.

To see if any other trends were present in the time-resolved data the decays of the aerated samples were analysed globally by linking the decay times. The outcome of this analysis is shown schematically in Fig. 1. In order to fit the data the sum of three exponentials was required. This clearly shows that the decay time of the major fluorescent component changes depending on which region of the phase diagram is observed. The longer-lived decay associated with the *o/w* region relates to pyrene situated closer to the polyoxyethylene head in the swollen micellar structures found in this region. This component becomes negligible (or non-existent) in the other regions. The shorter-lived fluorescence (major component in the *w/o* region) most likely relates to pyrene in bulk tetradecane. The small quantity found in the water rich region can be ascribed to pyrene deep in the micelle interior. In the bicontinuous region the major component expresses a different decay time, which can relate to movement away from the surfactant head, but because of the confinement of the surfactant the less polar environment of pure tetradecane is not achievable, except for approximately 20% of the emission ascribed to pyrene in the oil channels.

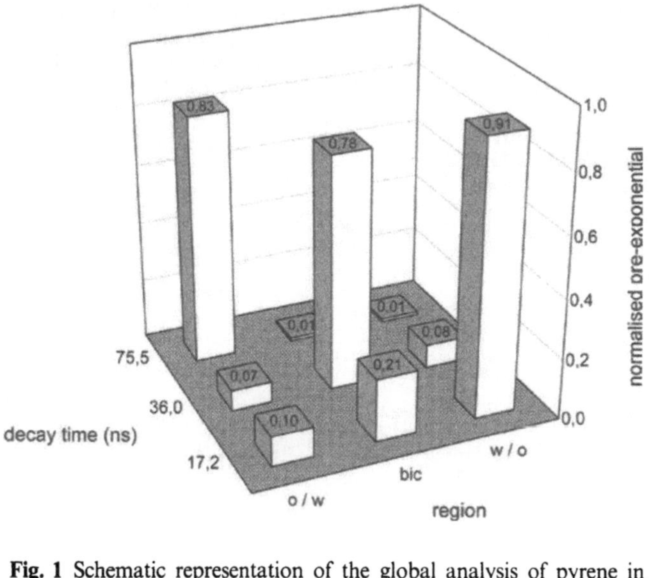

Fig. 1 Schematic representation of the global analysis of pyrene in $C_{12}E_5$/water/tetradecane showing the decay parameters. The global χ^2 is 1.13

Small angle neutron scattering measurements on the water rich region show that this consists of spherical $C_{12}E_5$ micelles swollen with oil [13]. It is reasonable to assume that normal three-dimensional quenching behaviour is observed in this system, such that the decay of excited pyrene can be represented by

$$^1Py^* - k_{fl} \rightarrow Py + h\nu$$

$$^1Py^* + O_2 - k_q \rightarrow Py + O_2$$

where k_{fl} contains both radiative and non-radiative components. Using the observed value for the fluorescence decay for the degassed system, the lifetime of the dominant component in the aerated *o/w* microemulsion (92.0 ns) and the quenching rate constant, taken as the rate of pyrene excimer formation in this phase [11], an oxygen concentration of 2.0 mM is estimated for the region where the pyrene is localised. This value is physically realistic, and lies between the oxygen solubility in aerated solutions of the ether tetrahydrofuran (2.1 mM) and the alkane dodecane (1.7 mM [12]). In contrast, using a similar treatment to calculate the oxygen concentrations in the other two microemulsion phases gives values 10.3 mM (*bic*) and 12.4 mM (*w/o*), which seem unrealistically high. A likely explanation is that there are differences in the kinetics of quenching of excited pyrene by oxygen and excimer formation. Studies on quenching of pyrene fluorescence by 3,4-dimethylbenzophenone in the L_3 phase of the $C_{12}E_5$ water binary system, which should have a very similar structure of the *w/o* microemulsions in the three component system, suggest the presence of regions where the probe and quencher are gathered together [8]. A similar situation may exist with pyrene excimer formation, whereas, with oxygen quenching three dimensional quenching behaviour may occur.

To provide data concerning the microviscosity of the different regions the well-known laser dye rhodamine 6G was used and the time-resolved anisotropy of this probe was measured in the three regions at different temperatures. Fig. 2 shows the outcome with the

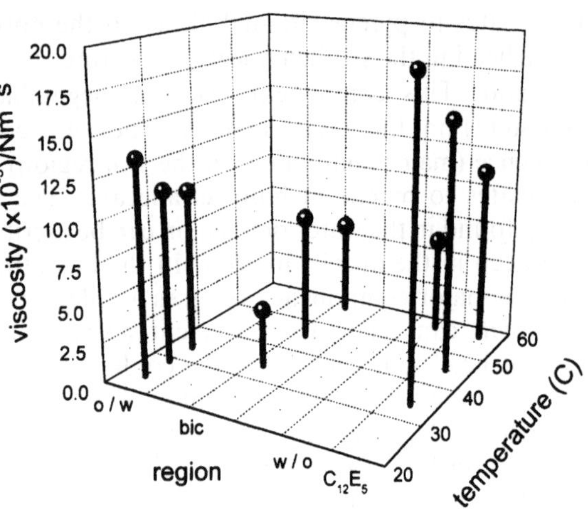

Fig. 2 The viscosity for the different ternary systems, along with that of pure $C_{12}E_5$ at different temperatures, obtained by the fluorescence depolarisation of rhodamine 6G

rotational correlation times (τ_R) converted to viscosities (η), ($\eta = \tau_R\, kT/V$, k is Boltzmann's constant, T absolute temperature and V is the effective volume [14]). The value obtained is probably an average microviscosity as the environment in practice is likely to be anisotropic. Generally there appears to be a decrease in viscosity with increasing the amount of tetradecane in the system. Interestingly for the bicontinuous region there is a change in viscosity with temperature passing from the L + O to the L + W phase via the L phase where the viscosity experienced by the dye increases. Also the viscosities experienced by the rhodamine 6G are never as high as those in pure $C_{12}E_5$. As rhodamine 6G was found to only be sparingly soluble in tetradecane it is most likely situated close to the surfactant head, as the rotational correlation times recovered were longer than could be expected in bulk water. Although not as informative as the pyrene probe, rhodamine 6G is also seen to provide useful information on the dynamics of these systems.

Conclusion

In this work we have shown that by using fluorescence it is possible to monitor the different phases present in $C_{12}E_5$/tetradecane/water systems. From both the steady state fluorescence of concentrated pyrene solutions observing excimer formation and from the global analysis of aerated dilute solutions it is possible to monitor the transition from a phase rich in water to one rich in oil via a phase both continuous in oil and water.

Acknowledgements Financial support from the *Fundação para a Ciência e a Tecnologia* through the *PRAXIS XXI* programme and *Sapiens* (POCTI/35415/QUI/2000) is acknowledged.

References

1. Olsson U, Shinoda K, Lindman B (1986) J Phys Chem 90:4083
2. Lichterfeld F, Schmeling T, Strey R (1986) J Phys Chem 90:5762
3. Leaver MS, Olsson U, Wennerstrom H, Strey R, Wurz U (1995) J Chem Soc Faraday Trans 91:4269
4. Scriven LE (1976) Nature 263:123
5. Olsson U, Wennerstrom H (1994) Adv Colloid Interface Sci 49:113
6. Real Oliveira MECD, Hungerford G, Miguel M da G, Burrows HD (2001) J Molecular Structure 563–564:443
7. Van der Auweraer M, Reekmans S, Boens N, De Schryver FC (1989) Chem Phys 132:91
8. Medhage B, Almgren M, Alsins J (1993) J Phys Chem 97:7753
9. Kalyanasundaram K, Thomas JK (1977) J Am Chem Soc 99:2039
10. Dong DC, Winnik MA (1984) Can J Chem 62:2560
11. Real Oliveira MECD, Hungerford G, Castanheira EMS, Miguel M da G, Burrows HD (2000) J Fluorescence 10:347
12. Murov SL, Carmichael I, Hug GL (1993) Handbook of photochemistry, 2nd edn. Marcel Dekker, New York, pp 289–293
13. Bagger-Jörgensen H, Olsson U, Mortensen K (1997) Langmuir 13:1413
14. Porter G, Sadkowski PJ, Tredwell CJ (1977) Chem Phys Lett 49:416

Progr Colloid Polym Sci (2004) 123: 5–7
DOI 10.1007/b11609

S. Siegel
D. Vollhardt

Phase behaviour and domain structure of 9-hydroxyhexadecanoic acid monolayers

S. Siegel · D. Vollhardt (✉)
Max-Planck-Institut für Kolloid- und Grenzflächenforschung,
14424 Potsdam/Golm, Germany

Abstract The temperature effect on the surface pressure (π)-molecular area (A) isotherms and the texture of the condensed phase domains of 9-hydroxyhexadecanoic acid monolayers are studied. The features of the monolayer are drastically changed by alkyl chain substitution by an OH group in the 9 position. The 9-hydroxyhexadecanoic acid monolayers show unusual temperature behaviour of the π-A isotherms and striking shape changes of the condensed phase domains at different temperatures. The morphological features indicate molecular packing of non-tilted alkyl chains in an orthorhombic and hexagonal lattice, respectively.

Keywords Hydroxyhexadecanoic acid monolayers · Brewster angle microscopy · Surface pressure

Introduction

Monolayers of hydroxy fatty acids are ideal candidates as bipolar model amphiphiles for studying the effect of an attached secondary polar group. Although measurements of the surface pressure (π)-area (A) isotherms of hydroxy fatty acid monolayers have provided some knowledge about the effect of position of the hydroxy groups on the thermodynamic features of the monolayers [1–4] nearly no information is available about the structure and texture of their condensed monolayer phases [5]. It was found that the position of the hydroxy group has a remarkable influence on the phase behaviour of the monolayers [2–4]. Therefore, it has been the objective of our current studies to obtain detailed information on the structure and texture properties of a homologous series of hydroxyhexadecanoic acids wherein the hydroxy group was positioned at the 2, 9 and 16 positions, respectively.

In this work we focus on the temperature dependence of the phase transition and demonstrate the first results of the domain structure of 9-hydroxyhexadecanoic acid monolayers.

Materials and methods

The 9-hydroxyhexadecanoic acid, purchased from Nu_check Prep Inc., Elysian Minnesota, was a gift of Dr. Cadenhead. The substance was dissolved in a 9:1 hexane-ethanol mixture and spread onto a 1 M aqueous NaCl subphase adjusted with HCl to pH 3. Under these conditions the slight solubility of the monolayer material in pure water at higher temperatures can be reduced. The monolayers were investigated at different temperatures using a thermostatted Langmuir film balance coupled with a Brewster angle microscope BAM 1+ (NFT Göttingen, Germany). The distortion caused by the angle of view was corrected by an image processing software. The condensed phase domains grown within the "plateau" region were recorded by BAM.

Results and discussion

The surface pressure (π)-area (A) isotherms show temperature-dependent "plateau" regions of the surface pressure (Fig. 1). Contrary to the most amphiphilic monolayers, 9-hydroxyhexadecanoic acid monolayers exhibit only a slight temperature dependence of the phase transition plateau pressure.

In the plateau region condensed phase domains are formed which were visualised by Brewster angle micros-

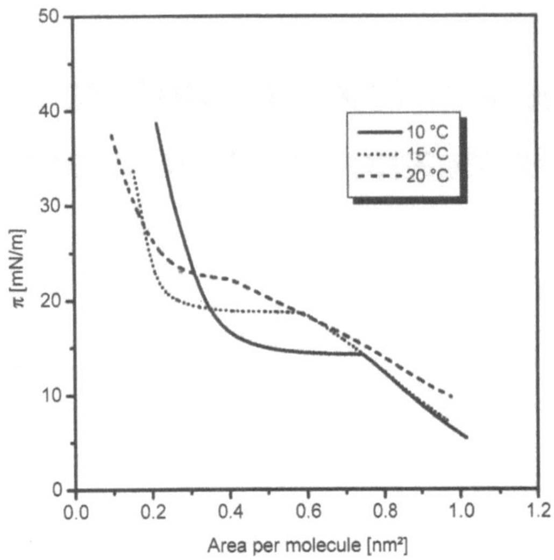

Fig. 1 Surface pressure-area isotherms of 9-hydroxyhexadecanoic acid at different temperatures

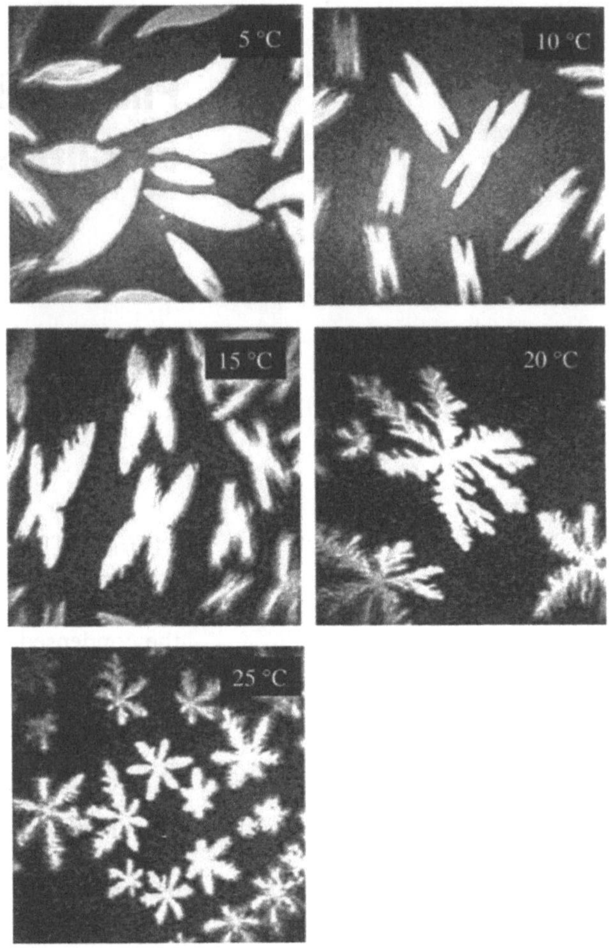

Fig. 2 BAM images of condensed phase domains in 9-hydroxyhexadecanoic acid monolayers at 5, 10, 15, 20, and 25 °C. The image size is 750 × 750 μm

copy (Fig. 2). Homogeneously reflecting domains are visualised for all temperatures indicating that no inner texture exists. However, the domain shape changes remarkably with the temperature. Grain-like domains grow at low temperatures (5 °C). It can be clearly seen that with increasing temperature there is the tendency to develop side arms. At first, four-arm structures are formed in the temperature region between 10 and 15 °C. The angles directly opposite have similar degree values. At 10 °C, two small acute angles and two large obtuse angles are seen, but. with increasing temperature the acute angles between the arms increase and the obtuse angles decrease. The increase of the acute angle between two arms of a four-arm domain with the temperature is shown in Fig. 3. According to a linear fit the acute angles increases in a straight line and can become larger than 60°. In the temperature region 16 °C< T< 25 °C, the development of domains with additional arms can be observed, as demonstrated in Fig. 2 for 20 °C. It can be seen that in this state the angles between the different domain arms are different. Finally, all angles approximate to 60 degrees and realise a six-fold symmetry at 25 °C.

A comparison with non-substituted hexadecanoic (palmitic) acid monolayers reveals that the attached polar OH group in the 9 position completely changes both the phase behaviour and the domain texture.

In palmitic acid monolayers the two phase coexistence region exists already at zero pressure after spreading. Correspondingly, the fluid-like condensed phase domains are irregularly shaped as they are mainly affected by the spreading conditions and prehistory of the monolayer (Fig. 4) [6]. Hence, both the π-A isotherms and the BAM images of 9-hydroxyhexadecanoic acid monolayers are completely different to those of the non-

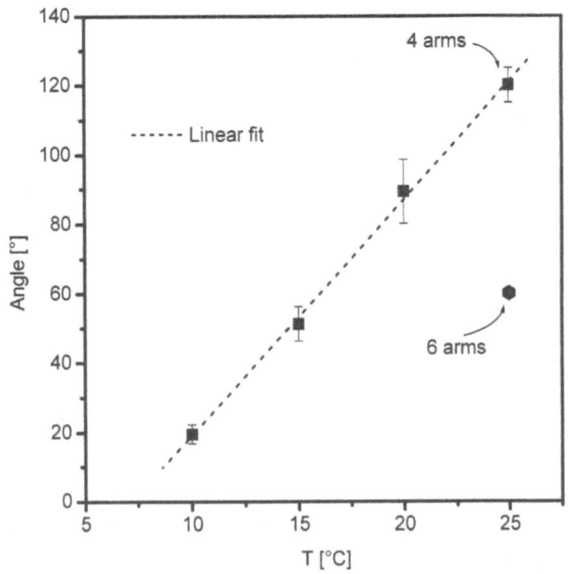

Fig. 3 Temperature dependence of the acute angle between the domain arms

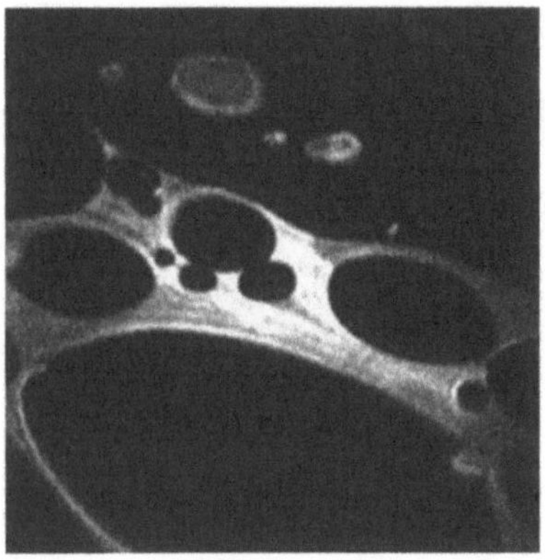

Fig. 4 BAM image of a hexadecanoic acid monolayer (pH 3, 20 °C, $\pi \approx 0$ mN/m)

substituted palmitic acid. The condensed phase domains of the 9-hydroxyhexadecanoic acid reveal a higher crystallinity and striking changes in the morphology with the temperature, The higher crystallinity of these domains can be correlated to the hydrogen bonding capability of the OH groups, see, e.g. [7].

Conclusions

The combination of π-A isotherms and BAM studies provides detailed information on the effect of alkyl chain substitution of fatty acid monolayers. Alkyl chain substitution by an OH group rises the temperature of the phase transition and changes drastically the features of the condensed phase domains. 9-Hydroxyhexadecanoic acid monolayers reveal an unusual temperature behaviour. The temperature effect on the phase transition pressure is comparably small.

The hydrogen bonding capability of the OH groups should be the reason for the higher two-dimensional crystallinity. Remarkable shape changes of the condensed phase domains of 9-hydroxyhexadecanoic acid monolayers are induced by different temperatures.

The morphological features of the 9-hydroxyhexadecanoic acid domains suggest a molecular packing of non-tilted alkyl chains in an orthorhombic and hexagonal lattice, respectively. Details of the molecular packing should be clarified by Synchrotron X-ray diffraction (GIXD) measurements at grazing incidence.

References

1. Tachibana T, Hori K (1977) J Colloid Interface Sci 61:398
2. Kellner BM, Cadenhead DA (1978) J Colloid Interface Sci 63:452
3. Kellner BM, Cadenhead DA (1979) Chem Phys Lipids 23:41
4. Matuo H, Rice DK, Balthasar DM, Cadenhead DA (1982) Chem Phys Lipids 30:367
5. Asgharian B, Cadenhead DA (2000) Langmuir 16:677
6. Gutberlet T, Vollhardt D (1995) J Colloid Interface Sci 173:429
7. Melzer V, Vollhardt D, Weidemann G, Brezesinski G, Wagner R, Möhwald H (1998) Phys Rev E 57:901

Progr Colloid Polym Sci (2004) 123: 8–11
DOI 10.1007/b11610

Marcus Wright
Ken-ichi Kurumada
Brian Robinson

Rates of incorporation of small molecules into pluronic micelles

Work first presented at the 15th ECIS Conference, Coimbra, Portugal, 2001.

M. Wright · K.-i. Kurumada
B. Robinson (✉)
School of Chemical Sciences,
University of East Anglia,
Norwich, Norfolk, NR4 7TJ, UK
e-mail: B.H.Robinson@uea.ac.uk

Present address: K.-i. Kurumada,
Department of Chemical Engineering,
Kyoto University, Japan

Abstract The kinetics of incorporation of the hydrophobic dye 1,6 diphenyl-1,3,5-hexatriene (DPH) into block copolymer micelles (F127) have been studied using fluorescence and spectrophotometric techniques. It appears that this incorporation process takes place over ~1 hour. After fitting the data to a first order transient, the fluorescence kinetics appear to be slower than those studied spectrophotometrically with rate constants changing from 8.4×10^{-4} to 2.4×10^{-4} s^{-1}, respectively, for the block copolymer F127 at a concentration of 1 g/L. For comparison purposes other block copolymers, e.g., F68 and PE6400, have also been investigated using the same methods of DPH incorporation and the kinetics are much quicker than those for F127. When F127 is mixed with bromo-cresol green, BCG, the kinetics are much quicker. It is thought that F127 is more hard-sphere-like in comparison with the other block-copolymers under investigation which may explain why the kinetics are slow. However, it is not really clear why the kinetics vary so much between F127, F68, and PE6400.

Keywords Pluronic · Block copolymer micelles · Dye incorporation · Entrapment · Kinetics

Introduction

Block co-polymers are made up of segments of different hydrophobicity. The block co-polymers under investigation in this study are Pluronic F127, Pluronic F68, and PE6400. Micellar aggregates are formed above a critical micellisation temperature (cmt). For a number of common systems, e.g., F127, the cmt is in the vicinity of room temperature. For F127, the micelle structure is made up of an essentially anhydrous polypropylene oxide core and a hydrated polyethylene oxide coat. Over a wide range of concentration conditions above the cmt, the micelles are consistent with hard-sphere systems.

Micelle dynamic studies have been carried out by a number of groups. For absorption of dyes into normal micelles such as sodium dodecyl sulphate, the kinetics of absorption is a fast process on the milli-second time scale, e.g., for acridine orange incorporation [1]. From temperature-jump studies on block co-polymer micelles, three relaxation processes have been detected, which have been associated with micelle-monomer entrapment, restructuring and micelle-micelle interactions [2]. Solubilisation of a number of small molecules (e.g., toluene, *p*-xylene, and pyrene) into block co-polymer systems has also been studied [3–5].

1,6-Diphenyl-1,3,5-hexatriene (DPH) as shown in Fig. 1 is a probe molecule which has been used to study the fluidity of membranes. The solubility in water is very low so it is normally prepared as a dilute solution in up to 2% methanol in water. DPH shows essentially no fluorescence in water; however, solubilisation in the presence of block co-polymer micelles leads to both a large change in the absorption spectrum

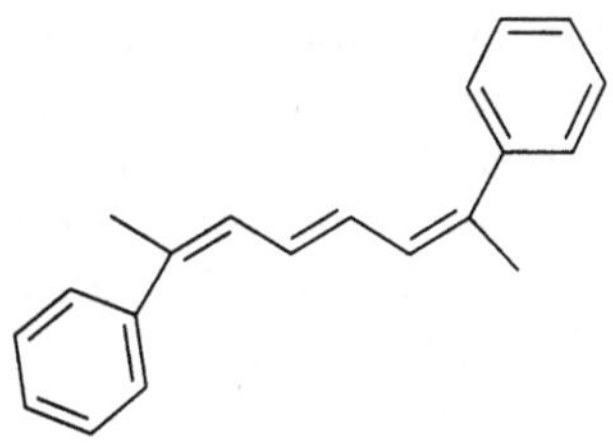

Fig.1 1,6 Diphenyl-1,3,5-hexatriene (DPH)

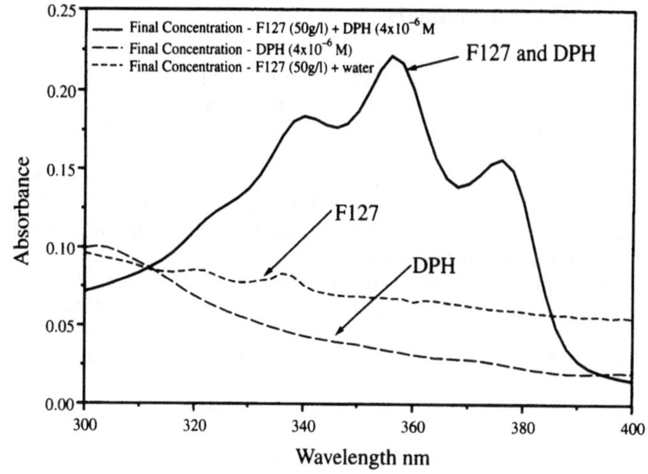

Fig. 2 Visible spectra of DPH, F127 and F127/DPH, T = 35 °C

in the region of 300–400 nm and an increase in fluorescence. The absorption spectra of DPH with and without pluronic F127 are shown in Fig. 2. We have also studied the incorporation of the dye molecule bromo cresol green (BCG) into F127 micelles for comparison purposes.

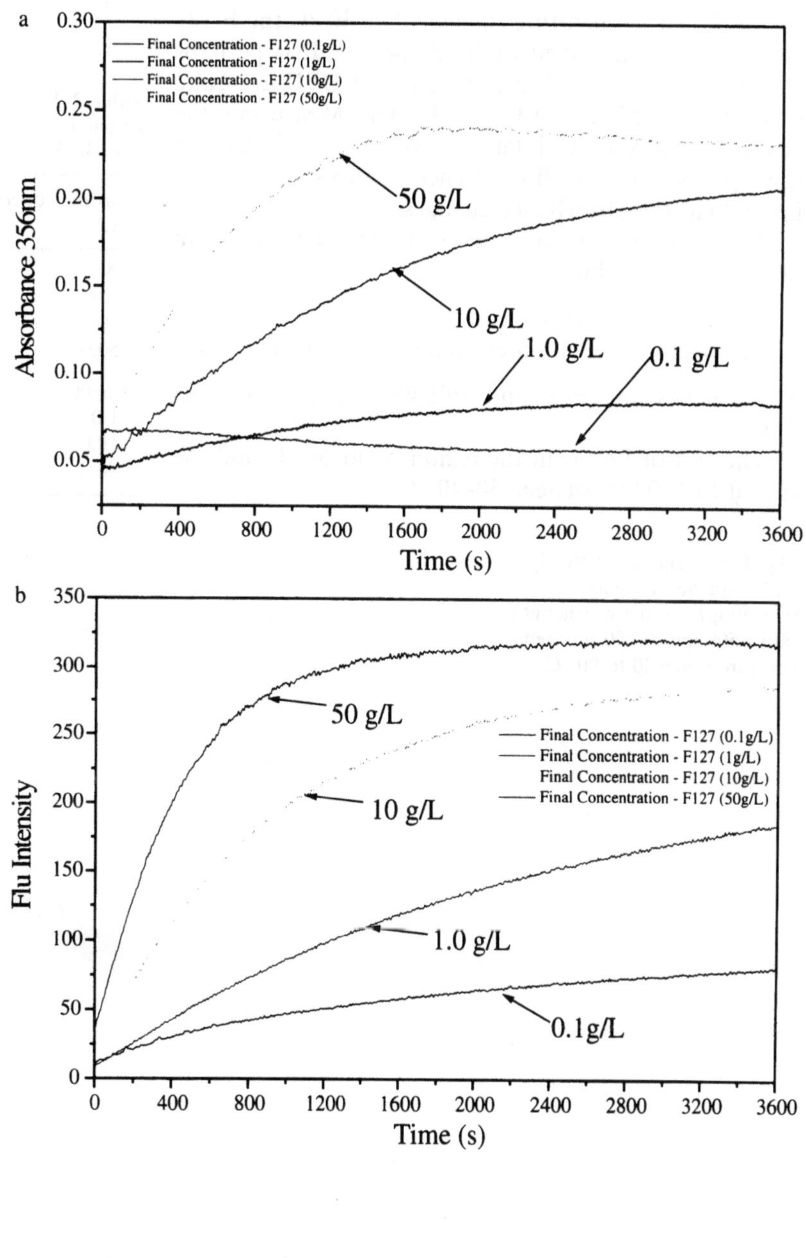

Fig. 3 Kinetic scans of F127/DPH at various concentrations. Wavelength 356 nm, T = 35 °C. **b** Fluorescence kinetic scans of F127/DPH at various concentrations. Emission wavelength 457 nm, excitation wavelength 350 nm, slit width 2.5 nm and T = 35 °C

Materials and methods

Pluronics F127, F68, and DPH were obtained from Sigma, Poole Dorset, UK. PE6400 was a gift from BASF, Cheadle Hulme, Cheshire, UK. BCG was obtained from Fisher Scientific Loughborough, UK.

The pluronics will have some polydispersity associated with the unimer structure which was as follows: F127 13,900 g mol^{-1}, F68 8,000 g mol^{-1}, and PE6400 2,900 g mol^{-1}. DPH and BCG were not subjected to further purification.

Absorption spectra were recorded on a Hewlett Packard HP8452A diode array spectrophotometer with a resolution of 2 nm. Fluorescence measurements were made using a Perkin Elmer Luminescence Spectrometer; thermostatting was to ± 0.5 °C using a Haake water bath.

Results and discussion

Fig. 3 shows results for DPH entrapment into F127 micelles above the cmt of F127 which is in the region 20–30 °C. At temperatures greater than 30 °C the F 127 system is well described by hard spheres. It is perhaps surprising that rates of incorporation of DPH are quite slow, taking of the order of 1 hour. Measuring the change in fluorescence intensity gives results of a similar time period but the fluorescence changes are always faster than the absorbance changes.

Analysing the data as a first order transient, the data in Table 1 were obtained.

For absorption: DPH concentration = 4×10^{-6} mol dm^{-3}.
For fluorescence: DPH concentration = 2.5×10^{-7} mol dm^{-3}.

The kinetics for F68 and PE6400 are presented in Table 2.

The cmt of F68 is in the region of 40–50 °C and the cmt of PE6400 is between 30–40 °C.

Comparing the results for F68 with those in Table 1, the rate of DPH entrapment appears quicker although the time for complete incorporation is still ~1 hour. It appears that the rates of incorporation for PE6400 are faster than those of F68 and F127. From the data the rates are in the following order:

PE6400>F68>F127

Fast ⟷ Slow

Table 1 Kinetic data for DPH incorporation into the block co-copolymer F127 as obtained from fluorescence and spectrophotometric techniques

Concentration F127 [g/L]	k_{Spec} [s^{-1}]	k_F [s^{-1}]
0.1	–	–
1.0	2.4×10^{-4}	8.4×10^{-4}
10.0	8.0×10^{-4}	1.2×10^{-4}
50.0	1.5×10^{-3}	2.1×10^{-3}

Table 2 Kinetic data for dye incorporation of DPH into F68 and PE6400 Block Copolymer Micelles; Temp = 70 °C (F68); 50 °C (PE6400)

Pluronic concentration [g/L]	k_{Spec} [s^{-1}]	k_{FL} [s^{-1}]
F68		
1.0	7.3×10^{-3}	2.1×10^{-3}
10.0	5.5×10^{-3}	5.4×10^{-3}
50.0	1.3×10^{-3}	2.5×10^{-2}
PE6400		
1.0	1.3×10^{-3}	1.7×10^{-3}
10.0	5.1×10^{-3}	5.1×10^{-3}
50.0	2.9×10^{-2}	7.8×10^{-3}

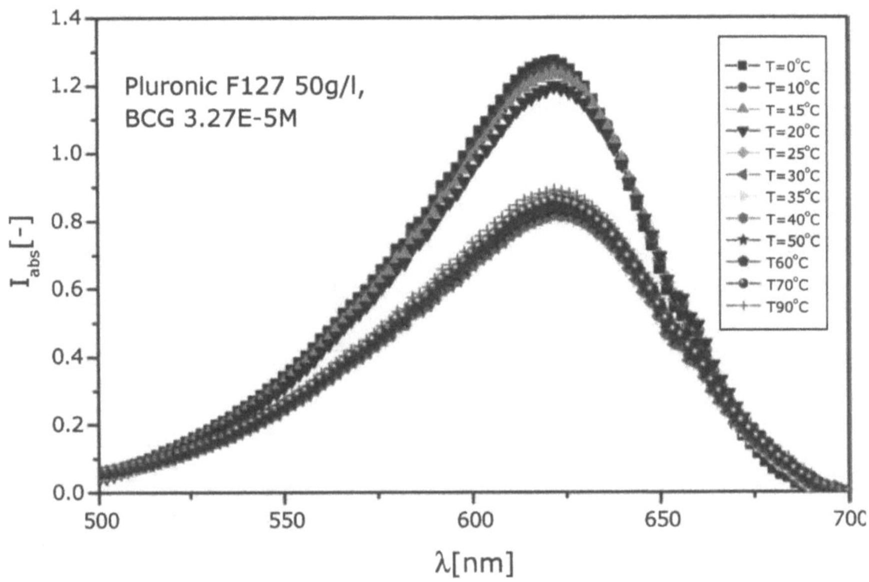

Fig. 4 Absorbance of BCG/F127 aqueous systems (C = 50 g/L) in the wavelength range 500 nm to 700 nm; temperature range 20 to 90 °C

For BCG, the situation is very different. The change in absorption spectrum as the temperature is increased through the cmt region is shown in Fig. 4. When the kinetics are studied, the change is very fast – in the msec time range.

The reasons for the very large difference in the rates are not clear at present but experiments with a number of other dyes are being performed in an attempt to clarify the situation.

References

1. Robinson BH, White NC, Mateo C (1975) Adv Mol Relax Proc 7:321
2. Kositza MJ, Bohne C, Holzwarth JF, et al. (1999) Macromolecules 32:5539–5551
3. Hurter PN, Hatton TA (1992) Langmuir 8:1291–1299
4. Gadelle F, Koros WJ, Schechter RS (1995) Macromolecules 28:4883–4892
5. Xing LF, Mattice WL (1997) Macromolecules 30:1711–1717

Progr Colloid Polym Sci (2004) 123: 12–15
DOI 10.1007/b11611

Ken-ichi Kurumada
Brian H. Robinson

Viscosity studies of pluronic F127 in aqueous solution

K.-i. Kurumada (✉) · B.H. Robinson
School of Chemical Sciences,
University of East Anglia, Norwich, UK
e-mail: kurumada@ynu.ac.jp

Present address: K.-i. Kurumada,
Graduate School of Environment &
Information Science,
Yokohama National University,
Yokohama, 240-8501, Japan

Abstract The viscosities of the triblock copolymer F127 [(poly(ethylene oxide))$_{106}$-(poly(propylene oxide))$_{70}$-(poly(ethylene oxide))$_{106}$] in water and the corresponding pluronic F127/water/SDS (sodium dodecyl sulphate) system have been studied as a function of concentration and temperature. The results are discussed in terms of the solution microstructures and transitions between the dissociated state at low temperatures and an associated state at high temperatures. Above 35 °C, the system is consistent with hard sphere dispersions. As the temperature is lowered below 35 °C, the viscosity data indicate a progressive weakening of the structure. This is consistent with a critical micellisation temperature in the region 20 to 30 °C, and a critical micellisation concentration in the region 25 to 50 °C of ca. 1.0 g/L. Addition of SDS (sodium dodecyl sulphate) leads to micellar softening and micellar dissociation at higher SDS concentrations. Data from dynamic light scattering support the viscosity observations.

Introduction

Block copolymers comprised of parts with different hydrophilicity can form micellar aggregates in aqueous solution [1]. The basic micellar structure in an aqueous environment is a predominantly dehydrated core of polypropylene oxide enclosed by a polyethylene oxide/water shell [2, 3]. When applications of the micellar states of pluronics are considered, an important phenomenon is the switching between the dissociated and associated (micellar) state as the temperature is increased. It has been found that there are two main factors responsible for the switching in pluronic aqueous solutions, i.e., the cmc (critical micellisation concentration) and the cmt (critical micellisation temperature) [4, 5]. Generally, the solution properties change markedly in the vicinity of the dissociation-association transition, from which the cmc and cmt can, in principle, be elucidated. Wanka et al. estimated the cmc by surface tension [4]. They also evaluated the cmt by DSC and light scattering measurements. Alexandridis et al. used a dye solubilisation method to detect the cmc and cmt; micelle formation is detected by the UV absorbance change when a dye indicator species becomes entrapped in the core of the micelle [5]. Meilleur et al. measured the specific volume, heat capacity and viscosity at 5 °C, 25 °C and 45 °C [6] and they suggested that the unimer-to-micelle transition was a gradual process. SANS measurements indicate the micelles are spherical [4], and the sizes are in reasonable agreement with cryo-TEM measurements of Lam et al. [7]. Schillen et al. have studied the related pluronic P123 system (PEO_{20}-PPO_{68}-PEO_{20}) using dynamic light scattering and DSC [8]. Micelles are again formed at ~25 °C and there is also a very significant effect of added SDS and CTAB.

Previously, Kurumada et al. reported that reverse-micellar systems of Na^+-AOT (Na^+-Aerosol-OT) in hexane basically behave like hard sphere dispersions [9]. In the present work, the viscosity will be discussed from the viewpoint of solution structure and the cmc and cmt concepts. Dynamic light scattering measurements are consistent with micelle formation above the cmt.

Besides concentration and temperature, addition of low-molecular-weight amphiphiles [e.g., SDS (sodium dodecyl sulphate)] significantly affects the dissociation-association transition [10, 11, 12]. By fluorescence and NMR methods, SDS has been reported to adsorb on pluronic, particularly on the PPO part, which results in uncoiling of the micelles [13, 14]. On the whole, SDS suppresses micelle formation of pluronic F127. PEO [poly(ethylene oxide)] has been reported to be quite interactive with SDS as shown by the viscosity measurements of Chari et al. [15]. Strong interactions between water-soluble polymers and low-molecular-weight amphiphiles are also indicated from Monte Carlo computations [16, 17].

Experimental

Pluronic F127 and SDS were purchased from SIGMA, and used without further purification. Water used for sample preparation was supplied from BDH. No buffers were required. The viscosity was measured using an Ostwald capillary viscometer immersed in a thermostatted water bath. Dynamic light scattering measurements were analysed using an ALV-5000 correlator (ALV Gesellschaft, Germany). The light source was a 400 mW YAG laser ($\lambda = 532$ nm) (Coherent, USA).

Results and discussion

According to Quemada [18], the relative viscosity η_r of hard sphere dispersions is given by

$$\eta_r = (1 - \Phi/\Phi_{max})^{-2} \quad (1)$$

where Φ and Φ_{max} represent the hard sphere volume fraction and the maximum value at which the viscosity of the dispersion diverges as a result of gelation [18]. A value of $\Phi_{max} = 0.64$ was determined by fitting the measured zero-shear-rate viscosity by de Kruif et al. for monodisperse silica microparticles in cyclohexane to Eq. (1) [19] as shown in Fig. 1.

For pluronic F127 in an aqueous solution, Fig. 2 shows the dependence of the relative viscosity η_r on $(1-C/C_{max})$ at 50 °C for various C_{max} values, where C [g/L] and C_{max} [g/L] denote the concentration and the presumed maximum concentration of pluronic F127.

At 50 °C, C_{max} is evaluated as 140 g/L from the best fit to the Quemada equation. From 35 °C to 50 °C, the measured viscosity dependence is also in accordance with the Quemada equation and C_{max} can be obtained at 130 g/L~140 g/L. At 30 °C and below, there are increasing deviations from the Quemada equation, and at temperatures below the cmt, there is no evidence for hard sphere interactions at any F127 concentration.

Above the cmt, a reasonable model for the micelle is an essentially dehydrated PPO core surrounded by a PEO corona which contains a large amount of water.

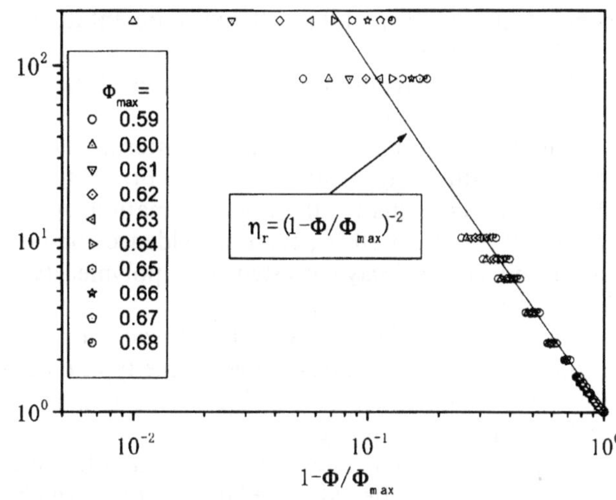

Fig. 1 Data fitting of the viscosity of hard sphere dispersions (monodisperse silica microparticles in cyclohexane) by de Kruif et al. [23] using the Quemada equation [18, 22] $\eta_r = (1-\Phi/\Phi_{max})^{-2}$ to determine the maximum volume fraction Φ_{max} at which viscosity divergence takes place. The symbols for various Φ_{max} values are shown in the figure; Φ_{max} is determined as 0.64

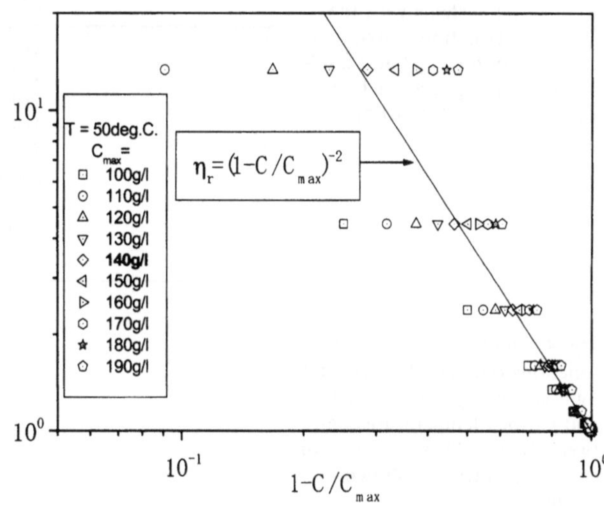

Fig. 2 η_r versus $1-C/C_{max}$ in pluronic F127/water systems at 50 °C with various values of C_{max} (symbols in the figure) fitted to the Quemada equation (solid line); C_{max} at 50 °C is determined as 140 g/L from the best fit to the Quemada equation

A calculation based on the assumption of a uniform density of 1 g/cm^3 suggests that the volume fraction of PEO in the corona is $vf_{PEO} = 0.15$, so that vf_{H_2O} is 0.85. In practice, it is not likely that there will be such a sharp distinction between the core and the corona, particularly, as the temperature is decreased to the region of the cmt. The micelles would also be expected to undergo some size fluctuations, partly as a consequence of the polydispersity of the unimer species.

At high temperatures, there is evidence for a cmc since the hard sphere like behaviour is only established at concentrations above 1.0 g/l. Some representative data are shown in Fig. 3.

Table 1 summarises the main properties of the system in the region from 1 to 50 °C.

Fig. 4 shows some dynamic light scattering data over the temperature range from 10 to 40 °C. For 30 to 40 °C, the data are consistent with micelle formation. At lower temperatures 10 to 20 °C, there is evidence for both a faster and slower decay consistent with polymer dynamics involving the unimer species.

The effect of added SDS on the dissociation-association process was also investigated. Some typical data at 50 °C for molar ratios of [SDS]/[pluronic F127] = 1, 10 and 100 are shown in Fig. 5.

The data are best interpreted in terms of, initially, formation of a softened micellar structure which, at higher concentrations of SDS, leads to disruption of the micellar structure. The divergence in the viscosity which is evident at molar ratio 0 and 1 is not so clear at higher relative concentrations of SDS. In fact, the system is

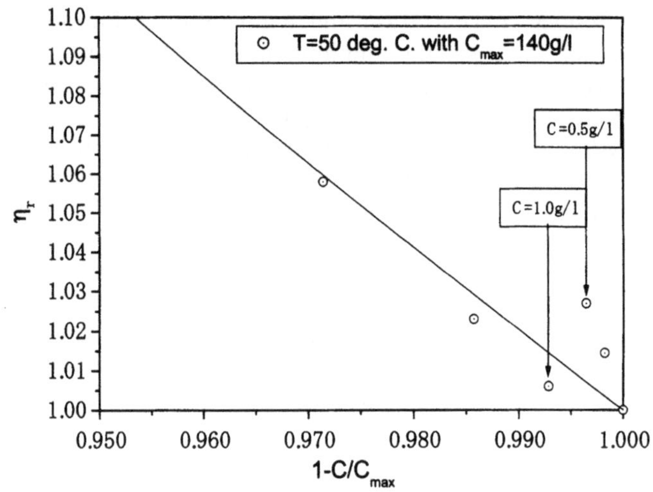

Fig. 3 Comparison with the Quemada equation (solid line) in the very dilute region for pluronic F127 at 50 °C with $C_{max} = 140$ g/L; it should be noted that η_r shows an abrupt decrease between C = 0.5 g/L and C = 1.0 g/L

Table 1 Fitted range of the measured viscosity of pluronic F127/water systems with the Quemada equation, maximum concentration (C_{max}) and fluidity at C = 150 g/L at each examined temperature from 1 °C to 50 °C

Temperature [°C]	Fitting with the Quemada equation	Maximum concentration: C_{max} [g/L]	State at C = 150 g/L
50	Totally Fitted	140	No Fluidity(Gel-Like)
45	Totally Fitted	140	No Fluidity(Gel-Like)
40	Totally Fitted	130	No Fluidity(Gel-Like)
38	Totally Fitted	130	No Fluidity(Gel-Like)
35	Totally Fitted	130	No Fluidity(Gel-Like)
30	Fitted in C≤50 g/L	120	No Fluidity(Gel-Like)
25	Fitted in C≤30 g/L	100	Fluid
5	Not Fitted	-	Fluid
1	Not Fitted	-	Fluid

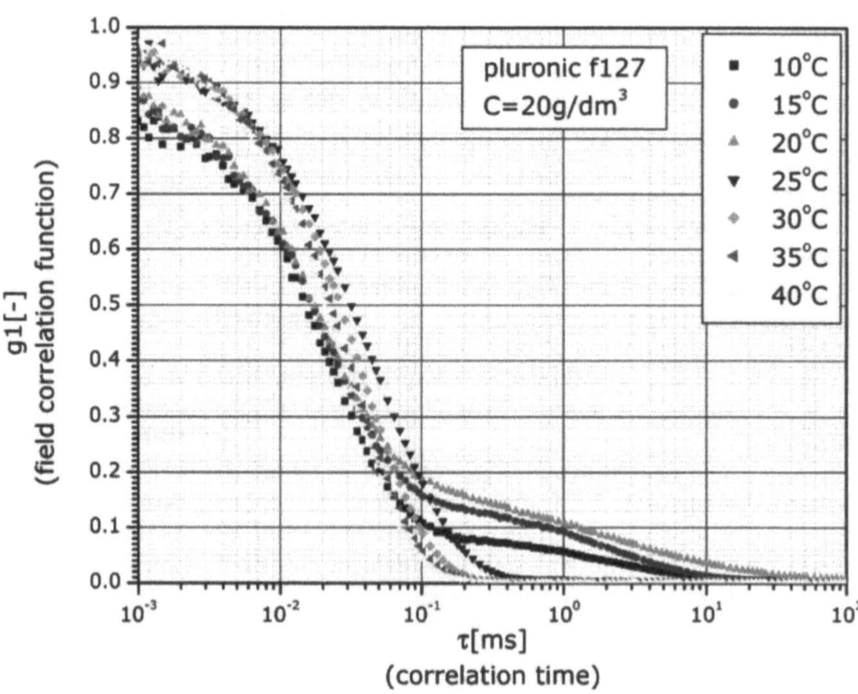

Fig. 4 Field autocorrelation functions obtained by dynamic light scattering at C = 20 g/L and various temperatures between 10 and 40 °C; symbol for each temperature is shown in the figure

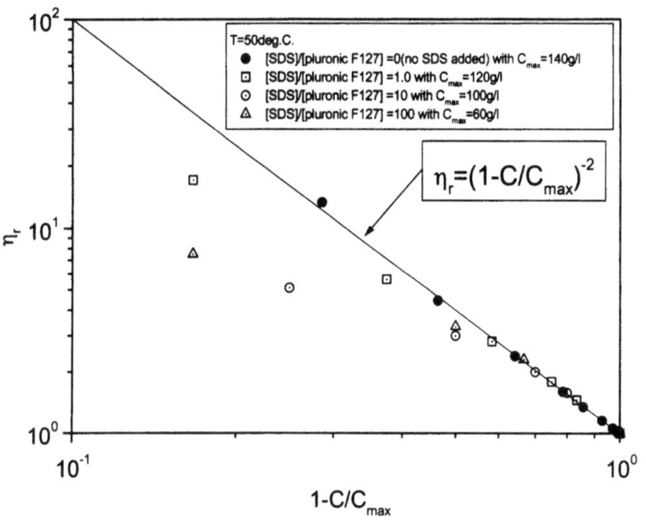

Fig. 5 η_r versus $1-C/C_{max}$ in pluronic F127/water/SDS systems; Comparison is made with the Quemada equation (solid line) at 50 °C; (●) at [SDS]/[pluronic F127] = 0 (no SDS added) with C_{max} = 140 g/L, (□) at [SDS]/[pluronic F127] = 1.0 with C_{max} = 120 g/L, (○) at [SDS]/[pluronic F127] = 10 with C_{max} = 100 g/L, (△) at [SDS]/[pluronic F127] = 100 with C_{max} = 60 g/L; C_{max} values were determined based on fitting to the Quemada equation in the dilute region for [SDS]/[pluronic F127] = 1.0, 10 and 100

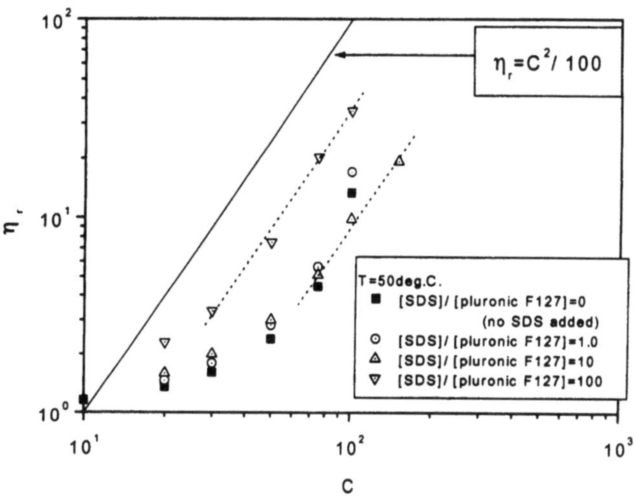

Fig. 6 η_r versus C for the full logarithmic scale in pluronic F127/water/SDS systems at 50 °C; (■) at [SDS]/[pluronic F127] = 0 (no SDS added), (○) at [SDS]/[pluronic F127] = 1.0, at [SDS]/[pluronic F127] = 10, (▽) at [SDS]/[pluronic F127] = 100; The lines are guides for the eye for $\eta_r \propto C^2$

tending to a state where the relative viscosity is proportional to C^2, which is characteristic of the formation of an open network structure in a polymer solution as shown in Fig. 6 [9]. According to Almgren and coworkers, SDS, which is a typical ionic surfactant, attractively adsorbs on pluronic molecules at sufficiently high concentrations [13, 14]. The above results can be ascribed to the uncoiling and bridging effect of SDS due to coverage of pluronic chains by SDS molecules, which can be clearly observed at the molar ratio 100.

Conclusions

Pluronic F127 in an aqueous medium forms hard-sphere micellar aggregates when the temperature exceeds 35 °C. As the temperature is lowered, there is a progressive breakdown of these micellar aggregates, such that the system is more like a polymer solution. Addition of SDS has a similar disrupting effect on the micellar structure.

Acknowledgements We would like to gratefully acknowledge the Daiwa Anglo-Japanese Foundation for support of this collaboration. We also appreciate stimulating discussions with Dr. D.C. Hone, Mr. M. Wright and Dr. M. Silbert.

References

1. Schmolka IR (1991) Poloxamers in the pharmaceutical industry. In: Tarcha PJ (ed.), Polymers for controlled drug delivery. CRC Press, Boston
2. Zhou Z, Chu BJ (1988) Colloid Interface Sci 126:171
3. Mortensen K, Pedersen JS (1993) Macromolecules 26:805
4. Wanka G, Hoffmann H, Ulbricht W (1994) Macromolecules 27:4145
5. Alexandridis P, Holzwarth JF, Hatton TA (1994) Macromolecules 27:2414
6. Meilleur L, Hardy A, Quirion F (1996) Langmuir 12:4697
7. Lam Y, Grigorieff N, Goldbeck-Wood G (1999) Phys Chem Chem Phys 1:3331
8. Jansson J, Silva RC sa, Olofsson G, Schillën K (2001) Presented at ECIS 2001, Coimbra, Portugal
9. Kurumada K, Shioi A, Harada M (1998) J Phys Chem:123:82
10. Hecht E, Hoffmann H (1994) Langmuir 10:86
11. Li Y, Xu R, Bloor DM, Holzwarth JF, Wyn-Jones E (2000) Langmuir 16:10515
12. Li Y, Xu R, Couderc S, Bloor DM, Wyn-Jones E, Holzwarth JF (2000) Langmuir 17:183
13. Almgren M, Brown W, Hvidt S (1995) Colloid Polym Sci 273:2
14. Almgren M, Stem J van, Lindbrad C, Li P, Stilbs P, Bahadur P (1991) J Phys Chem 95:5677
15. Chari K, Antalek B, Lin MY, Sinha SKJ (1994) Chem Phys 100:5294
16. Jennings DE, Kuznetsov YA, Timoshenko EG, Dawson KA (1998) J Chem Phys 108:1702
17. Jennings DE, Kuznetsov YA, Timoshenko EG, Dawson KA (2000) J Chem Phys 112:7711
18. Quemada D (1977) Rheol Acta 16:82
19. Kruif CG de, Iersel EMF van, Vrij A, Russel WB (1985) J Chem Phys 83:4717

Progr Colloid Polym Sci (2004) 123:16–22
DOI 10.1007/b11613

Magnus Bergström
Jan Christer Eriksson

Synergistic effects in binary surfactant mixtures

M. Bergström · J.C. Eriksson
Department of Chemistry,
Surface Chemistry, Royal Institute of Technology, Drottning Kristinas väg 51, 100 44 Stockholm, Sweden

M. Bergström (✉) · J.C. Eriksson
YKI, Institute for Surface Chemistry, Box 5607, 114 86 Stockholm, Sweden
Tel.: +46-8-7909905
Fax: +46-8-208998
e-mail: magnus.bergstrom@surfchem.kth.se

Abstract By considering the main contributions to the micellar free energy we have analysed the synergistic effect often seen on the CMC of a binary surfactant mixture. The synergistic effects are due mainly to the entropic free energy contributions related with the surfactant head groups. Several cases have been treated: (i) For a mixture of a monovalent ionic and a non-ionic surfactant in the absence of added salt we obtain, entirely because of electrostatic reasons, a negative deviation from the ideal behaviour corresponding to an interaction parameter $\beta \approx -1$. Upon adding an inert salt we found that the magnitude of the synergistic effect first increases, reaches a maximum and eventually decreases. (ii) For mixtures of two ionic surfactants with the same charge number but with different hydrocarbon moieties β-values as low as –10 may arise. (iii) For mixtures of an anionic and a cationic surfactant enormous effects are anticipated yielding $\beta \leq -20$ depending on the CMCs of respective pure surfactant. (iv) Synergistic effects due to different cross-section areas of the head groups are found to be rather small, with $0 > \beta > -1$, provided the difference in head group size is modest but can become more significant when the size difference is larger.

Introduction

Surface active substances (surfactants) self-assemble above a certain rather well-defined concentration, the critical micelle concentration (CMC), to form droplet-like aggregates (micelles). For mixtures of two only slightly differing surfactants CMC is found to obey an approximately linear behaviour with respect to the composition in a micelle and they are generally referred as to ideal mixtures. However, many binary surfactant mixtures cannot be accurately described with a linear relation with respect to surfactant composition and, in analogy with the theory for regular solution, it is frequently assumed that CMC may be written as

$$CMC(x) = x\exp\left[(1-x)^2\beta\right]CMC_1 + (1-x)\exp\left[x^2\beta\right]CMC_a \quad (1)$$

where CMC_1 and CMC_2 are the CMCs of pure Surfactant 1 and Surfactant 2, respectively, and x and $(1-x)$ are the corresponding mole fractions *in the aggregates* formed in a binary surfactant mixture. The non-ideal behaviour is taken into account by the parameter β whereas ideality is recovered as a special case from Eq. (1) when $\beta = 0$ giving the linear relation

$$CMC(x) = x\exp CMC_1 + (1-x)CMC_2 \quad (2)$$

Eq. (2) may be rewritten so as to relate *CMC* with *the overall surfactant concentration* (free + aggregated surfactant) y. By taking into account that, at CMC, the concentration of free surfactant is much larger than the concentration of aggregated surfactant, it is straightforward to show that

$$\frac{1}{CMC(y)} = \frac{y}{CMC_1} + \frac{(1-y)}{CMC_2} \quad (3)$$

Synergistic effects by definition are present when β assumes negative values giving a negative deviation from ideal behaviour, i.e., $CMC(x)$ is generally lower than expected from the ideal expression in Eq. (2), whereas antagonistic effects (positive deviations from linearity) occurs for $\beta > 0$.

The concentration of free surfactant (= CMC at CMC) may in principle be determined from the aggregate free energy through an equilibrium condition that the chemical potential of free and aggregated surfactant must be equal. In other words, a relation between aggregate free energy and CMC is obtained from equilibrium thermodynamics. As a result, it may be demonstrated that the free energy per aggregated surfactant of forming a surfactant aggregate must be written in the form

$$\varepsilon(x) = x\varepsilon_1 + (1-x)\varepsilon_2 + x\ln x + (1-x)\ln(1-x) + \beta x(1-x) \quad (4)$$

where ε_1 and ε_2 are constants with respect to x, in order to yield Eq. (1) [1]. In accordance with Eq. (3) a linear behaviour of ε is taken into account by the first terms $x\varepsilon_1+(1-x)\varepsilon_2$ whereas non-linear free energy contributions other than the entropy of mixing the two surfactants in the aggregate [$=x\ln x+(1-x)\ln(1-x)$] are taken into account by the 'pairwise molecular interaction' term $\beta x(1-x)$ with $\beta \neq 0$. Hence, it follows from Eqs. (1) and (4) that contributions to the free energy that are linear with respect to composition do not give rise to any synergism nor antagonism, i.e., a non-linear behaviour of $CMC(x)$. Only non-linear free energy contributions may contribute to a non-vanishing value of β and generate deviations of $CMC(x)$ from linearity. Incidentally, the free energy of mixing [$=x\ln x+(1-x)\ln(1-x)$] gives rise to the factor of x proportional to CMC_1 and the factor $(1-x)$ proportional to CMC_2 in the expression $CMC(x)= x\exp CMC_1+(1-x)CMC_2$.

In general, the free energy $\varepsilon(x)$ cannot be written in the form given in Eq. (4) and, as a result, the $CMC(x)$ cannot be accurately described with Eq. (1). However, a more general expression for CMC as a function of composition has recently been derived by the present authors from which $CMC(x)$ may be calculated from an arbitrary expression $\varepsilon(x)$ of the aggregate free energy. Hence [2],

$$CMC(x) = A(x)xCMC_1 + B(x)(1-x)CMC_2 \quad (5)$$

where

$$A(x) = \exp\left[\left(\varepsilon_{ex}(x) - \varepsilon_{ex}(x=1) + (1-x)\frac{d\varepsilon_{ex}(x)}{dx}\right)/kT\right] \quad (6)$$

and

$$B(x) = \exp\left[\left(\varepsilon_{ex}(x) - \varepsilon_{ex}(x=0) - x\frac{d\varepsilon_{ex}(x)}{dx}\right)/kT\right] \quad (7)$$

and $\varepsilon_{ex}(x) \equiv \varepsilon(x) - x\ln x - (1-x)\ln(1-x)$ is the excess free energy.

It is straightforward to demonstrate that Eq. (1) is recovered when Eq. (4) is inserted in Eqs. (5–7). We may also note that the functions in Eqs. (6) and (7) are related to the activities of aggregated Surfactant 1 and Surfactant 2, respectively, as $a_1(x)=A(x)x$ and $a_2(x)=B(x)(1-x)$.

Hence, synergistic (or antagonistic) effects may be calculated from Eqs. (5–7) for any appropriate free energy function $\varepsilon(x)$. Below synergistic effects in binary surfactant systems are investigated by means of evaluating $CMC(x)$ for several cases: Mixture of a monovalent ionic and a non-ionic surfactant, mixture of two ionic surfactants with different hydrocarbon tails, mixture of an anionic and a cationic surfactant and mixture of two non-ionic surfactants with inert rigid head groups.

Contributions to the free energy of a surfactant aggregate

The free energy of forming a surfactant aggregate can be written as a sum of several contributions related to either the tails or the head groups of the surfactants [3, 4]: The reduction of contact area between hydrocarbon and water as well as the conformational entropy due to packing restrictions of the hydrocarbon chains are related to the tails whereas electrostatics for a charged aggregate surface and its diffuse layer of counter-ions as well as other effects are related to the head groups.

Contributions due to the surfactant tails The driving force for the otherwise entropically unfavourable self-assembly of surfactant molecules is the hydrophobic effect, i.e., the reduction of hydrocarbon/water interfacial area as the hydrocarbon tails of the surfactants form the liquid-like cores of the aggregates [5]. This contribution can be calculated as the work of bringing free surfactants from the aqueous bulk solution to a hydrocarbon bulk phase ($= v - x\ln x_1^{free} - (1-x)\ln x_2^{free}$) [5] plus the hydrocarbon/water interfacial tension times the area per aggregated surfactant at the hydrocarbon/water interface ($\gamma_{hc/w} \times a$). This free energy contribution is linear with respect to the aggregate composition provided the structure of the aggregates is constant (since a is a function of aggregate structure) and, as a consequence, no synergistic nor antagonistic effects are obtained as a result of this free energy contribution as far as the structural change of the aggregates with x is small.

Moreover, it has been demonstrated that the contribution to a non-ideal behaviour of $CMC(x)$ due to hydrocarbon chain conformational entropy is small [2]. Hence, since we have assumed a constant (planar) structure of the aggregates with respect to surfactant composition throughout our calculations the contributions to synergistic effects from the tails are found to be negligible.

Contributions due to the surfactant head groups Electrostatics yield a large positive contribution to the aggregate free energy for mixtures consisting of at least one ionic surfactant. According to the Poisson-Boltzmann (mean field) description, this contribution is mainly due to the entropically unfavourable organisation of the counter-ions into a diffuse layer located outside the electrically charged surface of an aggregate, whereas energetic effects usually are much smaller.

For aggregates encompassing monovalent surfactants the electrostatic free energy per unit charge can be rather accurately calculated from the Poisson-Boltzmann theory, which for planar geometry gives a free energy per charge equal to

$$\varepsilon_{el}/kT = 2\left[\ln\left(S+\sqrt{S^2+1}\right) - \frac{\sqrt{S^2+1}-1}{S}\right] \tag{8}$$

Hence, the electrostatic free energy may be written as a function of one single parameter, the reduced charge density S, which is defined as

$$S = \frac{\sigma}{\sqrt{8c_t\varepsilon_0\varepsilon_r N_A kT}} \tag{9}$$

$\sigma = e_{el}/a_{charge}$ denotes the surface charge density, e_{el} the elementary charge and a_{charge} the area per charge at the aggregate surface. ε_0 and ε_r are the electric permeability in vacuum and the relative permeability, respectively, and N_A is the Avogadro constant.

Normally S is a rather strong function of the surfactant composition and, as a consequence, Eq. (8) is not a constant but depends on the aggregate composition. In accordance, the electrostatic free energy contribution gives rise to significant synergistic effects which will be treated below. In physical terms, these synergistic effects arise as a consequence of the dilution of counter-ions, and the subsequent increase in entropy of mixing counter-ions and water molecules, when, e.g., a non-ionic or an oppositely charged surfactant is admixed to an ionic surfactant solution.

The most important contribution associated with *non-ionic* surfactant head groups is due to the entropy of mixing head groups and solvent molecules. For a rather concentrated mixture of particles with a circular cross-section the following expression may be used

$$\varepsilon_{hg}(x) = \ln\eta + \frac{\eta(20-\eta)}{15(1-\eta)} - \frac{2}{3}\ln(1-\eta) \tag{10}$$

where the area fraction of head groups at the aggregate surface equals $\eta(x) = \left[xa_1^{hg} + (1-x)a_2^{hg}\right]/a$ [2]. a_1^{hg} and a_2^{hg} are the cross-section areas of the head groups of Surfactant 1 and 2, respectively, and a the area per aggregated surfactant. The derivation of Eq. (10) is based on the two-dimensional equation of state obtained by a Padé approximation from the known virial coefficients up to B_6 [6, 7]. It is evident that Eq. (10) is non-linear with respect to x and the resulting synergistic effects are treated below.

Other free energy contributions related to the surfactant head groups, including specific interactions between different head groups, are difficult to estimate quantitatively but are probably small. However, we cannot exclude that specific interactions may occasionally contribute non-negligibly to any observed synergistic effects.

Mixture of an ionic and a non-ionic surfactant

For mixtures consisting of a monovalent ionic surfactant and a non-ionic surfactant with an otherwise similar head group (same size and hydration number etc. so that only electrostatic effects differ between the two surfactants), the electrostatic free energy per aggregated surfactant can be written as follows

$$\varepsilon_{ex}(x) = x\varepsilon_{el} \tag{11}$$

where the free energy per charge ε_{el} is approximately given by Eq. (8) which is strictly valid for planar geometry. An approximate expression for ε_{el} can be obtained by means of considering the case where $S \gg 1$. This is a fairly good approximation for electrolyte concentrations c_t below about 0.2 M, for which $S \geq 10$, provided x assumes values not too far from unity. Moreover, the contribution from energetic effects to the electrostatic free energy is negligible in the regime of S values larger than about unity [8]. Hence, we may conclude that the synergistic effects due to electrostatics obtained in our analysis is entirely of an entropic origin (entropy of mixing counter-ions and solvent molecules).

In accordance with the assumption $S \gg 1$, we can simplify Eq. (8) as

$$\frac{\varepsilon_{el}}{kT} = 2\ln\left(\frac{2S}{e}\right) \tag{12}$$

Hence,

$$\varepsilon_{ex}(x) = x\varepsilon_{el} = 2x\ln\left(\frac{x}{CMC_1^m(x)+c_{salt}}\right) + const \tag{13}$$

where CMC_1^m is the concentration of free ionic monomers at CMC and c_{salt} is the molar concentration of any added inert salt giving a total electrolyte concentration $c_t = CMC_1^m + c_{salt}$.

An expression for $CMC(x)$ may be derived by combining Eqs. (5–7) and (13) for the case of no added salt ($c_{salt}=0$). The result is a differential equation with respect to $CMC(x)$ that is not analytically solvable but the following approximate expressions for $A(x)$ and $B(x)$ may be evaluated by iteration

$$A(x) = \exp\left[-(1-x)^2\right] \tag{14}$$

and

$$B(x) = \exp(-x^2) \tag{15}$$

which, incidentally, exactly corresponds to the expression for regular solutions in Eq. (1), with $\beta=-1$. A more accurate but less simple (and still approximate) solution to the differential equation is

$$\begin{aligned} \mathrm{CMC}(x) &= x^{3/2}\exp\left[\frac{(1-x)(1+x^2)}{4}\right]CMC_1 \\ &\quad + (1-x)\exp\left[-\frac{x(1+x^2)}{4}\right]CMC_2 \end{aligned} \tag{16}$$

which corresponds to synergistic effects close to $\beta=-1$ but, in contrast to the dependence of $CMC(x)$ according to Eqs. (5), (14) and (15), is asymmetrical about to $x=0.5$. The somewhat skewed plot of Eq. (16) has a minimum at mole fractions where the non-ionic surfactant is in excess (cf. Fig. 1).

The addition of salt influences the appearance of $\varepsilon_{ex}(x)$ in Eq. (13) and, hence, the synergistic effects as expressed by $CMC(x)$. For the special case $c_{salt} \gg CMC_1^m$ and $S \gg 1$ the explicit expressions

$$A(x) = x^2\exp(2-2x) \tag{17}$$

and

$$B(x) = \exp(-2x) \tag{18}$$

may be derived from Eqs. (5–7) and (13). Those expressions roughly corresponds to $\beta=-3$, i.e., considerably larger synergistic effects than in the case of no added salt. However, for large electrolyte concentrations the S parameter becomes significantly reduced and the assumption $S \gg 1$ no longer holds true.

The effect of a decreasing S is to reduce the magnitude of the synergistic effects, which are expected to vanish as $S \to 0$. Hence, two effects, one tending to increase and the other tending to decrease the synergistic effects are expected to be present as an inert electrolyte is added to an ionic/non-ionic surfactant mixture. As a result, according to calculations using Eqs. (8), (9) and (11) (without assuming $S \gg 1$) the magnitudes of the synergistic effects are observed to first increase, reach a maximum, and then decrease as an increasing amount of salt is added to the surfactant solution (cf. Fig. 1).

In physical terms, the synergistic effects observed in mixtures of an ionic and a non-ionic surfactant is mainly a result of the entropy of the diffuse layer of counter-ions outside the micelle interfaces. In particular, the dilution of counter-ions when adding a non-ionic surfactant to a ionic surfactant system raises the counter-ion entropy (in analogy with a gas that increases its entropy upon expansion). As a result, the chemical potential of aggregated surfactant is lowered giving rise to a negative deviation of $\varepsilon(x)$ from linear behaviour. In the absence of added salt, the resulting synergistic effects are, however, counteracted by the reduction in CMC, i.e., the reduction of electrolyte concentration, which raises the electrostatic free energy per aggregated surfactant giving synergistic effects of smaller magnitude than expected from the former effect. When, on the other hand, a small amount of salt is added these counteracting effects are reduced resulting in an increase in magnitude of the synergistic effects. Further increase of c_{salt} eventually reduces S to become close to unity and the synergistic effects begin to decrease in magnitude.

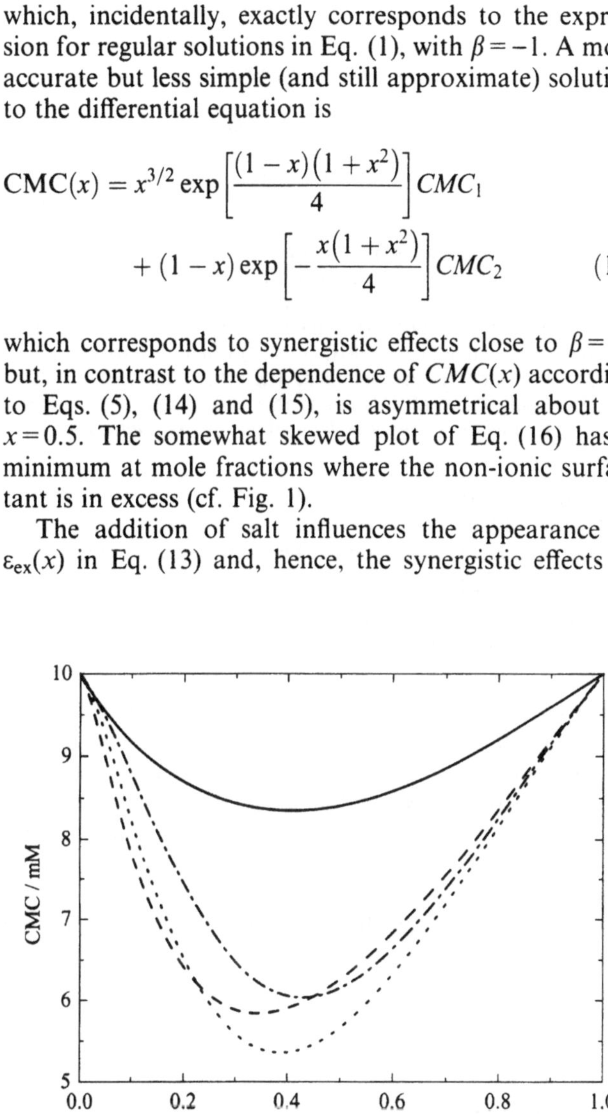

Fig. 1 CMC for a mixture of a monovalent ionic and an otherwise similar non-ionic surfactant plotted against the mole fraction of the ionic surfactant in the aggregates (x) in the absence of added salt (*solid line*, $\beta \approx -1$) and at concentrations of added salt 0.01 M (*dashed line*, $\beta \approx -2.1$), 0.1 M (*dotted line*, $\beta \approx -2.5$) and 1 M (*dashed-dotted line*, $\beta \approx -2.0$). The CMCs for the pure surfactants are set to $CMC_1 = CMC_2 = 10$ mM

Mixture of two ionic surfactants with identical head groups but different hydrocarbon tails

For mixtures of two monovalent surfactants with identical charge the free energy per aggregated surfactant may be written as

$$G(x) = \varepsilon_{el} \tag{19}$$

Since the electrostatic free energy is a function of electrolyte concentration, it must depend on the CMC_1 and CMC_2 of the two ionic surfactants. As a matter of fact, the following expressions for $A(x)$ and $B(x)$ in Eqs. (5–7) may be derived from Eqs. (12), (9) and (19) [2]

$$A(x) = \left(1 + \frac{c_{salt}}{CMC_1}\right) \frac{\exp[-(1-x)(1-\lambda)/(x+\lambda(1-x)+c_{salt}/CMC_1)]}{x+\lambda(1-x)+c_{salt}/CMC_1} \quad (20)$$

and

$$B(x)=\left(\lambda+\frac{c_{salt}}{CMC_1}\right)\frac{\exp[x(1-\lambda)/(x+\lambda(1-x)+c_{salt}/CMC_1)]}{x+\lambda(1-x)+c_{salt}/CMC_1} \quad (21)$$

In other words, $CMC(x)$ is completely determined by the ratio $\lambda \equiv CMC_2/CMC_1$ between the CMCs of respective pure surfactant as well as the concentration of any added salt c_{salt}. In Fig. 2 we have plotted $CMC(x)$ for three cases and it is seen that CMC increases with increasing difference between CMC_1 and CMC_2. For the special case $\lambda=1$ (or $CMC_1=CMC_2$) the synergistic effects vanish. It is also seen from Fig. 2 that the synergistic effects are most pronounced at compositions where the surfactant with the lower CMC is in excess. Moreover, the synergistic effects decrease monotonically in magnitude with increasing c_{salt} and vanish as $c_{salt} \gg CMC_1 \approx CMC_2$.

The synergistic effects obtained as a result of Eqs. (20) and (21) may be rationalised as a purely electrostatic effect. Because of the different CMCs the addition of the two ionic surfactants will influence the electrostatic free energy to different extents. If a small amount of the surfactant with the higher CMC is added to a solution with a composition rich in the surfactant with the lower CMC, most of the added surfactant will be located as free surfactant in the bulk solution. This means that it will mainly have the same effect as when an inert salt is added. Hence, the electrostatic free energy will be reduced and the chemical potential of aggregated surfactant (mostly surfactants with lower CMC) be reduced causing CMC to become lower than expected from the ideal linear behaviour. On the other hand, when the surfactant with lower CMC is added to a solution rich in the high CMC surfactant most of it will aggregate and the synergistic effects be significantly reduced. Since the effect is entirely electrostatic the synergistic effects vanish as an inert electrolyte is added and the electrostatic free energy reduced and eliminated.

Mixtures of an anionic and a cationic surfactant

The synergistic effects in mixtures of an anionic and a cationic surfactant have been observed to be much larger

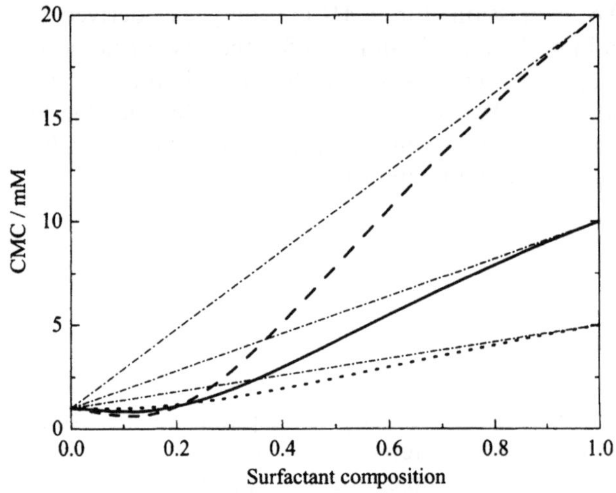

Fig. 2 CMC for mixtures of two monovalent ionic surfactants with identical head groups but with different CMCs ($CMC_1 > CMC_2$) plotted against the mole fraction of Surfactant 1 in the aggregates (x) in the absence of added salt according to Eqs. (5), (20) and (21). CMC_2 was fixed to 1 mM for the three cases whereas $\lambda \equiv CMC_2/CMC_1$ was set to 0.1 (*solid line*), 0.05 (*dashed line*) and 0.2 (*dotted line*). The corresponding curves for the ideal surfactant mixtures as obtained from Eq. (1) are indicated as *dashed-dotted lines*. The synergistic effects are most pronounced when the mixture is rich of the surfactant with the lowest value of the CMC (CMC_2) where distinct minima in the CMC vs. x curves are obtained. The synergistic effects increase with decreasing λ and at the minima they approximately correspond to $\beta=-2$ (*dotted line*), $\beta=-4$ (*solid line*) and $\beta=-6$ (*dashed line*).

in magnitude than, e.g., in mixtures of an ionic and a non-ionic surfactant. β-values well below -20 have frequently been observed. The enormous synergism may be rationalised as a result of the elimination of the unfavourable electrostatic free energy as oppositely charged surfactants aggregate giving an aggregate mole fraction $x=0.5$ in virtually the entire overall composition range y. As a matter of fact, overall surfactant compositions in the range $0.002 < y_1 < 0.998$ (shown in 3) correspond to compositions in the aggregates of $0.497 < x_1 < 0.503$. Since the unfavourable electrostatic free energy cancels at $x=0.5$ the CMC is brought down considerably in catanionic mixtures implying enormous synergistic effects.

As a matter of fact, the following expressions for $A(x)$ and $B(x)$ in the limit $x \rightarrow 0.5$ may be derived [9]

$$\lim_{x\rightarrow 0.5} A(x) = e^{-\varepsilon_{el}^1/kT} \quad (22)$$

and

$$\lim_{x\rightarrow 0.5} B(x) = e^{-\varepsilon_{el}^2/kT} \quad (23)$$

where ε_{el}^1 and ε_{el}^2 are the electrostatic free energy per aggregated charge of pure Surfactant 1 and Surfactant 2, respectively. Hence, in the special case $\varepsilon_{el}^1 = \varepsilon_{el}^2$, the simple expression

$$\beta \approx -4\varepsilon_{el}^1/kT = -4\varepsilon_{el}^2/kT \quad (24)$$

is obtained and, as a consequence, a plot of CMC against y using the regular solution expression with $\beta = -4\varepsilon_{el}^{1}/kT$ coincides with the calculated $CMC(y)$ curve at $y=0.5$ (cf. Fig. 3).

Mixtures containing non-ionic surfactants with rigid head groups

The most important contribution to the aggregate free energy associated with the head groups of non-ionic surfactants is due to the free energy of mixing the head groups with solvent molecules. For a rather dense mixture of rigid head-groups we can use Eq. (10) to calculate the corresponding free energy contribution. It immediately appears that this equation is non-linear with respect to x and, hence, the mixing of head groups gives rise to synergistic effects and a non-linear behaviour of CMC. The synergistic effects are highly dependent on the cross-section areas of both head groups as well as of the area per aggregated surfactant a and increases with increasing a_1^{hg} and a_2^{hg} and decreasing a, i.e., with increasing η. The deviation from linear behaviour increases with increasing difference between the head group cross-section areas a_1^{hg} and a_2^{hg}.

Examples of $CMC(x)$ as calculated from Eq. (10) for different values of a_1^{hg} and a_2^{hg} at fixed values of $(a_1^{hg}+a_2^{hg})$ and a are given in Fig. 4. For moderate differences in head group size the synergistic effects correspond to β values of rather small magnitude, i.e., less than unity. However, large differences in head group size may result in comparatively large synergistic effects.

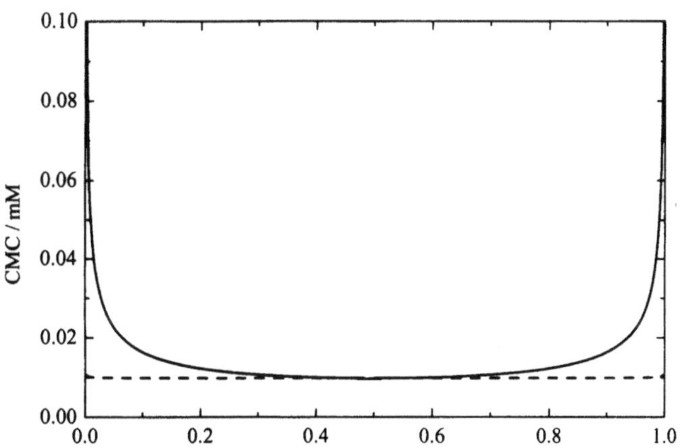

Fig. 3 CMC for a mixture of a monovalent anionic and an otherwise identical monovalent cationic surfactant plotted against the overall mole fraction of the anionic surfactant y_1 (*solid line*). The CMCs for the pure surfactant are set equal to $CMC_1 = CMC_2 = 10$ mM which correspond to $\varepsilon_{el} = 7kT$. As a result, the CMC curve according to Eq. (1) with $\beta = -4\varepsilon_{el}/kT = -28$ (*dashed line*) coincides with the former curve near $y=0.5$. The aggregates are assumed to have a planar geometry

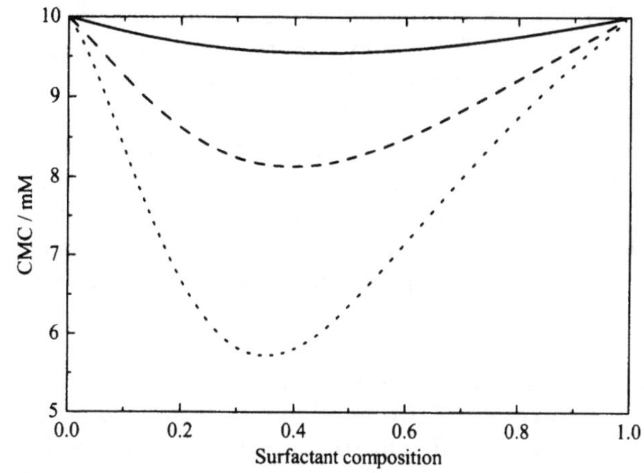

Fig. 4 CMC for mixtures of two non-ionic surfactants with rigid circular head groups of different size, as calculated using Eq. (10), plotted against the mole fraction of Surfactant 1 in the aggregates (x). For all three curves the area per aggregated surfactant was set to a constant value of $a=32$ Å^2 and the sum of the cross-section areas of the two head groups to $a_1^{hg}+a_2^{hg}=30$ Å^2. The difference in head group cross-section area is $\Delta a_{hg} \equiv a_2^{hg}-a_1^{hg}=5$ Å^2 (*solid line*, approximately corresponding to $\beta=-0.2$), $\Delta a_{hg} \equiv a_2^{hg}-a_1^{hg}=10$ Å^2 (*dashed line*, $\beta=-0.8$) and $\Delta a_{hg} \equiv a_2^{hg}-a_1^{hg}=20$ Å^2 (*dotted line*, $\beta=-2.2$). The minima of the

The electrostatic and residual head group contributions to synergistic effects are found to be additive in a first approximation [2, 9]. As a result, we may conclude that $\beta=-1$ makes a maximum value for mixtures of a monovalent ionic and a non-ionic surfactant and β is predicted to be somewhat lower (i.e., more negative) for mixtures where the surfactant head groups have different cross-section areas. Hence, our theoretical results agree very well with the general experimental observation that synergistic effects in mixtures of two non-ionic surfactants are for most cases studied small whereas β is always somewhat lower than -1 for mixtures of a monovalent ionic and a non-ionic surfactant [1, 10, 11].

Conclusions

Synergistic effects in mixed surfactant systems have been rationalised by investigating the composition dependence of the excess free energy per aggregated surfactant ε_{ex}. When ε_{ex} is linear with respect to the composition in the aggregates x, CMC will depend linearly on the composition. However, the inclusion of any additional non-linear free energy contribution gives rise to synergistic effects. It is found that if $\varepsilon_{ex}(x)$ has a negative deviation from linear behaviour so does $CMC(x)$ and vice verse.

Out of the various free energy contributions those related with the surfactant head groups are mainly responsible for generating synergistic effects. For mixtures of a monovalent ionic surfactant and an otherwise similar non-ionic surfactant in the absence of added salt, the electrostatic free energy gives rise to synergistic effects approximately corresponding to $\beta=-1$. When a small amount of salt is added the magnitude of β increases. Upon further addition of electrolyte a maximum in magnitude of the synergism is reached and β begin to approach zero at higher concentrations. With our calculation scheme we are also able to qualitatively reproduce the more pronounced synergistic effects that arise in mixtures of two identically charged surfactants having different CMCs as well as in mixtures of an anionic and a cationic surfactant.

For mixtures of two non-ionic surfactants with rigid head groups of different size, a negative deviation from linear behaviour arises from the entropy of mixing the rather concentrated head groups with surrounding water molecules at the aggregate surface. This contribution is fairly small, i.e., $0>\beta>-1$, for moderate differences between the two head groups but may be much larger if the two head groups become sufficiently different from each other or if they are densely concentrated. Moreover, the electrostatic and the residual head group contributions are found to be additive in a first approximation.

For most cases the regular solution theory is found to furnish a rather poor description of the CMC vs. x behaviour. This is not surprising considering that various entropic effects, rather than pairwise specific interactions among the surfactants, mainly contribute to the aggregate free energy. Qualitatively, our results agree with a host of experimental findings as well as with detailed model calculations performed by Blankschtein and co-workers [12].

Acknowledgements M.B. was financed by The Swedish Agency for Innovation Systems (VINNOVA) via the Competence Center for Surfactants Based on Natural Products (SNAP).

References

1. Holland PM (1992) In: Holland PM, Rubingh DN (eds), Mixed surfactant systems, Chap. 2. American Chemical Society, Washington, DC, p 40
2. Bergström M, Eriksson JC (2000) Langmuir 16:7173–7181
3. Eriksson JC, Ljunggren S, Henriksson U (1985) J Chem Soc Faraday Trans 2 81:833–868
4. Ljunggren S, Eriksson JC (1987) Prog Colloid Polym Sci 74:38–47
5. Tanford C (1980) The hydrophobic effect, Chap. 7, New York
6. Erpenbeck JJ, Luban M (1985) Phys Rev A 32:2920–2922
7. Nilsson U, Jönsson B, Wennerström H (1993) J Phys Chem 97:5654
8. Evans DF, Wennerström H (1994) The Colloidal Domain, Chap. 3, , New York
9. Bergström M (2001) Langmuir 17:993–998
10. Jönsson B, Lindman B, Holmberg K, Kronberg B (1998) Surfactant and polymers in aqueous solution, Chap. 5, Chichester
11. Liljekvist P, Kronberg B (2000) J Colloid Interface Sci 222:159
12. Shiloach A, Blankschtein D (1998) Langmuir 14:1618–1636

Progr Colloid Polym Sci (2004) 123:23–27
DOI 10.1007/b11614

A. Chittofrati
R. Pieri
F. D'Aprile
D. Lenti
P. Maccone
M. Visca

Perfluoropolyether carboxylic salts in micellar solution and O/W microemulsions

A. Chittofrati (✉) · R. Pieri
F. D'Aprile · D. Lenti · P. Maccone
M. Visca
Solvay Solexis, v.le Lombardia 20,
20021 Bollate, Milano, Italy
e-mail: alba.chittofrati@solvay.com

Abstract High purity laboratory-scale samples of $Cl(C_3F_6O)_nCF_2$-COOX salts are examined for their micellization in water, with **n** from 2 to 4 and NH_4^+ or Na^+ counterion. Well defined mixtures form O/W microemulsions with perfluoropolyether oils. Their strong dilution, outside the thermodynamic stability region, in pseudo-ternary phase diagrams, can lead to droplets with fairly high kinetic stability, as monitored by dynamic light scattering.

Keywords Perfluoropolyethers · Micellar solutions · O/W microemulsions

Introduction

This work aims firstly, to sketch the solution behaviour of high purity carboxylic perfluoropolyether (PFPE) salts with respect to the number of perfluoroisopropoxy units in their chain and, secondly, provide a "formulator approach" to obtain fluorinated droplets with fairly high kinetic stability in water, by strong dilution of O/W microemulsions outside their thermodynamic stability region in the pseudo-ternary phase diagrams.

In the past, fractions of PFPE surfactant mixtures by the Solvay Solexis (Former Ausimont) process [1] had shown packing analogies with double-tail hydrocarbon surfactants [2–5]. They had a narrow MW distribution of terms with the following general structure, with $n \gg m$, q and terminals Y either fully fluorinated or chlorine-containing [6].

$$Y\text{-}(C_3F_6O)_n\,(C_2F_4O)_m(CF_2O)_qCF_2COOX$$

Here, high purity samples, with 2 to 4 pefluoroisopropoxy units, n, having Y = Cl and X = Na^+, K^+ or NH_4^+, are used for the first time. They differ from other PFPE carboxylic salts, such as those described by Hoffmann [7], in the counterion, the tail terminal and the absence of an α-perfluoromethyl group, while sharing perfluoroisopropoxy units as a major packing feature.

Experimental

Materials

All the surfactant salts had purity of at least 99% with respect to the formula below, yet including two isomers. Purities of 99.8% and 99.5% were achieved with the n2 and n3 salts, respectively.

$$Cl(C_3F_6O)_nCF_2COOX$$

In all cases, the molecular weight, by titration and NMR, agreed within 5%, with the value calculated from the molecular structure. All the samples were free of any precursor and by-product within analytical sensitivities. Even in the worst case, the n4 salts, the global residue of fluorinated impurities, irregular species with different terminals or m, q chain-units, did not exceed 1% by mole. The sodium salts contained Na_2CO_3 up to 1–2 mg/g, while calcium ion was less than 0.02 mg/g. The water was MilliQ grade.

The PFPE oils were commercial Galden® oils, from Solvay Solexis, with the general formula below

$$CF_3\text{–}(O\text{–}\underset{\displaystyle CF_3}{\underset{|}{CF}}\text{–}CF_2)_N\text{–}(OCF_2)_M\text{-}OCF_3$$

and the bulk characteristics given in Table 1. Further information on PFPE oils have been reported by Marchionni [8].

Methods

The phase diagrams have been determined by visual inspection, at 25 °C, firstly by a titration method allowing 30 minutes equilibration

Table 1

Galden® oil	av. MW	Density (25 °C) g/cm^5	Kinematic Viscosity (25 °C) cSt
HT55	340	1.65	0.45
HT110	580	1.72	0.83
HT135	610	1.73	1.0
D02–TS	760	1.77	1.8
LS215	950	1.80	3.80

for each composition and then checked, with crossed-polaroids too, on Aged Independent samples. Liquid crystalline phases have been detected by optical microscopy in polarised light, after long-term equilibration. The equilibrium surface tension has been measured, within 0.2 mN/m, with a Lauda TE1C tensiometer, by the De Nouy ring method with Harkins-Jordan correction factors [9].

Dynamic light scattering (DLS) has been performed at 25 °C, with a BI 200SM goniometer, a BI2030 correlator from Brookhaven Instruments Co. and a Spectra Physics argon ion laser at 514.5 nm. Each sample was prepared by adding microemulsion to water, at a concentration sufficiently low for negligible interaction among droplets, to monitor the Stokes diameter on time since the dilution. Strictly comparative sets of measurements, with different oils, implied the use of O/W systems with the same initial microemulsion composition, but for the type of oil, and with the same final composition for the diluted system. The presence of nanosize oil droplets in similar systems, with less pure PFPE surfactants, had been assessed and has been industrially exploited in the past [10].

Results and discussion

Surfactant-water systems

In water, these surfactants display a concentration threshold for the appearance of liquid crystals which decreases by more than a order of magnitude per each perfluoropropoxy unit in the tail. For instance, at 25 °C, the ammonium and sodium salts of n2 have an L_1 region of up to 25 and 45% wt. respectively, against the 1–2% wt. of the n3 analogues, while no solution region could be detected, within 0.1% wt., with n4 salts at this temperature. Increasing concentration and allowing equilibrium to be attained, n2 and n3 salts undergo $L_1 \rightarrow L_1 + L_\alpha \rightarrow L_\alpha$ transitions, the lamellar phase spanning over a wide concentration range. Other phases form at higher concentration and the binary phase diagrams are under examination in collaboration with Monduzzi, as in a previous NMR study of a mixture [6].

For the mixtures reported in the past, rough estimations of molecular volume, by bulk density at 25 °C, suggested 400 to over 800 Å^3 in the same MW range as the present series. Recently, a molecular volume of 433 Å^3 for the present n2NH$_4$ in aqueous solution has been reported [11], along with an estimated tail volume of 384 Å^3, which gives a first idea of the lateral CF_3 contribution.

Very preliminary evaluations of Krafft points have been carried out only to ensure appropriate experimental conditions for cmc detection, while an accurate study of dissolution and dissociation equilibria is being undertaken by Kallay.

To exemplify cmc detection, Fig. 1 shows the equilibrium surface tension of the ammonium and sodium salts of the terms n2 and n3 in aqueous solutions.

To get a first feeling of cmc variation with the number of perfluoroisopropoxy units, the cmcs at 25 °C of the n2 and n3 salts have been compared to the value obtained with a less pure sample of n1 analogue. The run was repeated at 40 °C to include n4 salts. Fig. 2 summarises the results for the ammonium salt series, suggesting a cmc reduction of roughly 1.3–1.7 orders of magnitude per each unit. The present data do not allow further speculation to date.

In micellar solutions of NH_4^+ or K^+ salts of $n = 2$, aggregation numbers of 40–60 have been reported by Gambi [11], in an SANS study proposing a transition

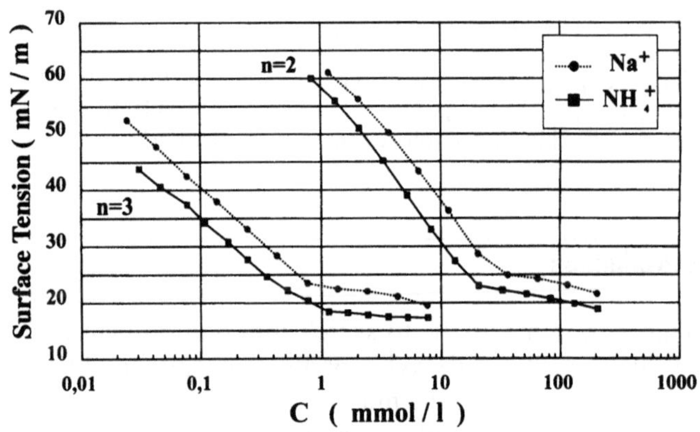

Fig. 1 Variation of the equilibrium surface tension, at 25 °C, with concentration, on log scale, for the ammonium or sodium salts of the n2 or n3 perfuoropolyether surfactants in aqueous solution

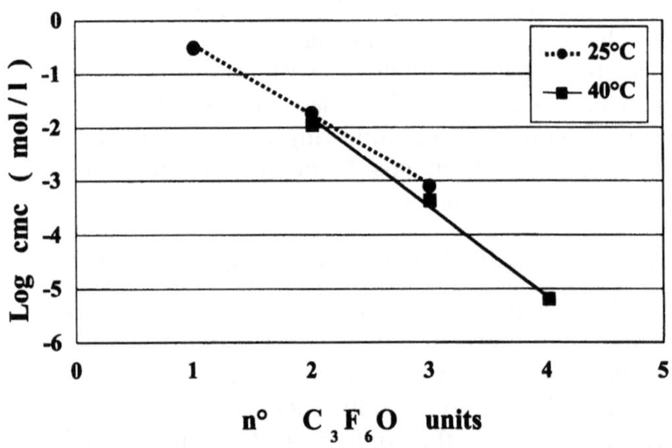

Fig. 2 Dependency of critical micellar concentration of the ammonium salts series from the number of perfluoroisopropoxy units n, at 25 °C for n1 to 3 and 40 °C for n2 to 4

from spherical to ellipsoidal micelles in 0.1–0.2 molar solutions.

Before describing surfactant mixtures in microemulsions, it is worth recalling that these analogues with a common ion have been suggested to ideally mix by the experimental cmcs of n2/n3-Na mixtures [12] in good agreement with the values calculated from the individual cmcs by the Rubingh equation [13].

O/W systems: phase diagrams

Some examples of the regions, in pseudo-ternary phase diagrams at 25 °C, where low viscosity, isotropic and clear systems form spontaneously, are now provided to illustrate the effect of the main composition parameters. Such regions include the relatively oil-rich portion where droplets occur.

Counterion

The areas obtained with Galden® D02-TS oil and the sodium, potassium or ammonium salts of an n2/n3 mixture, with molar ratio 2, are compared in Fig. 3.

The change of counterion from Na^+ to $NH_4{}^+$ reduces the extension and shifts the location of the monophasic region. This shift parallels the shift of the range for lamellar phase in pertinent S/W systems, supporting the link between these two kind of phases, similarly to other systems [14].

The maximum oil to surfactant ratio decreases in the order $NH_4{}^+ > K^+ > Na^+$, in agreement with the Hofmeister series for increasingly hydrated cations.

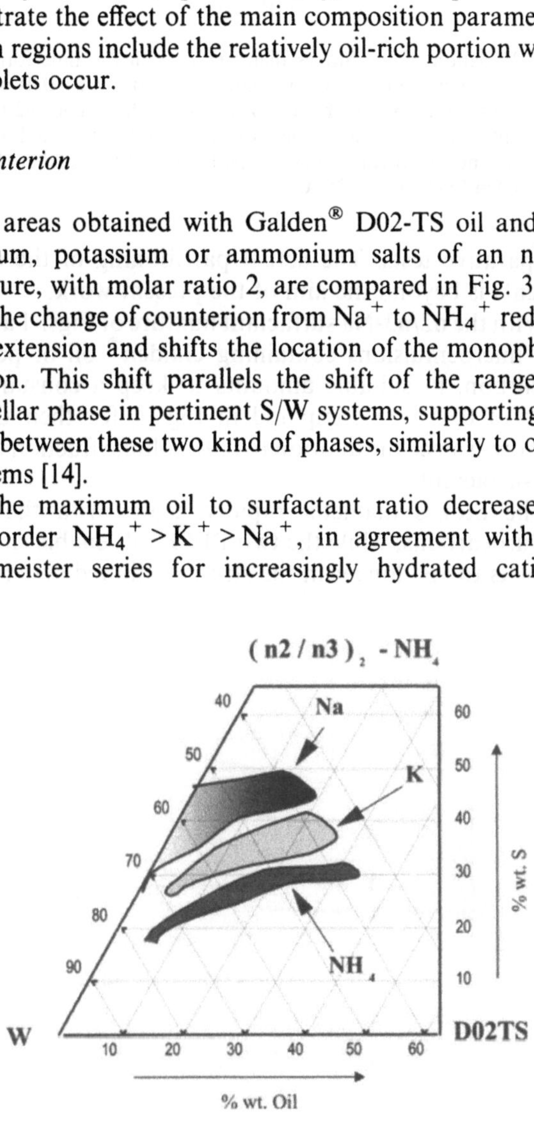

Fig. 3 Counterion effect on the region for the spontaneous formation, at 25 °C, of low viscosity, isotropic and clear monophasic systems, including O/W microemulsion area, with the ammonium, potassium and sodium surfactants salts of a binary surfactant mixture of n2/n3 (molar ratio = 2). The PFPE oil is Galden D02-TS (Table 1)

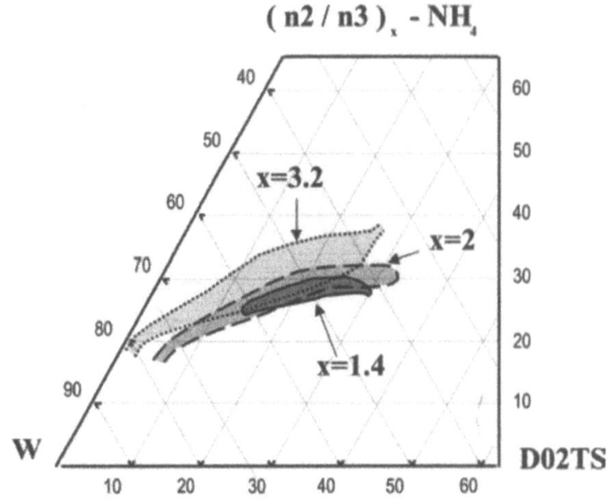

Fig. 4 Effect of the n2/n3 molar ratio, varied from 3.2 to 1.4, on the O/W region, at 25 °C, with ammonium binary surfactant salts n2/n3 and the same PFPE oil as in Fig. 3

Surfactant mixture

Figure 4 compares, in the same conditions of Fig. 3, the areas obtained with the ammonium salts upon variation of the molar ratio n2/n3. The area is progressively reduced with the decrease of n2 in the surfactant mixture. The pure n2 salt will be shown in the next section to form the same kind of systems. Although n3 and n4 terms give reverse systems only, their ternary surfactant mixtures with n2 have been ascertained to provide O/W microemulsions too.

PFPE oil

The binary surfactant mixture of Fig. 3 has been used, with sodium counterion, to exemplify the effect on the area of the average molecular size of two Galden® oils (Table 1). The HT135 oil, with average MW much smaller than the LS215 analogue, enables the largest area in Fig. 5.

The same trend, towards reduction of the area upon increase of oil size, is shown in Fig. 6 with the pure n2 ammonium salt with the same LS215 oil as the previous case compared to HT55, the Galden® of smallest average MW in the oil series.

We have assessed similar trends with binary and ternary PFPE surfactant mixtures, with either counterion, using not only the Galden® oils of Table 1 but also linear perfluorocarbons. For instance, the substitution of the Galden oils with C_6F_{14} and C_8F_{18} in the systems of Fig. 3, all other conditions kept strictly constant, enables further extension of the area, up to a maximum perfluorocarbon content exceeding 30% wt with C_6F_{14} and 25% wt with C_8F_{18}.

Selecting a set of relatively oil-rich microemulsions with the same composition, within the O/W area given by

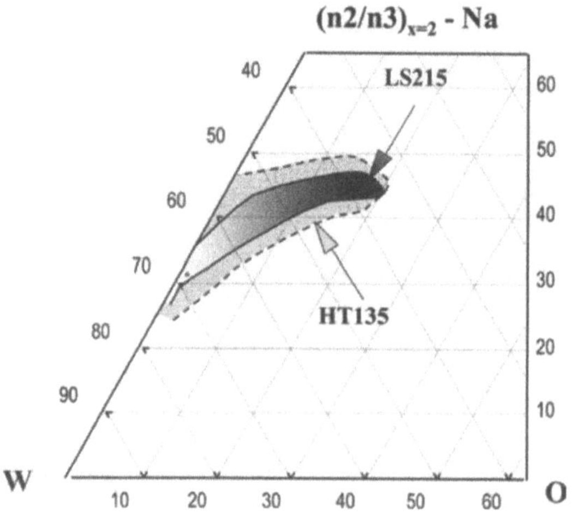

Fig. 5 Effect of the average MW of the PFPE oil (Galden HT135 and LS215, Table 3) on the O/W region, at 25 °C, with a binary surfactant mixture of n2/n3 (molar ratio = 2) sodium salts

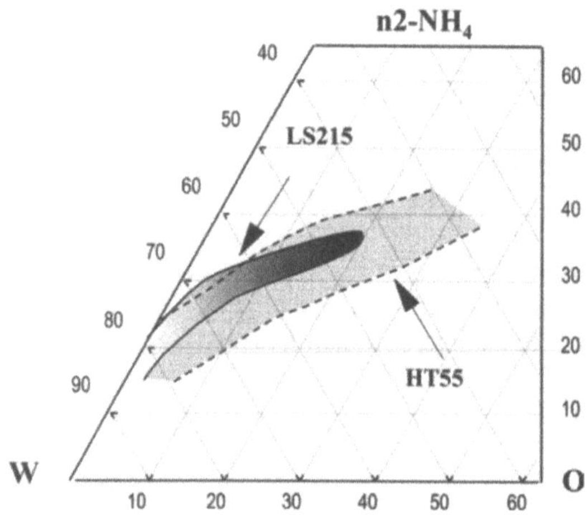

Fig. 6 Effect of the PFPE oil (Galden HT55 and LS215, Table 3) on the O/W region, at 25 °C, with a the sodium salt of n2 surfactant

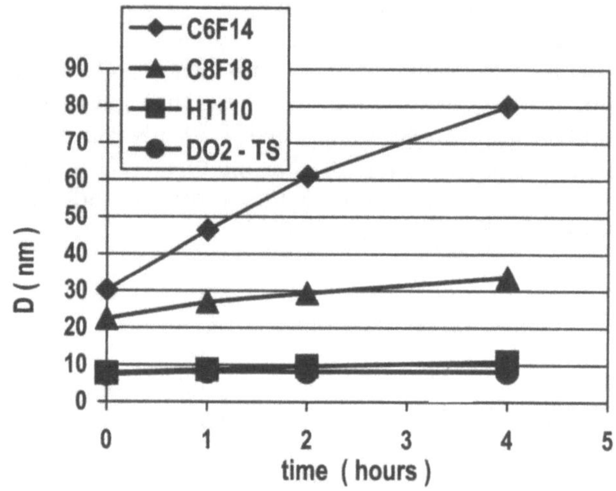

Fig. 7 Variation of the Stokes diameter D with the time form the dilution to volume fraction of fluorinated components $\Phi = 0.01$ of four O/W microemulsions with $\Phi_i = 0.45$, with the same n2/n3-Na surfactant mixture of Fig. 5: comparison of the kinetic stability with four different oils: perfluorohexane, perfluorooctane, Galden HT110 and Galden D02-TS. T = 25 °C

comparative tests. The actual partitioning of the components is beyond the aim of the present work.

With the n2/n3-Na surfactant mixture of Figs. 1 and 3, four microemulsions containing Galden® oil or perfluorocarbon, in suitable amounts to keep a constant Φ_i value of 0.45, are compared in Fig. 7, for their droplet growth since the dilution to 0.01 volume fraction of oil and surfactant.

The Stokes diameter rapidly grows with the two perflurocarbons, while the two PFPE oils enable smaller droplets to last longer in the strongly diluted system.

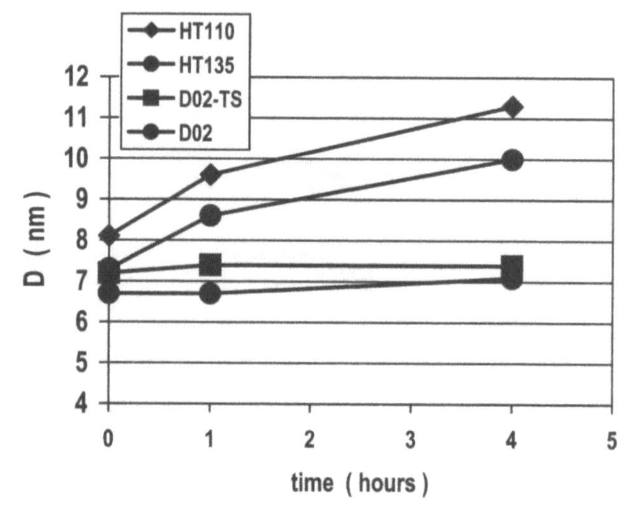

Fig. 8 Droplet growth, in the same experimental conditions of Fig. 7, for the dilution to $\Phi = 0.01$ of four O/W microemulsions ($\Phi_i = 0.35$) containing the same ternary surfactant mixture (n2/n3/n4 ammonium salts) and four different Galden oils having average MW increasing in the order HT110 < HT135 < D02TS < D02

the identical surfactant combined to different oils, it is then possible to compare their dilution behaviour, often of interest for application purposes.

Droplet growth by dilution outside the microemulsion region: kinetic stability

This section exemplifies the relatively high kinetic stability that can be achieved by nanosize fluorinated droplets in water, against ripening and surfactant repartitioning, upon dilution from an initial volume fraction of fluorinated components (Φ_i) around 0.4 to a final value of 0.01 for all the systems examined. The Stokes diameter by DLS has been used to monitor the droplets growth, versus time since the dilution, in strictly

The difference between the two Galden® oils cannot be appreciated in Fig. 7, but the average MW and MW distribution of the Galden® oil are important parameters to pursue kinetic stability, as shown in Fig. 8, where another set of microemulsions having Φ_i of 0.35, all with the same ternary mixture of ammonium surfactants but different Galden® oils, are compared in the same conditions as the previous test.

The two Galden® oils with relatively low average MW allow droplets which undergo a very limited growth, less than twice the initial diameter in few hours, but only in the case of the Galden® oils with relatively high average MW does the diameter remain constant within the experimental deviation.

Conclusion

The examination of a series of high purity PFPE carboxylic salts in solution has provided a first feeling for the contribution of each perfluoroisopropoxy unit to micellization. The same surfactants have than been used to obtain O/W microemulsions with PFPE oils and sketch the effects of counterion and relative size of oil and surfactant. Finally, the strong dilution of microemulsions in water has been shown to provide nanosize droplets with relatively high kinetic stability.

Acknowledgements Thanks are due to all the Solvay Solexis people involved in synthesis, purification and characterisation of the surfactant samples, particularly S. Fontana, G. Geniram and E. Barchiesi. E. Giannetti is thanked for his pioneering work in the exploitation of PFPE microemulsions in fluoropolymer manufacturing. AC is grateful to N. Kallay, M. Monduzzi, C. Gambi and P. Baglioni for their enthusiasm in the study of PFPE systems.

Note added in Proof: The surfactant preparation has been described by Tonelli et al. in J. Fluorine Chemistry (2002) 118: 107–121.

Further information on $n2NH_4$ micellar solutions, with focus on counterion association, have been recently reported by Kallay et al. in Colloid Surf. A (2003) 222: 95–101.

References

1. Sianesi D, Marchionni G, De Pasquale RJ (1994) In: Banks E, Smart BE, Tatlow JC (eds), Organofluorine chemistry: principles and commercial applications, chap. 20. Plenum Press, New York
2. Chittofrati A, Sanguineti A, Visca M, Kallay N (1992) Colloid Surf 63:219; Chittofrati A, Sanguineti A, Visca M, Kallay N (1993) Colloid Surf A 74:251
3. Monduzzi M, Knackstedt A, Ninham BW (1995) J Phys Chem 99:17772
4. Gambi CMC, Giri MG, Carlà M, Senatra D, Chittofrati A (1997) Phys Rev E 56:4356; Baglioni P, Gambi CMC, Giordano R (1997) Physica B 234–236:295
5. Chittofrati A, Visca M (1997) Chim Industr 79:30
6. Caboi F, Chittofrati A, Lazzari P, Monduzzi M (1999) Colloid Surf A 160:47
7. Wurtz J, Meyer J, Hoffmann H (2001) Phys Chem Chem Phys 3: DOI 10.1039/b102776j
8. Marchionni G, Ajroldi G, Pezzin G (1992) Rheology tribology engine oils (SP-936). SAE International, Warrendale, PA, pp 87–96
9. Harkins WD, Jordan HF (1930) J Am Chem Soc 52:1751
10. Giannetti E, Chittofrati A, Sanguineti A (1997) Chim Industr 79:22
11. Gambi C, Giordano R, Chittofrati A, Pieri R, Baglioni P, Texeira J (2002) Appl. Phys. A 74 [Suppl.] S436–S438 DOI 10.007/s003390201519
12. Lenti D, D'Aprile F, Chittofrati A, Visca M (1999) Communication at XIII ECIS Conference, Dublin
13. Rosen MJ (1989) Surfactants and interfacial phenomena, Chap. 3. Wiley, New York, pp 161–162 (and references thereupon)
14. Salanger JL, Anton R (1999) In: Kumar P, Mittal KL (eds), Handbook of microemulsion science and technology, Chap. 8. Marcel Dekker, New York, pp 247–280

Progr Colloid Polym Sci (2004) 123:28–30
DOI 10.1007/b11616

Abílio J.F.N. Sobral
Susana H. Lopes
António M. d'A. Rocha Gonsalves
M. Ramos Silva
A. Matos Beja
J.A. Paixão
L. Alte da Veiga

Synthesis and crystal structure of new phase-transfer catalysts based on 1,8-diazabicyclo[5.4.0]undec-7-ene and 1,5-diazabicyclo[4.3.0]non-5-ene

Abílio J.F.N. Sobral · Susana H. Lopes
António M. d'A. Rocha Gonsalves (✉)
Departamento de Química, FCTUC,
Universidade de Coimbra,
3049 Coimbra, Portugal

M. Ramos Silva · A. Matos Beja
J.A. Paixão · L. Alte da Veiga
CEMDRX, Departamento de Física,
FCTUC, Universidade de Coimbra,
3000 Coimbra, Portugal

Introduction

The efficient *N*-alkylation of 1,8-diazabicyclo[5.4.0] undec-7-ene (DBU) and 1,5-diazabicyclo[4.3.0] non-5-ene (DBN) with long-chain alkyl iodides opens the way to a new family of phase transfer catalysts.

The use of organic hindered amines such as 1,8-diazobicyclo[5.4.0]undec-7-ene (DBU) as catalysts when a strong non-nucleophilic base is required is a usual procedure. It is the case in the synthesis of 3-substituted pentane-2,4-diones [1]. The catalyst is particularly useful in the case of long-chain alkyl iodides due to its lower reactivity.

In the course of our own studies on the synthesis of pyrroles and porphyrins for the production of Langmuir-Blodgett films [2], we prepared 3-octadecylpentane-2,4-dione as a pyrrole precursor, through the C-3 alkylation of pentane-2,4-dione with 1-octadecyl iodide. When DBU was used as catalyst in that synthesis, we unexpectedly isolated the DBU iodide salt **1** (Scheme 1) as a secondary product, a stable, sharp melting point crystalline solid, in 15% yield.

Performing the reaction in the absence of pentane-2,4-dione gives exclusively the iodide salt of the *N*-alkylated DBU **1**, in 57% yield. An analogous result was obtained with 1,5-diazobicyclo[4.3.0]non-5-ene (DBN), furnishing in this case the iodide salt **2** (Scheme 1).

The new compounds **1** and **2** were characterised by ^{1}H-NMR, FT-IR and elemental analysis, giving spectroscopic and physical characteristics for the iminium salts **1** and **2**.[1] The full characterisation of these interesting compounds was definitive when the structure of salt **1** was solved by single crystal X-ray diffraction (Scheme 2).[2]

Except for a paper of 1982 that refers to the synthesis of some *N*-alkyl derivatives [3] of DBU and DBN, these nitrogen bases are considered to be very hindered non-nucleophilic bases, and are usually used taking their non-nucleophilicity as granted. Actually there are scarce

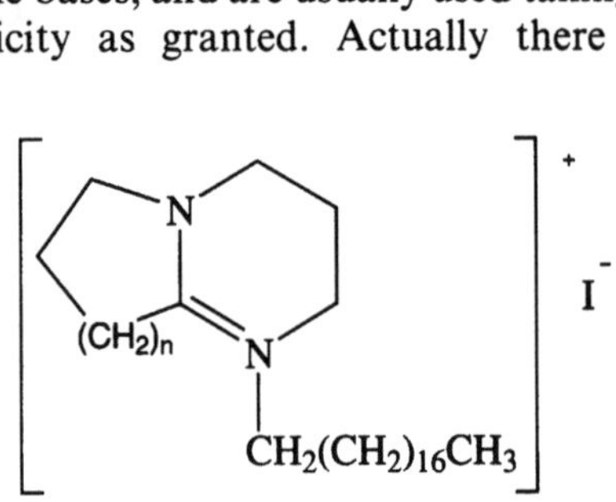

1 n = 3 DBU salt

2 n = 1 DBN salt

Scheme 1

Scheme 2 ORTEP diagram of the DBU salt **1**. Displacements ellipsoids are drawn at the 50% probability level.

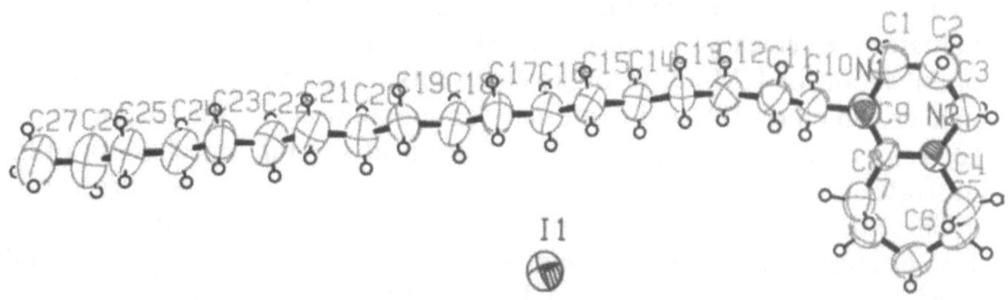

literature references to the nucleophilicity of DBU and DBN although they are always considered as unexpected. To explain those products, the delocalised positive iminium salts were suggested as intermediates for the reaction of DBU and/or DBN in the esterification of carboxylic acids with alkyl halides [4], on the reaction with bicyclic bromoketones [5], 1-halocyclopropane-1, 2-diesters [6], 1-bromo-4-benzoyloxyimino-1,2,3,4-tetrahydrophenanthrene [7] and more recently with phosphanes [8], 4-halo-3,5-dimethyl-1-nitro-1*H*-pyrazoles [9] and even with the large macrocycle of methyl pheophorbide *a* [10].

Although these results showed the nucleophilic character of DBU and DBN, the foregoing reactions have still been considered unexpected and the existence of a covalent bond between the DBU or DBN to the carbon N-substituents was only now confirmed by X-ray crystallography.

In salt **1** the bond distances between carbon C9 and nitrogen atoms are N1-C9 1.305(15) Å and N2-C9 1.330(14) Å, showing a delocalised character of the double bond and confirming the existence of a large delocalised iminium cation.

The amphiphilic nature of these salts prompted us to check their performance as phase-transfer agents. For these preliminary studies we chose the transfer of

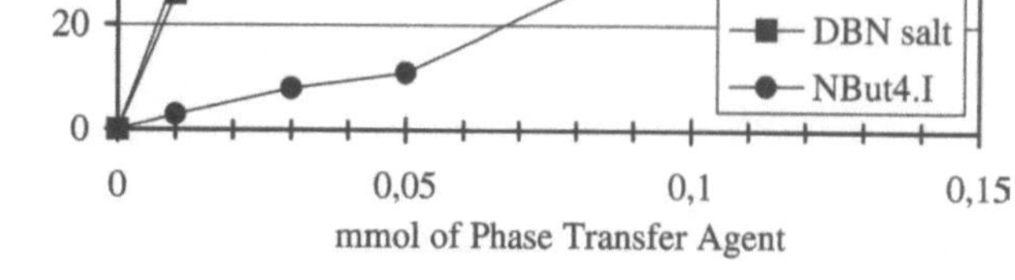

Fig. 1 Percentage of $KMnO_4$ transferred to benzene after extraction from water, with salts **1**, **2** and tetrabutylammonium iodide, presents as phase-transfer agents

[1] a) Synthesis of the DBU salt **1** (1-octadecyl-2,3,4,6,7,8,9,10-octahydropyrimido[1,2-a]azepin-1-ium; iodide): a mixture of octadecyl iodide (2 g, 5 mmol) and 1,8-diazobiciclo[5.4.0] undec-7-ene (DBU) (0,76 ml, 5 mmol) in 60 mL of dry acetone is placed in a 100-mL round-bottomed flask fitted with a reflux condenser and a silica guard tube. The mixture is stirred and heated under reflux for 5 hours. The required compound is extracted into dichloromethane/water and the organic phase is dried with anhydrous $MgSO_4$. Some remaining octadecyl iodide is removed by dissolution with ethyl ether and the desired product, which is insoluble in this solvent, is filtered off. The desired product undergoes crystallisation by slow evaporation of the solvent and is obtained with 57% yield. Melting point: 111–112 °C. ^{1}H-NMR (solvent: $CDCl_3$; internal reference: TMS): $\delta = 0.88$ (3H, t, $J = 6.7$ Hz, CH_3-$(CH_2)_n$), 1.25 (30H, m, CH_3-$(CH_2)_{15}$-CH_2), 1.64 (2H, s (broad), N-CH_2-CH_2), 1.84 (6H, s (broad), NCH_2-$(CH_2)_3$-CH_2), 2.18 (2H, m, NCH_2-CH_2-$(CH_2)_{15}$), 2.89 (2H, d, $J = 6.1$ Hz, C-CH_2-CH_2), 3.51 (2H, t, $J = 8.0$ Hz, N-CH_2-CH_2), 3.70 (6H, m, N-$(CH_2)_3$-N). Elemental analysis for $C_{27}H_{53}IN_2$: Required (C 60.88; H 10.03; N 5.26); Found: (C 60.38; H 10.05; N 5.27). FT-IR in KBr (cm^{-1};% T; group): (722.10, 73.64, $\gamma(CH_2)$); (1465.83, 60.91, $\delta(CH_2)$); (1626.55, 43.06, $\nu(C=N)$); (2849.08, 39.82, ν(C-H)); (2954.04, 52.49, ν(C-H)). b) Synthesis of the DBN salt **2** (1-octadecyl-2,3,4,6,7,8-hexahydro-pyrrolo[1,2-a]pyrimidin-1-ium; iodide): a mixture of octadecyl iodide (2 g, 5 mmol) and 1,5-diazabicyclo[4.3.0]non-5-ene (DBN) (0.65 mL, 5 mmol) in 60 mL of dry acetone is placed in a 100-mL round-bottomed flask fitted with a reflux condenser and a silica guard tube. The synthesis and isolation are as reported above for salt **1**, giving the desired product with 74% yield. Melting point: 76–78 °C. ^{1}H-NMR (solvent: $CDCl_3$; internal reference: TMS): $\delta = 0.88$ (3H, t, $J = 6.5$ Hz, CH_3-$(CH_2)_n$), 1.25 (28H, m, CH_2-$(CH_2)_{14}$-CH_2), 2.10 (2H, m, NCH_2-CH_2-CH_2N), 2.25 (4H, m, CH_2-$(CH_2)_2$-$(CH_2)_{14}$), 3.21 (2H, t, $J = 7.9$ Hz, N-CH_2-CH_2), 3.41 (2H, t, $J = 7.7$ Hz, NCH_2-CH_2-CH_2), 3.54 (4H, m, N-$(CH_2)_2$-CH_2), 3.85 (2H, t, $J = 7.4$ Hz, N-CH_2-CH_2). Elemental analysis for $C_{25}H_{49}IN_2$: required (C 59.59; H 9.79; N 5.55); Found: (C 59.19; H 9.78; N 5.68). FT-IR in KBr (cm^{-1};% T; group): (747.92, 74.15, $\gamma(CH_2)$); (765.60, 74.16, $\gamma(CH_2)$); (1505.55, 69.23, $\delta(CH_2)$); (1732.34, 68.08, $\nu(C=N)$); (3057.11, 60.84, ν(C-H)); (3075.91, 60.79, ν(C-H)).

[2] Crystal data: $C_{27}H_{53}N_2I$, M = 532.6, monoclinic a = 6.9488(5) Å, $b = 63.300(5)$ Å, $c = 7.0619(17)$ Å, $\beta = 108.751(14)°$, $V = 2941.38(8)$ Å^3, $T = 293(2)$ K, space group $P2_1/n$ (No. 14), $Z = 4$, $\mu(CuK\alpha) = 8.64$ mm^{-1}, 2860 reflections measured, 2596 unique ($R_{int} = 0.051$) which were used in the full matrix least-squares refinement. The final $R(F^2)$ was 0.079 (for $I > 2\,\sigma(I)$) and $wR(F^2)$ was 0.17 (for all reflections). Full crystal data has been deposited at the Cambridge Crystallographic Data Centre and allocated the deposition number CCDC 184814.

$KMnO_4$ from water to benzene [13], a standard system to evaluate the efficacy of phase-transfer agents. The solubilisation of $KMnO_4$ in organic solvents, aided by some crown ethers [11] and quaternary ammonium salts [12, 13], is crucial for the efficient oxidation of several substrates.

These results show that the iminium salts **1** and **2** are very promising materials. In Fig. 1 we see that the transfer of $KMnO_4$ from a water solution (0.05 mmol of $KMnO_4$ in 20 mL of water) to benzene (20 mL) is much faster with our salts than with tetrabutylammonium iodide, a classical phase-transfer agent.

The transfer of $KMnO_4$ to the organic layer is almost complete when we use the phase-transfer agent in an equimolar ratio to the inorganic salt, in opposition with the tetrabutylammonium iodide where a much higher ratio is required. Whether or not this good behaviour is related to the delocalised nature of the salts is a matter for future studies to fully interpret the phase transfer mechanism of these new compounds.

Studies are also under way to extend this *N*-alkylation reaction to other hindered nitrogen bases of the DBU family, to produce new nitrogen amphiphilic compounds.

Acknowledgements The authors would like to thank Prof. Hugh D. Burrows from the University of Coimbra for the useful discussions on the phase-transfer studies. Financial assistance from FCT (Sapiens POCTI/QUI/42536) and Chymiotechnon, Portugal, is also acknowledged.

References

1. Price R, Johnson AW, Markham E (1962) Org Synth 42:75; Clark JH, Miler JM (1977) J Chem Soc Perkin Trans I1743; Raban M, Yamamoto G (1977) J Org Chem 42:2549
2. Richardson T, Smith VC, Johnstone RAW, Sobral AJFN, d'A. Rocha Gonsalves AM (1998) Thin Solid Films 327–329:315; Ramos Silva M, Matos Beja A, Paixão JA, Alte da Veiga L, Sobral AJFN, d'A. Rocha Gonsalves AM (2000) Acta Cryst C56:1263
3. Alder RW, Sessions RB (1982) Tetrahedron Lett 23:1121
4. Ono N, Yamada T, Saito T, Tanaka K, Kaji A (1978) Bull Chem Soc Jpn 51:2401
5. House HO, DeTar MB, Vanderveer D (1979) J Org Chem 44:3793
6. McCoy LL, Mal D (1981) J Org Chem 46:1016
7. Juneja TR, Garg DK, Schafer W (1982) Tetrahedron 38:551
8. Reed R, Reau R, Dahan F, Bertrand B (1993) Angew Chem Int Ed Engl 32:399
9. Lammers H, Choen-Fernandes P, Habraken CL (1994) Tetrahedron 50:865
10. Ma L, Dolphin D (1996) Tetrahedron 52:849–860
11. Weber WP, Shepherd JP (1972) Tetrahedron Lett 4907
12. Sam DJ, Simmons HE (1972) J Am Chem Soc 94:4024
13. Herriott AW, Picker D (1974) Tetrahedron Lett 4907

Progr Colloid Polym Sci (2004) 123:31–35
DOI 10.1007/b11617

Isabelle Berlot
Yves Chevalier
Liliane Coche-Guérente
Pierre Labbé
Jean-Claude Moutet

Interfacial and micellar behaviour of pyrrole-containing surfactants

Y. Chevalier (✉)
Laboratoire des Matériaux Organiques à Propriétés Spécifiques,
UMR 5041 CNRS-Université de Savoie,
BP 24, 69390 Vernaison, France
e-mail: Yves.Chevalier@lmops.cnrs.fr
Tel.: +33-4-78022271
Fax: +33-4-78027187

I. Berlot · L. Coche-Guérente · P. Labbé
J.-C. Moutet
Laboratoire d'Électrochimie Organique et de Photochimie Rédox, UMR 5630
CNRS – Université de Grenoble 1,
BP 53, 38041 Grenoble, France

Abstract The physicochemical properties of new electropolymerisable cationic surfactants having a pyrrolyl group attached and unusual counterions have been studied in aqueous solutions and at the air-water interface. The tetrafluoroborate and tosylate anions behave as quite hydrophobic counterions as compared to the conventional bromide. The pyrrolyl group of moderate polarity has a dual behaviour: it behaves as a hydrophobic substituent when it is attached close to the polar head of the surfactants, but its low polarity manifests when it is attached to the end of the hydrophobic chain. Thus, the presence of the pyrrolyl group at the chain end does not affect the *cmc* value. The pyrrole ring was found located at the micellar surface in the dilute regime; the resulting folding of the hydrophobic chain induces a strong curvature of the interface; small and spherical micelles are formed. A concentrated regime is reached where the interfacial curvature is reduced: the micelles progressively grow in size and change their shape into elongated ellipsoids. The increasing lateral interactions at the level of the headgroups expel the pyrrolyl groups into the hydrophobic micellar core.

Keywords Cationic surfactant · Pyrrole · Tosylate · Micelles · Adsorption

Introduction

Electropolymerisable pyrrole-containing cationic surfactants allow the synthesis of water-swollen cationic gels at the surface of electrodes by means of their in situ electrochemical polymerisation [1, 2]. Thin layers are easily obtained at the surface of electrodes by a simple electropolymerisation from an aqueous solution [3]. But thick layers can be prepared as well since the monomer is allowed to penetrate and diffuse inside the water-swollen polymerised materials [3]. On the contrary, a water-insoluble polymer is obtained by polymerisation of the water-soluble pyrrole; the thin waterproof layer of polypyrrole formed at the electrode surface prevents the electropolymerisation to go on. The cationic gels are used as an immobilisation matrix for various redox species including redox enzymes such as glucose oxidase or polyphenol oxidase [4]. Lastly, the amphiphilic polycationic gel forms a structured layer which influences the course of redox reactions of entrapped species [1, 2]. The modified electrodes with the entrapped redox enzymes are the primary units for the elaboration of electrochemical devices used as chemical sensors [5].

The cationic surfactants used for the electrochemical polymerisation are quite different from the usual ones. Firstly, an electroactive group, the pyrrole in the present case, is attached to the surfactant molecules. Secondly, the counterions should not interfere with the electropolymerisation. Usual counterions such as chloride or bromide anions are oxidised at the potentials used for the polymerisation of pyrrole. Nitrate, tetrafluoroborate or tosylate anions which are often chosen for that purpose

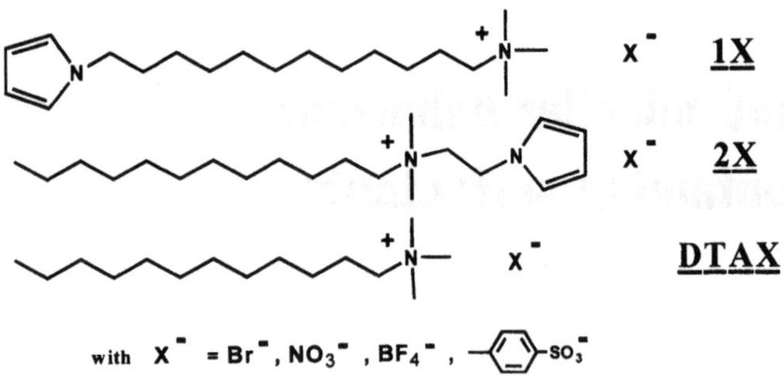

Fig. 1 General chemical formulae of the cationic surfactants studied

are less common in the field of surfactant science and this fact deserves some investigation into their interfacial properties [6].

In the present work, the influence of these structural peculiarities of the electropolymerisable surfactants is investigated: the presence of the pyrrolyl group and the substitution of the bromide for the anions of electrochemists. A series of electropolymerisable cationic surfactants was studied and compared to the common cationic surfactant DTABr as a reference. Thus, the pyrrolyl group was attached either at the end of the alkyl chains in the 1X series, or at the level of the cationic headgroup in the 2X series. The counterions X^- were NO_3^-, BF_4^- or tosylate (OTs^-) and were compared to Br^- (Fig. 1).

This paper is divided into three parts dealing with the properties of the unusual counterions used in electrochemistry, the influence of the presence of the pyrrolyl group attached to the surfactant molecules and a detailed study of the 1OTs surfactant which shows bistability behaviour.

Influence of the nature of the counterions

The nature of the counterions affects the properties of surfactants because of the contribution of non electrostatic interactions. Thus, on the ground of electrostatic interactions only, every monovalent counterion should have identical properties. Their adsorption in the electrical double layer at the surfactant interface should follow the Poisson-Boltzmann equation. This is far from reality. There is a counterion specificity following the Hofmeister series. Some anions such as hydroxide or acetate bind very weakly to cationic surfactants, leaving strong electrostatic repulsions between the surfactant headgroups at the interface; the consequence is a strong curvature of the surface, a large *cmc* value and a small micellar size [7]. On the contrary, more "hydrophobic" anions such as iodide or salicylate bind very strongly, the *cmc* is small, the resulting interface becomes electrically quasi-neutral and of moderate curvature; large elongated micelles form and viscoelastic behaviour can be observed in solutions of very long cylindrical micelles [8].

The basic properties of the surfactants (Table 1), namely the *cmc*, the surface tension lowering and the area per molecule at the air-water interface, allow us to sort the counterions with respect to their increasing "hydrophobicity". The *cmc* values are particularly well suited for that purpose. The same order was found in the DTAX and 1X series: $\mathbf{Br^- \approx NO_3^- < BF_4^- < OTs^-}$. The area per molecule (a_0) and surface tension lowering at the *cmc* (γ_{cmc}) follow the same trend, but less clearly. The stronger adsorptions of the more "hydrophobic" counterions which decrease the electrostatic repulsions between headgroups, gave a lower *cmc*, a denser packing of the surfactant at the air-water interface (lower a_0) and a larger effect on the surface tension (lower γ_{cmc}). The values of a_0 and γ_{cmc} pertaining to the BF_4^- counterion depart from the overall trend, but the measurements were carried out at 50 °C because of the high Krafft temperature and the completion of ion exchange may also be questioned. The full 2X series could not be studied because of the too high Krafft temperatures of $2BF_4$ and 2OTs.

Table 1 Basic properties of the DTAX, 1X and 2X surfactants as a function of the type of counterion. They were determined by means of surface tension and electrical conductivity measurements (conductivity alone for the entries where γ_{cmc} and a_0 values are lacking)

Compound	Krafft temperature	Critical micelle concentration, *cmc*	Surface tension at the cmc, γ_{cmc}	Area per surfactant molecule, a_0
DTABr	<0 °C	15 mmolL^{-1}	39 mN/m	68 Å^2
$DTANO_3$	<0 °C	11.2 mmol·L^{-1}		
$DTABF_4$ (at 55 °C)	50 °C	8 mmol·L^{-1}	28 mN/m	50 Å^2
DTAOTs	<0 °C	5 mmol·L^{-1}	31 mN/m	62 Å^2
1Br	10 °C	13 mmol·L^{-1}		
$1NO_3$	<0 °C	10 mmol·L^{-1}		
$1BF_4$ (at 55 °C)	44 °C	5 mmol·L^{-1}	38 mN/m	160 Å^2
1OTs	13 °C	4.4 mmol·L^{-1}	41 mN/m	92 Å^2
2Br (at 50 °C)	40 °C	4 mmol·L^{-1}	33 mN/m	176 Å^2
$2NO_3$ (at 45 °C)	40 °C	3.5 mmol·L^{-1}	31 mN/m	87 Å^2
$2BF_4$	95 °C			
2OTs	95 °C			

Influence of the pyrrolyl group

Pyrrolyl attached to the headgroup

In the 2X series, a (2-pyrrolyl)ethyl group was attached to the ammonium nitrogen atom in place of a methyl group of the reference DTAX. For both the Br^- and NO_3^- counterions, the *cmc* were lower by a factor of 3–4 in the 2X series (Table 1). The pyrrolyl group acts as an hydrophobic substituent when it is attached to the headgroup. The same effect was observed when one *N*-methyl group was substituted for a short *n*-alkyl chain in the DTABr series [9]. The (2-pyrrolyl)ethyl group is found equivalent to a pentyl residue [9] on the basis of the *cmc* values. The same *cmc* ratio was also observed between dodecyldimethylbenzylammonium chloride and the corresponding DTAC reference [10]. The pyrrolyl substituent acts as a second short hydrophobic chain.

Pyrrolyl attached at the end of the hydrophobic chain

The properties of the pyrrolyl group appear quite different when it is attached at the end of the hydrophobic chain. Thus, the *cmc* of the 1X series are close to the *cmc* of the DTAX reference (Table 1). The substitution of the hydrogen atom of the terminal methyl of DTAX for pyrrole does not change the overall balance between hydrophobic and hydrophilic contributions. This is obviously a compensation of opposing effects since differences are observed when looking at other properties. Thus, the mean interfacial area per molecule at the air-water interface (a_0) is significantly larger for 1OTs than for DTAOTs although their *cmc* are identical (Fig. 2). This supports the idea that the pyrrolyl group contributes to the steric hindrance at the level of the polar layer where the headgroups and the counterions are. The pyrrolyl group is located at the vicinity of the headgroups in the monolayer. The average conformation of the surfactants is a loop as shown in Fig. 2.

These observations show that the pyrrole has a dual character because of its medium polarity. It behaves as a polar group in the 1X series but as a hydrophobic substituent in the 2X series.

Dual behaviour of the pyrrolyl group

The micellar behaviour of the 1OTs micelles shows two distinct regimes as a function of the surfactant concentration. A clear evidence is the Debye plot of a classical

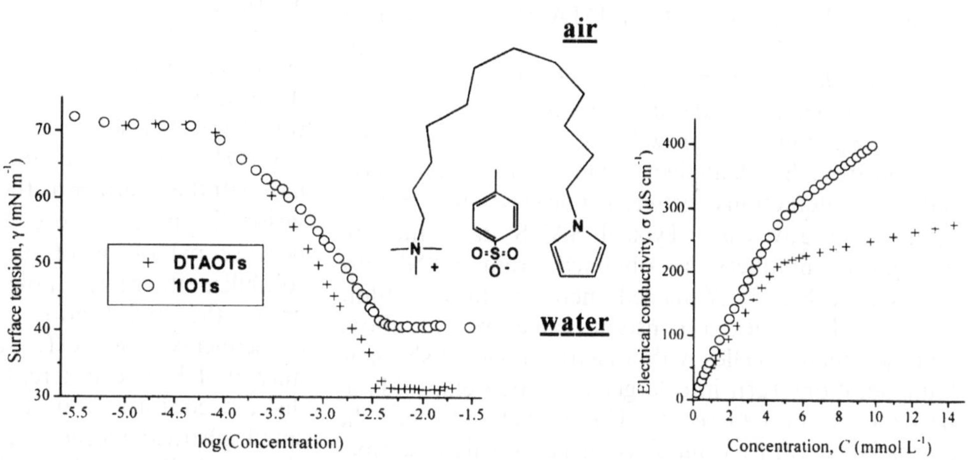

Fig. 2 Comparison of the tensioactive properties of 1OTs and DTAOTs. Surface tension (*left*) and electrical conductivity (*right*) measurements show that their *cmc* are the same. But the slope of γ vs. $\log(C)$ is less for 1OTs, indicating a larger area a_0. A sketch of the conformation of 1OTs at the air-water interface is presented in the middle of the figure

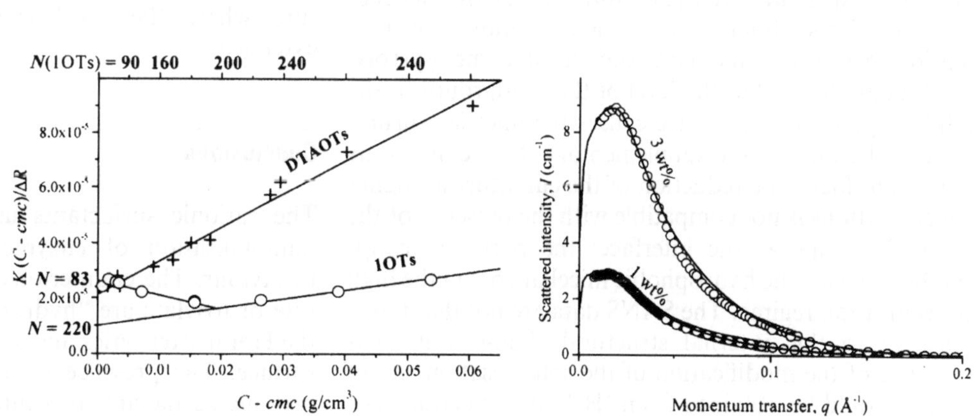

Fig. 3 Debye plot of classical light scattering (*left*) for 1OTs (○) and DTAOTs (+) and typical SANS data (*right*) for 1OTs 1 wt % and 3 wt %. The solid lines in the SANS plot are the best fits to the experimental data for charged ellipsoids with the aggregation numbers reported on top of the Debye plot

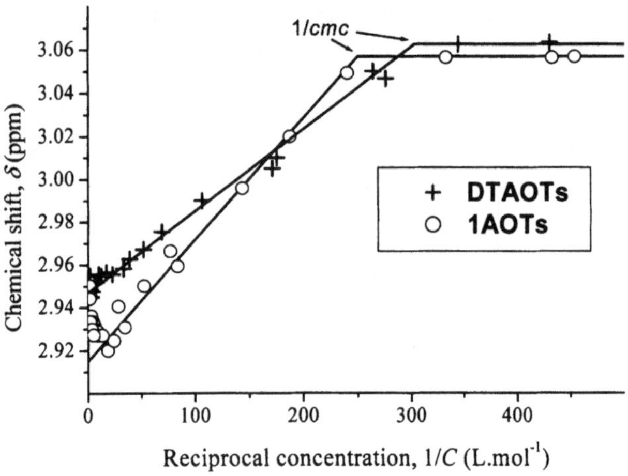

Fig. 4 ^{1}H-NMR chemical shift of the $N(CH_3)_3$ protons of 1OTs (○) and DTAOTs (+) as a function of the inverse concentration $1/C$. The solid lines correspond to the linear relationship

light scattering experiment at the fixed angle of 90° (Fig. 3). A concentrated regime above 2–3 wt % looks similar to the data for DTAOTs: the slope is positive indicating strong repulsive interactions between ionic micelles and the extrapolation of the data to infinite dilution gives an aggregation number $N=220$. But the data depart from the classical behaviour in the dilute regime for 1OTs (but not for DTAOTs): the slope is negative and the extrapolation to infinite dilution indicates smaller micelles with $N=83$. The negative slope indicates either attractive interactions between micelles or a steep micellar growth as a function of the concentration. Small angle neutron scattering data show that the interactions between micelles are strongly repulsive in any case. Typical SANS data and the aggregation numbers as obtained from SANS are reported in Fig. 3. N indeed increases in the dilute regime and reaches a constant value of 240 above 3 wt %. Small micelles with a nearly spherical shape in dilute solutions turn into larger elongated ellipsoids as the concentration increases. This transition where the interfacial curvature is modified, is necessarily accompanied by a structural reorganisation inside the micelles because of the different packing constraints. Surface tension measurements have shown that the pyrrolyl residues are located at the level of the headgroups in the dilute regime; their presence causes the micellar surface to curve because of the supplementary steric constraints at the interface. The reduction of the curvature at higher concentrations is not compatible with the presence of the pyrrolyl groups at the interface: the pyrrolyl groups solubilise inside the hydrophobic micellar core in the high concentration regime. The SANS data are not that much sensitive to this internal structural change but clear evidence of the modification of the solubilisation site of pyrrole could be obtained from ^{1}H-NMR experiments.

The ^{1}H-NMR resonances of the $N(CH_3)_3$ protons of 1OTs and DTAOTs shift upfield as the surfactant concentration increases above the *cmc*. The upfield shift is caused by the strong adsorption of the tosylate counterions at the micellar surface; the presence of the aromatic counterion shifts upfield the ^{1}H-NMR line of the neighbouring protons. This effect is well documented for different types of aromatic counterions [11] as well as for neutral aromatic solutes [12]. It is observed for both 1OTs and DTAOTs. Since micellised surfactant is in equilibrium with a residual concentration of non-micellised surfactant (monomeric) equal to the *cmc*, the chemical shift δ varies linearly as a function of the reciprocal concentration of the surfactant $1/C$ [13]:

$$\delta = x_{\mathrm{mono}}\delta_{\mathrm{mono}} + x_{\mathrm{mic}}\delta_{\mathrm{mic}} = \delta_{\mathrm{mic}} + (\delta_{\mathrm{mono}} - \delta_{\mathrm{mic}})\frac{cmc}{C}$$

The ^{1}H-NMR data of DTAOTs follow this linear relationship, showing that there exists a unique micellar state in the concentration domain studied. But the data pertaining to 1OTs depart from this two-states model (monomeric and micellar), indicating a second micellar state at high concentrations. The extrapolation to infinite concentration ($1/C=0$) gives the value of the chemical shift in the micellar state. The upfield shift of 1OTs in the dilute micellar regime is larger than that of DTAOTs because of the presence of the pyrrolyl groups at the interface. But the ^{1}H-NMR line of 1OTs shifts back downfield above 0.1–0.2 $mol{\cdot}L^{-1}$ and tends to the same extrapolated value as that of DTAOTs. This shows that the environment of the $N(CH_3)_3$ protons in the concentrated micellar solutions of 1OTs is identical to that in DTAOTs, that is, free of pyrrole. The pyrrolyl groups have left the interface as the micelles grew and solubilised inside the micellar core. Such a behaviour is possible because pyrrole is of medium polarity, it is able to solubilise either in the polar environment of the interface or in the hydrophobic core of the micelles. As a consequence, the 1OTs surfactant is bistable [14]. It may exist in two different states corresponding to two distinct conformations of surfactant molecules: either small spherical micelles with a loop-like conformation, or elongated ellipsoidal micelles of lower interfacial curvature where the alkyl chain conformation is more extended.

Conclusions

The cationic surfactants used for the electrochemical immobilisation of enzymes have quite a particular behaviour. The counterions used such as tetrafluoroborate or tosylate are "hydrophobic" anions according to the Hofmeister series and strongly adsorb at the micellar surface. The presence of a pyrrole group attached to cationic surfactant molecules is quite complex because pyrrole can solubilise both in polar and non-polar environments. Depending on its solubilisation site, its contributions to the surfactant properties are opposite. Pyrrole acts as an hydrophobic substituent when it is attached to the surfactant headgroup because it is forced to solubilise in the polar interfacial medium. But it acts as a polar group when it is attached at the end of the hydrophobic chain of the surfactant. Small changes of the external constraints such as the surfactant concentration are enough for switching the role of pyrrole from a polar-like substituent to a hydrophobic substituent. The origin of the bistability is the medium polarity of pyrrole.

References

1. Coche-Guérente L, Deronzier A, Galland B, Labbé P, Moutet J-C, Reverdy G (1991) J Chem Soc, Chem Commun 386; Coche-Guérente L, Deronzier A, Galland B, Moutet J-C, Labbé P, Reverdy G, Chevalier Y, Amrhar J (1994) Langmuir 10:602
2. Collard DM, Fox MA (1991) J Am Chem Soc 113:9494; Collard DM, Stoakes MS (1994) Chem Mater 6:850
3. Berlot I, Labbé P, Letellier P, Moutet J-C, Turmine M (1997) J Electroanal Chem 431:57; Berlot I, Labbé P, Moutet J-C (2000) Langmuir 16:5814
4. Cosnier S, Labbé P (1993) In: Guilbault GG, Mascini M (eds) Uses of immobilized biological compounds. NATO ASI Ser, Ser E, Applied Sciences, Kluwer Academic Pub 252, p 231
5. Jaffrezic N, Souteyrand É, Martelet C, Cosnier S, Labbé P, Pijolat C (1996) Les capteurs chimiques. CMC2, Lyon (available from CMC2, École Centrale de Lyon, http://www.ec-lyon.fr/associations/cmc2)
6. Berlot I, Chevalier Y, Labbé P, Moutet J-C (2001) Langmuir 17:2639
7. Lianos P, Zana R (1983) J Phys Chem 87:1289; Ninham BW, Evans DF, Wei GJ (1983) J Phys Chem 87:5020
8. Gravsholt S (1976) J Colloid Interface Sci 57:575; Ulmius J, Wennerström H, Johansson LB-Å, Lindblom G, Gravsholt S (1979) J Phys Chem 83:2232; Hoffmann H, Platz G, Rehage H, Schorr W (1981) Ber Bunsenges Phys Chem 85:877; Hoffmann H, Kalus J, Thurn H, Ibel K (1983) Ber Bunsenges Phys Chem 87:1120; Olsson U, Söderman O, Guéring P (1986) J Phys Chem 90:5223
9. Zana R (1980) J Colloid Interface Sci 78:330; Zana R, Levy H (1995) J Colloid Interface Sci 170:128
10. Rózycka-Roszak B (1990) J Colloid Interface Sci 140:538
11. Rao URK, Manohar C, Valaulikar BS, Iyer RM (1987) J Phys Chem 91:3286; Bachofer SJ, Simonis U (1996) Langmuir 12:1744
12. Eriksson JC, Gillberg G (1966) Acta Chem Scand 20:2019
13. Muller N, Platko FE (1971) J Phys Chem 75:547; Odberg L, Svens B, Danielsson I (1972) J Colloid Interface Sci 41:298; Drakenberg T, Lindman B (1973) J Colloid Interface Sci 44:184; Chevalier Y, Chachaty C (1984) Colloid Polym Sci 262:489
14. Cantù L, Corti M, Del Favero E, Raudino A (1997) J Phys Condens Matter 9:5033; Raudino A, Cantù L, Corti M, Del Favero E (2000) Curr Opin Colloid Interface Sci 5:13

Progr Colloid Polym Sci (2004) 123: 36–39
DOI 10.1007/b11618

Gerd Persson
Håkan Edlund
Göran Lindblom

Phase behaviour of the 1-monooleoyl-*rac*-glycerol/*n*-octyl-β-D-glucoside/water system

G. Persson (✉) · H. Edlund
Department of Natural and Environmental Sciences, Mid Sweden University, 851 70 Sundsvall, Sweden
e-mail: Gerd.Persson@kep.mh.sem
Fax: +46-60-148802

G. Lindblom
Department of Biophysical Chemistry, Umeå University, 901 87 Umeå, Sweden

Abstract Obtaining high-quality crystals for X-ray diffraction from membrane proteins has proven to be a difficult task. One recently presented method utilises the cubic phases formed by 1-monooleoyl-*rac*-glycerol (MO). Removing the proteins from their native environment requires the use of surfactants. One commonly used surfactant is *n*-octyl-β-D-glucopyranoside (OG). Using NMR techniques and visual observations, the ternary phase diagram of MO/OG/^2H_2O was outlined at 25 °C. The preliminary data show that all phases present in the binary systems at this temperature are also found in the ternary. Further, at the OG-rich side, an additional phase that appears to be hexagonal occurs. Addition of minor amounts (≈1.5 wt/wt %) of OG converts the cubic phases of MO to a lamellar structure, while the OG-rich cubic phase is able to dissolve about 15 wt/wt % MO. OG in water forms a large micellar solution phase. Increasing the MO concentration at constant water content leads to a series of two- and three-phase areas in which one or two phases are in equilibrium with almost pure water.

Introduction

A complete structural determination of membrane proteins is one of the cornerstones in the understanding of their function in the biological cell. The most important method used is X-ray-diffraction, which require high-quality crystals, and one of the main issues is concerned with the problem to obtain such crystals. Water-soluble proteins are relatively easy to crystallise, which can be seen from the amount of structures so far determined [1]. Membrane proteins are far more difficult to work with, mainly due to the need to remove them from their native environment. The solubilisation process may lead to denaturation of the proteins, thus destroying them. A new approach to the problem was introduced in 1996 [2]. This method includes the use of a bicontinuous cubic phase as the crystallisation medium. The general idea behind this approach was to introduce the proteins into an environment that was similar to their native one [2]. The bicontinuous cubic phases formed by 1-monooleoyl-*rac*-glycerol (MO) [3, 4] meets the requirements. The exact mechanisms involved in the process are yet to be elucidated, and since the introduction of this method, to this day only two membrane proteins have been successfully crystallised [2, 5, 6]. Briefly, this method can be described as follows: mix appropriate amounts of MO and protein solution, add a precipitant (for example salt) and wait for a couple of months for the crystals to form. This has even been accomplished with a membrane dispersion of native membranes naturally enriched with a protein, thus omitting any potentially denaturating detergent-depending procedures [7]. In order to fully utilise this method several crucial issues need to be solved. From a liquid crystalline point of view, knowledge of the microstructure of the liquid crystalline phases (LC) involved, and the effect different additives may have on these LC, are important. In the literature, the most frequently used surfactant is *n*-octyl-β-D-glucopyranoside (OG), but to our knowledge, only a few attempts to investigate its effects on the stability of the MO cubic phases have been published [8, 9]. In reference [9] only one composition was investigated, and the focus lied on the formation of the lamellar phase. In a recent study by Ai and Caffrey the effect of *n*-dodecyl-*β*-D-maltoside (DM) on the stability of the MO cubic phases was investigated [10]. This work shows that addition of DM converts the *Pn3m* cubic phase to a lamellar structure via the *Ia3d* structure.

In this report, we present the preliminary data of the ternary MO/OG/^2H_2O phase diagram studied at 25 °C by NMR.

Materials and methods

1-Monooleoyl-*rac*-glycerol (MO) (>99% purity) and *n*-octyl-*β*-D-glucopyranoside (OG) (>98% purity) were purchased from Sigma Aldrich Chemie Gmbh, Germany and used without further purification. ^2H_2O (99.9% ^2H) was purchased from Cambridge Isotope Laboratories, USA.

Sample preparation. The first series of samples were prepared in 8-mm glass test tubes, which were sealed with removable caps. The entire phase diagram was scanned by successive addition of ^2H_2O followed by a short equilibrium time (1 week). All samples were inspected both visually and between crossed polaroids to check homogeneity and birefringency. All anisotropic samples were investigated by ^2H-NMR.

Equilibrium samples were prepared by weighing the appropriate amounts of each substance into 8-mm glass tubes, which were flame sealed. The samples were homogenised by rotation and equilibrated at 25 °C. The ^2H-NMR experiment was repeated at regular intervals until no further change could be detected. All samples were inspected both visually and between crossed polaroids to check homogeneity and birefringency.

^1H-NMR samples were prepared in 8-mm test tubes and sealed with removable caps. After an incubation time of 7 days in 25 °C they were placed into 5-mm NMR tubes. When more than one phase was present, the samples were separated into the constituent phases.

Methods

^2H-NMR. The ^2H-NMR quadrupolar splittings were obtained using a Bruker Avance DPX 250 NMR spectrometer. This method is a well established [11] non-destructive method for investigating multiphase samples.

1H-NMR. The high-resolution ^1H-NMR spectra were obtained using a Bruker Avance DPX 250 NMR spectrometer.

Results

The phase diagram was outlined using ^2H-NMR, visual observations as well as observations in polarised light. In the main phase diagram (Fig. 1) no two- and three-phase areas are indicated. They are observed, but neither their exact locations nor the tie-lines have been determined. The phase boundaries of the single-phase areas are exact within 2 wt % except for the lamellar phase. This boundary is based on the first, approximate, series of samples. Since no SAXS results are available, the boundary between the two MO-rich cubic phases may be less accurately determined.

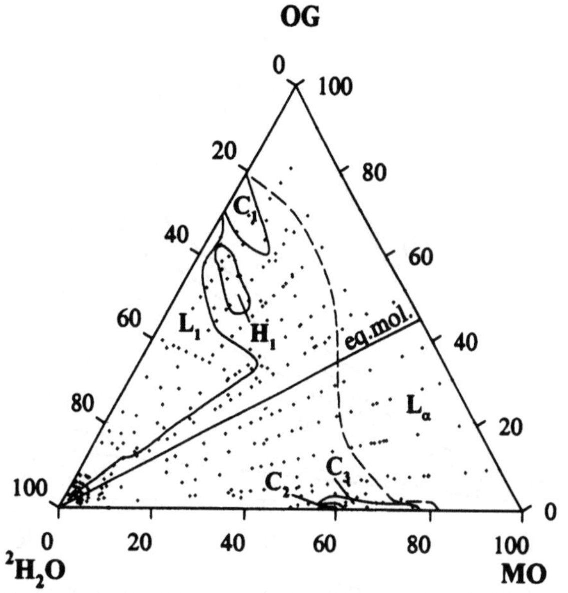

Fig. 1 Isothermal ternary phase diagram of the MO-OG-^2H_2O system at 25 °C. *Abbreviations:* L_I, isotropic solution phase; H_I, OG-rich normal hexagonal phase; C_1, OG-rich cubic phase; C_2, MO-rich cubic phase of space group *Pn3m*; C_3, MO-rich cubic phase of space group *Ia3d*; L_α, lamellar phase. The equimolar line is also indicated

OG-rich side

The preliminary data indicate that, except from all the phases present in the binary system [12] at this temperature, an additional phase occurs at high concentration of OG in the ternary system. The appearance of this phase indicates that it is hexagonal. The OG-rich cubic phase is able to dissolve about 15 wt/wt % MO. At this stage we do not know whether incorporation of MO affects this cubic structure. At low water content a phase, which appears to be lamellar, stretches across the entire phase diagram.

MO-rich side

The structure of the cubic phases has not been determined, thus we cannot tell whether addition of OG transforms the *Pn3m* cubic phase to the lamellar structure via the *Ia3d* structure, or if it converts to the lamellar structure directly. Previous results for DM [10] show this trend though, and therefore it is plausible for it to occur also in our case. Addition of minor amounts (≈1.5 wt/wt %) of OG to the *Ia3d* cubic phase of MO converts it to a lamellar structure.

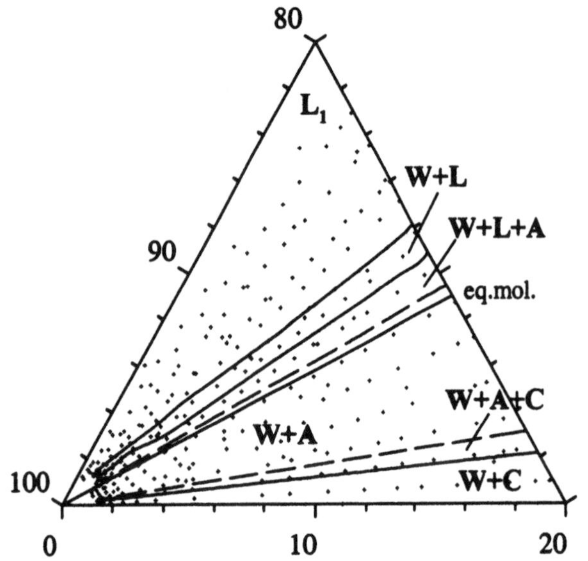

Fig. 2 Aqueous corner. *Abbreviations:* L_1, isotropic solution phase; W, water-rich isotropic liquid; L, water-depleted isotropic liquid; A, white substance as described in the text; C, cubic phase. The equimolar line is also indicated.

Table 1 Samples studied by ^{1}H-NMR. *Abbreviations:* L_1: isotropic liquid; W: water-rich, dense, isotropic liquid containing very small amounts of OG; L: water-depleted, low-density, viscous, isotropic liquid containing the majority of MO and OG; A: white, non-transparent, viscous, anistropic substance; C: MO-rich cubic phase

Sample	Wt % $MO/OG/^2H_2O$	X_{MO}/X_{OG}	Appearance
A	0.30/1.70/98.00	0.22	1 ϕ, L_1
B	0.62/1.49/97.89	0.51	1 ϕ, L_1, bluish
C	0.73/1.45/97.82	0.61	2 ϕ, W+L
D	0.82/1.33/97.85	0.75	2 ϕ, W+L
E	0.93/1.08/97.99	1.05	3 ϕ, W+L+A
F	1.05/1.06/97.89	1.21	2 ϕ, W+A
G	1.34/0.75/97.91	2.18	2 ϕ, W+A
H	1.76/0.58/97.66	3.70	2 ϕ, W+A
I	1.96/0.29/97.75	8.24	3 ϕ, W+A+C
J	1.89/0.10/98.02	23.59	2 ϕ, W+C

Isotropic solution phase

OG in water forms a large region of a micellar solution phase, and at a water content range of 43 to 75 wt/wt %, MO can be solubilised in this phase up to a molar ratio (MO/OG) of ca. 1:1.5. The aggregate structures have not been studied in detail, so far. Visual observations of sample viscosity do not, however, imply enhanced micellar growth. Upon a further increase in the water content, this ratio decreases to ca. 1:1.8. Increasing the MO concentration beyond this ratio leads to separation into two liquid phases. The denser phase contains almost pure water, while the remainder contains mainly MO and OG as determined by ^{1}H-NMR. Close to the two-phase region in the solution phase the samples appears bluish. This is probably due to the onset of the phase separation, similar to the cloudpoint-phenomenon.

Further addition of MO yields a series of two and three-phase areas (Fig. 2 and Table 1). In all these multiphase areas, except the one of highest MO concentration, a white, non-transparent, anisotropic and viscous substance constitutes one of the phases. Preliminary microscopy results indicate that this white substance consists of anisotropic particles. Furthermore, besides the large singlet, splittings are obtained by ^{2}H-NMR for these samples. Thus, we believe that this region of the phase diagram consists of a lamellar dispersion in equilibrium with a dilute OG solution. This is also consistent with the results obtained by Angelov et al. [9].

Discussion and conclusions

In this report, we have investigated the entire ternary phase diagram of $MO/OG/^2H_2O$ at 25 °C by NMR. These preliminary results show that only small amounts (1 OG per 28 to 14 MO) of OG are sufficient to transform the cubic structure to a lamellar one. At this stage of our study we are unable to conclude whether this transformation goes via the *Ia3d* phase, in analogy with previous results published by Ai and Caffrey for a similar system [10].

The observed phase behaviour of the ternary system may play an important role in the crystallisation process of membrane proteins. The structures of these two cubic phases have been proposed to be described by infinite periodic minimal surfaces (IPMS). The lipid forms a bilayer and the IMPS constitutes the dividing surface between the two monolayers [13]. If the proteins are reconstituted in the lipid bilayer, the large protein molecules may affect the phase behaviour in such a way that a more planar lipid film is created. However, the effect may not be large enough to change the phase behaviour of the entire sample, since the total protein concentration is quite low, but locally the effect may be dramatic.

The OG-rich cubic phase can incorporate about 1 MO per 6 OG. At lower concentrations of OG, addition of MO leads to the formation of a phase which most likely is hexagonal.

As can be seen from the phase diagram, the OG micellar aggregates can accommodate substantial amounts of MO. In the concentrated region of the L_1 phase 1 MO per 1.5 OG molecules can be solubilised. Upon an increase in the water content this ratio decreases to 1:1.8. Increasing the MO to OG ratio in this dilute regime leads to a complex phase behaviour involving several multi-phase areas in which surfactant-rich phases are in equilibrium with almost pure water. This is currently being studied in more detail in our laboratories.

References

1. The protein data bank: http://www.rcsb.org/pdb
2. Landau E, Rosenbusch J (1996) Proc Natl Acad Sci USA 93:14535
3. Larsson K (1989) J Phys Chem 93:7304
4. Qiu H, Caffrey M (2000) Biomaterials 21:223
5. Kolbe M, Besir H, Essen LO, Oesterhelt D (2000) Science 288:1390
6. Pebay-Peyroula E, Rummel G, Rosenbusch JP, Landau, EM (1997) Science 277:1676
7. Nollert P, Royant A, Pebay-Peyroula E, Landau E (1999) FEBS Lett 457:205
8. Landau E, Rummel G, Cowan-Jacob SW, Rosenbusch JP (1997) J Phys Chem B 101:1935
9. Angelov B, Ollivon M, Angelova A (1999) Langmuir 15:8225
10. Ai X, Caffrey M (2000) Biophys J 79:394
11. Lindblom G (1996) Nuclear magnetic resonance spectroscopy and lipid phase behaviour and lipid diffusion. In: Christie WW (ed), Advances in lipid methodology, vol. 3. Oily Press, Dundee, Scotland, pp 133–209
12. Nilsson F, Söderman O (1996) Langmuir 12:902
13. Lindblom G, Rilfors L (1989) Biochim Biophys Acta 988:221

Progr Colloid Polym Sci (2004) 123: 40–43
DOI 10.1007/b11619

Krister Thuresson
Filipe E. Antunes
Maria G. Miguel
Björn Lindman

The association between a non-ionic microemulsion and hydrophobically modified PEG. A rheological investigation

K. Thuresson · F.E. Antunes
M.G. Miguel · B. Lindman (✉)
Physical Chemistry 1,
Center for Chemistry and
Chemical Engineering, Lund University
P.O. Box 124, 221 00 Lund, Sweden
e-mail: Krister.Thuresson@fkem1.lu.se
Fax: +46-46-2224413

F.E. Antunes · B. Lindman
Department of Chemistry,
Coimbra University, 3005–535 Coimbra,
Portugal

Abstract The thickening effect of a hydrophobically modified polymer in an O/W microemulsion is investigated. The hydrophobically modified polymer is a triblock copolymer, alkyl end-capped poly(ethylene glycol) and the microemulsion is based on a non-ionic surfactant, pentaethylene oxide dodecyl ether ($C_{12}E_5$) and decane. The rheological properties vary strongly with microemulsion droplet volume fraction, with temperature and with end-caps of the polymer. Particularly interesting are a maximum in viscosity as a function of droplet volume fraction, a decrease in cross-link life-time and very strong temperature dependences. These can be understood on the basis of interdroplet distances and interactions.

Introduction

Hydrophilic (homo)polymers have traditionally found use as rheology modifiers in water-based systems but during the past years hydrophobically modified polymers (HM-P), or water-soluble associative polymers (WSAP), have extensively replaced them. These polymers show larger thickening effects but also increase the compatibility with dispersions by adsorbing to the surfaces of colloidal particles. However, the hydrophobic attraction appeared also to be responsible for problems that were encountered when HM-P polymers were mixed in formulations. For instance, instead of the desired increase in stability, phase separation may be induced due to bridging-related mechanisms.

Despite draw-backs, HM-Ps are today found in many technical formulations [1]. Such formulations generally have in common a complex composition, and they may contain a range of liquids, particles and colloids, and surface-active compounds. Water-based paints may serve as a representative example. To be able to understand the underlying mechanisms that determine a sometimes unwanted, and seemingly mysterious behaviour, the investigated mixtures have to be simple and the number of components therefore has to be reduced. In the present investigation we have used a well characterised system that is based on a thoroughly investigated microemulsion [2], with a non-ionic surfactant pentaethylene oxide dodecyl ether ($C_{12}E_5$), water and decane, and rather mono disperse end-capped poly(ethylene glycol) polymers. In the temperature and compositional ranges where we have chosen to work, the microemulsion forms spherical droplets with a diameter of approximately 160 Å [3]. To refer back to the case of a water-based paint formulation, the microemulsion droplets may represent the binder particles (latex) and the end-capped polymer the thickener. We have investigated effects of varying the hydrophobic end-group as well as of a variation of the length of the mid-block of the polymers, and of a changed microemulsion concentration. Questions that we seek answers to are how micro-structure and dynamics are influenced by these variations, and to obtain such information we have performed rheological investigations. This report gives a preliminary account of some intriguing observations, while the complete study will be presented later.

Experimental

Materials

A range of hydrophobically end-modified polyethylene glycols (HM-PEG), were obtained as kind gifts from Akzo Nobel Surface Chemistry Stenungsund, Sweden. The average molecular weight of the PEG polymers that were used was either 12,000 or 20,000. This corresponds to 280 or 450 repeating oxyethylene units in the mid-block, respectively. The hydrocarbon part had an average composition of C_{21} or C_{24}. The microemulsion with which the different HM-PEG polymers were mixed is composed of a non-ionic surfactant pentaethylene oxide dodecyl ether ($C_{12}E_5$), water and decane at a constant surfactant to oil ratio of 52/48. $C_{12}E_5$ was obtained from Nikko Chemicals Co. Ltd., Tokyo, Japan, and decane (> 99%) was obtained from Sigma. All chemicals were used without further purification and for all samples water that had been passed through a Millipore™ purification unit was used.

Methods

The rheological measurements were performed by using a Physica UDS 200 rheometer with automatic gap setting. A 5 cm, 1° cone and plate geometry with a solvent trap was used for all investigated samples, and the temperature in the measuring geometry was controlled to within ± 0.1 °C by a Peltier system. All measurements were performed with the instrument in the oscillatory shear mode, and to ensure that all determinations were performed in the linear viscoelastic regime, each rheological determination was preceded by stress-sweep measurements.

Results and discussion

The microemulsion that we have chosen is well characterised; it forms spherical microemulsion droplets in the concentration and temperature ranges that we have used [2, 3]. The formed droplets can be viewed as oil-swollen micelles and they have a hydrocarbon core radius of about 75 Å. We also know from an earlier investigation that the droplets are virtually insensitive to an addition of HM-PEG polymers [4]. This is an observation that is important for interpretation of our results below. Due to hydrophobic attraction an HM-PEG polymer is expected to adsorb to the microemulsion droplets via their end-groups, while the mid-block is expected to be in the aqueous environment. These requirements are fulfilled in a situation where the polymer chains form loop conformations and both stickers from a HM-PEG polymer are attached to the same microemulsion droplet. This is likely to be the case when the average distance between microemulsion droplets is large compared to the length of the PEG-block of the polymers. When the average distance between micelles is small compared to the length of the mid-block of the polymer chains the situation is different. Under such conditions an HM-PEG polymer may also form a bridge between two microemulsion droplets. Since the latter enables a network formation, a transformation between the loop and bridge conformations is expected to have large consequences for the rheological properties of the system. This is also what was observed.

In Fig. 1 is shown that the microemulsion concentration has a pronounced influence on the Newtonian viscosity, and at low concentrations (5–8%) the viscosity increases strongly. This is in sharp contrast to the weak concentration dependence that was observed for the same microemulsion but without added HM-PEG [5]. In the latter case the concentration dependence followed a hard sphere model. A strong concentration dependency has also been found in other systems. Examples are semi-dilute systems of flexible polymers of high molecular weight under theta solvent conditions [6], and aqueous solutions containing only an HM-PEG polymer where

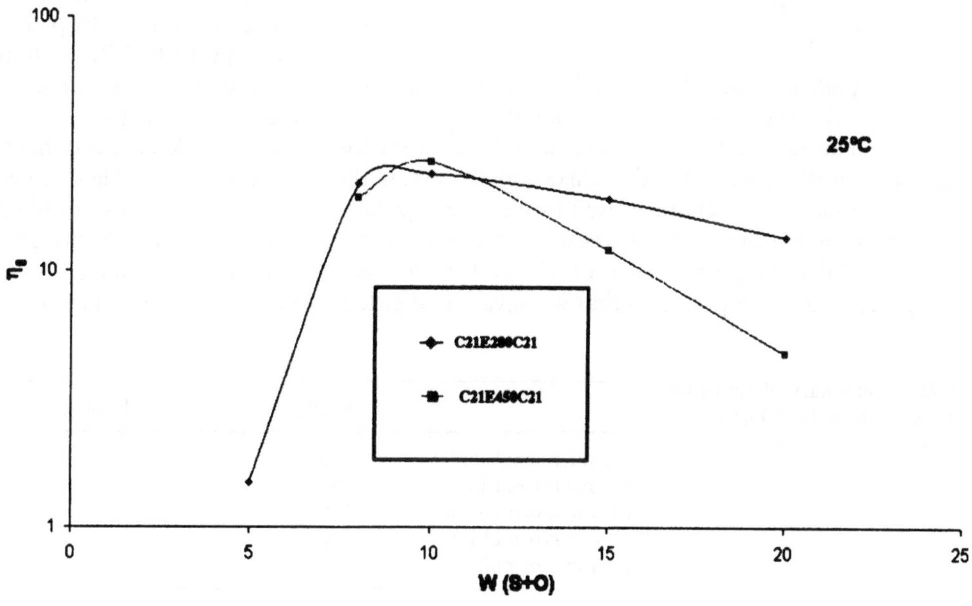

Fig. 1 The Newtonian viscosity as a function of the microemulsion concentration. The average ratio between HM-PEG polymers and microemulsion droplets was held constant to 4

the viscosity was found to increase approximately according to the power law $c \propto c^5$ [7]. The large exponent was believed to be due to attraction between polymer chains as a result of poor solvent conditions. In fact we believe that the generally accepted mechanism that is responsible for the strong concentration dependence in solutions containing end-modified polymers [8] is closely related to what happens also in the present system. At low concentrations, HM-polymers form polymeric micelles, often referred to as flowers. A large fraction of the HM-polymers within these flowers are in a loop conformation. This is mainly due to the fact that the concentration of polymeric micelles is low, and the average distance between flowers is long. When the concentration increases the average distance between adjacent flowers decreases, and a bridge conformation becomes more likely. The bridges provide connectivity in the system and give rise to a strong attraction between micelles [9, 10]. Apart from favouring phase separation, the attraction also influences dynamics, and in particular the viscosity is expected to increase strongly as a result from network formation.

Since the polymer concentration is rather low in the present investigation, and the contribution to polymer dynamics from chain entanglements therefore is expected to be negligible, a simple rheological behaviour can be anticipated [4, 7]. Indeed, the Maxwell model that is the simplest model of a viscoelastic fluid gives a fair description of the dynamic moduli as a function of frequency. Within the framework of this model, that has one elastic component (spring) and one viscous component (dashpot) in series, the storage, G', and loss, G'', moduli become

$$G'(\omega) = G_\infty \frac{\tau^2\omega^2}{1+\tau^2\omega^2} \tag{1a}$$

$$G''(\omega) = G_\infty \frac{\tau^2\omega}{1+\tau^2\omega^2} \tag{1b}$$

G_∞ is the plateau value of G' at high angular frequencies, $\omega = 2\pi f$, and τ is the characteristic time of the relaxation process. The fact that this simple model describes the data quite well suggests that the relaxation process, in the terminal zone, can be characterised by a single exponential relaxation time. Fitting these equations to the experimental data gives values of G_∞ and τ for each composition and temperatures that we have investigated. G_∞ is proportional to the number density, n, of rheologically active chains [11]

$$G_\infty \cong nk_BT \tag{2}$$

where k_B is the Boltzmann factor, and T the absolute temperature. τ is related to the characteristic lifetime of a cross-link, which depends on the activation energy for the relaxation process, E.

$$\tau \propto \exp\left(\frac{E}{k_BT}\right) \tag{3}$$

From Arrhenius' plots, which use the temperature dependency of τ, E can be obtained. The fact that straight lines are obtained in the Arrhenius' representation may be taken to suggest that the structure of the solution is not influenced by the variation in temperature. This was also the conclusion in a previous report where the same microemulsion system was investigated [4]. Since E is likely to be coupled to the process of transferring a polymer hydrophobic tail from a microemulsion droplet (oil environment) to an aqueous environment it is not surprising that the numbers obtained from the Arrhenius' plots are similar for polymers with the same hydrophobic end-groups, Table 1. From this it is also expected that the $C_{24}E_xC_{24}$ polymers should have higher E values than the $C_{21}E_xC_{21}$ polymers. This is indeed what is found, see Table 1. The absolute values seem however to be high and E is affected by the concentration of the microemulsion; we come back to the E values below.

In addition to the high (apparent) activation energy there are two other interesting and intriguing observations, the decrease in the viscosity at higher droplet concentrations and the decrease in the cross-link lifetime; the latter are of course interconnected but we believe that all three effects have a common cause. The increase in viscosity as the droplet volume fraction increases we attribute as discussed above to the relation between the interdroplet distance and the extension of the unperturbed PEG chains. For low volume fractions, the distance is too large to allow extensive cross-linking and the end-caps of a given polymer chain are to a considerable extent connected to the same droplet, they form loops. As the droplet volume fraction increases to very high values another phenomenon sets in: The droplets are close enough so that transfer of alkyl end-caps does not have to proceed via a purely aqueous environment but via palisade layers rich in ethylene oxide

Table 1 Summary of activation energies (in units of k_BT) in the different systems

	$C_{21}E_{280}C_{21}$	$C_{24}E_{280}C_{24}$	$C_{21}E_{450}C_{21}$	$C_{24}E_{450}C_{24}$
5% microemulsion	65	–	–	–
8% microemulsion	95	–	92	–
10% microemulsion	106	117	96	115
15% microemulsion	140	–	146	–
20% microemulsion	155	–	165	–

groups, with a significantly lower polarity than water; this facilitated transfer of alkyl chains between microemulsion droplets decreases the life-time of cross-links and, as a consequence, also the viscosity. This transfer is for hard-sphere microemulsion droplets mainly determined by the volume fraction of droplets but will be modified if there is an attractive interaction. It is well established that poly(ethylene oxide) chains repel each other at low temperatures. However, as temperature increases there is a progressively weaker repulsion which turns into an attraction [12]. This explains the strong temperature dependence and the high apparent energies of activation.

Conclusions

In this investigation we have used well-defined end-capped polymers with a low polydispersity index, and added them to a microemulsion containing spherical droplets also with a small size distribution. We have studied effects on structure and dynamics by using rheology. We found that the structure was sensitive to variations in the microemulsion concentration and in the length of the mid-block of the HM-PEG polymers, while the characteristic relaxation time was sensitive to variations in the polymer hydrophobic moieties. The results could be understood by translating the microemulsion concentration to a mean distance between microemulsion droplets and comparing this length with the end-end distance of the unmodified parent polymer. When the microemulsion concentration was decreased and the average distance between droplets became significantly longer than the end-end distance of the unmodified parent polymer a loop-conformation was favoured over a rheologically active link-formation. Apart from the concentration of active links which could be controlled by the microemulsion concentration and by the length of the HM-PEG mid-block, the viscosity also depended on the characteristic relaxation time of the system. This relaxation time was found to be largely influenced by the length of the hydrophobic moiety of the HM-PEG polymers, the concentration of the microemulsion, and by the temperature.

Acknowledgements We thank Leif Karlson at Akzo Nobel Surface Chemistry, Stenungsund, Sweden, for supplying HM-PEG polymers and for valuable discussions. KT also thanks the Centre for Amphiphilic Polymers (CAP) at Lund University for financial support. The stay of F.A. in Lund was financed from a grant from the Swedish Science Research Council. The Colloid Group in Coimbra is supported by a grant from the Fundação para Ciência e Tecnologia (FCT) (project Sapiens PCTI/99/QUI/35415), and from University of Coimbra.

References

1. Glass JE (ed) (1989) Polymers in Aqueous Media. American Chemical Society, Washington, DC, vol 223
2. Olsson U, Schurtenberger P (1993) Langmuir 9:3389
3. Bagger-Jörgensen H, Olsson U, Mortensen K (1997) Langmuir 13:1413–1421
4. Bagger-Jörgensen H, Coppola L, Thuresson K, Olsson U, Mortensen K (1997) Langmuir 13:4204–4218
5. Leaver M, Olsson U (1994) Langmuir 10:3449–3454
6. Takahashi Y, Isono Y, Noda I, Nagasawa M (1985) Macromolecules 18:1002
7. Thuresson K, Nilsson S, Kjøniksen A-L, Walderhaug H, Lindman B, Nyström B (1999) J Phys Chem B 103:1425–1436
8. Winnik MA, Yekta A (1997) Curr Opin Colloid Interface Sci 2:424–436
9. Semenov AN, Joanny J-F, Khokhlov AR (1995) Macromolecules 28:1066–1075
10. Witten TA (1988) J Phys 49:1055–1063
11. Green MS, Tobolsky AV (1946) J Chem Phys 14:80–92
12. Jönsson B, Lindman B, Holmberg K, Kronberg B (1997) Surfactants and polymers in aqueous solution. Wiley, London

Progr Colloid Polym Sci (2004) 123: 44–47
DOI 10.1007/b11621

K. Esumi

Adsolubilization by mixtures of ionic and non-ionic surfactants

Abstract The adsolubilization of 2-naphthol by surfactant mixtures of ionic and non-ionic surfactants has been investigated for three systems; sodium dodecyl sulfate(SDS)-hexaoxyethylene decyl ether($C_{10}E_6$)-alumina, hexadecyltrimethylammonium bromide (HTAB)-$C_{10}E_6$-silica, and 1,2-bis(dodecyldimethylammonio)-ethane dibromide (2RenQ)-$C_{10}E_6$-silica. For three systems, the interaction parameters estimated from mixed critical micelle concentrations are −3.4 for both SDS-$C_{10}E_6$ and HTAB-$C_{10}E_6$ and −1.5 for 2RenQ-$C_{10}E_6$, suggesting that the interaction between SDS and $C_{10}E_6$ as well as that between HTAB and $C_{10}E_6$ is stronger than that between 2RenQ and $C_{10}E_6$ in mixed micelle formation. In the SDS-$C_{10}E_6$-alumina system where only SDS shows an appreciable adsolubilization, the adsolubilization becomes greater with an increase in the SDS content in the initial mixtures. On the other hand, the ratios of adsolubilized amount to surfactant adsorbed amount decrease when the SDS content in the initial mixtures increases. In the cationic surfactant-$C_{10}E_6$-silica systems where each surfactant such as HTAB, 2RenQ, or $C_{10}E_6$ shows an appreciable adsolubilization, the ratios of adsolubilized amount to surfactant adsorbed amount are greater at some mixture compositions compared to those of single surfactants. In particular, the effect of chemical structure of cationic surfactant on the adsolubilization is also discussed.

Keywords Adsolubilization of 2-naphthol · Binary surfactants · Silica · Alumina

K. Esumi
Department of Applied Chemistry and Institute of Colloid and Interface Science, Science University of Tokyo, Kagurazaka, Shinjuku-ku, Tokyo 162–8601, Japan

Introduction

Surfactant adsorbed layers formed on particles by adsorption of surfactants exhibit hydrophobic properties which have been characterized by many techniques including ESR and fluorescence spectroscopy [1]. Accordingly, water-insoluble compounds can be incorporated into the surfactant adsorbed layers, which is called adsolubilization [2]. Factors influencing adsolubilization behavior are mainly as follows: (a) surfactant structure; (b) kind of water-insoluble compound; (c) kind of particles. Until now, adsolubilization has been studied for many systems consisting of different combinations of water-insoluble compounds and particles using single surfactants [3–12].

Adsorption from surfactant mixtures onto solids has been studied to understand complex interfacial phenomena at solid/liquid interfaces. In particular, it is known [13] that for adsorption from binary mixtures of ionic and non-ionic surfactants, the adsorption of one surfactant is often enhanced by the addition of a small amount of the other surfactant. Surfactant mixtures provide several advantages over single surfactants, because the adsorption of surfactants onto particles can be controlled using appropriate surfactants and solution properties. It is anticipated that the adsolubilization can be enhanced using a minimum amount of binary mixtures of ionic and non-ionic surfactants.

In this study, adsolubilization of 2-naphthol, which has been used as a model compound for a number of adsolubilization experiments, has been studied using three binary surfactant systems. They are sodium dodecyl sulfate (SDS)-hexaoxyethylene decyl ether ($C_{10}E_6$)-alumina, hexadecyltrimethylammonium bromide (HTAB)-$C_{10}E_6$-silica, and 1,2-bis(dodecyldimethylammonio)ethane dibromide (2RenQ)-$C_{10}E_6$-silica.

Experimental

Materials

SDS and HTAB were obtained commercially and used after several recrystallizations from ethanol and acetone, respectively. 2RenQ was synthesized and purified by recrystallization from mixtures of hexane and ethanol. $C_{10}E_6$ was supplied by Nikko Chemicals Co. and used as received. α-Alumina was supplied by Showa Denkou Co.; its specific surface area and average particle size were 30.2 m^2 g^{-1} and 0.3 μm, respectively. Silica was obtained from Nippon Shokubai Kogyo Co., and its specific surface area and particle size were 50 m^2 g^{-1} and 0.03 μm, respectively. Water used was purified with the use of a Mill-Q Plus system (Millipore). The other chemicals used were of analytical grade.

Methods and Measurements

Adsolubilization of 2-naphthol was carried out as follows. A series of mixed surfactant solutions was prepared, containing fixed concentration of 2-naphthol (0.4 mmol dm^{-3}) and NaCl or NaBr (10 mmol dm^{-3}). The solutions were then added to alumina or silica in glass vials with caps. All suspensions were equilibrated at 25 °C for 24 h in a shaker-water bath. For SDS-$C_{10}E_6$-alumina system the suspensions were adjusted to about pH 3.5 with HCl, while the pH of the suspensions was found to be about 6 for both HTAB-$C_{10}E_6$-silica and 2RenQ-$C_{10}E_6$-silica systems. After equilibration, the solids were separated by centrifugation and the supernatants were analyzed for 2-naphthol (at 328 nm using a UV detector) and for surfactants (using an RI detector of a high-performance liquid chromatography). A mixture of methanol and water (85:15 in volume) containing 0.4 mol dm^{-3} NaCl was used as the mobile phase for the HPLC assays.

The surface tensions of surfactant mixtures in the presence of 10 mmol dm^{-3} NaCl or NaBr were measured at 25 °C using a Kruss K12 tensiometer.

Results and discussion

Since aqueous properties of surfactants affect the adsorption behavior of surfactants onto particles, firstly mixed critical micelle concentrations for the three systems were determined by measuring surface tensions of mixtures. Figure. 1 shows the mixed cmc vs. molar fraction of ionic surfactant in the mixtures. It is found that the mixed cmcs are very similar to those of the surfactant with lower cmc for the three systems. In addition, the interactions between two surfactants in mixed micelles can be estimated using the regular solution theory [14]. In Fig. 1, the interaction parameter, β of the SDS-$C_{10}E_6$ mixed system is about −3.4, while those of the HTAB-$C_{10}E_6$ and 2RenQ-$C_{10}E_6$ systems are −3.4 and −1.5, respectively. It is suggested that the mutual phobicity between the hydrocarbon chains, as

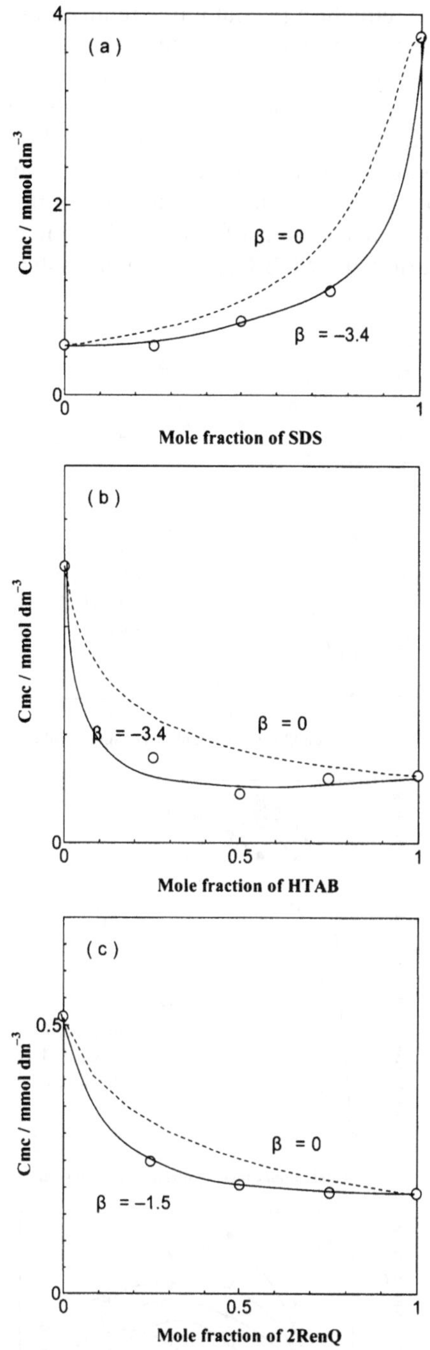

Fig. 1 Cmc's in for (**a**) SDS-$C_{10}E_6$, (**b**) HTAB-$C_{10}E_6$, and (**c**) 2RenQ-$C_{10}E_6$ mixtures in 10 mmol dm^{-3} salt at 25 °C. The plotted points are experimental results; the solid lines are the prediction of the regular solution theory; and the dotted lines are the prediction for ideal mixing

well as the reduction in coulombic repulsion between the headgroups, dominates the interactions between ionic surfactants and $C_{10}E_6$. Interestingly, the comparison of β values between HTAB-$C_{10}E_6$ and 2RenQ-$C_{10}E_6$ systems indicates that the interaction between HTAB and $C_{10}E_6$ is much stronger than that between 2RenQ and $C_{10}E_6$.

Figure. 2 shows the adsolubilized amount of 2-naphthol as a function of total surfactant equilibrium concentration for SDS-$C_{10}E_6$-alumina system and as a function of total initial surfactant concentration for HTAB-$C_{10}E_6$-silica and 2RenQ-$C_{10}E_6$-silica systems. Because the adsolubilization of 2-naphthol is not observed onto alumina and silica without surfactants, it is apparent that 2-naphthol is incorporated into the surfactant adsorbed layer that exhibits a hydrophobic property. For the SDS-$C_{10}E_6$-alumina system the adsolubilized amount of 2-naphthol is very low by the adsorption of $C_{10}E_6$ alone, but becomes greater with an

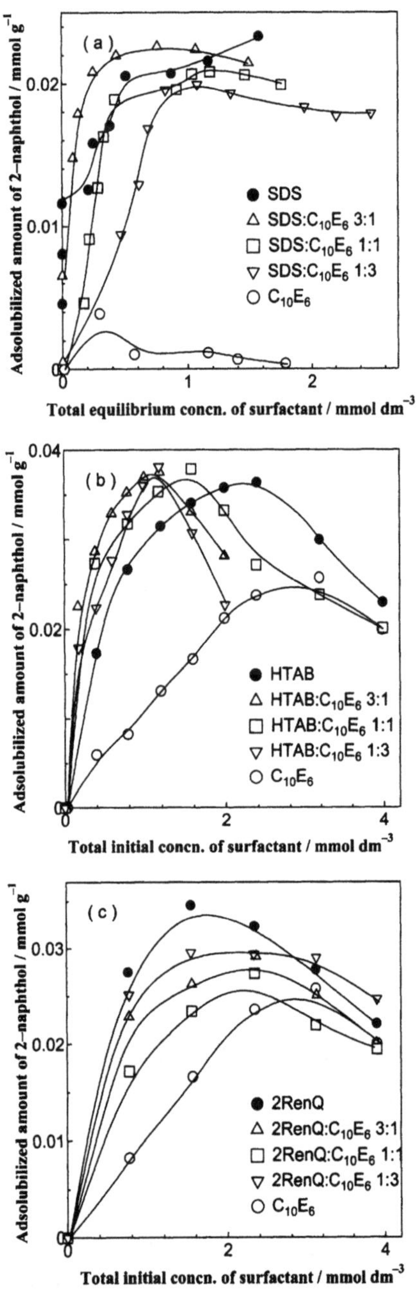

Fig. 2 Adsolubilization of 2-naphthol on **(a)** alumina for SDS-$C_{10}E_6$, **(b)** silica for HTAB-$C_{10}E_6$, and **(c)** silica for 2RenQ-$C_{10}E_6$ mixtures as function of total surfactant concentration; 25 °C, 0.4 mmol dm^{-3} 2-naphthol, 10 mmol dm^{-3} salt

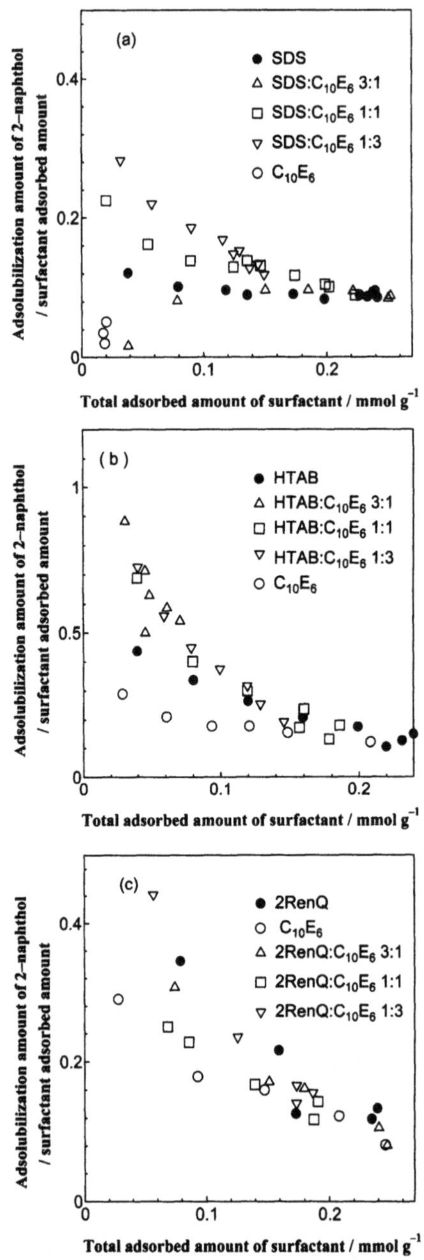

Fig. 3 The relationship between adsolubilized amount/surfactant adsorbed amount and total adsorbed amount of surfactant for SDS-$C_{10}E_6$-alumina, **(b)** HTAB-$C_{10}E_6$-silica, and **(c)** 2RenQ-$C_{10}E_6$-silica systems

increase in the SDS content of the initial mixtures, from SDS:$C_{10}E_6$ = 1:3 to 3:1 [15]. The adsolubilized amount of 2-naphthol by the adsorption of SDS alone ranges between that of SDS:$C_{10}E_6$ = 1:1 and 3:1. In the case of HTAB-$C_{10}E_6$-silica system the adsolubilization of 2-naphthol is appreciably observed by the adsorption of $C_{10}E_6$ alone and the maximum of 2-naphthol adsolubilization is almost the same for all three different mixed compositions and is a slightly greater than that for HTAB alone. In addition, the maximum occurs at the lowest initial concentration for HATAB:$C_{10}E_6$ = 3:1 and 1:3. On the other hand, for 2RenQ-$C_{10}E_6$-silica system, the change in the adsolubilization with the total initial concentration of surfactant from all three different mixed compositions is similar to that of 2RenQ alone, although the magnitude in the adsolubilization is not directly proportional to the mixed compositions of 2RenQ in the mixtures.

To elucidate the effect of mixed surfactant compositions on the adsolubilization of 2-naphthol, the ratios of adsolubilized amount to adsorbed amount of surfactant are plotted with the total adsorbed amount of surfactant in Fig. 3. For the SDS-$C_{10}E_6$-alumina system, the proportions by the adsorption of SDS alone are about 0.1 in a whole SDS adsorption concentration, whereas those for $C_{10}E_6$ alone range between 0.02 and 0.06, suggesting that the efficiency of adsolubilization is considerably different between SDS and $C_{10}E_6$ single adsorption. This difference occurs due to the adsorption state of surfactants that SDS adsorbs onto positively charged alumina surface by electrostatic attractive force as a monolayer and then an SDS bilayer is formed by the hydrophobic interaction between SDS, while $C_{10}E_6$ interacts weakly with the surface of alumina. On the other hand, the ratios for the three different mixed compositions are greater than those for SDS alone below 0.2 mmol g^{-1} of adsorption. This result can be explained by a view that the mixed surfactant adsorbed layer is more compact because of a shield of electrostatic repulsion of SDS adsorbed by incorporation of $C_{10}E_6$, in particular for the case of SDS:$C_{10}E_6$ = 1:3. It is interesting to note that the ratio of adsorbed amount of SDS to $C_{10}E_6$ onto alumina is about 1:1 for the SDS:$C_{10}E_6$ = 1:3. In the cases of HTAB-$C_{10}E_6$-silica and 2RenQ-$C_{10}E_6$-silica systems, the adsolubilized amount of 2-naphthol to adsorbed amount of surfactant for HTAB alone is greater than that for $C_{10}E_6$ alone. Since it is conceivable that HTAB adsorbs onto negatively charged silica surface by electrostatic attractive force at low HTAB concentration and as bilayer due to hydrophobic interaction between HTAB at high HTAB concentration, while $C_{10}E_6$ adsorbs onto negatively charged silica surface by hydrogen bonding with surface silanol groups, the difference in the ratios may attributed to the alkyl chain length between HTAB and $C_{10}E_6$. The ratios for all the three different compositions are greater than that for HTAB alone below 0.15 mmol g^{-1} of surfactant adsorption. On the other hand, the ratios for all the three different compositions of 2RenQ-$C_{10}E_6$-silica system are smaller than that for 2RenQ alone. This suggests that the mixed adsorbed layer consisting of 2RenQ and $C_{10}E_6$ is not so compact to enhance the adsolubilization of 2-naphthol. From the interaction parameter of the mixed surfactant systems it is also deduced that the interaction between HTAB and $C_{10}E_6$ in the adsorbed layer is greater than that between 2RenQ and $C_{10}E_6$.

Thus, the above results lead to a conclusion that a proper surfactant selection for binary mixtures of ionic and non-ionic surfactants is required to enhance the adsolubilization of 2-naphthol at low surfactant mixture concentrations.

References

1. Esumi K (1997) In: Esumi K, Ueno M (eds), Structure-performance relationships in surfactants, Chap. 11 (and references therein). Marcel Dekker
2. Wu J, Harwell JH, O'Rear EA (1987) Langmuir 3:531–537
3. Esumi K, Sakamoto Y, Nagahama T, Meguro K (1989) Bull Chem Soc Jpn 62:2502
4. Zhu B-Y, Xhao X, Gu T (1988) J Chem Soc Faraday Trans 184:3951–3960
5. Monticone V, Mannebach MH, Treiner C (1994) Langmuir 10:2395–2398
6. Nayyar SP, Sabatini DA, Harwell JH (1994) Environ Sci Technol 28:1874–1881
7. Schieder D, Dobias B, Klumpp E, Schwuger MJ (1994) Colloids Surf 88:103–111
8. Esumi K, Matoba M, Yamanaka Y (1996) Langmuir 12:2130–2135
9. Esumi K, Goino M, Koide Y (1996) J Colloid Interface Sci 183:539–545
10. Favoriti P, Treiner C (1998) Langmuir 14:7493–7502
11. Esumi K (2001) Colloids Surf 176:25–34
12. Rosen MJ, Fang Li (2001) J Colloid Interface Sci 234:418–424
13. Holland PM, Rubingh DN (1992) In: Holland PM, Rubingh DN (eds), Mixed surfactant systems, Chap. 1. Washington DC, American Chemical Society
14. Holland PM (1992) In: Holland PM, Rubingh DN (eds), Mixed surfactant systems, Chap. 2. Washington DC, American Chemical Society
15. Esumi K, Maedomari N, Torigoe K (2000) Langmuir 16:9217–9220

Progr Colloid Polym Sci (2004) 123:48–51
DOI 10.1007/b11622

Patrick Oliger
Arnaud Fischer
Marc Hebrant
Christian Tondre

Probe entrapment by vesicular systems in relation with the properties of the amphiphilic film

P. Oliger · A. Fischer · M. Hebrant
C. Tondre (✉)
Laboratoire de Chimie Physique Organique et Colloïdale, Unité Mixte de Recherche CNRS-UHP (UMR 7565),Université Henri Poincaré-Nancy 1, B.P. 239,
54506 Vandoeuvre-les-Nancy Cedex, France
e-mail: Christian.Tondre@lesoc.uhp-nancy.fr
Tel.: +33-3-83912189
Fax: +33-3-83912532

Abstract Two different vesicular systems were investigated as regards their ability to substantially retain a hydrophilic probe. The first system was a series of phospholipid analogues, and its glucose encapsulation was revealed to be much dependent on the temperature, in reference to the transition temperature – from the gel to the liquid crystalline phase – of the system. The second system consisted of different catanionic mixtures, whose entrapment efficiency was investigated as a function of the original conditions (glucose concentration, surfactant proportion in water. Dialysis experiments were also conducted in order to assess the permeability of the vesicular bilayer and, thus, to settle the issue of long-term encapsulation.

Keywords Vesicles · Amphiphilic bilayer · Glucose entrapment · Transition temperature · Permeation kinetics

Introduction

The encapsulation of biologically active agents by liposomes or vesicles – in order to design drug-delivery systems or simple nanoscale containers for other active substances – is a topic of growing interest. Cancer and gene therapy, but also delivery of agriculturally active substances (herbicides, pesticides, etc.), are known examples of applications. Whatever the purpose of the considered encapsulation, a compromise must be found in order to ensure not only a good entrapment efficiency, but also a possibility of drug release. The control of the permeability is not an easy problem. We report here the behaviour of two types of synthetic vesicles regarding the entrapment/release of glucose, which was selected as a neutral hydrophilic probe.

Two largely different systems were investigated: System I was constituted by a series of newly synthesised phospholipid analogues (dialkyl *N*-(phosphonoacetyl)-L-aspartate) which present a potential anti-tumoural activity. The thermosensitivity of the entrapment efficiency of the vesicles prepared from these molecules was studied as a function of the bilayer thickness. System II was a so-called "catanionic" system consisting of a mixture of two surfactants, the first one (cetyltrimethylammonium tosylate) being cationic, and the second one (sodium dodecyl benzenesulphonate) being anionic. These systems are obviously more labile than true phospholipid vesicles and their long-term ability to retain entrapped molecules remains an unanswered question.

System I: vesicles formed by dialkyl *N*-(phosphonoacetyl)-L-aspartate

The above-mentioned compounds were synthesised as part of a work conducted within the framework of a collaboration with P. Coutrot and C. Grison (University of Nancy-1) and intended to experiment a new concept of drug delivery, so as to increase the penetration of an anti-cancer agent [*N*-(phosphonacetyl)-L-aspartate, which is also known as PALA] [1, 2] through cell membranes, and, thus, to improve its bioavailability [3, 4].

The ability of such derivatives to form vesicular aggregates was examined. In order to prove the formation of

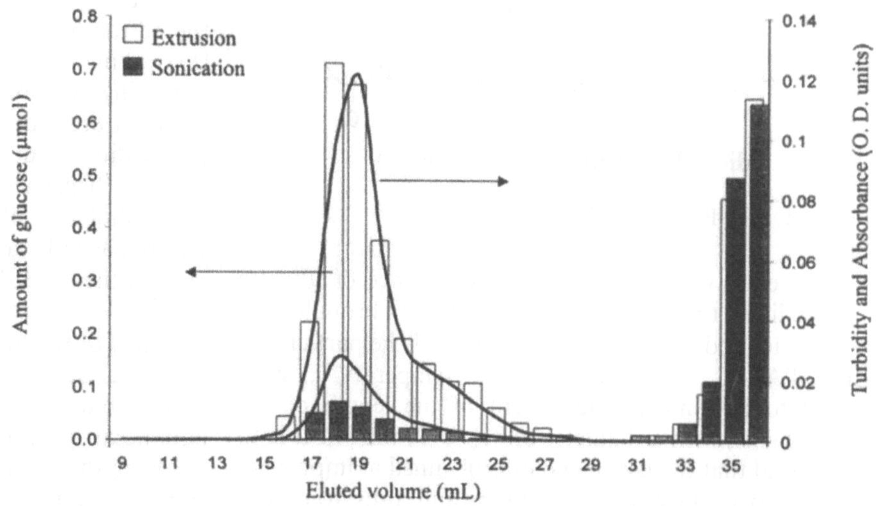

Fig. 1 Example of a size exclusion chromatogram. The current figure presents the case of glucose-containing vesicles of di-C_{14}-PALA obtained by extrusion or by sonication. [di-C_{14}-PALA]$_0$ = 7 × 10^{-3} mol/L; [Glucose]$_0$ = 0.5 mol/L

close structures and not only of bilayer fragments, we studied the encapsulation of glucose by the surfactant dispersions. Dispersions of either di-C_n-PALA, phosphatidylcholines (PC), or mixtures of di-C_n-PALA and PC, were prepared in the presence of glucose.

Glucose is partially trapped in the aqueous core of the vesicles. This encapsulated probe can be separated from the rest – which is free in solution – by size exclusion chromatography (SEC) (see Fig. 1), and the amount of trapped glucose can be quantified by an enzymatic titration [5]. As reported in Fig. 1, the entrapment is much less important when vesicles are obtained by sonication (0.1 to 0.2%) than when they are obtained by extrusion (0.7 to 1.1%). This result is perfectly consistent with the cryo-TEM experiments which were conducted in parallel [4], and which revealed that sonication was mainly likely to yield small size vesicles and fragmentary objects. Contrary to the preparation procedure, the alkyl chain length of the PALA derivatives did not seem to have much influence on encapsulation.

The phase transition temperature T_m of a surfactant bilayer characterises the transition between the β (gel) and the α (liquid crystalline) phases. This value provides interesting information as concerns the physical state of the bilayer, which can influence the stability of the vesicles. T_m was determined for di-C_n-PALA, lecithins, and mixtures of di-C_n-PALA and lecithins with different alkyl chain lengths by measuring the fluorescence polarisation factor of diphenylhexatriene at different temperatures [6].

The values obtained with phosphatidylcholines (PCs) proved similar to those encountered in the literature. As far as the PALA derivatives are concerned, their T_m turned out to be 12 to 18 °C (depending on the chain length) higher than those of the PCs for a similar alkyl chain length. Mixtures of di-C_n-PALA with PCs or di-C_n-PALA with different alkyl chain lengths thus proved likely to allow a modulation of the phase transition temperature. Indeed, a given T_m can be reached by mixing these derivatives under different ratios. Glucose entrapment was consequently studied as a function of temperature. A considerable drop in the encapsulation ratio was detected when elution was performed at a temperature which was higher than the phase transition temperature (Table 1). Reversely, as proved by further

Table 1 Percentage of encapsulated glucose as a function of the elution temperature for different di-C_n-PALA- and phosphatidylcholine-derivative mixtures. DPPC stands for dipalmitoyl-phosphatidylcholine

Composition: 7 mM surfactant extruded dispersion	T_m [°C]	Percentage of encapsulated glucose					
		Elution temperature					
		15 °C	25 °C	35 °C	45 °C	55 °C	65 °C
di-C_{12}-PALA	25	–	0.67	–	–	–	–
di-C_{14}-PALA	41	0.99	1.07	0.98	0.09	–	–
di-C_{16}-PALA	54	–	0.81	0.66	0.74	0.09	–
di-C_{18}-PALA	64	–	1.1	–	–	–	0.07
di-C_{16}-PALA + DPPC (molar ratio 2/1)	54/39	–	1	0.93	0.2	–	–
di-C_{16}-PALA + Cholesterol (molar ratio 2/1)	–	–	–	1.54	1.34	0.57	0.23
DPPC	39	–	1.85	0.85	0.99	–	–
DPPC + Cholesterol (molar ratio 2/1)	–	–	2.8	2.3	1.4	1.9	–

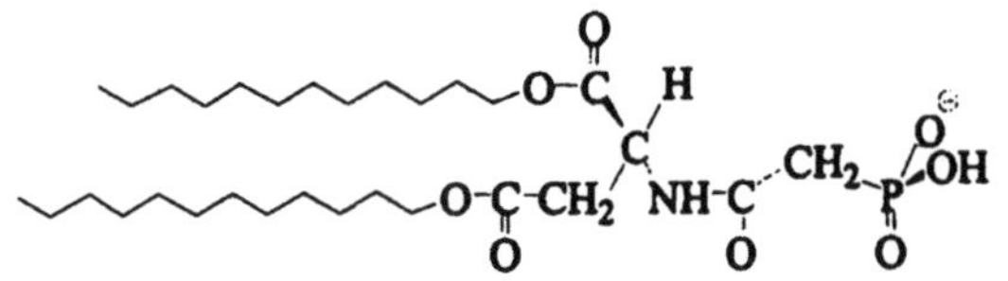

Scheme 1 di-C_n-PALA (n = 12, 14, 16 or 18)

experiments, vesicles formed in the absence of glucose proved likely to incorporate the probe when brought into its contact and left at $T > T_m$. The size of the particles of di-C_n-PALA, measured at room temperature, were found to be almost independent of the length of the alkyl chains, with a diameter of 81(± 2) nm [4]. It is to be noticed that the use of cholesterol aimed at improving the rigidity of the bilayer membrane and, thus, to reduce permeation.

As for classical phospholipids, the thermosensitivity of di-C_n-PALA regarding the entrapment efficiency was clearly shown to be correlated with the bilayer thickness and with the T_m value. The longer the chain length in di-C_n-PALA, the higher is the temperature at which leakage occurs.

System II: encapsulation in catanionic vesicles

A significant number of papers have dealt with the so-called "catanionic" vesicles for the last two decades [7–12]. Such systems are spontaneously obtained when both an anionic and a cationic surfactant are simultaneously dissolved in aqueous medium. Unlike other vesicular architectures, these objects are likely to form by simple solution-mixing, and no external energy – be it extrusion or sonication – is necessary to make them reach their final structure. A quantitative investigation concerning the entrapment ability of such systems was undertaken [13], taking into account the fact that preliminary studies regarding the sodium octyl sulphate/cetyltrimethylammonium bromide/water system had revealed that probe permeation occurred [14]. Octyl sulphate was replaced by dodecyl benzenesulphonate in order to increase the hydrophobicity of the anionic surfactant, and bromide was replaced by a tosylate counterion, which eventually yielded the SDBS/CTAT/water system [7, 8].

Vesicle formation is observed in two very restricted domains of the SDBS/CTAT/water ternary diagram [8], i.e., one SDBS-rich (V^-) and one CTAT-rich (V^+) region, but also when simple surfactants are mixed in equimolar proportions without any counterion, thus allowing the formation of an ion-pair amphiphile (IPA). Each of the three types of systems was studied as regards its glucose retention ability. Glucose encapsulation was quantitatively assessed by enzymatic titration after size exclusion chromatography.

An increase in the concentration of glucose in which the mixture is initially prepared was shown to bring about an increase in the final average glucose concentration in the vesicles, even though osmotic pressure tends to reduce this effect at higher initial glucose concentrations. The influence of the overall surfactant weight percentage in solution was also investigated. Even if the V^- and IPA systems allow a better entrapment, all three systems tend to encapsulate all the more glucose as the initial surfactant weight percentage is high, which is logical since this triggers bigger or more numerous vesicles.

Since the SDBS/CTAT/water system had proved to be able to encapsulate a hydrophilic probe, it was decided to enquire about the remnant character of this retention, and, thus, to investigate the dynamic phenomena which

Fig. 2 Release of glucose: ratio between vesicular current and initial glucose concentration as a function of dialysis duration. $[Glucose]_0 = 0.3$ mol/L; overall surfactant weight proportion in water for V^+ and V^-: 2%; $[IPA] = 1.5 \times 10^{-2}$ mol/L. A blank experiment without vesicles is also reported

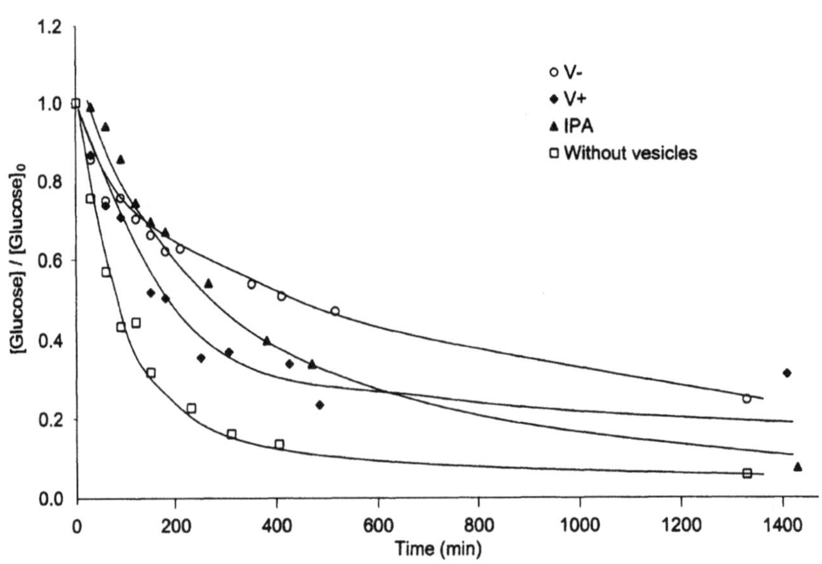

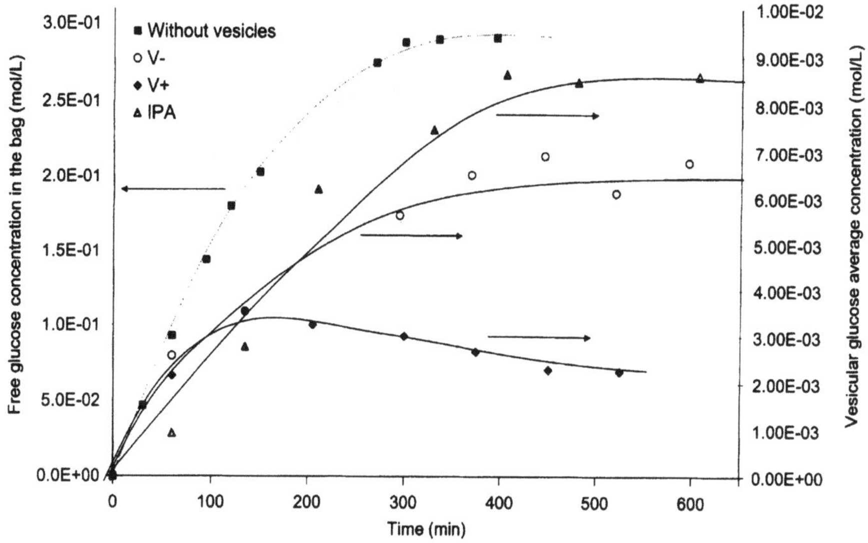

Fig. 3 Penetration of glucose: vesicular glucose concentration as a function of dialysis duration. This value represents the amount of glucose which can be found in the vesicles expressed in reference to one litre of the original dispersion. The blank experiment without vesicles ascertains that there is no glucose limitation by external diffusion. External initial glucose concentration = 0.315 mol/L; Overall surfactant weight proportion in water for V^+ and V^-: 2%; [IPA] = 1.5×10^{-2} mol/L

ruled glucose-exchange through the vesicular bilayer. Dialysis experiments were conducted on SEC-isolated glucose-containing vesicles, so as to assess the release kinetics of vesicle-entrapped probe. Symmetrical experiments provided further information about the penetration dynamics, via the titration of the glucose found inside SEC-isolated vesicles which had been prepared in the absence of the probe. As reported in Fig. 2, long-term entrapment seems to be quite low, except for the V^- system. According to Kaler et al. [7, 8], the mean diameter of the V^+ particles is generally larger than that of the V^-, which proves that, in the present case, the difference in particle size is not the only parameter which influences encapsulation.

Glucose penetration was also studied, and proved rather easy (Fig. 3), thus conveying the feeling that the vesicular bilayer was relatively permeable, even though a significant improvement in retention could be observed compared with the sodium octyl sulphate/cetyltrimethylammonium bromide/water system. The low probe retention ability compared with that obtained in the case of lipidic systems is likely to be due to a larger dynamics of the involved amphiphilic molecules.

Acknowledgments P. Coutrot and C. Grison (Laboratory of Organic Biomolecular Chemistry, UMR CNRS-UHP n°7565, University Henri Poincaré – Nancy-1) are greatly acknowledged for making available the di-C_n-PALA derivatives, which were originally synthesised in their laboratory.

References

1. Tsuboi KK, Edmunds HN, Kwong LK (1977) Cancer Res 37:3080
2. Sharma A, Straubinger NL, Straubinger RM (1993) Pharm Res 10:1434
3. Coutrot P, Oliger P, Grison C, Joliez S, Hébrant M, Tondre C (1999) New J Chem 23:981
4. Oliger P, Schmutz M, Hébrant M, Grison C, Coutrot P, Tondre C (2001) Langmuir 17:3893
5. Huggett AS^{t}G, Nixon DA (1957) Biochem J 66:12
6. Andrich MP, Vanderkoi JM (1976) Biochemistry 15:1257
7. Kaler EW, Murthy AK, Rodriguez BE, Zasadzinski JAN (1989) Science 245:1371
8. Kaler EW, Herrington KL, Murthy AK, Zasadzinski JAN (1992) J Phys Chem 96:6698
9. Zhao GX, Yu WL (1995) J Colloid Interface Sci 173:159
10. Kondo Y, Uchiyama H, Yoshino N, Nishiyama K, Abe M (1995) Langmuir 11:2380
11. Regev O, Khan A (1996) J Colloid Interface Sci 182:95
12. Yatcilla MT, Herrington KL, Brasher LL, Kaler EW, Chiruvolu S, Zasadzinski JAN (1996) J Phys Chem 100:5874
13. Fischer A, Hebrant M, Tondre C (2002) J Colloid Interface Sci 248:163
14. Caillet C, Hebrant M, Tondre C (2000) Langmuir 16:9099

Progr Colloid Polym Sci (2004) 123: 52–55
DOI 10.1007/b11623

H.D. Burrows
A.A. Kharlamov

About energy and electron transfer processes in C_{60}/phthalocyanine films

H.D. Burrows (✉), A.A. Kharlamov
Department of Chemistry,
University of Coimbra,
3004–535 Coimbra, Portugal
e-mail: burrows@ci.uc.pt

Abstract In the present work absorption, photoluminescence excitation and emission spectra of fullerene C_{60}, and a zinc derivative of phthalocyanine (ZnPc) in solutions and in multiple-layers have been investigated. In solution, the absorption spectra of the mixtures were identical with the superposition of the absorption of components, suggesting no interaction in the ground state under the dilute concentration range used in this study. Photoluminescence emission spectra of the ZnPc solution and a mixture of C_{60} and ZnPc in a non-polar solvent showed a tendency for ZnPc aggregation. The films were prepared by casting the solution onto substrates, and were tested from layer to layer. The C_{60} film thickness dependence of photoluminescence, in the C_{60}/ZnPc system was studied from the initial stages of deposition to the formation of the thick film. A series of the prepared structures was characterised. These exhibited photoluminescence bands associated with the interaction between C_{60} and ZnPc films within the bilayer structure. The photoluminescence features of these very thin C_{60} films on a ZnPc surface may be analysed in the context of a symmetry reduction at the interface, and can be discussed in terms of energy and electron transfer between C_{60} and the chromophore.

Keywords Fullerene C_{60} · Visible chromophores · Binary systems · Electron transfer · Mesoscopic effect

Introduction

Since the discovery of fullerenes, much attention has been devoted to C_{60} film formation over different surfaces and to characterising their electronic and optical properties. Particular interest is focused on fullerene films that exhibit such outstanding properties as photoconductivity [1–3] and luminescence [4–6]. Binary systems of fullerenes with visible chromophores, such as metal phthalocyanines and porphyrins, are of current interest for their potential applications in molecular electronic devices, such as photovoltaic cells [7–9]. However, these long-term goals require considerable basic research on the effects of interaction of fullerene C_{60} systems with visible chromophores on their structure, spectroscopy and photophysics. Photoinduced electron transfer from zinc phthalocyanine (ZnPc) [10] and zinc porphyrin (ZnTPP) [11] to fullerenes (C_{60} and C_{70}) in a polar solvent has been confirmed. In non-polar solvent, energy transfer from excited triplet states of fullerenes (C_{60} and C_{70}) to phthalocyanines occurs predominantly as reported in [10]. In the present work absorption, photoluminescence excitation and emission spectra of binary systems of the fullerene C_{60} with the zinc derivative of phthalocyanine (ZnPc) in solutions and in multiple-layers have been investigated.

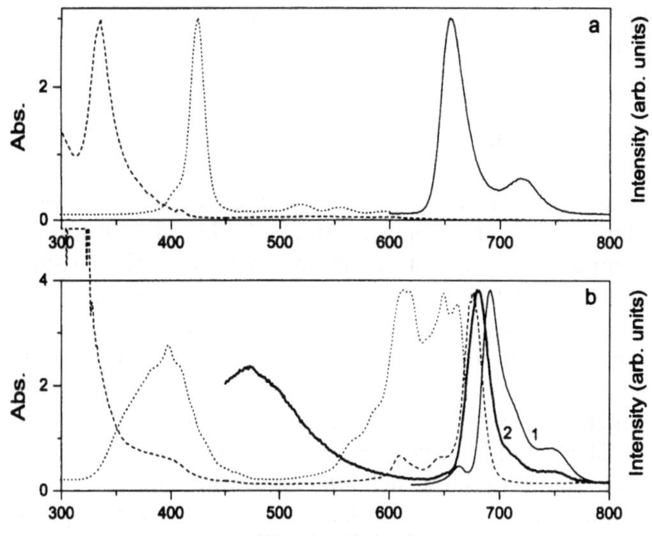

Fig. 1 Absorption (*dash*), photoluminescence excitation (*dot*) and emission (*solid*) spectra of (**a**) C_{60} solution in benzene ($\lambda_{em} = 655$ nm and $\lambda_{exc} = 424$ nm) and (**b**) ZnPc fresh solution in benzonitrile ($\lambda_{em} = 690$ nm and $\lambda_{exc} = 420$ nm). Emission spectra ZnPc (*curve-1*) correspond to a new solution 2×10^{-4} M and after 20 days (*curve-2*)

Experimental

Materials

The C_{60} (99.95%) powder, obtained from MER Corporation (Tucson, USA), and zinc tetra-*tert*-butyl phthalocyanine (ZnPc) from Kodak Corporation were used.

Preparation

The single films and multilayer structures have been sublimed in vacuum onto transparent glass and quartz substrates. An Alpha Step 200 (Tencor Instruments) profilometer was used to estimate the thickness (within ± 10 nm). The films thicknesses were estimated as 20–80 (± 10) nm and 300–500 (± 10) nm for ZnPc; and 20–70 (± 10) nm and 150–200 (± 10) nm for C_{60}. Films were also prepared by casting and spinning the benzene or benzonitrile solutions onto substrates by adding drops in 20–30 μL increments (liquid deposition), and were tested from layer to layer.

Characterization

Photoluminescence emission and excitation spectra were measured at room temperature with a SPEX DM 3000 F and Jobin Yvon-SPEX Fluorolog-3 spectrofluorimeters with 150 W and 450 W xenon arc lamps, respectively, as excitation sources. Excitation wavelengths (± 2 nm) were isolated by appropriate monochromators with filter combinations. Photoluminescence spectra of films were also measured with a Renishaw Image Confocal Microscope-2000 model, using He-Ne laser (Spectra-Physics-127) for excitation, with a power of 2.5–25 mW at the source. The absorption spectra were measured at room temperature between 300–800 nm, at 2 nm resolution, with a Shimadzu 2100 spectrophotometer. The spectral components were extracted from the data using Microcal Origin-6 (Microcal Software, Inc.).

Results and discussion

Single systems

Optical absorption and photoluminescence spectra of C_{60} and ZnPc in solution and in individual films are

Fig. 2 *Left*: absorption and photoluminescence emission spectra of C_{60} (40 nm) and ZnPc (20 nm) films. *Right*: Raman and photoluminescence spectral contributions from C_{60} and ZnPc single films; on the top: the Raman scattering of C_{60} film (excitation at $\lambda_{exc} = 632.8$ nm, the wavelength of an He-Ne laser)

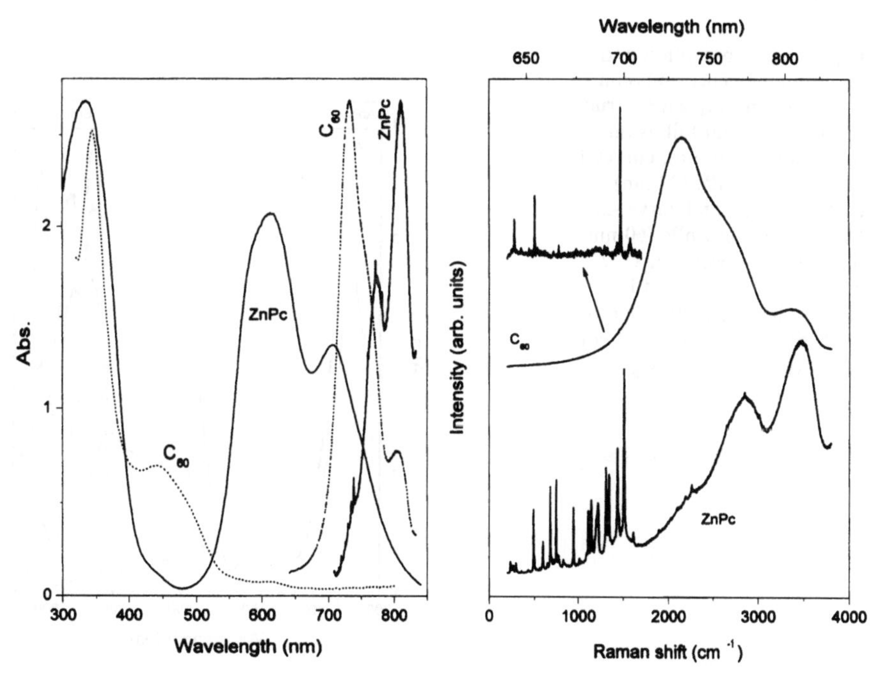

shown in Fig. 1a,b and on the left in Fig. 2. These spectra of C_{60} show the main features already known [1–6]. Absorption, excitation ($\lambda_{em}=655$ nm) and emission ($\lambda_{exc}=424$ nm) spectra of C_{60} solution in benzene (Fig. 1a) demonstrate main bands at 335, 424 and 655 nm, respectively, typical for individual isolated C_{60} molecules in solution within the dilute concentration range. In Fig. 1b absorption and excitation ($\lambda_{em}=690$ nm) spectra of a ZnPc fresh solution show similar spectral contributions in the 340–460 nm and 550–670 nm regions. ZnPc in a polar solvent (benzonitrile) at a short time after preparation of the solution shows a blue shift in the main emission band from 691 to 680 nm and the presence of a new broad band at 480 nm ($\lambda_{exc}=420$ nm). These spectral transformations can be explained by a strong tendency of ZnPc molecules in solution to aggregation. In solid C_{60}, the ground state absorption has been attributed to interaction between C_{60} molecules that in turn enhance the forbidden bands at 450 and 620 nm [12]. The most obvious feature of the C_{60} film luminescence is that the main peak at 732 nm has a large fraction of the total emission intensity and a relatively small vibronic contribution. Bands at 774 and 810 nm dominate the photoluminescence spectrum of ZnPc films as has been reported in [13]. The Raman spectra with photoluminescence contributions of C_{60} and ZnPc films are also presented on the right in Fig. 2. The Raman spectrum of the fullerene film exhibits the ten Raman active vibrations found previously, which is dominated by a peak at 1469 cm^{-1} [2–6]. Very strong Raman bands have been observed in ZnPc film that can be assigned to resonance conditions [14, 15] under excitation at $\lambda_{exc}=632.8$ nm, the wavelength of a He-Ne laser.

Binary systems

In solution, the absorption spectra of the mixtures were identical to the superposition of the absorption of components, suggesting no interaction in the ground state within the dilute concentration range used in this study. The results obtained from individual solution and mixtures demonstrate a tendency of ZnPc to aggregation, while spectra of mixtures can be interpreted as a superposition of C_{60} and ZnPc contributions.

The C_{60} film thickness dependencies of absorption and photoluminescence in C_{60}/ZnPc bilayers were studied. Fig. 3 (on the left) gives the typical optical absorption and photoluminescence spectra of the C_{60}/ZnPc bilayers: C_{60} films were prepared by physical vapour deposition onto ZnPc films. The absorption spectra of bilayers were interpreted as resulting from the superposition of components, while the C_{60} forbidden band intensity around 450 nm was markedly weaker in the spectra of very thin C_{60} films prepared on the ZnPc film surface. The possibility exists that the decrease in the C_{60} forbidden band intensity results from the anisotropy of the substrate associated with the ZnPc film [16].

Fig. 3 (on the right) shows also the selected spectra of C_{60}/ZnPc bilayers prepared by casting the C_{60} solution in benzene onto a ZnPc thick film (liquid deposition). The spectra obtained from these samples cannot be interpreted as a pure superposition of C_{60} and ZnPc

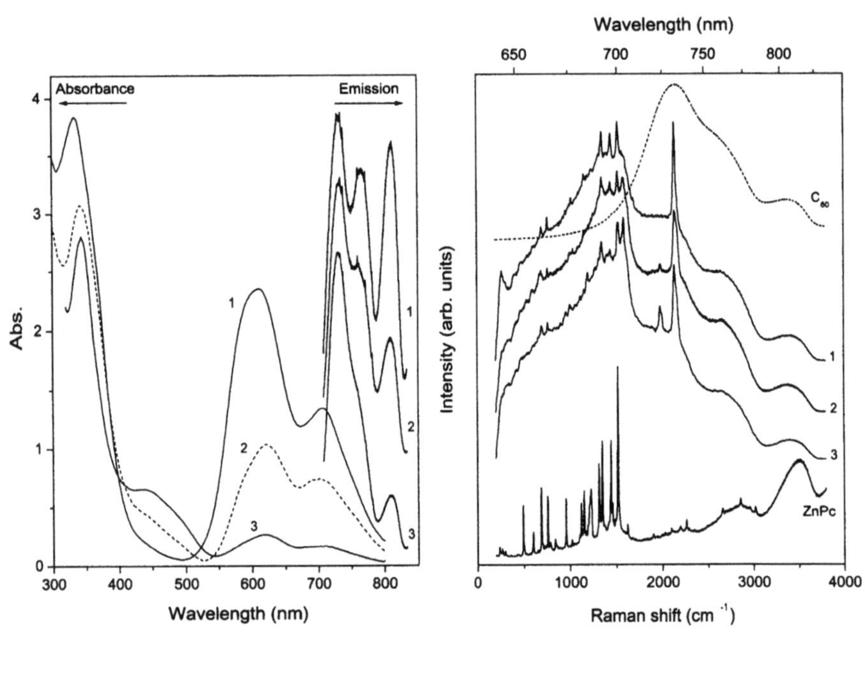

Fig. 3 *Left*: absorption and photoluminescence emission spectra of the C_{60}/ZnPc structures: C_{60} (40 nm) films on thick ZnPc (450 nm), curves-1, and on thin ZnPc (20 nm), curves-3, films and curves-2: C_{60} (70 nm) on ZnPc (60 nm) films. *Right*: spectra of C_{60}/ZnPc bilayers prepared by casting the C_{60} solution onto ZnPc film, curves-1, -2, -3 correspond to the initial C_{60} monolayers

independent components. In fact, there are significant differences between the shapes of photoluminescence spectra of C_{60}/ZnPc structures prepared with the previously discussed procedure (vacuum deposition) and by liquid deposition. These exhibited photoluminescence bands associated with the interaction between C_{60} and the zinc phthalocyanine within the bilayer structure. The interesting features of the spectra of these very thin C_{60} films ("monolayers") on ZnPc surface have been analysed in the context of the electron transfer from ZnPc to C_{60} and interface-induced aggregation in the initial stages of deposition (so-called mesoscopic condition). Charge transfer between C_{60} and ZnPc was also suggested recently in [17] for co-deposited C_{60}/ZnPc films. From a structural point of view, a possible explanation of these photo luminescence spectra of the C_{60} "monolayer" on the large effective area of ZnPc polycrystalline surface may be found in a mechanism of symmetry reduction at the interface and weaker interaction between C_{60} molecules in monolayer fragments, due to an electron transfer process from ZnPc.

Conclusions

In the present work results are presented on the absorption, photoluminescence excitation and emission spectra of the fullerene C_{60}, and ZnPc in solutions and in multiple-layers. Films and multilayer structures have been prepared using various methods and have been analysed with respect to the preparation method. The C_{60} film thickness dependence of photoluminescence and absorption, in binary systems of C_{60}, with visible chromophore were studied. A series of the structures was characterised. These exhibited photoluminescence bands associated with the interaction between C_{60} and the ZnPc within the bilayer structure that can be analysed in the context of a symmetry reduction of C_{60} at the C_{60}/ZnPc interface.

Acknowledgements The authors wish to thank the FCT (Portugal) for the award of a fellowship (BPD/1525/2000) to A.A.K.

References

1. Minami N (1991) Chem Lett Chem Soc Jpn 10:1791
2. Kiser M, Reichenbach J, Byrne HJ, Andres J, Maser W, Roth S, Zahab A, Bernier P (1992) Solid State Commun 81:261
3. Kazaoui S, Ross R, Minami N (1994) Solid State Commun 90–10:623
4. Sinha A, Menendez J, Hanson RC, Adams GB, Page JB, Sankey OF, Lamb LD, Huffman DR (1991) Chem Phys Lett 186:2
5. Reber C, Yee L, McKiernan J, Zink JI, Williams RS, Tong WM, Ohlberg DA, Whetten RL, Diederich F (1991) J Phys Chem 95:2127
6. Sariciftci NS, Smilowitz L, Heeger AJ, Wudl F (1992) Science 258:1474
7. Sauvajol JL, Hricha Z, Coustel N, Zahab A, Aznar R (1993) J Phys Condens Matter 5:2045
8. Smilowitz L, Sariciftci NS, Wu R, Gettinger C, Heeger AJ, Wudl F (1993) Phys Rev B 47:3835
9. Hiromitsu I, Kitano M, Shinto R, Ito T (2000) Solid State Commun 113:165
10. Nojiri T, Alam M A, Konami H, Watanabe A, Ito O (1997) J Phys Chem A 101:7943
11. Nojiri T, Watanabe A, Ito O (1998) J Phys Chem A 102:5215
12. Wang YM, Kamat PV, Patterson LK (1993) J Phys Chem 97:8793
13. Siebentritt S, Günster S, Meissner D (1991) Synthetic Metals 41/43:1173
14. Larina LL, Melnik NN, Poponin VP, Shevaleevskii OI, Kalachev AA (1995) Mater Sci Forum 173/174:231
15. Kharlamov A, Rainho J, Giullo G, Iannotta S (1998) EPS Europhys Conf 22G:27
16. Rainho JP, Santos LM, Kharlamov AA (1998) Mat Res Soc Symp Proc 488:933
17. Toccoli T, Boschetti A, Iannotta S (2001) Synthetic Metals 122-1:229

Progr Colloid Polym Sci (2004) 123: 56–60
DOI 10.1007/b11629

Masakatsu Hato
Hiroyuki Minamikawa
Rajesh A. Salkar
Sanae Matsutani

Phase behavior of phytanyl-chained akylglycoside/water systems

M. Hato (✉) · H. Minamikawa
R.A. Salkar · S. Matsutani
Bionanomaterials and Surface Interactions Group, Nanotechnology Research Institute, AIST, Tsukuba Central-5, Higashi 1-1-1, Tsukuba, Ibaraki, 305–8565, Japan
e-mail: m.hato@aist.go.jp
Tel.: +81-29-861-9324
Fax: +81-29-861-6243

Present address: H. Minamikawa
Nanoarchitectonics Research Center, AIST, Tsukuba Central-5, Higashi 1-1-1, Tsukuba, Ibaraki, 305–8565, Japan

Present address: R.A. Salkar
Department of Physics, Astronomy & Mathematics, University of Central Lancashire, Corporation Street, Preston, Lancashire, UK

Abstract The aqueous phase behavior of novel alkylglycosides, AGs, that have a 3,7,11,15-tetramethylhexadecyl (phytanyl) group as their hydrophobic part was examined as a function of the headgroup type, i.e., glycerol, xylose, glucose, and maltose in order of the increasing number of hydroxy groups of the headgroup. The aqueous phase structures at full hydration are well correlated with the headgroup size; an H_{II} phase for the glycerol-headgroup, a cubic phase of crystallographic space group *Pn3m*/*Pn3* at lower temperatures and an H_{II} phase at higher temperatures for the xylose-headgroup, and an L_α phase for the glucose- and the maltose-headgroup. Krafft eutectic temperatures, T_K, of the phytanyl-chained AGs are significantly lower than those of conventional AGs, although the total number of carbon atoms in the hydrophobic group is as large as 20. Their low T_K values afford greater control of the aqueous phase structures at low temperatures.

Keywords Phytanyl-chained akylglycoside · Akylglycoside · Phase diagram · Cubic phase Krafft eutectic temperature

Introduction

Alkylglycosides are recently becoming important from the ecological and industrial view points [1, 2]. This is because these surfactants can be synthesized from renewable resources and are generally non-toxic and biodegradable. Moreover, biological functions of the sugar groups such as molecular recognition [3] or stabilizing effects of protein functions and structures [4] make them particularly attractive as a new material for biotechnology [5–9].

Values of T_K of conventional AGs are unusually high for "non-ionic" surfactants [10, 11]. For example, the T_K values of C_{12}-glucosides are already 55 °C (*n*-dodecyl-α-D-glucoside) and 36 °C (*n*-dodecyl-β-D-glucoside) [12]. As further extension of the alkyl chain will result in AGs with T_K significantly higher than room temperatures, the conventional AGs have relatively short alkyl chains mainly in a range C_8–C_{12}. This fact should not be underestimated. Microemulsions based on the short-chain AGs, for example, require high surfactant concentrations that are not acceptable in many technical applications [13]. Furthermore, the aqueous phases formed in their phase diagrams are of the normal-type [13–15]. A biologically important lamellar phase (that can form stable vesicles) and in particular inverted liquid crystalline phases [16, 17] could not be obtained at room temperatures. This seriously limits the usefulness of the AGs in many technical applications. We have recently developed novel AGs that largely overcome these difficulties [18]. In this communication, we report their phase diagrams in order to understand how the aqueous phase structures are controlled by modification of the sugar headgroups.

Materials and methods

Figure 1 shows the chemical structures of the AGs examined in this communication. They were synthesized by similar procedures as

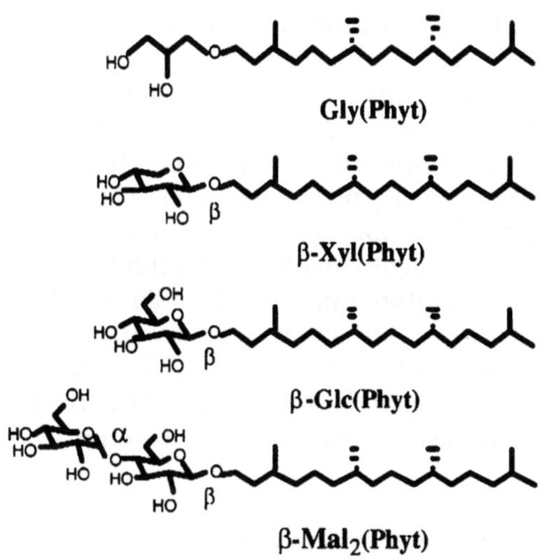

Fig. 1 Chemical structures of phytanyl-chained alkyl glycosides. Gly(Phyt): glyceryl phytanyl ether. *β*-Xyl(Phyt): 1-*O*-phytanyl-*β*-D-xyloside. *β*-Glc(Phyt): 1-*O*-phytanyl-*β*-D-glucoside. *β*-Mal_2(Phyt): 1-*O*-phytanyl-*β*-D-maltoside. Phyt: 3,7,11,15-tetramethylhexadecyl group (phytanyl group). Mal_N: maltooligosaccharide that consists of N glucose residues that are linked by α-1,4-*O*-glucosidic bonds

previously reported [19, 20]. The hydrophobic chain is a phytanyl group (Phyt) that contains 16 backbone carbon atoms and 4 methyl groups at the 3, 7, 11, and 15 positions (a total of 20 carbon atoms). The headgroups are systematically altered, i.e., maltose (Mal_2) where 2 glucose residues are linked by α-1,4-O-glucosidic bonds, glucose (Glc), xylose (Xyl) and glycerol (Gly) in order of the decreasing number of hydroxy groups of the headgroup. The purity of the surfactants was checked by NMR and TLC and was at least 98%.

The phase diagram of each surfactant/water system was determined by small angle X-ray scattering (SAXS) measurements. When needed, we performed an optical examination of surfactant/water mixtures with an Olympus BHS-751-P polarizing microscope to substantiate the SAXS results [21]. The phase boundary was estimated from a deflection in a d-spacing vs. surfactant concentration curve. The SAXS measurements were performed with Ni-filtered CuKα radiation (wavelength = 0.154 nm) generated by a Rigaku RU-200 X-ray generator (40 kV, 100 mA) with a double pinhole collimator (0.5 mm ϕ × 0.3 mm ϕ). The hydrated AG was flame-sealed into a quartz capillary (Glas, Berlin, 1.5 mm ϕ in outer diameter) after repeated cycles of freezing and thawing. The sample-loaded capillaries were then incubated at 4 °C for at least 24 h. The first measurements were carried out at 4 °C followed by measurements at temperatures ramped in 10 °C increments. At each temperature we allowed the sample to equilibrate for 30 min∼1 h. When the temperature approached a phase transition temperature, the temperature was ramped at a 3 °C interval and 5 h–48 h of incubation were allowed to assure the equilibrium. After the measurements at highest temperature, we performed a second measurement at 25 °C to confirm that the second measurement gave practically identical results to those of the first measurement. The sample temperature was controlled to within ± 0.5 °C by a Mettler FP82HT hotstage. Exposure time was 30 min–1 h at a sample to film distance of 195 mm. The values of T_K for fully hydrated surfactants were estimated by a Seiko SSC/560U differential scanning calorimeter.

Results

Gly(Phyt)/water system (Fig. 2a)

Below 65 °C, the Gly(Phyt)/water system was characterized by three diffraction peaks whose spacing ratio is $d_1:d_2:d_3 \approx 1:1/\sqrt{3}:1/2$ that is consistent with a two-dimensional hexagonal lattice (Table 1). The value of the first-order diffraction at full hydration, $d_1 = 4.1_5$ nm (at 4 °C), decreased slightly with temperature: about 0.01 nm/°C over the temperature range from 4 °C to 60 °C. By analogy with other lipid/water systems [22], the low hydration capacity of the mesophase indicates that the structure of this phase is of type H_{II}. On dilution below about 85 wt % surfactant, excess water, W, separates from the H_{II} phase to form a two-phase W + H_{II} region. At higher temperatures, the H_{II} phase transforms into a fluid isotropic phase, I, which gave a broad peak around 3.7 nm. Above about 96% surfactant, the system no longer formed a liquid crystalline phase. An H_{II} + I coexistence region seemed very narrow and could not be clearly detected.

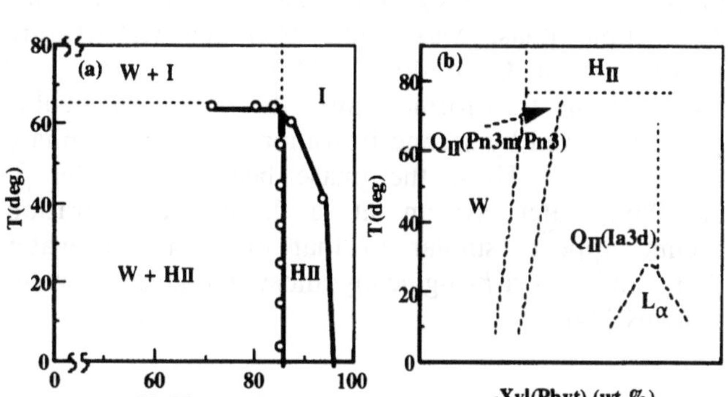

Fig. 2 Partial phase diagrams of phytanyl-chained alkyl glycosides. (**a**) Gly(Phyt)/water system. (**b**) *β*-Xyl(Phyt)/water system. (schematic). (**c**) *β*-Glc(Phyt)/water system. (**d**) *β*-Mal_2(Phyt)/water system. W: excess water. I: a fluid isotropic phase. H_{II}: an inverted hexagonal phase. Q_{II}(Pn3m/Pn3): an inverted cubic phase of *Pn3m*/*Pn3* symmetry. Q_{II}(Ia3d): an inverted cubic phase of *Ia3d* symmetry. L_α: a lamellar phase

β-Xyl(Phyt)/water system (Fig. 2b)

The β-Xyl(Phyt)/water system exhibited rich phase behavior involving two different cubic phases at lower temperatures and an H_{II} phase at higher temperatures. At higher concentration (≥80%) and at lower temperatures (< 35 °C), extra peaks that could be ascribed to the presence of unknown phases complicated the identification of the phase structures. Thus the phase diagram presented here is only a schematic one. Nevertheless, the data so far obtained can provide us with useful information to characterize the β-Xyl(Phyt)/water system. The SAXS diffraction peaks observed for an isotropic phase in equilibrium with W gave 6 peaks (e.g., 6.3_8, 5.2_0, 4.4_9, 3.6_7, 3.1_8, 2.9_9 nm for 50 wt % surfactant at 25 °C), with spacing ratios of $\sqrt{2}$, $\sqrt{3}$, $\sqrt{4}$, $\sqrt{6}$, $\sqrt{8}$, $\sqrt{9}$, that were consistent with a cubic phase of the crystallographic space group *Pn3m*/*Pn3*. This cubic phase was stable at least from 4 °C to 74 °C. Above 76 °C, the *Pn3m*/*Pn3* cubic phase transformed into an H_{II} phase that persisted at least up to 95 °C. With increasing the surfactant concentration to 70%, a second set of SAXS diffraction peaks (e.g., 5.6_1, 4.9_7, 3.6_1, 3.4_6, 3.0_9, 2.9_5 nm at 35 °C) were observed with the spacing rations of $\sqrt{6}$, $\sqrt{8}$, $\sqrt{14}$, $\sqrt{16}$, $\sqrt{20}$, $\sqrt{22}$. This indicated a cubic phase of the crystallographic space group *Ia3d*. This *Ia3d* cubic phase was stable over the temperature range from 4 °C to 40 °C. Above 45 °C, the *Ia3d* cubic phase transformed into the *Pn3m*/*Pn3* cubic phase followed by second transition into the H_{II} phase above 75 °C. Thus, the phase behavior of the β-Xyl(Phyt)/water system at a lower concentration regime appears similar to that of monoolein/water [23], and alkyl-β-D-glucopyranosyl-rac-glycerol/water systems [24].

β-Glc(Phyt)/water system (Fig. 2c)

The β-Glc(Phyt)/water system gave diffractions at positions in the ratio 1:1/2:1/3, corresponding to first-, second- and third-order diffractions from a bilayer periodicity of an L_α phase (Table 1). The value of the first-order diffraction line, $d_1 = 4.3_8$ nm (at 25 °C) changed only slightly with temperature: 0.001 nm/°C over the temperature range 25 °C to 80 °C. On dilution below ~80 wt % surfactant, excess water separated to form a two-phase region, $W + L_\alpha$. The L_α phase extended at least up to 98 wt % surfactant that corresponds to β-Glc(Phyt)1/2H_2O and stable at least up to 100 °C.

β-Mal_2(Phyt)/water system (Fig. 2d)

The phase behavior of the β-Mal_2(Phyt)/water system is very similar to the β-Glc(Phyt)/water system, indicating that addition of one glucose residue to the glucose headgroup does not appreciably affect the phase behavior. The observed X-ray diffraction spacing in the ratio of 1:1/2:1/3 corresponds to an L_α phase (Table 1). The value of the first order diffraction line, $d_1 = 4.9_8$ nm (at 2 °C) decreased only slightly with temperature: 0.002 nm/°C over the temperature range 2 °C to 70 °C. On dilution below about 75 wt % surfactant, excess water separated from the L_α phase to form an $L_\alpha + W$ two-phase region. The L_α phase extended at least up to 83 wt % surfactant (maximum concentration examined) and stable at least up to 100 °C. The major difference between the β-Glc(Phyt)/water and the β-Mal_2(Phyt)/water system lies in a maximum amount of hydration of the L_α phase: about 8 moles of water per β-Glc(Phyt) and about 13 moles of water per β-Mal_2(Phyt), respectively.

Table 1 Phase behavior of the phytanyl-chained AGs at full hydration (at 25 °C)

Surfactant	T_K [°C]	Phase sequence	$(n_w/n_L)_{max}$ [mol/mol]	d_l [nm]	d_2	d_3	Reference
Gly(Phyt)	< 0	W-H_{II}-I	4	3.9_3	2.2_7	1.9_8	this work
β-Xyl(Phyt)	~10	W-Q_{II}(*Pn3m*/*Pn3*)-Q_{II}(*Ia3d*)-L_α	–	(see text)			this work
β-Glc(Phyt)	< 0	W- L_α	8	4.3_8	2.1_9	1.4_6	this work
β-Mal_2(Phyt)	< 0	W-L_α	13	4.9_4	2.46	1.6_2	this work
β-$GlcC_{12}$[a]	36	L_I (phase separation)-L_α[e]		–			12
β-Mal_2C_{12}[b]	≤0	L_I-H_I-(M_α)-S[e]		–			34
β-$GlcC_{18}$[c]	~55	W-L_α-?[e]		–			27
β-Mal_2C_{18}[d]	~40	?-H_I-L_α-?[e]		–			28

$(n_w/n_L)_{max}$: maximum swelling of the phase (mole of H_2O per mole of surfactant). W: excess water. I: a fluid isotropic phase
L_I: a normal micellar solution. H_I: a normal hexagonal phase. M_α: a centered rectangular phase. S: a hydrated solid
[a] β-$GlcC_{12}$: *n*-dodecyl-β-glucoside
[b] β-Mal_2C_{12}: *n*-dodecyl-β-maltoside
[c] β-$GlcC_{18}$: *n*-octadecyl-β-D-glucoside
[d] β-Mal_2C_{18}: *n*-octadecyl-β-maltoside
[e] The phase sequence at temperatures above T_K

Discussion

Table 1 summarizes the aqueous phase behavior of the phytanyl-chained AGs at full hydration at 25 °C. The phase data available for the AGs with *n*-dodecyl and *n*-octadecyl chains are also listed. The data display several important features for the AG/water systems.

First, the phase structures of the phytanyl-chained AGs are well correlated with the size (cross-section area) of the headgroups; the H_{II} phase for the smallest glycerol headgroup, and the L_α phase for the glucose and maltose headgroups. Xylose differs from glucose in that the hydroxymethylene group, CH_2-OH, at carbon C5 of glucose is replaced by a hydrogen atom. Thus, the cross-section area of the xylose headgroup would be intermediate between the glucose and the glycerol headgroup. The *β*-Xyl(Phyt)/water system in fact gave phase behavior intermediate between that of Gly(Phyt) and *β*-Glc(Phyt). This indicates that carbohydrate chemistry is a powerful tool for fine tuning of the aqueous phase structures of AGs. In this context, it is interesting to speculate what will happen when one further increases the number of glucose residues (N) in the maltooligosaccharide headgroup of Mal_N(Phyt)? Let us consider surfactants with double *n*-dodecyl chains: 1,3-di-*O*-dodecyl-2-*O*-(*β*-glycosyl)glycerols bearing a series of maltooligosaccharide headgroups, $Mal_N(C_{12})_2$ [10, 25, 26]. Their liquid crystalline phases at full hydration shift from an H_{II} phase to an H_I phase as N increases from 1 to 7, i.e., an H_{II} phase (N = 1), an L_α phase (N = 2, 3), and an H_I phase (N = 7), indicating that the "hydrophilicity" of the surfactant is constantly enhanced as N increases. With a maltoheptaosyl headgroup (N = 7), even a surfactant with strongly "hydrophobic" di-dodecyl chains can form normal micelles with a CMC of less than 5×10^{-6} M [26]. In analogy with $Mal_N(C_{12})_2$, an increase in N of the Mal_N(Phyt) would eventually result in normal micelle formation. This suggest that the phytanyl-chained AGs can form a nearly full range of aqueous phase structures at low temperatures, from the strongly hydrophilic normal-type to the strongly hydrophobic inverted-type.

Second, the T_K values of the phytanyl-chained AGs are significantly lower than those of the conventional AGs [27, 28], although the total number of carbon atoms in the phytanyl chain is as large as 20. This indicates that the phytanyl group (and other isoprenoid-type hydrophobic groups) is suited to design AGs with a large-enough hydrophobic group while keeping the values of T_K sufficiently low [10, 11].

Third, the maximum amount of water molecules incorporated in the liquid crystalline phases, $(n_w/n_L)_{max}$, is low and temperature-insensitive: about two water molecules per OH group in the headgroups. The average area per AG molecule in the L_α phase that was estimated from the SAXS data is 0.46 nm^2 for *β*-Glc(Phyt) and 0.53 nm^2 for *β*-Mal_2(Phyt). These values are only marginally larger than the cross section area of about 0.4 nm^2 for anhydrous glucose and maltose (estimated from the Allinger's MM2 calculation) [25]. This implies that apart from strongly bound primary hydrated water molecules, one or two water layers can exist in a "hydration shell" of the sugar group. The strong interlamellar attractions [29, 30] arise in part from the low hydration of the headgroups and explain ion-induced aggregation of aqueous glycolipid vesicles [30, 31]. The low hydration of sugar-headgroup has also been reported for other surfactants such as *n*-octyl-*β*-D-glucoside [32], mono-glucosyldiglyceride [33], and digalacosyldiacylglycerol [34].

Acknowledgements R.A.S. is grateful for the STA-fellowship. Financial support of AIST is highly acknowledged (Subject: Structures and functions of organized solutions). This work was also performed as a part of the International Joint Research Program 2001 supported by NEDO.

References

1. Hill K, Rybinski W von, Stoll G (1997) Alkyl polyglycosides. VCH, Weinheim
2. Söderman O, Johansson I (2000) Curr Opin Colloid Interface Sci 4:391
3. Hakomori S (1991) Pure Appl Chem 63:473
4. Franks F, Hatley RHM (1993) In: Tweel WJJ van den, Harder A, Buitelaar RM (eds) Stability and stabilization of enzymes. Elsevier, New York
5. Helenius A, McCaslin DR, Fries E, Tanford C (1979) Methods Enzymol 56:734
6. Kühlbrandt W (1992) Quart Rev Biophys 25:1
7. Rigaud J-L, Pitard B, Levy D (1995) Biochim Biophys Acta 1231:223
8. Rummel G, Hardmeyer A, Widmer C, Chiu ML, Nollert P, Locher KP, Pedruzzi I, Landau EM, Rosenbush J (1998) J Struc Biol 121:82
9. Baba T, Minamikawa H, Hato M, Motoki A, Hirano M, Zhou D, Kawasaki K (1999) Biochem Biophys Res Commun 265:734
10. Hato M, Minamikawa H, Tamada K, Baba T, Tanabe Y (1999) Adv Colloid Interface Sci 80:233
11. Hato M (2001) Current Opinion Colloid Interface Sci 6:268
12. Boyd BJ, Drummond CJ, Krodkiewska I, Grieser F (2000) Langmuir 16:7359
13. Stubenrauch C (2001) Curr Opin Colloid Interface Sci 6:160
14. Sakya P, Seddon JM, Vill V (1997) Liq Cryst 23:409
15. Nilsson F, Söderman O, Hansson P (1998) Langmuir 14:4050
16. Larsson K (1989) J Phys Chem 93:7304
17. Lindblom G, Rilfors L (1992) Adv Colloid Interface Sci 41:101
18. Hato M, Minamikawa H, Salkar RA, Matsutani M (2002) Langmuir 18:3425
19. Minamikawa H, Murakami T, Hato M (1994) Chem Phys Lipids 72:111

20. Minamikawa H, Hato M (1997) Langmuir 13:2564
21. Rosevear FB (1954) J Am Oil Chem Soc 31:628
22. Larsson K (1994) Lipids – molecular organization, physical functions and technical applications. The Oily Press, Glasgow
23. Hyde ST, Andersson S, Ericsson B, Larsson K (1984) Z Kristallogr 168:213
24. Turner DC, Wang Z-G, Gruner SM, Mannock DA, McElhaney N (1992) J Phys II France 2:2039
25. Hato M, Minamikawa H (1996) Langmuir 12:1658
26. Hato M, Seguer JB, Minamikawa H (1998) J Phys Chem B 102:11035
27. Vill V, Minden HM von, Koch MHJ, Seydel U, Brandenburg K (2000) Chem Phys Lipids 104:75
28. Minden HM von, Brandenburg K, Seydel U, Koch MHJ, Garamus V, Willumeit R, Vill V (2000) Chem Phys Lipids 106:157
29. Waltermo Å, Manev E, Pugh R, Claesson PM (1994) J Disp Sci Tech 15:273
30. Korchowiec BM, Baba T, Minamikawa H, Hato M (2001) Langmuir 17:1853
31. Baba T, Zheng L-Q, Minamikawa H, Hato M (2000) J Colloid Interface Sci 223:235
32. Nilsson F, Söderman O, Johansson I (1996) Langmuir 12:902
33. Wieslander A, Ulmius J, Lindblom G, Fontell K (1978) Biochim Biophys Acta 512:241
34. McDaniel RV (1988) Biochim Biophys Acta 940:158

Progr Colloid Polym Sci (2004) 123:61–64
DOI 10.1007/b11632

David Llères
Jean-Pierre Clamme
Emmanuel Dauty
Jean-Paul Behr
Yves Mély
Guy Duportail

Oxidisable cationic detergent for gene therapy: condensation of DNA and interaction with model membranes

D. Llères · J.-P. Clamme · Y. Mély
G. Duportail (✉)
Laboratoire de Pharmacologie et Physicochimie des Interactions Cellulaires et Moléculaires, UMR 7034 du CNRS, Faculté de Pharmacie, Université Louis Pasteur, B.P. 24, 67401 Illkirch Cedex, France
e-mail: duportai@aspirine.u-strasbg.fr
Tel.: +33-390244260
Fax: +33-390244312

E. Dauty · J.-P. Behr
Laboratoire de Chimie Génétique, UMR 7514 du CNRS, Faculté de Pharmacie, Université Louis Pasteur, B.P. 24, 67401 Illkirch Cedex, France

Abstract Cationic amphiphile-mediated delivery of plasmid DNA is the non-viral gene transfer method most often used. In the present work, we considered a new cysteine-detergent, ornithinyl-cysteinyl-tetradecylamide (C_{14}-CO), able to convert itself, via oxidative dimerisation, into a cationic cystine-lipid. By using fluorescence techniques, we first characterised the structure of complexes of plasmid DNA with C_{14}-CO molecules either kept as monomers, or oxidised into dimers. Both forms are able to condense DNA, with the formation of hydrophobic micelle-like domains along the DNA chain. Domains with a larger molecular order were obtained with dimeric C_{14}-CO/DNA complexes. In a second step, the interactions of these complexes with lipid vesicles considered as membrane models were investigated. In the presence of vesicles, we observed a decondensation of the DNA involved in complexes obtained with C_{14}-CO monomers. With anionic vesicles, the DNA is released into the bulk solution, while with neutral vesicles, it remains bound to the vesicles via electrostatic interactions with inserted C_{14}-CO molecules. In sharp contrast, the complexes with C_{14}-CO dimers are unaffected by the addition of either neutral or anionic vesicles and show no interaction with them. These results may partly explain the low transfection efficiency of these complexes at the +/− charge ratios used in this study.

Keywords DNA condensation · Cationic detergents · Non-viral vectors · Transfection · Lipid vesicles

Introduction

Gene therapy is based on the use of nucleic acids for the treatment of both genetic and acquired diseases. As nucleic acids hardly cross cell membranes, the feasibility of gene therapy depends strongly on the use of vectors able to transport the gene into the target cells. Viral vectors show high gene transfer efficiency, but suffer from drawbacks of immunogenicity and mutagenicity. These problems can be circumvented by using non-viral vectors like cationic polymers or cationic liposomes [1, 2]. Although efficient for transfecting cells in culture, these complexes aggregate into particles larger than 1 μm, thus limiting their ability to cross membranes. Aggregation may be overcome by the use of cationic detergents able to condense DNA in discrete particles containing a single nucleic acid molecule but unable to transfect cells per se [3]. With these considerations in mind, it was appealing to try to combine the respective favourable features of cationic detergents and cationic lipids. In this respect, Dauty et al. [4] have developed a new type of cysteine-based detergent with a hydrophobic chain of 14 carbon atoms and an ornithine polar head, ornithinyl-cysteinyl-tetradecylamide (C_{14}-CO) (Fig. 1, insert) able to convert itself, via oxidative dimerisation into a cationic cystine-lipid on the template DNA [5].

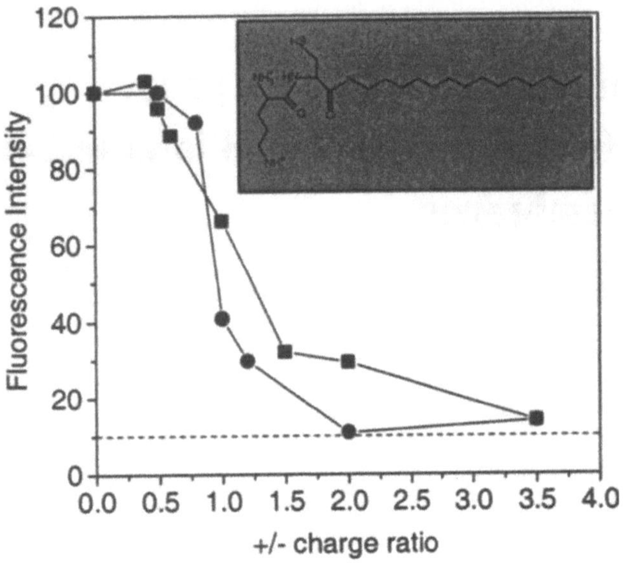

Fig. 1 Plasmid DNA condensation induced by the addition of the oxidisable cationic surfactant (C_{14}-CO). The condensation is monitored by the fluorescence intensity of EtBr added at a 1:50 molar ratio of probe to nucleotide. (●) C_{14}-CO monomer; (■) C_{14}-CO dimer. The dotted line corresponds to the fluorescence intensity of free EtBr. The chemical structure of C_{14}-CO is given in the insert

In this context, our objective was to characterise, by fluorescence spectroscopy, the structure of the complexes of plasmid DNA with C_{14}-CO molecules either kept as monomers, or oxidised into dimers. In a second step, we investigated the interactions of these complexes with lipid vesicles considered as membrane models in order to explain the low transfection efficiency of these complexes at the +/− charge ratios used to form single DNA-containing particles.

Materials and methods

Egg yolk phosphatidylcholine (EYPC) and phosphatidyl-DL-glycerol (EYPG) were from Sigma. Ethidium bromide (EtBr), YOYO-1, 1,6-diphenylhexatriene (DPH) and 2-[3-(diphenylhexatrienyl)propanoyl]-1-hexadecanoyl-*sn*-glycero-3-phosphocholine (DPHpPC) were from Molecular Probes. pCMV-luc plasmid was produced and purified as described [7]. C_{14}-CO was prepared as described [4] and stored under argon at −80 °C in neat ethanol.

The complexes were prepared in Hepes 15 mM, pH 7.4 buffer by mixing pCMV-luc plasmid DNA at a final concentration of 10 μM phosphate with an appropriate amount of detergent to reach a given +/− charge ratio, *r*, assuming that the two positive charges of a C_{14}-CO molecule neutralise two DNA phosphate groups. The mixture was kept at room temperature for 4 h to allow for complete C_{14}-CO oxidation. The studies with the C_{14}-CO monomer form were performed in the same buffer, but previously degassed and containing 10 mM dithiothreitol to avoid detergent oxidation.

Fluorescence measurements were performed with a thermostatted SLM 48000 spectrofluorimeter as described [6, 7].

Results

Characterisation of C_{14}-CO/DNA complexes Condensation of DNA occurs spontaneously when the DNA phosphate charges are neutralised by 90% during the formation of complexes with cationic molecules [8]. The condensation of DNA that results from the binding of both forms (monomer and dimer) of C_{14}-CO molecules is followed by the fluorescence decrease of the intercalating fluorophore ethidium bromide (EtBr) [9]. In presence of C_{14}-CO, either in its monomer or dimer form, the EtBr fluorescence does not change significantly at low *r* (Fig. 1). In contrast, a steep decrease of fluorescence intensity is observed, for both forms, at *r* close to 1 and this intensity reaches, at *r* between 2 and 3, the level of EtBr in the absence of DNA. Fluorescence lifetime measurements (data not shown) revealed that this sharp fluorescence decrease is due to an ejection of EtBr from its intercalation sites and not to some environmental changes [10]. Thus, it appears from Fig. 1 that the dimerisation of the C_{14}-CO molecules does not significantly modify the condensation of DNA.

Upon addition of increasing amounts of C_{14}-CO to plasmid DNA, the appearance of hydrophobic domains resulting from the binding of both forms of C_{14}-CO on the DNA was followed by the fluorescence intensity increase of the hydrophobic probe DPH (data not shown) [6, 11]. The molecular order of the micelle-like domains was characterised by the mobility of DPH as estimated by the rotational correlation time, ρ, related to the average fluorescence lifetime, $\langle\tau\rangle$, the DPH fundamental anisotropy, $a_0 = 0.362$ [12], and the steady-state fluorescence anisotropy, a, by Perrin-Weber's equation:

$$\left(\frac{a_0}{a} - 1\right) = 3\frac{\tau}{\rho}.$$

The fluorescence parameters of DPH are summarised in Table 1. The rotational correlation times are relatively large for both C_{14}-CO/DNA complexes. With the monomer form of C_{14}-CO, the ρ value approaches that of obtained with egg yolk phosphatidylcholine vesicles (14.5 ns), while with the dimer form it is even higher ($\rho = 20$ ns). Accordingly, the detergent form (monomer or dimer) is of critical importance for the structure of the micelle-like domains of surfactant/DNA complexes.

Table 1 Fluorescence parameters of DPH in C_{14}-CO/DNA complexes[a]

	a	τ [ns]	ρ [ns]
C_{14}-CO reduced (monomer)	0.165 (±0.006)	4.9 (±0.4)	12 (±2)
C_{14}-CO oxidised (dimer)	0.211 (±0.006)	4.9 (±0.4)	20 (±3)

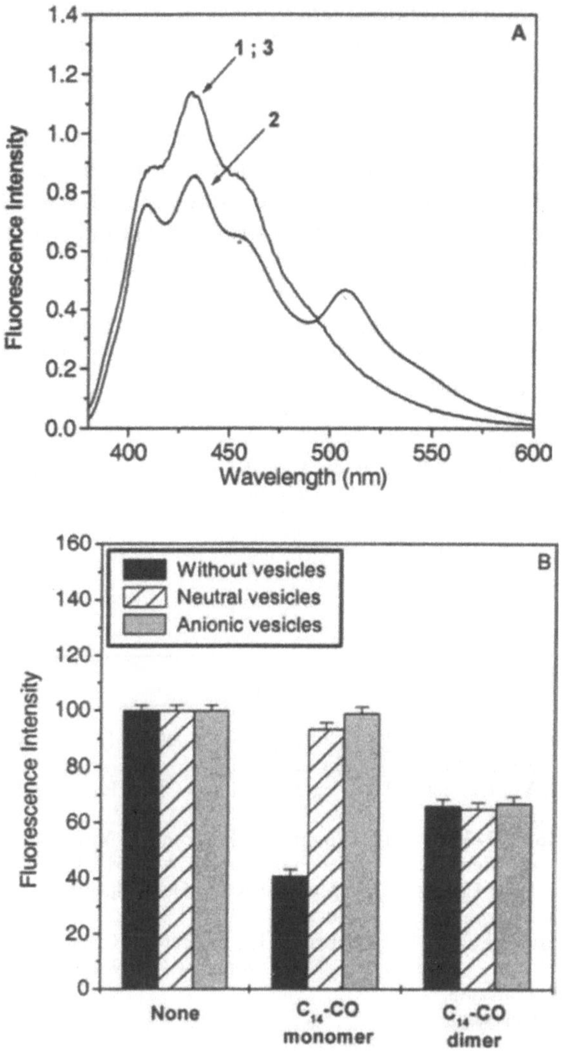

Fig. 2 Interaction of C_{14}-CO/DNA complexes with model membranes. **A** Fluorescence spectra of DPHpPC-labelled EYPC vesicles in the presence of YOYO-labelled C_{14}-CO/DNA complexes. Spectra 1 to 3 correspond to EYPC vesicles either in the absence of complexes or in the presence of monomer C_{14}-CO/DNA and dimer C_{14}-CO/DNA complexes, respectively. The excitation wavelength is 360 nm. **B** Interaction followed by EtBr intercalation. The DNA concentration is 10 μM DNA and $r=1$

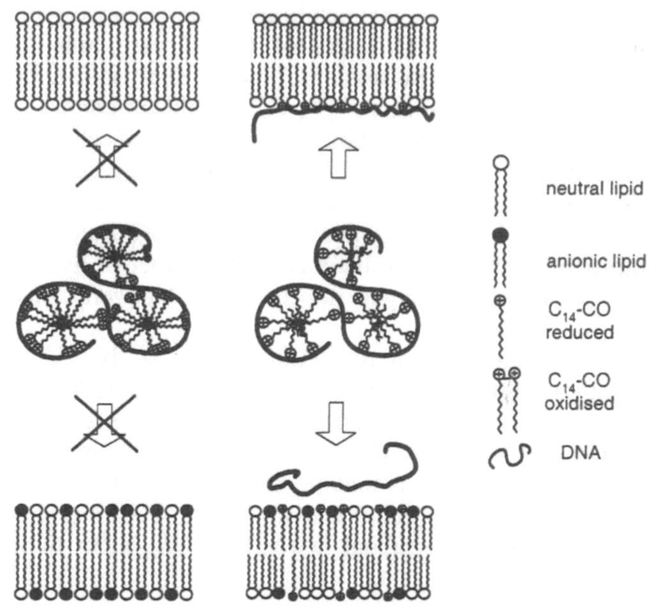

Fig. 3 Proposed mechanism for the interaction of C_{14}-CO/DNA complexes with vesicles. The mechanism is described for dimeric (oxidised) C_{14}-CO/DNA complexes (*left*) and monomeric (reduced) C_{14}-CO/DNA complexes (*right*) with either neutral (*top*) or anionic (*bottom*) vesicles

Interaction of C_{14}-CO/DNA complexes with lipid vesicles The interaction of the detergent/DNA complexes with plasma membrane can be modelled by using neutral and anionic vesicles composed of EYPC, or of an equimolar mixture of EYPC and EYPG, respectively [13].

To evidence these interactions, we first measured, by FRET, the proximity between an acceptor probe linked to the DNA and a donor probe covalently linked to a phospholipid. The acceptor is YOYO-1, a bis-intercalating agent characterised by a very high affinity for DNA [14]. The labelled phospholipid used as a donor is DPHpPC. The fluorescence spectra of C_{14}-CO/DNA complexes in the presence of neutral vesicles (EYPC) are shown in Fig. 2A. For the complexes with monomeric C_{14}-CO, an important fluorescence energy transfer is observed. In sharp contrast, there is no evidence of energy transfer when using complexes with C_{14}-CO dimer. In the case of anionic vesicles, no energy transfer could be observed with either type of complexes. It results that a close contact between YOYO-labelled DNA and DPHpPC-labelled vesicles only occurs with monomeric C_{14}-CO/DNA complexes in the presence of neutral EYPC vesicles.

Additional information is obtained with the fluorescence of intercalated EtBr. With monomeric C_{14}-CO/DNA complexes, the addition of either neutral or anionic vesicles leads to an almost complete recovery of EtBr fluorescence (Fig. 2B). This suggests that the interaction with the vesicles induces a re-exposure of EtBr intercalation sites that may be consecutive to DNA decondensation. In sharp contrast, the addition of both types of vesicles to dimeric C_{14}-CO/DNA complexes is unable to induce any recovery of EtBr fluorescence suggesting that DNA remains condensed.

Conclusion

This work allowed us to characterise the formation of complexes between plasmid DNA and a new cationic surfactant, C_{14}-CO, either reduced (detergent monomer) or oxidised (lipid-like dimer), and to study their behaviour in the presence of model membrane vesicles. Since both monomer and dimer forms of C_{14}-CO are able to efficiently expel EtBr from its binding sites, it can be inferred that both forms of C_{14}-CO are able to condense DNA. The high molecular order of the micelle-like domains for the C_{14}-CO dimers strongly suggests that dimerisation further stabilises the micelle-like domains. This may be linked to a tighter packing of the micelle-like domains resulting from the increased number of hydrophobic interactions per molecule of detergent that follows dimerisation.

Furthermore, from the results of the study of the complexes interacting with model membrane vesicles, we propose the scheme depicted in Fig. 3. In the case of monomeric C_{14}-CO/DNA complexes, the interaction with neutral vesicles is thought to induce an opening of the micelle-like domains and an insertion of C_{14}-CO molecules in the external layer. However, after their incorporation, the detergent molecules are probably still able to bind the unfolded DNA as revealed by the FRET data. In the presence of anionic vesicles, the C_{14}-CO monomers also insert into the lipid bilayer, but probably form neutral ion pairs with anionic lipids. Consequently, positive charge are no more available for DNA binding, and the unfolded DNA is released into the bulk solution.

In sharp contrast, due to the high stability of the micelle-like domains, the dimeric C_{14}-CO/DNA complexes do not interact with both types of vesicles and remain condensed. This may explain the low transfection efficiency of these complexes at the (+/−) charge ratio used in this study.

Acknowledgements This work was supported by ARC, Ligue Régionale du Bas-Rhin et du Haut-Rhin contre le Cancer, and FRM. D.L., J-P.C. and E.D. are recipient of doctoral fellowships from MENRT, the Ligue Régionale du Bas-Rhin et du Haut-Rhin contre le cancer and AFLM, respectively.

References

1. Miller AD (1998) Angew Chem Int Ed 37:1768–1785
2. Godbey WT, Wu KK, Mikos AG (1999) Biochemistry 32:149–160
3. Mel'nikov SM, Sergeyev VG, Yoshikawa K (1995) J Am Chem Soc 117:2401–2408
4. Dauty E, Remy JS, Blessing T, Behr JP (2001) J Am Chem Soc 123:9227–9234
5. Blessing T, Rémy J-S, Behr JP (1998) Proc Natl Acad Sci USA 95:1427–1431
6. Clamme J-P, Bernacchi S, Vuilleumier C, Duportail G, Mély Y (2000) Biochim Biophys Acta 1467:347–361
7. Llères D, Dauty E, Behr J-P, Mély Y, Duportail G (2001) Chem Phys Lipids 111:59–71
8. Bloomfield VA (1996) Curr Opinion Struct Biol 6:334–341
9. Gershon H, Ghirlando R, Guttman SB, Minsky A (1993) Biochemistry 32:7143–7151
10. Le Pecq J-B, Paoletti C (1967) J Mol Biol 27:87–106
11. Lentz BR (1989) Chem Phys Lipids 50:171–190
12. Shinitzky M, Barenholz Y (1978) Biochim Biophys Acta 515:367–394
13. Devaux PF (1991) Biochemistry 30:1163–1173
14. Hirons GT, Fawcett JJ, Crissmann HA (1994) Cytometry 15:129–140

Progr Colloid Polym Sci (2004) 123: 65–68
DOI 10.1007/b11633

Malin Ardhammar
Per Lincoln
Bengt Nordén

Orientation of ruthenium dipyridophenazine complexes in liposome membranes sensitively controlled by ligand substituents

Abstract Liposomes can be deformed to ellipsoidal shapes in a shear flow, and the orientation of chromophores in the lipid bilayers can then be examined with polarised light (linear dichroism, LD). The linear dichroism distinguishes between chromophore transition moments oriented along the lipid chains (negative LD), and parallel to the membrane surface (positive LD). A series of ruthenium(II) complexes comprising differently substituted dipyridophenazine (dppz) ligands has been examined with respect to their orientation when bound to phosphatidylcholine liposome bilayers. The polarisations, energies and absorption overlap of the electronic transitions of the ruthenium complexes, known from previous work, were used to characterise the orientation of liposome bound complexes. From luminescence lifetime studies further information about the localisation of the chromophores was obtained. Monomers with unsubstituted dppz as well as alkyl-substituted dppz ligands tend to dip the dppz ligand down into the bilayer, whereas nitrile and amide substituents contribute to the alignment of the dppz group parallel to the surface. The findings give insights into mechanisms that govern orientation of membrane solutes, thus providing useful leads, for example, for the composition of membrane molecular devices.

Keywords Membrane orientation · Linear dichroism · Shear-deformed liposomes · Ruthenium(II) dipyridophenazine complexes · Luminescence lifetimes

M. Ardhammar (✉) · P. Lincoln
B. Nordén
Department of Physical Chemistry,
Chalmers University of Technology,
412 96 Gothenburg, Sweden
e-mail: maar@phc.chalmers.se
Tel.: +46-31-7723857
Fax: +46-31-7723858

Introduction

Studies of membrane transport and membrane protein function call for a thorough knowledge of molecular interactions in the membrane, between the lipids themselves and between lipids and other species (proteins, drugs, ions). To this aim, the locations and orientations of molecules bound to the membrane can give important information. Linear dichroism of molecules bound to the bilayer of shear-deformed liposomes is one of few direct methods available for the study of orientation of membrane guest molecules. The only requirement to be met is that the molecules of interest have significant absorption in the visible and near-UV regions [1]. The choice of a series of ruthenium complexes with phenanthroline ligands, in another context studied with respect to their remarkable binding and photophysical properties upon interaction with DNA [2–8], was prompted by the observation that a dimeric ruthenium complex dislocates through a liposome membrane with a surprising efficiency, despite its large size [9]. In order to shed light on the interaction of such complexes with artificial lipid membranes, a number of differently substituted ruthenium complexes has been studied with respect to their membrane orientation in flow-oriented liposomes.

Materials and methods

Liposome preparation

L-α-Phosphatidylcholine type II-S, phosphatidylcholine content 10–20%, was purchased from Sigma. Chloride salts of ruthenium complexes (Fig. 1) were synthesised and purified as described elsewhere [2, 7]. Liposomes for LD measurements were prepared by dissolving lipids in buffer (5 mM phosphate buffer, pH 7) containing dissolved ruthenium complex. The solution was subjected to 10 minutes of forceful shaking, frozen (liquid N_2) and thawed (40 °C) five times, and extruded 11 times using a syringe extruder [10] through a polycarbonate filter with 100 nm-sized pores to give liposomes of an average size of approx. 120 nm, verified by small-angle light scattering. Total concentrations of ruthenium complex in the vesicle solutions ranged from 10 μM to 100 μM, lipid concentrations were 5 mg/mL, corresponding to approximately 8 mM lipids. For the luminescence lifetime experiments, liposomes were prepared in buffer and ruthenium complexes added subsequently. In these experiments, the lipid concentration was 0.5 mg/mL (approx. 0.8 mM lipids).

Linear dichroism measurements

Linear dichroism was measured on a Jasco-720 instrument equipped with an Oxley prism, using a Couette cell to orient the sample in a shear flow. The probing light beam enters the Couette cell radially; the horizontally polarised light is thus parallel to the flow direction, and the vertically polarised light is perpendicular to the flow. The initially spherical liposomes will be deformed by a sufficiently high shear into an ellipsoid, prolate shape that will orient its long axis preferentially parallel to the flow direction. A chromophore present in such a sample may thus show linear dichroism if it in turn is bound to, and oriented by, the lipid bilayer. Approximately, a prolate liposome can be thought of as a cylindrical part capped by to half-spheres. It can be shown that, at wavelength regions where only a single transition moment oriented contributes to the absorption, the shape of the LD spectrum will be identical to the isotropic absorbance, and the reduced linear dichroism, $LD^r(\lambda) = LD(\lambda)/A_{isotropic}(\lambda)$, will be wavelength-independent:

$$LD^r = 3S(1 - 3\cos^2\alpha_i)/4$$

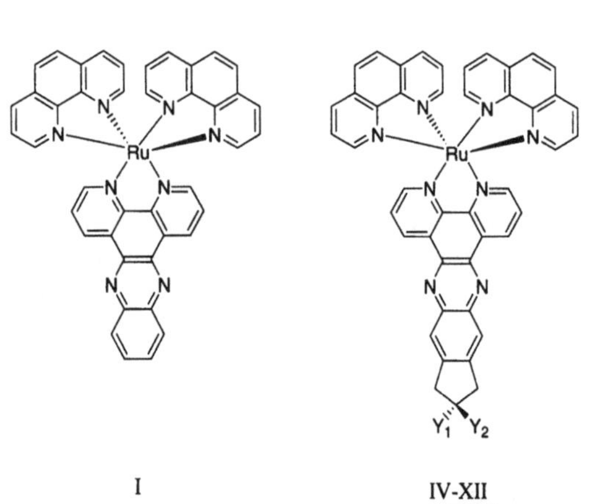

I

IV-XII
R(Y1)Y2
See Table 1

Fig. 1 Ruthenium complexes in this study. For structures of compounds II–XVI, see Table 1

Thus, the reduced linear dichroism can be maximum +0.75S, in case the transition moment is oriented parallel to the surface of the bilayer, and minimal −1.5S, in case the transition moment is oriented parallel to the lipid hydrocarbon chains, i.e., normal to the surface of the bilayer.

The visible spectrum of DNA-bound compound I (Fig. 1), Δ-$[Ru(phen)_2dppz]^{2+}$ (phen = 1,10-phenanthroline, dppz = dipyrido [*a*:2,3-*c*:3′,2′]phenazine), can be resolved into four polarised components [2]. The four polarised components are:

- A, transition moments directed along the long-axis of the dppz ligand,
- B(E), transition moments perpendicular to A and directed approximately along the line joining the middle rings of the two phenanthroline ligands,
- B(A2), transition moments perpendicular to A and at an angle of approx. 80° to B(E), and
- B(sh), transition moments directed along the in-plane short-axis of the dppz ligand.

Although the overlap of the differently polarised transitions is extensive, and the orientation factor remains unknown, impeding a quantitative analysis of the data with Eq. (1), inspection of the LD spectra obtained with liposomes gives a qualitative picture of the relative size of the angles between the transition moments and the membrane.

Results

Figure 2 shows the linear dichroism of compounds I, VII and XII after subtracting the scattering linear dichroism resulting from the deformed liposomes (compare with insert in Fig. 2a, showing actual measurements at 3100 s^{-1}). Dotted line spectra show the isotropic absorption of the samples, after subtracting the scattering from the liposomes and scaling the absorption to the same size as the dichroism. All the main features of the linear dichroism spectra observed are consistent with the resolved spectral components (see Materials and Methods), and may be accounted for, qualitatively, as follows: Since positive linear dichroism results from transitions with moments located predominantly along the membrane surface (perpendicular to the hydrocarbon chains), and a negative linear dichroism from transitions along the hydrocarbon chains, it can be concluded that compounds II–V, which show the same LD features as compound I (with negative peaks in the regions of A transitions and positive in the regions of lipid chains as in Fig. 3a, whereas compounds VIII–XI, showing the same LD features as compound XII (positive A components, negative B components), orient the dppz axis mainly along the membrane surface, as in Fig. 3c. Compound VII constitutes a special case; the A transitions being weakly positive, B(A2) and B(sh) also positive but B(E) slightly negative, indicating another type of surface-associated orientation, shown in Fig. 3b. The diester seems to represent a distribution of along-lipid and along-surface orientations. The three different orientations referred to above, summarised in Table 1, are

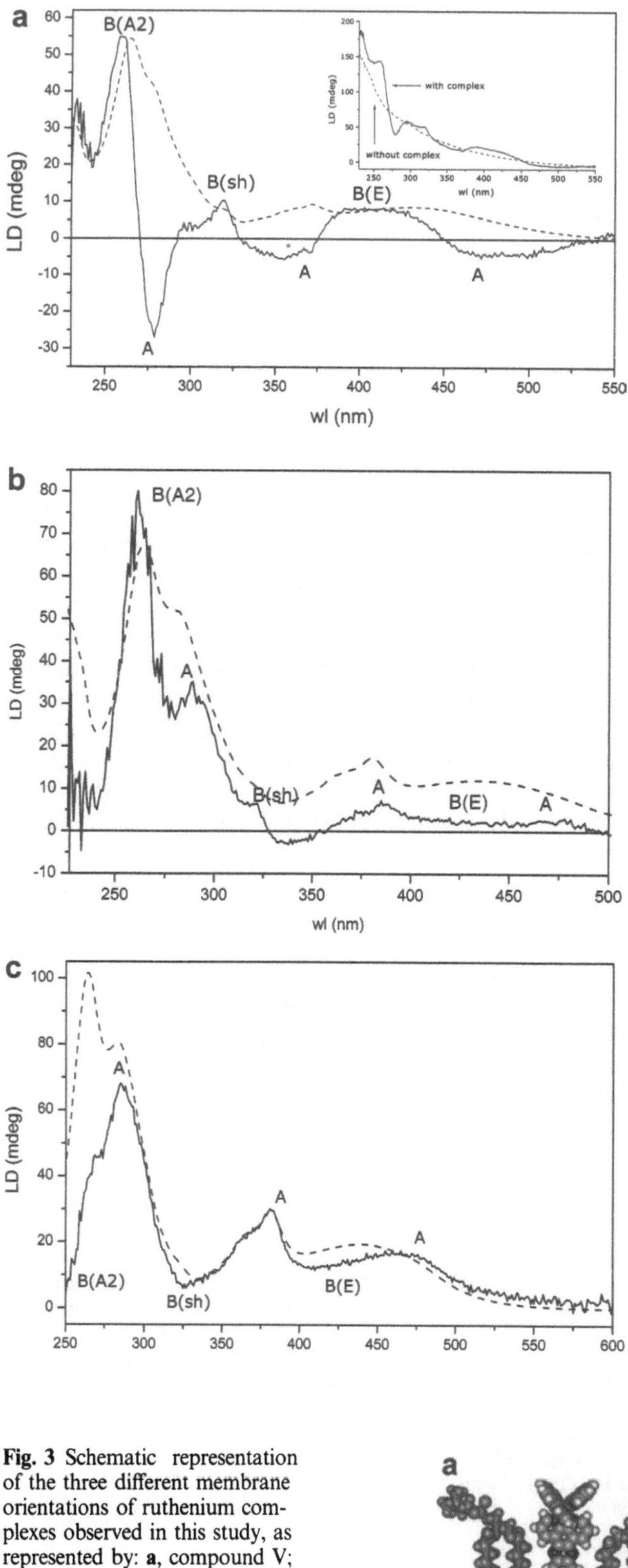

◀

Fig. 2 Linear dichroism – compounds I, VII and XII, with the location of the main transition moments A, B(sh), B(A2) and B(E) indicated (see Materials and Methods section). The dotted line indicates the shape of the absorption spectrum. The insert in Fig. 2a shows the actual measurement (solid line) before subtraction of the linear dichroism due to scattering from the liposomes (dotted line)

Table 1 Structures for compounds I–XII. For structures of Ru $(phen)_2dppz$ and $R(Y_1)Y_2$, see Fig. 1. Compound III has an extra benzene ring on the dppz moiety (bdppz). Letters a, b and c refer to the orientation modes represented in Fig. 3

Compound	Formula	Type of orientation
I	$[Ru(phen)_2dppz]^{2+}$, see Fig. 1	a
II	$[Ru(phen)_2dppz]^{2+}Me_2$	a
III	$[Ru(phen)_2bdppz]^{2+}$	a
IV	$R(H)COOCH_3$	a
V	$R(H)COO(CH_2)_7CH_3$	a
VI	$R(COOCH_3)COOCH_3$	a + b
VII	$R(CN)COOCH_3$	b
VIII	$R(CN)CONH(CH_2)_4NHCOCH_2Br$	c
IX	$R(CN)CONH(CH_2)_4NH_3^+$	c
X	$R(CN)CONH(CH_2)_4NH$-$CO(CH_2)_{14}CH_3$	c
XI	$[R(CN)CONHCH_2\text{-}]_2$	c
XII	$[R(CN)CONHCH_2CH_2\text{-}]_2$	c

shown schematically in Fig. 3a–c, represented by compounds V, VII and IX: the long-axis parallel to the lipid chains (3a, compound V), or the dppz long-axis parallel to the membrane surface with the short-axis either parallel (3b, compound VII) or perpendicular (3c, compound IX) to the surface.

Table 2 reports measurements of luminescence lifetimes and quantum yields for compounds I, II, IV, VI, VI and XII, made in the presence of liposomes, compared with quantum yields from steady-state measurements. Since the luminescence is very efficiently quenched by water for $Ru(phen)_2dppz$ derivatives, excited state lifetimes and emission quantum yields for this chromophore depend sensitively on water accessibility [8]. Four separate types of lifetimes can be discerned. The very short lifetime (< 5 ns) is caused by the free complex, and is only observed for the dimer, which can assume a folded conformation (protected from water quenching). The second lifetime is in the range

Fig. 3 Schematic representation of the three different membrane orientations of ruthenium complexes observed in this study, as represented by: **a**, compound V; **b**, compound VII; **c**, compound IX.

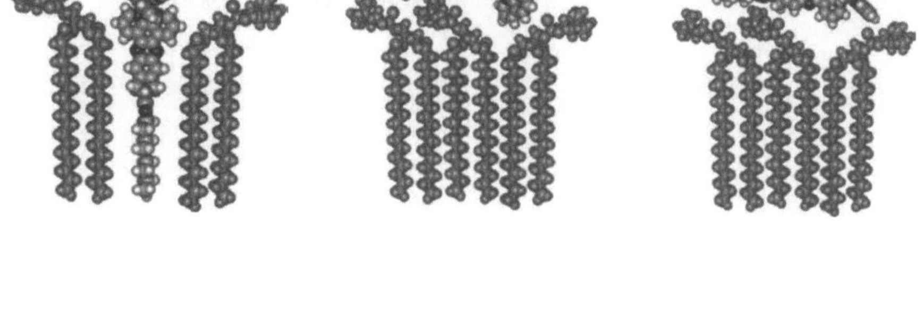

Table 2 Pre-exponential factors (α), lifetimes (τ, in ns) and relative dynamic quantum yields ($\Phi = (\alpha_i\tau_i)/\Phi$(compound II)) of ruthenium complexes I, II, IV, V, VI and XII in vesicle solutions

	α_1	τ_1	α_2	τ_2	α_3	τ_3	α_4	τ_4	Φ
Ru(...) (compound I)	–	–	0.61	97	0.39	20	–	–	0.26
Ru(...) Me_2 (compound II)	0.81	320	–	–	0.19	33	–	–	1.0
Methyl monoester (compound IV)	0.65	270	0.08	150	0.27	21	–	–	0.73
Octyl monoester (compound V)	0.57	300	0.14	100	0.29	12	–	–	0.73
Methyl diester (compound VI)	0.18	250	0.24	120	0.58	20	–	–	0.33
$[Ru(...)cp]_2C_4$ (compound XII)	–	–	0.05	130	0.42	26	0.53	3.2	0.07

from 10 to 40 ns, and can be assigned to complexes adsorbed to the surface of the liposome bilayer. The two longer lifetimes, one in the interval 100–150 ns, the other ranging from 250 to 320 ns, are presumably caused by complexes embedded to different extents in the bilayer, with the longer lifetime representing the most lipophilic environment. Lifetimes of similar size have been observed earlier for compound I in semipolar solvents such as pyridine, CH_2Cl_2 and DMSO [11]. The proportions of longer life-times agree well with the established orientations; the compounds with larger amounts of long life-times are the ones with the lipid-embedded orientation (Fig. 3a).

Discussion

Compounds I–X are structurally quite similar, the only difference being the variations of end groups of the dppz moiety. It is therefore interesting to note that they tend to take either of three distinct orientations in the membrane bilayer, as concluded from LD and lifetime results (see Fig. 3): embedded and with the dppz long-axis parallel to the lipid chains (*Mode a*) or aligned parallel with the surface, with the dppz short axis either aligned parallel with the surface (*Mode b*) or dipping down into the bilayer (*Mode c*). The two latter orientations (*b* and *c*) could only arise as a result of interaction with the surface and its polar environment, for example, as a consequence of attraction of both "bottom" and "top" of the complex to the polar head groups of the lipid molecules. It is possible to identify some factors that govern this behaviour: charge and polarity of the end group appear to be important factors, as is the flexibility of the end group, which may permit the ruthenium moiety to adopt a variety of different orientations, more or less independent of the orientation of the end group itself.

We have shown how flow linear dichroism and emission lifetime measurements, in conjunction with detailed knowledge of transition moment directions, may provide conclusions about location and orientation of solute molecules in a lipid bilayer with a lipid composition resembling the one found in living plants and animals. Our approach does not depend on the introduction of additional probes, which could affect the interaction with the membrane. More specifically, it is ideally suited for following sensitively the effects of systematic substitutions or other structural variations of the membrane solute molecules subject to study.

References

1. Ardhammar M, Mikati N, Nordén B (1998) J Am Chem Soc 120:9957
2. Lincoln P, Broo A, Nordén B (1996) J Am Chem Soc 118:2644
3. Choi SD, Kim MS, Kim SK, Lincoln P, Tuite E, Nordén B (1997) Biochemistry 36:214
4. Hiort C, Nordén B, Rodger A (1990) J Am Chem Sob 112:1971
5. Önfelt B, Lincoln P, Nordén B (1999) J Am Chem Soc 121:10846
6. Lincoln P, Tuite E, Nordén B (1997) J Am Chem Soc 119:1454
7. Hiort C, Lincoln P, Nordén B (1993) J Am Chem Soc 115:3448
8. Olson EJC, Hu D, Hormann A, Jonkman AM, Arkin MR, Stemp EDA, Barton JK, Barbara PF (1997) J Am Chem Soc 119:11458
9. Ardhammar M, Nordén B, Nielsen PE, Malmström BG, Wittung-Stafshede P (1999) J Bio Str Dyn 17:33
10. Lasic DD (1993) Liposomes: from physics to applications. Elsevier, Amsterdam
11. Nair RB, Cullum BM, Murphy CJ (1997) Inorg Chem 36:962

Progr Colloid Polym Sci (2004) 123: 69–72
DOI 10.1007/b11634

Marta Airoldi
C. Andrea Boicelli
Giuseppe Gennaro
Marcello Giomini
Anna Maria Giuliani
Mauro Giustini
Enrico Paci

Cationic microemulsion hosting polynucleotides: effect of NaCl on host and guest

M. Airoldi · G. Gennaro
A.M. Giuliani (✉)
Inorganic Chemistry Department,
University of Palermo, Viale delle Scienze,
Parco d'Orléans, 90128 Palermo, Italy
e-mail: denebdue@unipa.it
Tel.: +39-091-590246
Fax: +39-091-427584

C.A. Boicelli
Animal Biology Department,
University of Pavia, Italy

M. Giomini · M. Giustini · E. Paci
Chemistry Department,
University "La Sapienza", Roma, Italy

Abstract The structural features of the quaternary water-in-oil microemulsion CTAB/*n*-hexane/*n*-pentanol/water in the presence of fairly high concentrations of NaCl and of relatively high molecular weight polynucleotides have been determined. Even in these severe conditions, the hosting system can still be depicted as formed by water droplets stabilised by a surfactant/cosurfactant layer. Moreover, the time stability of the host/guest system has been evaluated and the phase behaviour of the hosting system in the presence of increasing concentrations of NaCl determined.

Keywords Quaternary microemulsions · Conductivity · Polynucleotides · Sodium chloride · Spectroscopy

Introduction

Water-in-oil microemulsions, and in particular reverse micelles, have been proposed as suitable systems where the properties of DNA or model molecules in a compartment of limited dimensions can be studied [1–3]. The use of such systems to investigate the properties of biological macromolecules solubilised in the aqueous core is of interest, since in vivo highly condensed forms of DNA are found in biological structures that have dimensions far smaller than those of the native polynucleotide considered as a stiff coil [4–6].

It has been reported that, while pairing of the complementary single strand polynucleotides polydeoxyadenylic acid and polydeoxythymidylic acid occurs to the same extent in microemulsion and in solution, dilution of native DNA and duplex polydeoxyadenylic-thymidylic acid (polyAT) has a hyperchromic effect absent in solution [1].

Solubilisation of duplex polyAT in the quaternary water-in-oil microemulsion hexadecyltrimethylammonium bromide (CTAB)/*n*-hexane/*n*-pentanol/water has been shown to suppress the helix-to-coil thermal transition and to induce the formation of condensed forms of the polynucleotide in conditions of NaCl concentration that cause no such phenomena in solution [2]. These compaction effects, leading to the formation of the $\psi(-)$ form [7, 8], have been ascribed not only to the limited dimensions of the water core of the microemulsion and to its low relative permittivity [9, 10], but also to the positive charge of the micellar wall, that appears to play the same role as cationic polymers in the condensation process of DNA [11–13].

In consideration of the essential role of the structure and the composition of the microemulsion in conditioning the behaviour of the solubilised polynucleotide, in the presence of different concentrations of NaCl, it seemed significant to ascertain whether and how the presence of the guest polymer and of the salt would alter the characteristics of the host system, that are fully described in their absence [14]. We report here the preliminary results of such a study.

Materials and methods

Chemicals

CTAB was from Fluka and was purified as described elsewhere [14].

PolyAT, sodium salt, molecular weight (0.9–1.9) × 10^6 Da, and 2-amino-2-(hydroxymethyl-1,3-propanediol) (TRIS) buffer were from Sigma.

NaCl was a Merck Suprapur product and was dried for 5 hours at 383 K and stored under vacuum over silica gel before use.

Twice distilled water was always employed; UV spectroscopy grade *n*-hexane and *n*-pentanol, from Fluka, were used without further purification.

Preparation of samples and microemulsions

The polyAT samples were prepared as elsewhere described [2] from the purchased vials of 50 units (1 unit yields an absorbance of 1.0 at 260 nm, when dissolved in 1.0 mL of water in a 1.0 cm optical pathlength cuvette).

The CTAB concentration was always 0.10 M; the aqueous TRIS solution was 1.0 mM at pH 8.0 ± 0.3, while the NaCl concentration varied from 0 to 1 M, as required; W_o = [water]/[surfactant] and P_o = [cosurfactant]/[surfactant] were 15 and 8.50, respectively.

The "empty" (without polyAT) and "filled" (with polyAT) microemulsions were prepared according to the procedure described in [2].

The actual concentration of the polynucleotide in each sample was determined, just before the measurements, from the absorbance at 262 nm ($\varepsilon = 6650$ mol^{-1} L cm^{-1}) [15].

Instrumentation

Conductivity measurements were made with a Radiometer CDM-83 conductometer at 73 Hz, using a thermostatted microcell (cell constant 1.00 cm^{-1} at 298.0 ± 0.2 K) suitably modified to accept a Pt-100 ceramic resistance (DEGUSSA-GR-2105) to monitor the sample temperature. The variable temperature measurements could not be extended beyond 336 K because of the boiling point of *n*-hexane (342 K).

The CD spectra were recorded with a Jasco J715 spectropolarimeter equipped with a 150 W xenon lamp under nitrogen flux, connected to a Julabo F10 thermostat for the temperature control (± 0.1 K). All the spectra are baseline corrected [2]. The characteristics of the microemulsive system restrict the wavelength range to values higher than 230 nm.

UV and near infrared (NIR) spectra were acquired with a Varian Cary 5E double-beam, double-monochromator spectrophotometer; quartz Hellma cells were used and the cell holder was thermostatted (Haake F3/K thermostat) at 298.0 ± 0.1 K.

Results and discussion

It is well assessed that, in the absence of any guest molecule, the quaternary system CTAB/*n*-hexane/*n*-pentanol/water consists of discrete spherical aggregates behaving like hard spheres without hydrodynamic interactions [14, 16]. This structure might be modified by the presence of the guest species object of this study, i.e., NaCl and/or polyAT. To check whether modifications do indeed occur, the conductivity behaviour of the microemulsion in the presence of both NaCl and polyAT has been investigated at different temperatures

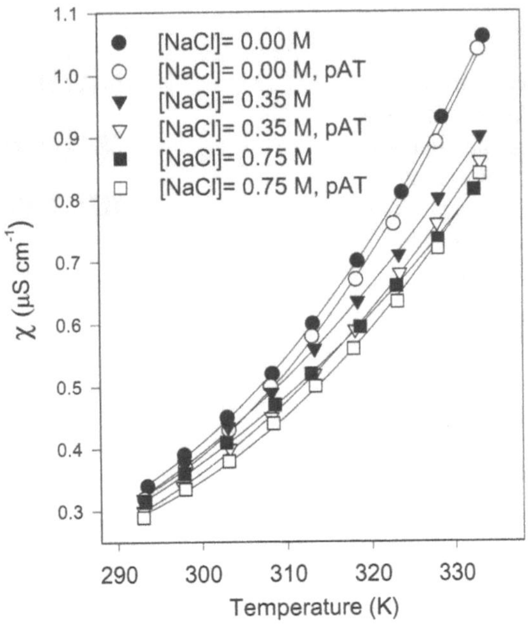

Fig. 1 Conductivity behaviour of CTAB/*n*-hexane/*n*-pentanol/water microemulsion as a function of temperature and guest species ([CTAB] = 0.10 M, W_o = 15, P_o = 8.50)

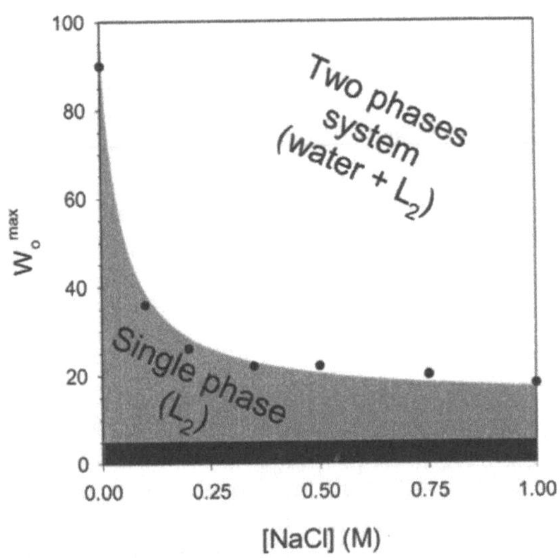

Fig. 2 Phase diagram at 298 K of the system CTAB/*n*-hexane/*n*-pentanol/water as a function of NaCl concentration in the water pool ([CTAB] = 0.10 M, P_o = 8.50). Below Wo = 5 (*black region*) the microemulsion does not exist

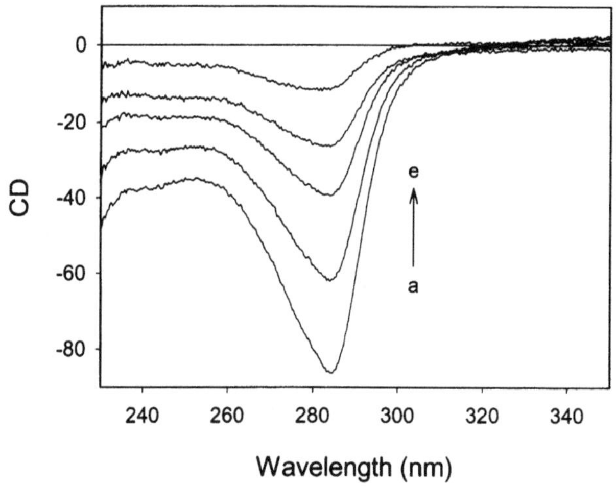

Fig. 3 CD spectra of polyAT in CTAB/*n*-hexane/*n*-pentanol/water microemulsion at 298 K at different times after preparation ([CTAB] = 0.10 M, $W_o = 15$, $P_o = 8.50$, [NaCl] = 0.75 M): a. 2 h; b. 2 h45′; c. 3 h50′; d. 5 h; e. 22 h

(293–333 K). Conductivity is a fast, non-invasive and extremely sensitive technique able to distinguish between discrete droplets and bicontinuous structures, that can be present in a quaternary microemulsion. Indeed, conductivity values range between 10^{-2} and 1 μS cm^{-1} for reverse micelles [17] and between 10^3 and 10^4 μS cm^{-1} for bicontinuous structures [18].

The data presented in Fig. 1 show that for our systems the conductivity values are always well below those typical of bicontinuous structures. It can thus be concluded that, even in the presence of polyAT and high concentrations of NaCl in the water pool, the system can still be described as formed by closed aqueous nanodomains diffusing in the bulk continuous oil phase, in the entire explored temperature range. On the other hand, it has to be pointed out that the NaCl concentration strongly affects the extension of the L_2 phase region, as shown in Fig. 2.

The presence of the guest polymer might, however, alter the time stability of the microemulsion, since it has been reported [2] that, at high NaCl concentrations, solubilised polyAT, which is in the ψ(−) form, is progressively expelled from the aqueous core (Fig. 3).

As can be seen from this figure, the CD spectra reduce their intensity with time, while retaining the ψ(−) features induced by the solubilisation in reverse micelles (negative band centred at ca. 283 nm). This sedimentation could occur either with or without simultaneous going out of endomicellar water.

To verify if the sedimentation of the guest destabilises the microemulsive system, we have followed the time behaviour of a microemulsion ([CTAB] = 0.10 M, $W_o = 15$, $P_o = 8.50$, T = 298 K, [NaCl] = 0.75 M) by recording in parallel the NIR band of water at ca. 1950 nm and the UV spectrum of solubilised polyAT (ca. 260 nm). The intensity and form of the water signal remain unchanged (Fig. 4B) through the 24 hours during which, instead, the absorbance of the polymer decreases to nearly zero (Fig. 4A).

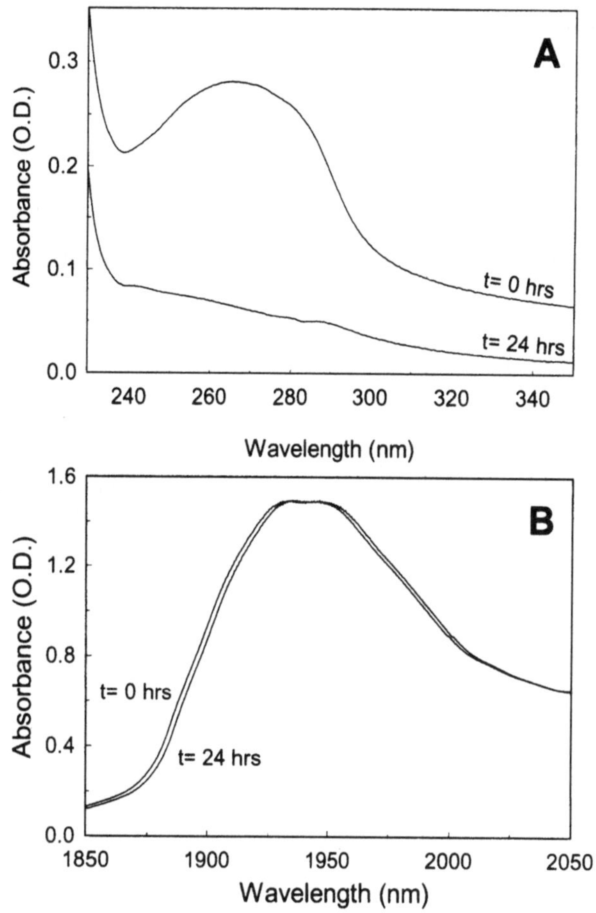

Fig. 4 UV (**A**) and NIR (**B**) spectra of polyAT in CTAB/*n*-hexane/*n*-pentanol/water microemulsion at 298 K at different times after preparation ([CTAB] = 0.10 M, $W_o = 15$, $P_o = 8.50$, [NaCl] = 0.75 M). To increase figure readability, the NIR spectrum acquired after 24 hours (see **B**) has been mathematically red-shifted by 2 nm

The above discussed results allow to conclude that, while prominent effects, modulated by the concentration of solubilised NaCl, are caused by the microemulsive system to the guest polyAT, this has practically no influence on the structure, properties or time stability of the host. On the other hand, increasing concentrations of NaCl in the aqueous core of the microemulsion produce a reduction of the region of existence of the L_2 phase to the benefit of Winsor II phase separation [19] and a less marked temperature dependence of the conductivity.

Acknowledgements We wish to thank Italian MURST – Cofin99 for financial support.

References

1. Balestrieri E, Giomini M, Giustini M, Giuliani AM, Ceglie A (1999) Progr Colloid Polym Sci 112:89
2. Airoldi M, Boicelli CA, Gennaro G, Giomini M, Giuliani AM, Giustini M (2000) Phys Chem Chem Phys 2:4636
3. Battistel E, Imre EV, Luisi PL (1989) Solubilization and structural properties of nucleic acids in reverse micelles. In: Rosoff M (ed), Controlled release of drugs: polymers and aggregate systems. VCH, Weinheim, pp 225–276
4. Doty P, Bunce BH (1952) J Am Chem Soc 74:5029
5. Harpst JA, Krasna AI, Zimm BH (1968) Biopolymers 6:595
6. Jolly DJ, Campbell AM (1972) Biochem J 128:569
7. Shin YA, Eichhorn GL (1984) Biopolymers 23:325
8. Lerman LS (1971) Proc Natl Acad Sci USA 68:1886
9. Wong M, Thomas JR, Novak T (1977) J Am Chem Soc 99:4730
10. Wells MA (1974) Biochemistry 13:4937
11. Carroll D (1972) Biochemistry 11:421
12. Maestre MF, Reich C (1980) Biochemistry 19:5214
13. Phillips CL, Mickols WE, Maestre MF, Tinoco I Jr (1986) Biochemistry 25:7803
14. Giustini M, Palazzo G, Colafemmina G, Della Monica M, Giomini M, Ceglie A (1996) J Phys Chem 100:3190
15. Gennis RB, Cantor CR (1972) J Mol Biol 65:381
16. Colafemmina G, Palazzo G, Balestrieri E, Giomini M, Giustini M, Ceglie A (1997) Progr Colloid Polym Sci 105:281
17. Lagourette B, Peyrelasse J, Boned C, Clausse M (1979) Nature 281:60
18. Clausse M, Boned C, Peyrelasse J, Lagourette B, McClean VER, Sheppard RJ (1981) In: Shah DO (ed), Surface phenomena in enhanced oil recovery. Plenum Press, New York, pp 199–228 and references therein
19. Quemada D, Langevin D (1989) A viscosity model of Winsor microemulsions. In: Mittal KL (ed), Surfactant in solution, vol 10. Plenum Press, New York, pp 123–145

Progr Colloid Polym Sci (2004) 123: 73–77
DOI 10.1007/b11645

M. Soledade C. S. Santos
Sara M. V. Lacerda
Ester F. G. Barbosa

Interactions of selected flavonoids with NaDS micelles

M. Soledade C.S. Santos (✉)
S.M.V. Lacerda · E.F.G. Barbosa
Departamento de Química e Bioquímica
e CECUL, Faculdade de Ciências da
Universidade de Lisboa,
Campo Grande 1749–016 Lisboa,
Portugal
e-mail: mssantos@fc.ul.pt
Tel.: +351-21-7500896
Fax: +351-21-7500088

Abstract Critical micelle concentrations (cmc) and degree of counterion dissociation (β) of sodium dodecyl sulphate micelles, in the presence of two flavonoids type additives, were determined at 298.15 ± 0.01 K using electrical conductivity. The dependences of the cmc on additive concentration follow a generalised type Setchenow equation and allowed the determination of Setchenow micellisation constants (K_M) for each solute. K_M values revealed the high hydrophobicity of these compounds, and suggest that structural factors are determinant for the partition towards the micellar phase.

Keywords Flavonoid · Hydrophobicity · Membrane mimetic system · Critical micelle concentration · Setchenow micellisation constant

Introduction

Flavonoids are secondary plant metabolites of polyphenolic type, present in 0.5 to 1.5%. Ever since their discovery, several epidemiological studies have suggested correlations between flavonoid-rich diets and a reduced risk of diseases like arteriosclerosis, diabetes and cancer [1, 2]. Several in vitro studies of these compounds supported a strong antioxidant activity, however numerous irregularities have been observed in structure-performance relationships. These results have been attributed to the wide range of processes involved in their radical scavenging ability, namely the power to act as hydrogen donors, the metal chelation capabilities and differences in the hydrophobicity of the compounds within the family [3].

The flavonoid chemical structure is characterised by a C_6–C_3–C_6 skeleton, flavone or flavanol-like, illustrated in Fig. 1, where the three-carbon bridge between the phenyl groups (rings A and B) is commonly cyclised with oxygen (forming a C ring). There are over 5000 flavonoids species that result from different unsaturation and oxidation degrees of this segment, and from a wide range and number of substituents such as hydroxy, sugars, oxygen atoms or methyl groups. The flavonoids studied were catechin and rutin, a flavonol and a flavone with IUPAC names *trans*-3,3′,4′,5,7-pentahydroxyflavanol and 3-rutinoside-3′,4′,5,7-tetrahydroxyflavone, respectively [1, 2].

Flavonoids are characteristic constituents of green plants with the possible exception of algae and hornworts, and those studied here can be found in fruits and vegetables such as pears, grapes, apples, peaches, and green tea with rutin also present in kale, spinach, onions, parsley, endives and citrus fruits [2].

A clear picture about the molecular mechanisms of flavonoid protection in living systems involves knowledge about the partition/adsorption of these compounds to the cell membranes. To pursue this goal NaDS micelles were chosen as a mimetic system due to the anionic character of the hydrophilic moiety of phospholipid membranes. Careful determinations of the critical micelle concentration (cmc), combined with a rigorous error analysis lead to the calculation of Setchenow micellisation constants, K_M, as an evaluation parameter of the partition/adsorption to the micelle [4, 5]. Both flavonoids present high hydrophobicity with rutin exhibiting twice the affinity of catechin towards the model membrane.

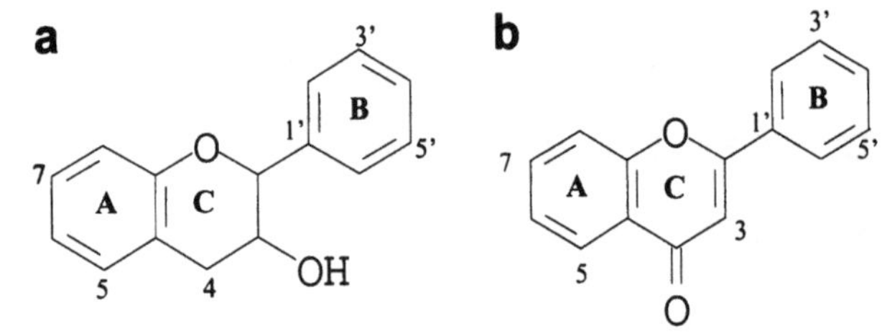

Fig. 1 General chemical structure of a flavonol (**a**) and a flavone (**b**)

Experimental

Materials and methods

All reagents were supplied by Sigma, and were used without further purification: sodium dodecyl sulphate (NaDS) was Sigma Ultra – GC grade (purity >99%) and the flavonoids: (+)-catechin hydrate (min. 98%) and rutin hydrate (min. 95%) were also supplied by Sigma.

The extremely low solubility of the flavonoids inhibited the preparation of accurate solutions by weight. Therefore the solids were weighted, in an A & D Instruments analytical balance to $\pm 2 \times 10^{-5}$ g, the solutions prepared volumetrically with analytical type I water produced either by a MilliQ System from Millipore ($\rho = 18.2$ MΩ cm) or WasserLab ($\rho = 18.3$ MΩ cm), and finally the density of the flavonoids solution determined.

The densities were measured, at 298.15 ± 0.01 K, using an Anton Paar 02D vibrating tube densimeter, calibrated with air and Millipore water, the density of the flavonoid solutions being indistinguishable from the one of the solvent ($\rho = 0.9970_{47}$ g cm^{-3} [6]).

Determination of the critical micelle concentration (cmc)

The cmc values were determined by electrical conductivity measurements, at 298.15 ± 0.01 K, using a Radiometer MeterLab CMD 230 Conductivity Meter and a Radiometer Copenhagen Conductivity Cell CDC 641 T, comprising 2 platinised platinum electrodes, and a temperature sensor. The conductivity cell was calibrated at 298.15 ± 0.01 K, before each run, with Radiometer Copenhagen conductivity standards of 0.01 D (1408 μS/cm ± 0.5%) and 0.1 D (12.85 mS/cm ± 0.35%) of KCl.

Determination of the cmc for aqueous NaDS. The experimental set-up used involved a titration type procedure. A concentrated micellar NaDS solution was added stepwise, with a 10 mL ± 1 μL Dosimate 665 automatic burette from Methrom, to 20.0 ± 0.038 mL of water. The titration was performed in a thermostatic beaker under constant stirring and the electrical conductivity was measured after each addition.

The dependence of the specific electrical conductivity on the surfactant concentration shows an abrupt change that allows the calculation of the cmc from the intersection of the two straight lines that best fit the experimental data, below and above the cmc. The identification of the set of experimental data points that unambiguously present pre- or post-micellar behaviour was based on the scattering pattern of the regression residual plots.

Determination of the cmc of NaDS in presence of flavonoid. The determinations in the presence of additive were performed using an identical experimental set-up. In order to maintain a constant additive concentration, during the titration with the surfactant solution, a flavonoid solution was introduced initially in the thermostatic beaker and simultaneously surfactant and concentrated flavonoid solution were added to the thermostatic vessel. The concentrated flavonoid solution was added using an automatic burette from Radiometer Copenhagen – ABU 901 Autoburette, of 10 mL ± 2 μL and the specific conductivity measured after the addition of both solutes (NaDS and flavonoid).

The uncertainty associated with the cmc determination was estimated in terms of the expanded uncertainty, U, from the intersection of the upper and lower regression bands of the linear fits for the pre- and post-micellar region, applying the following expression to both fits [7].

$$y_i = (bx_i + a) \pm t_{n-2,95\%} S_{y/x} \left[1 + \frac{1}{n} + \frac{(x_i - \bar{x})^2}{\sum (x_i - \bar{x})^2} \right]^{\frac{1}{2}} \quad (1)$$

where y_i and x_i are the conductivity and the concentration values, b is the slope and a is the intercept of the conductivity versus surfactant concentration, $t_{n-2,95\%}$ is the Student t value for a 95% confidence level and $n-2$ degrees of freedom, $S_{y/x}$ is the regression standard deviation, n is the number of experimental points and $\bar{x}$ is the mean of these x_i values. In Fig. 2 the linear fits as well as the corresponding upper and lower regression bands, for an experimental run, are plotted on a restricted concentration range around the cmc, illustrating the calculation of the corresponding uncertainty for a 95% confidence level, U.

Results and discussion

The overall effect of the presence of the flavonoids in the specific conductivity of the surfactant solutions shows an

Fig. 2 Determination of the expanded uncertainty of the cmc

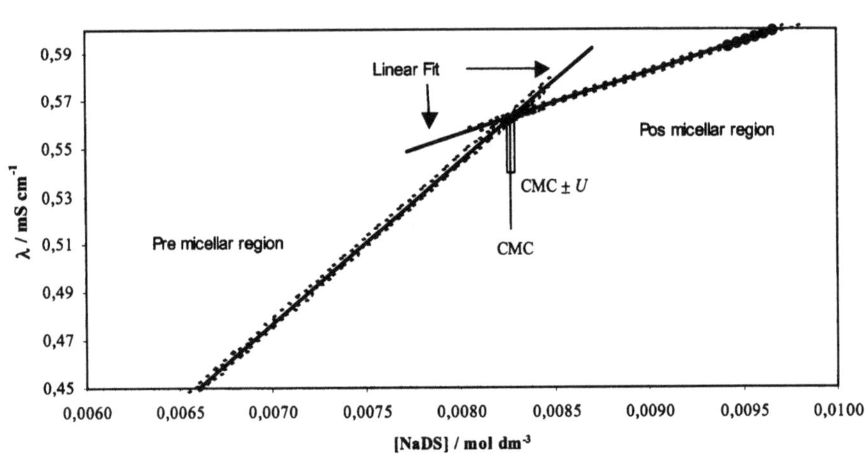

identical pattern for both additives, characterised by a small increase in the slopes below the cmc, S_1, that increases for the slopes above it, S_2. In Table 1 average values for the slopes obtained in independent runs, N_{exp}, in the absence and in the presence of additive are presented and these values clearly indicate a stronger interaction of the flavonoids with the surfactant aggregates than with its monomers.

The experimental results obtained for the cmc in the absence and presence of flavonoid are also included in Table 1. These values are weighed averages of several experimental runs and despite the differences registered in the cmc, in the absence of additive, for the two NaDS batches used, the values obtained agree well with literature data [8–10].

The reproducibility and accuracy of the experimental data obtained clearly showed a decrease in the cmc with flavonoid addition that is steeper for rutin than for catechin. The low solubility of both flavonoids in water restricted this study to a relatively narrow additive concentration range, but one must bear in mind that the estimated daily uptake and physiological concentrations (0.1–1 μmol dm^{-3}) [2, 11] of these compounds in the human body are even below these values. The observed cmc decrease in the presence of neutral polar additives is frequently attributed to the micellar solubilisation of the additive, accompanied by a decrease in the surface charge density and an increase in the entropy of mixing of the micelles [12–14]. This hypothesis was checked by resorting to the calculation of the degree of counterion dissociation, β,

$$\beta = \frac{n-m}{n} \qquad (2)$$

That can be estimated using Evans [5] equation [Eq. (3)]

$$(1000S_1 - \lambda_c)\frac{(n-m)^2}{n^{4/3}} + \lambda_c\left(\frac{n-m}{n}\right) - 1000S_2 = 0 \qquad (3)$$

where S_1 and S_2 are the slopes of the conductivity versus surfactant concentration below and above the cmc, n and m are, respectively, the number of monomers in the aggregate and the numbers of counterions bound to the micelle, and λ_c the equivalent conductivity of the counterion (Na^+). In these calculations λ_{Na} was approximated to the limiting value at infinite dilution, 50.10 $cm^2\ \Omega^{-1}\ eq^{-1}$ at 298.15 K [15] and the aggregation number for NaDS taken as 64 [9, 16] and considered independent of the concentration of additive as the concentration range is limited and the effect of n on the calculated β values is known to be small [5].

Calculated values for degree of counterion dissociation for both solutes were also included in Table 1 and, for both flavonoids, β values display a progressive increase that levels off or even decreases slightly for higher concentrations. This type of dependence of β on additive concentration has already been reported for several 1-alkanols [13] and may be associated with solute induced micelle destabilisation. It is worth mentioning that the decrease of β with additite concentration is sharper for rutin than for catechin.

Over the last 20 years several attempts have been made to relate the dependence of the cmc on solute concentration to either a Setchenow micellisation constant, K_M, or a corresponding partition coefficient (P) between the micelles and the solvent [5], parameters which are very promising in terms of the construction of hydrophobicity/hydrophilicity scales and also in terms of their relations to the classical octanol-water partition coefficient (P_{ow}) frequently used in medicinal, pharmacological and environmental studies.

Treiner's approach to the micellar pseudophase model [4, 5] which allows the calculation of K_M, directly from the cmc dependence on the concentration of flavonoids, in the Setchenow region, is used here and accordingly

$$\log\frac{CMC_W}{CMC_{W,N}} = K_M m_N \qquad (4)$$

where m_N is the molality of the neutral polar additive.

A comparison of these data with literature data apart from allowing a quantitative measure of their hydrophobicity can shed some light into the major

Table 1 Effect of flavonoids concentration on pre- and post-micellar slopes, S_1 and S_2, critical micelle concentration, cmc, degree of counterion dissociation, β, and number of independent experimental runs considered, N_{exp} (T = 298.15 K)

Flavonoid	[Flavonoid] ± μmol dm^{-3}	N_{exp}	S_1	S_2	CMC ± $U_{(t=95\%)}$/mol dm^{-3}	β ± $U_{(t=95\%)}$
	0	3	$67.3_0 \pm 0.3$	$24.5_5 \pm 0.2$	$8.32_2 \times 10^{-3} \pm 1 \times 10^{-5}$	$0.22_3 \pm 2 \times 10^{-3}$
	$3.01_2 \times 10^{-6} \pm 4 \times 10^{-8}$	4	$67.6_4 \pm 0.1$	$25.55_6 \pm 0.07$	$8.278_7 \times 10^{-3} \pm 8 \times 10^{-6}$	$0.226_4 \pm 8 \times 10^{-4}$
Catechin	$6.02_5 \times 10^{-6} \pm 1 \times 10^{-8}$	4	$67.65_4 \pm 0.09$	$26.0_6 \pm 0.2$	$8.271_5 \times 10^{-3} + 7 \times 10^{-6}$	$0.22_8 \pm 2 \times 10^{-3}$
	$1.00_4 \times 10^{-5} \pm 2 \times 10^{-8}$	4	$68.0_3 \pm 0.2$	$25.7_6 \pm 0.2$	$8.220_9 \times 10^{-3} \pm 8 \times 10^{-6}$	$0.22_5 \pm 2 \times 10^{-3}$
	$1.50_6 \times 10^{-5} \pm 4 \times 10^{-8}$	4	$68.1_6 \pm 0.3$	$26.0_6 \pm 0.1$	$8.16_7 \times 10^{-3} \pm 1 \times 10^{-5}$	$0.22_6 \pm 2 \times 10^{-3}$
	$2.00_8 \times 10^{-5} \pm 5 \times 10^{-8}$	5	$68.6_9 \pm 0.3$	$26.6_3 \pm 0.2$	$8.18_6 \times 10^{-3} \pm 2 \times 10^{-5}$	$0.22_6 \pm 2 \times 10^{-3}$
	0	4	$65.9_6 \pm 0.1$	$24.10_6 \pm 0.07$	$8.38_4 \times 10^{-3} \pm 1 \times 10^{-5}$	$0.224_9 \pm 4 \times 10^{-4}$
	$3.08_3 \times 10^{-6} \pm 1 \times 10^{-8}$	2	$66.0_2 \pm 0.2$	$25.1_5 \pm 0.4$	$8.33_7 \times 10^{-3} \pm 2 \times 10^{-5}$	$0.22_9 \pm 5 \times 10^{-3}$
Rutin	$4.62_3 \times 10^{-6} \pm 3 \times 10^{-8}$	1	$66.46_0 \pm 0.06$	$26.01_2 \pm 0.08$	$8.28_4 \times 10^{-3} \pm 1 \times 10^{-5}$	$0.23_4 \pm 5 \times 10^{-3}$
	$6.16_5 \times 10^{-6} \pm 1 \times 10^{-8}$	4	$65.9_4 \pm 0.5$	$25.7_4 \pm 0.2$	$8.262_8 \times 10^{-3} \pm 3 \times 10^{-6}$	$0.23_5 \pm 4 \times 10^{-3}$
	$7.70_7 \times 10^{-6} \pm 2 \times 10^{-8}$	3	$66.8_6 \pm 0.3$	$25.4_0 \pm 0.5$	$8.301_6 \times 10^{-3} \pm 1 \times 10^{-5}$	$0.23_2 \pm 2 \times 10^{-3}$

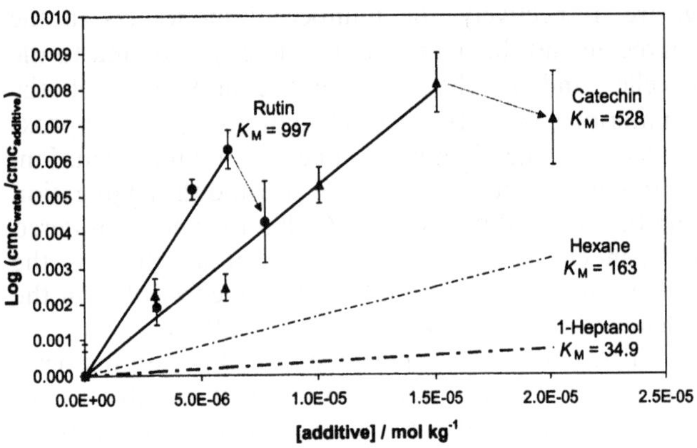

Fig. 3 Setchenow plots for catechin and rutin as well as calculated trendlines for hexane and 1-heptanol

factor influencing the hydrophilic/lipophilic balance in such complex molecular structures.

In Fig. 3 plots of the generalised form of Setchenow equation and calculated flavonoids micellisation constants as well as trendlines, based on literature data [17, 18], for hexane and 1-heptanol are plotted for the sake of comparison.

The very narrow linearity range for the compounds studied stands out and points their hydrophobic character. An analysis in terms of the magnitude of K_M reveals the extreme affinity of the flavonoids for the micelle [K_M(catechin) = 5.28×10^2; K_M(rutin) = $9._{97} \times 10^2$], showing values 20 to 40 times larger than the one reported for 1-heptanol ($K_M = 34.9$ [18]), and 200 to 400 times larger than the K_M value reported for phenol ($K_M = 2.36$ [18]). To the best of our knowledge these are the first values reported for this family of compounds thus rendering impossible comparisons with compounds with identical structures, further assessment resorting to other polyphenolic type compounds not being possible as data is extremely scarce.

Within the flavonoid family one of the major structural differences between the two compounds studied is the presence of a disaccharide structure in rutin, that is known to render hydrophilic character [2], thus one would expect a lower K_M for this compound. The experimental results show the opposite, namely that rutin is more hydrophobic than catechin, an order that agrees with the data obtained in terms of the octanol-water partition coefficients, K_{ow}(catechin) = 0.6_3 and K_{ow}(rutin) = 1.5_4 [19]. Such an order cannot be assigned to the presence of an extra hydroxy group in position 3 of the C ring instead of the rutinoside substituent. Therefore other factors, not considered so far, must play an important role in terms of the partition towards the model membrane.

Conclusions

In this work K_M values for two flavonoids were determined: K_M(catechin) = $5._{28} \times 10^2$ and K_M(rutin) = $9._{97} \times 10^2$, setting a hydrophobicity scale, for two compounds within the family, that parallels independently determined P_{ow} values.

Correlations between structure and solubilisation in a micellar phase, measured in terms of K_M or P have been reported before [17, 20] for other families of solutes like 1-alcohols, ketones, nitriles and aldehydes. Accordingly these experimental data indicate that the partition towards the micellar phase of rutin is more favourable and, simultaneously, this compound also shields more efficiently the counterions, suggesting its localisation on the micelle palisade layer.

A closer look at the flavonoids' three-dimensional structure evidences that the flexibility of the C ring in catechin may hinder its penetration in the palisade layer of the micelle whereas such constraints are less likely for the rigid and almost coplanar three rings of the flavone. In rutin the bulky rutinoside structure may also obstruct its embedment in the micelle, however the specific interactions between the aggregate anionic head groups and the hydrophilic glycoside substituent can stabilise its localisation on the interfacial double layer region. This hypothetical positioning of the flavonoids on the micelle is supported by the differences observed for the dependence of β on additive concentration, and agrees with the proposal of Saija et al. [21] about the establishment of a reversible chemical bond between flavonoids and the polar head groups of the biological membranes.

These results suggest that the partition towards the micellar phase, within the flavonoid family, is mainly determined by the C ring steric constraints associated with the aromaticity interruption, when going from a flavone (rutin) to a flavonol (catechin), thus rendering the former more hydrophobic. Furthermore the evaluation of additive effects on the cmc within this family of compounds is crucial for the establishment of structure-partition/solubilisation correlations that can be a very useful predictive tool in the hydrophobicity evaluation of the most insoluble compounds within the family.

Acknowledgements The authors wish to thank FCT for financial support of project PRAXIS 2/2.1/QUI/255/94.

References

1. Jovanovic SV, Steenken S, Tosic M, Marjanovic B, Simic MG (1994) J Am Chem Soc 116:4846
2. Robards K, Antolovich M (1997) Analyst 112:11R
3. Halliwell B, Gutteridge JMC (1999) Free radicals in biology and medicine, 3rd edn. Oxford University Press, pp 225–245
4. Treiner C (1982) J Colloid Interface Sci 90:444
5. Treiner C (1995) The partitioning of neutral solutes between micelles and water as deduced from critical micelle concentration determinations. In: Christian SD, Scamehorn JF (eds), Solubilization in surfactant aggregates: surfactant science series. vol 53. Marcel Dekker, New York, pp 383–428
6. Riddick JA, Bunger WB, Sakano TK (1986) In: Weissberger A (ed), Techniques of Chemistry, vol. II. Wiley-Interscience, New York, p 74
7. Miller JN (1991) Analyst 116:3
8. Goddard ED, Benson GC (1957) Can J Chem 35:986
9. Manabe M, Shirahama K, Koda M (1976) Bull Chem Soc Jpn 49:2904
10. Jönsson B, Lindman B, Holmberg R, Kronberg B (1998). Surfactants and polymers in aqueous solution. Wiley, Chichester, p 37
11. Hertog M, Feskens E, Hollman P, Katan M, Kromhout D (1993) The Lancet 349:1007
12. Kaneshina S, Kamaya H, Ueda I (1981) J Colloid Interface Sci 83:589
13. Zana R, Yiv S, Strazielle C, Lianos P (1981) J Colloid Interface Sci 80:208
14. Manabe M, Kawamura H, Kondo S, Kojima M, Tokunaga S (1990) Langmuir 6:1596
15. Robinson RA, Stokes RH (1959) Electrolyte solutions. Butterworths, London, p 463
16. Aniansson EAG, Wall SN, Almgren M, Hoffmann H, Kielmann I, Ulbricht W, Zana R, Lang J, Tondre C (1976) J Phys Chem 80:905
17. Treiner C, Mannebach M-H (1987) J Colloid Interface Sci 118:243
18. Abu-Hamdiyyah M, El-Danab C (1983) J Phys Chem 87:5443
19. Santos MS, Lacerda SMV, Silva LM, Barbosa EFG, to be published
20. Treiner C (1983) J Colloid Interface Sci 93:33
21. Saija A, Scalese M, Lanza M, Marzullo D, Bonina F, Castelli F (1995) Free Radical Biol Med 19:481

Progr Colloid Polym Sci (2004) 123: 78–82
DOI 10.1007/b11646

A. Di Biasio
F. Bordi
C. Cametti

Salt-induced aggregation in cationic liposome suspensions

A. Di Biasio
Dipartimento di Matematica e Fisica,
Universita' di Camerino, Camerino, Italy

F. Bordi · C. Cametti (✉)
Dipartimento di Fisica,
Universita' di Roma "La Sapienza",
Piazzale A. Moro 5,00185 Rome, Italy

A. Di Biasio · F. Bordi · C. Cametti
Istituto Nazionale per la Fisica della
Materia (INFM), Unita' di Roma 1,
Rome, Italy

Abstract The simple salt-induced aggregation of small unilamellar dioleoyltrimethylammoniumpropane [DOTAP] vesicles is investigated by measuring the change in the effective radius with time, using dynamic light scattering techniques. At small salt concentration (lower than 0.5–0.6 mol/L), an aggregation mechanism results in the formation of stable liposome structures of moderate size, before that the usual irreversible coagulation prevails, at higher salt concentration. The steady-state size reached by these aggregates, after their initial growth, is governed by binding counterions to liposome surfaces, resulting in a screening effect and in a reduction of electrostatic repulsive forces. These liposomal structures with hydrodynamic radius three or four times larger than that of the initial liposomes are stable as colloidal dispersions. These structures may be potentially useful to promote efficient DNA transfection of animal cells in tissue cultures.

Introduction

Liposomes are spherical structures composed by a closed lipid bilayer that encompasses an aqueous core disjoined from an external continuous medium [1, 2]. Liposomes in aqueous suspension are an interesting system not only as a model colloidal system in fundamental research concerning self-assembling molecules but also as a drug delivery vehicle, where the enclosed water core can be used to solubilise active substances and the biocompatible bilayer can be used as carrier to join the site of infection or disease [3, 4, 5].

In particular, liposomes, built up from cationic lipids, have been widely used for cell transfection in vitro and are being investigated in gene therapy and genetic engineering for the delivery of genes into mammalian cells.

The success of such a method depends on the stability of the liposomes and, in recent years, a number of amphiphile systems has been developed, each of them shows a different efficiency, depending on various parameters such as the composition of the incubation medium (pH, ionic strength), the nature of the lipid component, the incubation time before transfection, and so on.

Despite their widespread use, the critical factors determining the transfection activity of these systems are not clear yet and, although the structure of the resulting aggregates should play an important role, up until now, little effort has been put into understanding the more fundamental aspects concerning the structure and the relationship between the transfection efficiency and the morphological characteristics of the aggregates.

In order to elucidate the mechanism governing the formation and the stability of cationic liposome aggregates of various sizes, induced by electrostatic interactions, we have investigated the conformational behaviour of liposomes in the presence of simple uni- and divalent salt electrolyte solutions, at various charge ratios. We have monitored the formation of aggregates and their change in size over time by means of the dynamic light scattering technique, because of its unique feature to study the structure and the dynamics of colloidal-size aggregates in solution. We studied the early-stage formation of liposome aggregates, 100–1000 nm in diameter, using a cationic lipid (dioleoyltrimethylammoniumpropane, DOTAP) in the presence of NaCl, $CaCl_2$, $MgCl_2$ electrolyte solutions at different concentrations and have examined the time evolution of the hydrodynamic radius of the resulting aggregates from the initial value up to a steady-state, before the irreversible coagulation. The salt concentration investigated (in the range from 0.2 to 0.6 mol/L) is well below the critical value yielding the irreversible coagulation, so we observe in the initial process of aggregation, small clusters of particles that act on an individual basis.

Experimental

Dioleoyltrimethylammoniumpropane [DOTAP] was purchased from Avanti Polar Lipids (Alabaster, Al) and used without further purification. The amount of lipid component (10 mg/mL) was dissolved in chloroform-methanol (1:1, vol/vol) in a suitable vial. The solvent was then evaporated to give a dry lipid film and then resuspended in pure water (electrical conductivity less than 10^{-6} ohm/cm). The formation of liposomes was induced by ultrasonication and the resulting mixture was extracted through a polycarbonate membrane filter (pore size 100 nm) using an extruder (Lipex Biomembranes) until a homogeneous liposomal suspension of unilamellar vesicles was obtained.

The hydrodynamic radius of the diffusing particle aggregates in the aqueous suspension (at an initial concentration of about 10^{12} particle/mL) was measured by means of the dynamic light scattering method [6, 7, 8]. The intensity-intensity correlation function was obtained from a standard laboratory-built spectrometer equipped with an He-Ne laser operating at 10 mW and 632.8 nm wavelength. Measurements were collected at a scattering angle of 90°. The normalised field autocorrelation function $g^{(1)}(t)$, obtained from the intensity autocorrelation function through the Siegert relationship [8], has been expanded according to the method of cumulants [9] as

$$g^{(1)}(t) = \exp(-<\Gamma>t + 1/2\mu_2 t^2 - \cdots) \quad (1)$$

where $<\Gamma>$ is the average decay rate and μ_2 characterises the width of the size distribution.

The hydrodynamic radius R was obtained from the Stokes-Einstein equation

$$R = \frac{K_B T q^2}{6\pi\eta <\Gamma>} \quad (2)$$

with q the scattering vector, K_BT the thermal energy and η the solvent viscosity. The polydispersity index can be obtained from the ratio of the second to the first cumulant

$$P - \frac{\sqrt{\mu_2}}{<\Gamma>} \quad (3)$$

Characterisation of size and size distribution of liposome suspensions carried out by means of the dynamic light scattering technique has been discussed in detail elsewhere [10, 11]. Vesicle aggregation was induced by mixing equal volumes of liposome suspension and a simple salt solution (NaCl, $CaCl_2$, $MgCl_2$) to a final concentration varying from 0.2 to 0.6 mol/L and monitoring the change in radius upon time. Each measurement took typically between 30 and 60 s and the sample time was adjusted between measurements in order to obtain the optimal autocorrelation function. All measurements were performed at the temperature of 25.0 ± 0.1 °C.

Results and discussion

Thanks to the fine balance of the different forces existing between two approaching lipid bilayers, most often the liposomal structure is stable over an extended period of time. Among these forces, a crucial role is played by the attractive van der Waals forces and the repulsive electrostatic forces, the mechanism of their mutual interactions being described within the DLVO theory [12, 13]. Further contributions may derive from attractive hydrophobic interaction, steric repulsion, when flexible polymers are adsorbed onto the lipid bilayer, and repulsive effects of the head group hydration [14]. Alteration of one component of this intricate balance makes the liposomes unstable and they tend to aggregate in a process known as flocculation or to coagulate in an infinite cluster leading to the partial or total sedimentation of the liposomal structures.

Before salt addition, all the liposome suspensions investigated were stable over time, consisting in vesicles about 140 nm in radius dispersed in the aqueous phase with a typical polydispersity between 0.1 and 0.2.

Figs. 1 to 3 show the observed changes in the normalised radius derived from the first cumulant [Eq. (2)] as a function of time induced by addition of NaCl, $CaCl_2$, $MgCl_2$ salt solutions, at different concentrations, respectively. The systems exhibit aggregation at salt concentration in the range of 0.3–0.6 mol/L. The

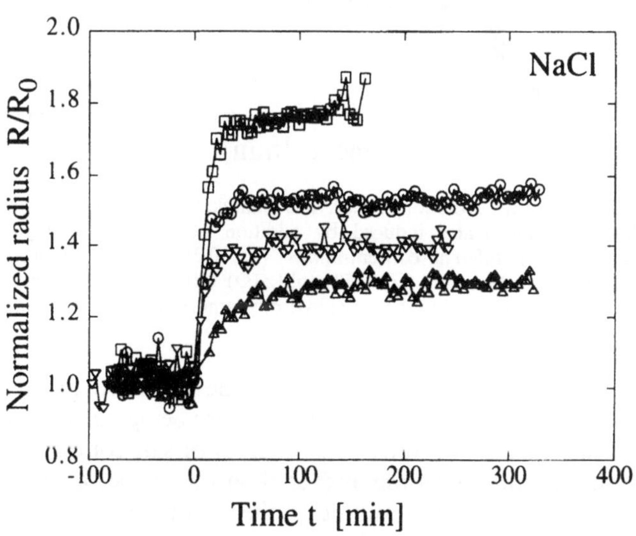

Fig. 1 The time evolution of the hydrodynamic radius of the resulting liposomal aggregates induced by addition of NaCl electrolyte solutions, at different concentrations: (△): 0.35 mol/L; (▽): 0.45 mol/L; (○): 0.50 mol/L; (□): 0.55 mol/L. The values have been normalised to the initial hydrodynamic radius (R = 140 nm)

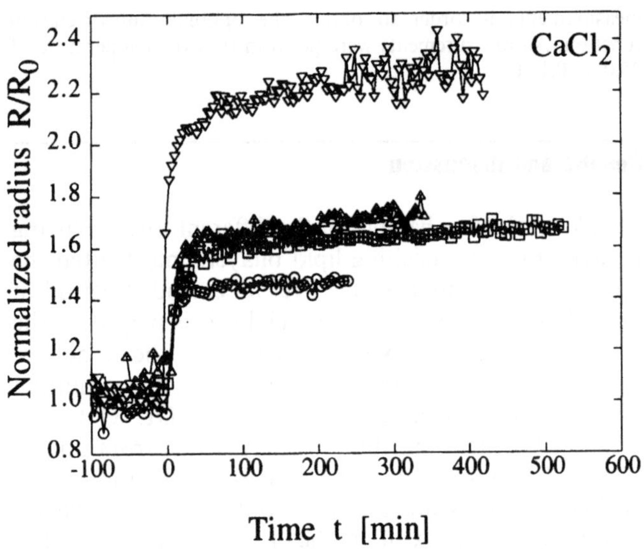

Fig. 2 The time evolution of the hydrodynamic radius of the resulting liposomal aggregates induced by addition of $CaCl_2$ electrolyte solutions, at different concentrations: (○): 0.25 mol/L; (□): 0.30 mol/L; (△): 0.35 mol/L; (▽): 0.40 mol/L. The values have been normalised to the initial hydrodynamic radius ($R = 140$ nm)

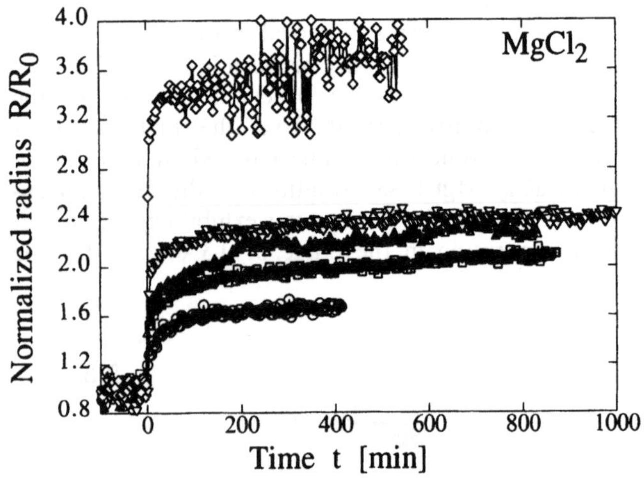

Fig. 3 The time evolution of the hydrodynamic radius of the resulting liposomal aggregates induced by addition of $MgCl_2$ electrolyte solutions, at different concentrations: (○): 0.25 mol/L; (□): 0.30 mol/L; (△): 0.40 mol/L; (▽): 0.50 mol/L; (◇): 0.60 mol/L. The values have been normalised to the initial hydrodynamic radius ($R = 140$ nm)

kinetic aggregate growth seems to be restricted to this interval, the aggregation behaviour being observed neither below nor above this interval of salt concentration. At concentrations higher than this value, a rapid coagulation occurs, resulting, at the end of the process, in an approximately complete phase separation.

In the systems investigated, in the presence of small amount of added salt, the aggregation process starts with the formation of structures of increasing size, but after an initial stage, the rate of change of the radius continually decreases, until an approximately steady-state is reached and the aggregation ends. It is noteworthy that, after this aggregation, the system undergoes a new stationary condition with particles (or aggregates of particles) of radius three or four times larger than their value at the beginning of the process. Fig. 4 shows the ability of monovalent and divalent salts to promote liposome aggregation in the condition of the experiment. As can be seen, divalent salts are able to induce larger aggregates than monovalent salts do, at the same molar concentrations. However, when the steady-state hydrodynamic radius is plotted as a function of the ionic strength, these differences disappear and mono- and divalent salts induce the same aggregation effect. This observation confirms that the important parameter in determining the salt-induced flocculation is the counterion concentration present in the aqueous phase, i.e., in this case, the concentration of chloride ions that bind the cationic DOTAP. The reduction of the electrostatic repulsion between vesicles through binding counterions to the lipid

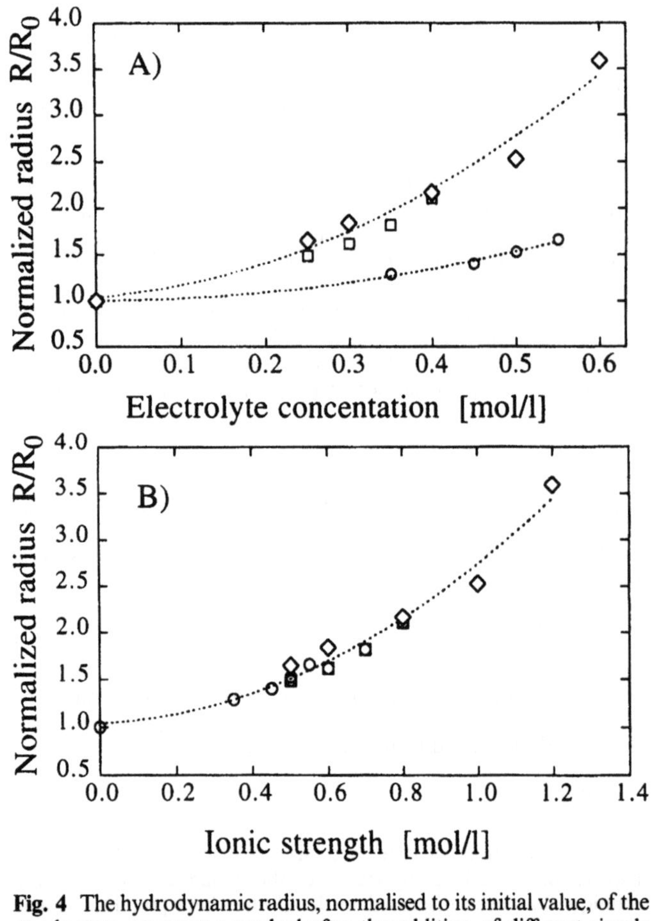

Fig. 4 The hydrodynamic radius, normalised to its initial value, of the steady-state aggregates reached after the addition of different simple salt electrolyte solutions: (○): NaCl; (□): $CaCl_2$; (◇): $MgCl_2$. Data are plotted as a function of the molar concentration (**A**) and as a function of the ionic strength of the electrolyte solution (**B**). The dotted lines are the second order polynomial best fit

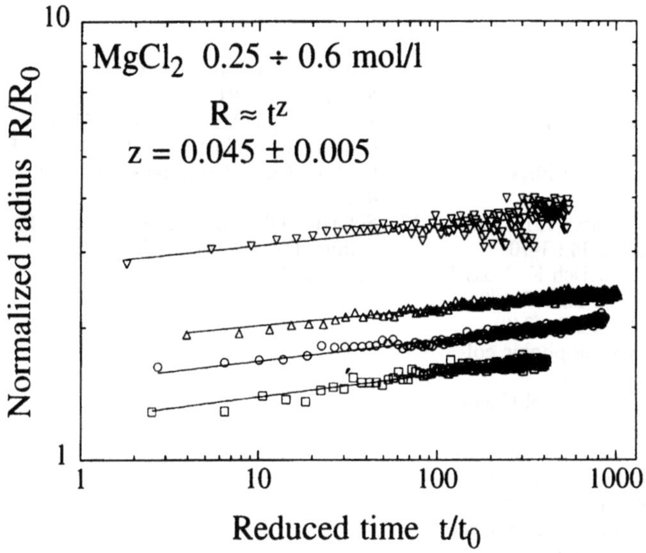

Fig. 5 The normalised radius of liposome suspensions in $MgCl_2$ electrolyte solutions at different concentrations between 0.25 and 0.6 mol/L. The data exhibit a power-law behaviour with an exponent of about $z = 0.045$, largely independent of salt concentration

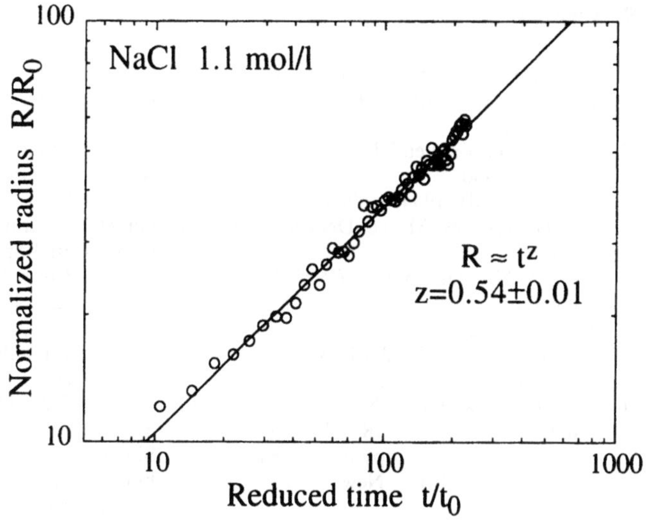

Fig. 6 The hydrodynamic radius of liposome aggregates during NaCl salt-induced aggregation. The salt concentration in the aqueous phase is 1.1 mol/L. The data exhibit the asymptotic behaviour expected for the DLCA regime and the full line gives the dependence upon time with an exponent $z = (0.54 \pm 0.01)$

surface induces aggregation and/or fusion, giving rise to stable structures in the same size range. This behaviour can offer new possibilities in biomedical applications because the tendency of such liposomes to associate with one another, maintaining separate structures, can be controlled by varying the ionic concentration of the appropriate counterion in the external medium.

In the usual salt-induced aggregation of colloidal suspensions, the kinetics of aggregation is generally described by dynamical scaling using the fractal morphology of the clusters [15, 16, 17]. Within this context, the average hydrodynamic radius of the aggregates in the diffusion limited cluster aggregation (DLCA) regime obeys a power-law behaviour $R(t) \approx R_0 t^z$ with $z = 1/D_f$, the inverse of the fractal dimension of the clusters. The exponent z characterises the aggregation mechanism.

In the systems investigated, however, the cluster growth kinetics can be described by a power-law for all the conditions and salts we have studied, with an exponent $z = 0.045$, largely independent of the electrolyte species and concentration. A typical example is shown in Fig. 5 in the case of liposome suspensions in $MgCl_2$ electrolyte solution.

The value of the exponent z rules out the possibility that this aggregation process may be considered as the initial stage of a diffusion limited aggregation, as described by scaling laws.

It must be noted, however, that the dynamic scaling law is strictly valid in the large cluster limit, whereas, in the present case, the aggregates involve only few monomers (or a confined number of monomers).

However, when the salt concentration is increased over a critical value and the coagulation occurs, the usual behaviour is observed. Fig. 6 shows the log-log representation of the growth of the hydrodynamic radius of liposome suspensions as a function of time in the presence of 1.1 mol/L NaCl electrolyte solution. The full line indicates the power-law region with an exponent $z = (0.54 \pm 0.01)$, to which corresponds a fractal dimension $D_f = 1.85$. This finding is in agreement with an aggregation governed by a diffusion regime, as expected for the DLCA regime.

The main result obtained in this work is the existence of an aggregation mechanism giving rise to stationary structures of relatively small size, before the usual coagulation of charge stabilised colloidal suspensions prevails, at higher ion concentrations. Moreover, the salt dependence of this liposome aggregation reflects the binding of counterions to vesicle surfaces. The existence of these relatively small structures and the possibility to modulate their size by counterion concentration could favour biological applications of cationic lipid-DNA complexes in gene delivery.

References

1. Lasic DD (1993) Liposomes: from physics to applications. Elsevier, Amsterdam
2. Hunter DG, Frisken BJ (1998) Biophys J 74:2990–3002
3. Kreuter J (ed) (1994) Colloidal drug delivery systems. Marcel Dekker, New York
4. Zuidam NJ, Barenholz Y (1998) Biochim Biophys Acta 1386:115–128
5. Pedroso de Lima MC, Simoes S, Faneca H, Duzgunes N (2001) Adv Drug Delivery Rew 47:277–294
6. Cumming HZ, Pusy PN (1977) In: Cummings ZH, Pike ER (eds), Photon correlation spectroscopy and velocimetry. Plenum Press, New York
7. Chu B (1974) Laser light scattering. Plenum Press, New York
8. Pecora R (1985) Dynamic light scattering. Plenum Press, New York
9. Koppel DE (1972) J Chem Phys 57:8414–8420
10. Hallet FR, Craig T, Marsh J, Nickel B (1989) Can J Spectrosc 34:63–70
11. Jin AJ, Huster D, Gawrisch K, Nossal R (1999) Eur Biophys J 28:187–199
12. Verwey EJW, Overbeek JT (1948) Theory and stability of liophobic colloids. Elsevier, Amsterdam
13. Ninham BW (1981) Pure Appl Chem 53:2153–2147
14. Silvander M (1999) Structure and stability of liposomes: interactions with micelle-forming surfactants. Uppsala Dissertation N. 455, Faculty of Science and Technology, Acta Universitatis Upsaliensis, Upsala
15. Lin MY, Lindsay HM, Weitz AA, Klein R, Ball RC, Meakin P (1990) J Phys Cond. Matter 2:3093–3113
16. Cametti C, Codastefano P, Tartaglia P (1989) J Colloid Interface Sci 131:409–422
17. Schaefer DW, Martin JE, Wiltzius P, Cannel PS (1984) Phys Rev Lett 52:2371–2374

Progr Colloid Polym Sci (2004) 123:83–87
DOI 10.1007/b11647

M. Céu Rei
P.J.G. Coutinho
E.M.S. Castanheira
M.E.C.D. Real Oliveira

$C_{12}E_7$-DPPC mixed systems studied by pyrene fluorescence emission

M. Céu Rei · P.J.G. Coutinho (✉)
E.M.S. Castanheira
M.E.C.D. Real Oliveira
Departamento de Física,
Universidade do Minho,
Campus de Gualtar,
4710–057 Braga, Portugal
e-mail: pcoutinho@fisica.uminho.pt
Tel.: +351-253-604321
Fax: +351-253-678981

Abstract The lipid/surfactant mixed interactions between the lipids dipalmitoylphosphatidylcholine (DPPC) or egg phosphatidylcholine (EggPC) and the non-ionic surfactant $C_{12}E_7$ [$C_{12}H_{25}(OCH_2CH_2)_7OH$] were studied by the use of the fluorescence properties of pyrene, namely the excimer to monomer emission intensity ratio, I_E/I_M. Previously, the behaviour of the $C_{12}E_7$/water system was also monitored. It was found to exhibit a significant preassociation of pyrene in ground state, which is more pronounced in micelles than in premicellar aggregates. In mixed systems, pyrene has proved to detect the changes from mixed bilayers to mixed micelles. The temperature influence in lipid/surfactant interactions was also studied. It was found that the pyrene I_E/I_M ratio is sensitive to the phase transition of DPPC. Pyrene microcrystallites are probably present in the gel phase region, justifying the enhancement of I_E/I_M in the DPPC/$C_{12}E_7$ system at low temperatures.

Keywords Non-ionic surfactants · Dipalmitoylphosphatidylcholine · Egg phosphatidylcholine · Lipid/surfactant interactions · Pyrene emission · Ground-state aggregation

Introduction

The surfactant micelles have an ability to solubilise insoluble or only sparingly soluble materials in aqueous media by incorporating them into the micellar interior with the formation of mixed micelles. Phospholipids, and other constituents of biomembranes, can also be solubilised by surfactant micelles. Due to this property, surfactants are widely used as molecular tools in membranology [1]. The applications of surfactants in membranology are based on the transformation from vesicles to mixed micelles (or reverse direction) occurring in aqueous surfactant/phospholipid mixtures [2, 3]. Understanding of the transformation phenomenon should be helpful to achieve these practical purposes and, hence, great efforts have been developed so far to elucidate the pathway and mechanism of the transformation between vesicles and mixed micelles [4, 5]. The surfactant action on the phase transition of vesicle membranes has been studied by the use of several techniques for various surfactant and phospholipid species [6, 7].

Among the entire range of biophysical and spectroscopic methods, several techniques have been used to elucidate the properties of lipid vesicles and vesicle/surfactant interactions: cryotransmission electron microscopy [4], differential scanning calorimetry [6], light scattering [5], and absorption and fluorescence spectroscopy [8]. Fluorescence spectroscopy is probably the technique with the highest sensitivity for the study of lipid vesicles, biomembranes and lipid/surfactant interactions.

Since lipids are not fluorescent, study of the fluorescence of lipid vesicles is possible by introducing a fluorescence probe into the lipid environment. Among all probes used so far, pyrene (and its derivatives) stands unique, owing to its useful and versatile properties. These include the sensitivity of the emission spectrum's vibronic structure to the polarity of the environment [9, 10] and its ability to form excimers, observed for the first time in 1954 [11].

In this work, we use pyrene's spectroscopic properties (excimer formation and the ratio of first to third peak emission intensity, I_1/I_3) to obtain information about the structural changes induced by different concentration ratios of the phospholipid dipalmitoylphosphatidylcholine (DPPC) and the non-ionic surfactant $C_{12}E_7$ [$C_{12}H_{25}(OCH_2CH_2)_7OH$]. The influence of temperature in these structures is also investigated.

Experimental

Materials Samples of polyoxyethylene 7 lauryl ether ($C_{12}E_7$) and of dipalmitoylphosphatidylcholine (DPPC) from Sigma were used as received. Pyrene (Koch Light, >99% pure) was zone refined (100 steps). Solutions were prepared using Milli-Q grade water.
Sample preparation The samples for the surfactant/water system were prepared by addition to water of the required amount of surfactant. Pyrene (2×10^{-7} M) was introduced by injection of a stock solution in ethanol. The samples were placed in an ultrasonic bath for mixing and left to stabilise.

DPPC was deposited by evaporation at 50 °C of a stock solution in ethanol. Then, the required amount of surfactant solution in water was added, followed by the injection of the probe. The samples were placed in an ultrasonic bath for mixing and left to stabilise.
Fluorescence measurements Steady-state excitation and emission spectra were recorded using a Spex Fluorolog 212 Spectrofluorimeter. The spectra were corrected for the instrumental response of the system. The temperature was maintained (± 0.2 °C) using a recirculating water supply connected to a water jacket on the cuvette holder.

Results and discussion

Before studying the influence of the phospholipid DPPC in mixed phospholipid/$C_{12}E_7$ systems, the behaviour of the fluorescent probe was monitored in the surfactant/water system.

For pyrene 2×10^{-7} M, it was found a pronounced decrease in the I_1/I_3 ratio (first to third peak monomer emission intensity ratio) until the critical micellar concentration (6.9×10^{-5} M) [12] is attained (Fig. 1a), resulting from the polarity decrease [9] experienced by the hydrophobic pyrene molecules as premicellar aggregates are formed, as already observed in other micellar systems [10, 13]. The cmc value corresponds to the inflection point of I_1/I_3 vs. log[$C_{12}E_7$] plot [10].

The excimer (500 nm) to monomer (372 nm) fluorescence intensity ratio, I_E/I_M, plotted in Fig. 1b, exhibits a significant increase, followed by a pronounced decrease. The peak in the I_E/I_M vs. log[$C_{12}E_7$] plot is located slightly after the cmc value. It should be noted that the pyrene concentration is the same in all the surfactant solutions. As the surfactant concentration is increased

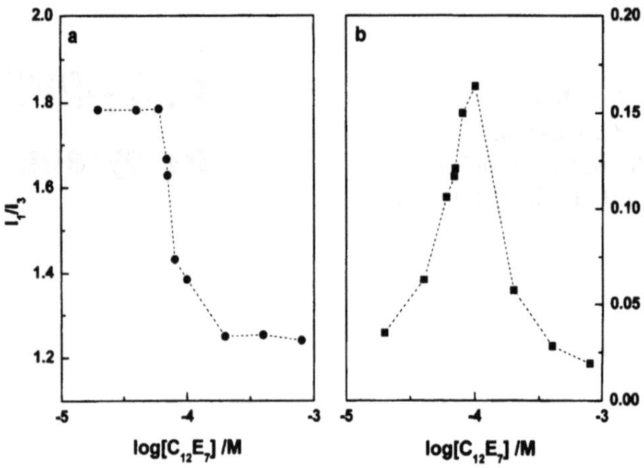

Fig. 1 (**a**) Fluorescence intensity ratio of pyrene first and third vibronic bands, I_1/I_3 (λ_{exc} = 337 nm), as a function of $C_{12}E_7$ concentration. (**b**) Excimer (500 nm) to monomer (372 nm) emission intensity ratio, I_E/I_M, of pyrene (2×10^{-7} M) as a function of $C_{12}E_7$ concentration (λ_{exc} = 337 nm)

through the cmc transition, the number of full-sized micelles increases. In this case, pyrene moves from premicellar aggregates to full-sized micelles, causing a rise in average occupancy of probe molecules in micelles, increasing therefore the I_E/I_M ratio [13]. After cmc, the number of full-sized micelles increases even further. Then, the average occupancy lowers and also the probability of excimer formation, decreasing I_E/I_M.

Assuming a radius of 40 Å for full-sized micelles [14], and a surface area per headgroup [12] of 61 Å^2, an aggregation number of 330 can be obtained (number of heads necessary to fill the micelle surface). From this value, pyrene average local concentrations of 6×10^{-3} M and 5.1×10^{-4} M can be estimated for surfactant solutions with concentrations 6.9×10^{-5} M (cmc) and 8×10^{-4} M, respectively. The probability (from a Poisson distribution) of having two or more pyrenes per micelle is 0.248 for cmc and 0.0032 for [$C_{12}E_7$] = 8×10^{-4} M. For premicellar aggregates ([$C_{12}E_7$] = 2×10^{-5} M), considered as spheres of a few surfactant molecules and water, a radius of 8 Å and a surface area per "hydrated" head of 80 Å^2 were estimated, giving an aggregation number of 10. In these conditions, the pyrene average local concentration is 0.078 M and the probability of having two or more probes per aggregate is 0.0047. Therefore, the I_E/I_M variation with [$C_{12}E_7$] agrees with the calculated probability of having two or more pyrenes per aggregate.

The temperature influence in the behaviour of this system was monitored by the variation of the pyrene excimer to monomer emission ratio, I_E/I_M, for several $C_{12}E_7$ concentrations. It was found that the I_E/I_M ratio decreases monotonically with temperature (Fig. 2a) for low surfactant concentrations (below 10^{-4} M) and does not follow the usual curve observed in low-viscosity

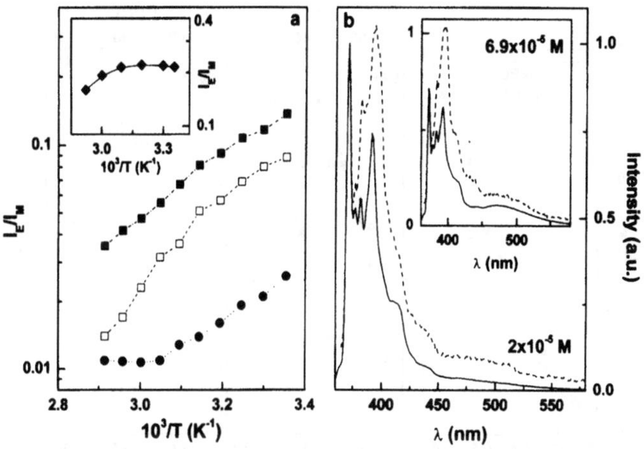

Fig. 2 (**a**) Excimer (500 nm) to monomer (372 nm) emission intensity ratio, I_E/I_M, for pyrene (2×10^{-7} M) as a function of reciprocal temperature ($\lambda_{exc} = 337$ nm), for several $C_{12}E_7$ concentrations: (●) 2×10^{-5} M; (□) 6.9×10^{-5} M; (■) 1×10^{-4} M. Insert shows I_E/I_M for pyrene in tetradecane. (**b**) Fluorescence spectra of pyrene (2×10^{-7} M) in a 2×10^{-5} M $C_{12}E_7$ solution, for $\lambda_{exc} = 337$ nm (—) and for $\lambda_{exc} = 350$ nm (- - -). Insert shows similar spectra for a 6.9×10^{-5} M $C_{12}E_7$ solution (room temperature). The spectra were normalised at 372 nm

solvents [15], with a maximum near or above room temperature (insert of Fig. 2a). The I_E/I_M ratio for pyrene in alkanes [16] shows the expected behaviour, resulting from the classical Birks scheme [15],

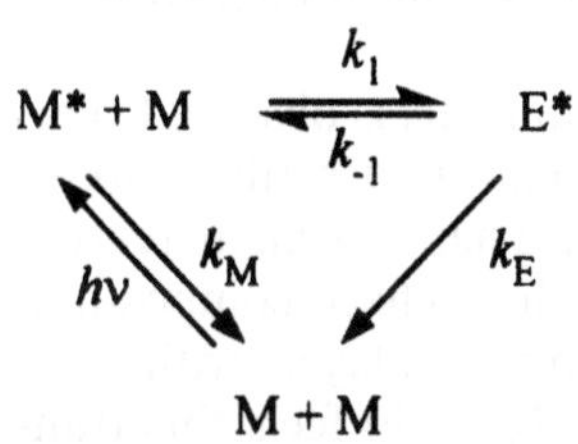

$$\frac{I_E}{I_M} \propto \frac{k_1}{k_{-1} + k_E} \qquad (1)$$

where k_1 and k_{-1} are the rate constants for excimer formation and dissociation, k_M and k_E are the rate constants for monomer and excimer deactivation, respectively. In the high-temperature region ($k_E << k_{-1}$), this scheme predicts a linear behaviour of $\ln(I_E/I_M)$ with $1/T$ [15].

For the $C_{12}E_7$/water system, it is clear that other photophysical processes involving pyrene molecules are playing an important role.

Fig. 2b shows the emission spectra of pyrene 2×10^{-7} M for the solution 2×10^{-5} M in $C_{12}E_7$, at room temperature, for two excitation wavelengths (337 nm and 350 nm). Insert of Fig. 2b displays the same spectra for $[C_{12}E_7] = 6.9 \times 10^{-5}$ M (cmc). In the spectra obtained with $\lambda_{exc} = 350$ nm, we can see the emission from an excited dimer preassociated in ground state. This ground state dimer absorbs a larger fraction of the excitation light at 350 nm, but not at 337 nm. In fact, the pyrene dimer to monomer relative absorption is higher at $\lambda_{exc} = 350$ nm, since the pyrene $S_1 \leftarrow S_0$ transition in the long wavelength region is symmetry forbidden, with a very low molar absorption coefficient. The Raman scattering of water (which should appear at 398 nm by excitation at 350 nm) is too small to justify the rise in emission intensity around 390 nm. This ground state aggregation was already observed for pyrene in several systems, especially in aqueous media [17–19]. The presence of a preassociated excited dimer strongly affects the I_E/I_M values and their variation with temperature, justifying the rather different behaviour observed in surfactant/water systems, compared to the one in homogeneous solvent. From spectra in Fig. 2b, it is clear that the emission from the dimer is stronger in the more concentrated $C_{12}E_7$ solution.

In order to investigate what happens well above the cmc, the I_E/I_M variation with temperature was recorded for a solution 8×10^{-4} M in $C_{12}E_7$ and in pure surfactant (Fig. 3a). The behaviour of I_E/I_M with temperature for these two systems is completely different from that observed in low $C_{12}E_7$ concentrations, the I_E/I_M now rising with temperature.

Fig. 3b shows the emission spectra for $\lambda_{exc} = 337$ nm and 350 nm for these two solutions. For $[C_{12}E_7] = 8 \times 10^{-4}$ M, a very strong dimer emission (relative to monomer) can be detected by excitation at 350 nm. This fact shows that, in this case, the excited dimer suffers a much slower conversion to excimer, indicating a lower mobility of the probe in micelles than in premicellar aggregates. This fact reflects the higher compactness of the surfactant molecules in micelles. Therefore, the major part of the excimers seems to come from the usual diffusion controlled process, and the I_E/I_M plot approaches that in a homogeneous solvent [15]. With

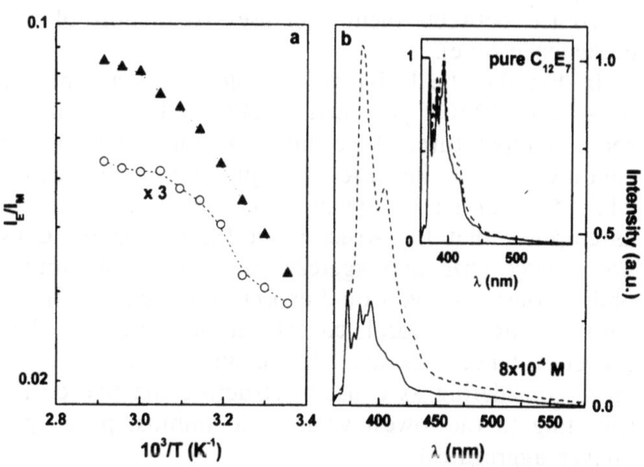

Fig. 3 (**a**) Excimer (500 nm) to monomer (372 nm) emission intensity ratio, I_E/I_M, for pyrene (2×10^{-7} M) as a function of reciprocal temperature ($\lambda_{exc} = 337$ nm), in a 8×10^{-4} M $C_{12}E_7$ solution (○) and for pyrene (10^{-3} M) in pure surfactant (▲). (**b**) Fluorescence spectra of pyrene (2×10^{-7} M) in a 8×10^{-4} M $C_{12}E_7$ solution, for $\lambda_{exc} = 337$ nm (—) and for $\lambda_{exc} = 350$ nm (- - -). Insert shows similar spectra for pyrene (10^{-3} M) in pure $C_{12}E_7$ (room temperature). The spectra were normalised at 372 nm

increasing temperature, the significant rise of the excimer dissociation rate (much higher than that of the excimer formation rate) and the highly compact medium surrounding pyrene in micelles may contribute to the occurrence of pyrene geminate pairs, which can recombine faster than the initial excited monomers [20]. This phenomenon will contribute to a further rise in I_E/I_M values with temperature.

In pure surfactant (insert of Fig. 3b), almost no dimer emission is observed. It is expected that some aggregation in the ground state is present, but the dimer suffers a fast conversion to excimer, with the corresponding decrease of the dimer emission. Therefore, the I_E/I_M plot with temperature exhibits almost no influence of pyrene aggregation, with a predictable maximum at high temperature, due to the significant viscosity of $C_{12}E_7$.

For the characterisation of lipid/micelle interactions, an important parameter is the surfactant molar ratio, x_s [7]:

$$x_s = \frac{[C_{12}E_7]}{[C_{12}E_7] + [DPPC]} \tag{2}$$

We will consider the usual three-stage model of solubilisation of liposomes by surfactants:

1. Mixed bilayers exist up to a critical surfactant concentration, x_{sat};
2. Between x_{sat} and x_{sol}, there are saturated bilayers and saturated mixed micelles with, respectively, x_{sat} and x_{sol} surfactant content;
3. Above x_{sol}, there are only mixed micelles.

Heerklotz et al. [21] obtained values of $x_{sat} = 0.45$ and $x_{sol} = 0.75$ for the POPC/$C_{12}E_7$ system at 25 °C. These two quantities both increase with temperature, while the difference between them decreases (reduction of the coexistence zone).

In Fig. 4a, the I_E/I_M ratio is plotted as a function of x_s for DPPC/$C_{12}E_7$ and EggPC/$C_{12}E_7$ systems at room temperature. According to our experimental conditions, as x_s increases the lipid content decreases, while the surfactant remains constant. Thus, on the assumption that the structure of the aggregates does not change (size and aggregation number of amphiphilic molecules), we would expect an increase of I_E/I_M with x_s, as the local concentration increases. The expected tendencies are plotted in Fig. 4a as solid lines, considering as constant structure the one corresponding to the lowest value of x_s (almost pure lipid bilayer aggregates).

For $x_s < 0.3$, it can be seen that there are no detectable changes in structure for both systems. In this region, the lipid bilayers are incorporating surfactant molecules, with a corresponding rise in the structure size. This would lead to a decrease in I_E/I_M, but there is an opposing factor: the fluidisation of the aggregate. In fact, surfactant molecules have a lower tail volume than the

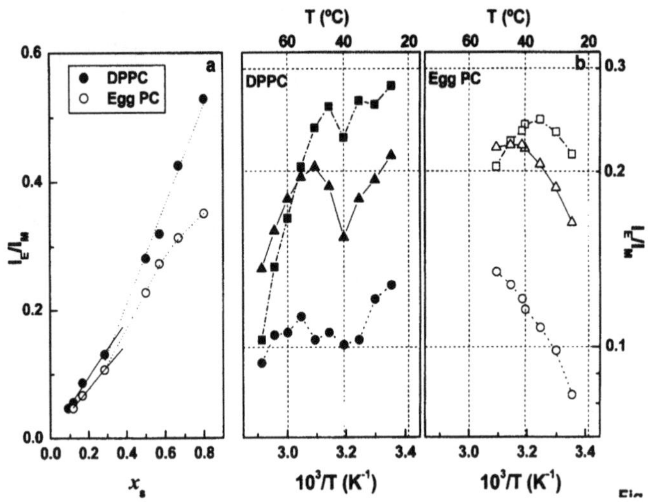

Fig. 4 (**a**) Excimer (500 nm) to monomer (372 nm) pyrene emission intensity ratio, I_E/I_M, in lipid/$C_{12}E_7$ mixed systems, as a function of surfactant molar ratio ($\lambda_{exc} = 337$ nm). (**b**) Excimer (500 nm) to monomer (372 nm) pyrene emission intensity ratio, I_E/I_M, as a function of reciprocal temperature ($\lambda_{exc} = 337$ nm), for DPPC/$C_{12}E_7$ and EggPC/$C_{12}E_7$ systems. DPPC: (■) $x_s = 0.67$; (▲) $x_s = 0.5$; (●) $x_s = 0.29$. EggPC: (□) $x_s = 0.67$; (△) $x_s = 0.5$; (○) $x_s = 0.29$.

lipid ones, but the surfactant headgroup is larger. Thus, the accommodation of $C_{12}E_7$ molecules in the lipid layer results in a lower compactness of the structure in the region below the headgroup [21]. A value for x_{sat} near 0.3 can be estimated.

For $x_s > 0.3$, a huge increase in I_E/I_M is observed. This corresponds to the second situation described above (between x_{sat} and x_{sol}). The system has now an increasing population of mixed micelles with x_{sol} fixed composition and a decreasing population of mixed bilayers with a x_{sat} surfactant composition. The I_E/I_M ratio feels this transition, as both the average local concentration and the probability of having two or more pyrene molecules per aggregate increase.

In DPPC the amount of excimer is always higher than for EggPC. At room temperature, DPPC bilayers are in the gel phase, whereas EggPC is in the liquid crystalline phase. Galla and Sackmann [22] observed that when DPPC vesicles change from the liquid crystalline to gel phase, a marked increase in I_E/I_M occurs. This behaviour was interpreted by considering that the pyrene solubility in the gel phase is very low, which results in the formation of microcrystallites with high yield of excimer formation.

The effect of temperature on these mixed aggregates was also investigated using pyrene monomer/excimer fluorescence. In Fig. 4b we can see that in the EggPC/$C_{12}E_7$ system there is a normal behaviour of I_E/I_M vs. temperature. The pronounced shift of the maximum with the lipid content is in agreement with an increase in the viscosity of the bilayer [16]. As EggPC is always in the liquid crystalline phase, it will mix almost homogeneously with $C_{12}E_7$, making a "fluid" of intermediate viscosity, irrespective of the aggregate structure (mixed micelle or mixed bilayer).

For the DPPC/$C_{12}E_7$ system, it is expected that the DPPC phase transition influence the results. In fact, near 40 °C a minimum in the I_E/I_M variation with temperature occurs. This compares well with the known value of 41 °C for the phase transition of pure DPPC [22]. After the phase transition, similar trends to those of the EggPC/$C_{12}E_7$ system are observed, with slight shifts, which can be attributed to changes in the fluidity of the surfactant/lipid "fluid". Further, when the lipid content is higher ($x_s = 0.5$ and $x_s = 0.29$), the excimer formation is more efficient in the EggPC/$C_{12}E_7$ than in DPPC/$C_{12}E_7$ system. This can also be explained by the higher viscosity of the DPPC bilayer. In the phase transition region (30–40 °C), the increase in I_E/I_M with decreasing temperature is again understandable by the presence of pyrene microcrystallites. When the value of x_s is 0.67, the presence of mixed micelles plays an important role. The DPPC content in these micelles should be ~20% ($1-x_{sol}$). Considering 400 molecules of $C_{12}E_7$ per micelle, we estimate that there are 100 molecules of DPPC per mixed micelle. It is expected that the packing of the lipid molecules in the micelle would be more favourable if they form small domains, because of the significant difference in tail volume and headgroup surface area of DPPC and $C_{12}E_7$. These domains would also exhibit a phase transition, which can be observed in our experimental results (Fig. 4b). Regarding ground-state pyrene association in these systems, a very weak emission from a preassociated dimer is observed in fluorescence spectra (with long wavelength excitation). Furthermore, the amount of dimer does not change appreciably with the phase transition of the lipid.

Acknowledgements The authors would like to thank Fundação para a Ciência e a Tecnologia (FCT) for financial support (Project 32901/99). M.C. Rei acknowledges FCT for a grant under the same project.

References

1. Sujatha J, Mishra AK (1997) J Photochem Photobiol A: Chem 104:173
2. Edwards K, Almgren M (1992) Langmuir 8:824
3. Miguel MG, Eiddelman O, Ollivon M, Walter A (1989) Biochemistry 28:8921
4. Edwards K, Gustafsson J, Almgren M, Karlsson G (1993) J Coll Interface Sci 161:299
5. Edwards K, Almgren M (1991) J Coll Interface Sci 147:1
6. Inoue T, Fukushima K, Shimozawa R (1988) Bull Chem Soc Jpn 61:1565
7. Inoue T (1996) In: Vesicles – surfactant science series, vol 62. Marcel Dekker, New York, pp 151–195
8. Duportail G, Lianos P (1996) In: Vesicles – surfactant science series, vol 62. Marcel Dekker, New York, pp 295–392
9. Dong DC, Winnik MA (1984) Can J Chem 62:2560
10. Kalyanasundaram K, Thomas JK (1977) J Am Chem Soc 99:2039
11. Förster T, Kasper K (1954) Z Phys Chem (NF) 1:275; (1955) Z Electrochem 59:976
12. Meguro K, Ueno M, Esumi K (1987) Micelle formation in aqueous media. In: Nonionic surfactants, vol 23, Surfactant science series. Marcel Dekker, New York, pp 109–183
13. Kawaguchi S, Yekta A, Duhamel J, Winnik MA, Ito K (1994) J Phys Chem 98:7891
14. Jönsson B, Lindman B, Holmerg K, Kronberg B (1998) Surfactants and polymers in aqueous solution. Wiley, New York
15. Birks JB (1970) Photophysics of aromatic molecules. Wiley-Interscience, London, pp 301–371
16. Real Oliveira MECD, Hungerford G, Castanheira EMS, Miguel M da G, Burrows HD (2000) J Fluorescence 10:347
17. Winnik FM (1993) Chem Rev 93:587 and references therein
18. Ilharco LM, Martins CI, Fedorov A, Martinho JMG (1997) Chem Phys Lett 277:51
19. Castanheira EMS, Martinho JMG, Duracher D, Elaïssani A, Charreyre MT, Pichot C (1999) Langmuir 15:6712
20. Reis e Sousa AT, Castanheira EMS, Fedorov A, Martinho JMG (1998) J Phys Chem A 102:6406
21. Heerklotz H, Binder H, Lantzsch G, Klose G, Blume A (1997) J Phys Chem 101:639
22. Galla HJ, Sackmann E (1974) Biochim Biophys Acta 339:103

Progr Colloid Polym Sci (2004) 123: 88–93
DOI 10.1007/b11648

A.L.F. Baptista
P.J.G. Coutinho
M.E.C.D. Real Oliveira
J.I.N. Rocha Gomes

Lipid interaction with textile fibres in dyeing conditions

A.L.F. Baptista · J.I.N. Rocha Gomes
Departamento de Engenharia Têxtil,
Universidade do Minho,
Campus de Azurém,
4810–258 Guimarães, Portugal

P.J.G. Coutinho
M.E.C.D. Real Oliveira (✉)
Departamento de Física,
Universidade do Minho,
Campus de Gualtar,
4710–057 Braga, Portugal
e-mail: beta@fisica.uminho.pt
Tel.: +351-253-604325
Fax: +351-253-678981

Abstract There is an increasing interest in the textile industry in eco-friendly textile processing, in which the use of naturally occurring materials such as phospholipids becomes important. In previous work [1] we have studied the effect of microencapsulation of dyes with soybean lecithin liposomes, in the dyeing of polyamide and cotton. We also found that, even if the dye was not encapsulated, there is an effect on the dyeing rate due to the lipid itself. In order to understand this effect we studied the interactions between phospholipids (labelled and non-labelled) with the fibres. We used a combination of reflectance, fluorescence, FTIR spectroscopy and electronic microscopy characterisation techniques. Our main conclusion is that phosphatidylcholine has more affinity for polyamide than for cotton and that cotton seems to interact more strongly with phosphatidylinositol.

Keywords Lecithin · Textile fibres · Fluorescence spectroscopy · SEM · FTIR

Introduction

Auxiliaries, as their name implies, are used to assist in dyeing, by wetting, levelling or when necessary by dispersing dyes of low solubility. Their interaction with fibres plays a very important role in achieving level dyeing or in controlling dye adsorption by fibres.

The aim of our work consists in the use of natural products, as soybean lecithin, as auxiliary in the dyeing of polyamide and cotton, instead of chemical products so as to improve the effluent characteristics [2].

In previous work [1, 2] we have studied the influence of the microencapsulation of acid dyes, with soybean lecithin liposomes, in the dyeing of polyamide and we have concluded that the liposomes have a retarding effect. This effect is very important to achieve level dyeing. It was interesting to observe that some of this same effect (to a lesser extent) is also present when we process the dyeing in the presence of lecithin liposomes and non-microencapsulated dyes.

Later we studied the influence of microencapsulation of reactive dyes, in soybean lecithin liposomes, in the dyeing of cotton and we concluded that these dyes have better affinity than the non-microencapsulated ones. In the same way, we have dyed cotton in the presence of lecithin liposomes with non-microencapsulated dyes, and the effect was nearly the same, i.e., the affinity of the dyes was improved.

The interactions between lipids or surfactants and textile fibres are not completely established.

Maza and his coworkers [3] reported a number of investigations on unilamellar or multilamellar vesicles of egg phosphatidylcholine as vehicles of dyes or oxidative reagents for dyeing and finishing wool. The role of the liposomes can be attributed to the fact that bilayer structure of lipids from the cell membrane complex of wool is similar to that of the liposomes. Kim and colleagues [4] proposed to use double tailed surfactants for preparing synthetic vesicles in disperse dyeing of polyamide to improve dyebath exhaustion and colour uniformity. They concluded that there might be fairly strong interactions between those surfactants and the fibre by virtue of electrostatic and hydrophobic interactions.

In the present work, the interactions between lecithin and textile fibres (polyamide and cotton) are investigated by reflectance, fluorescence and FTIR spectroscopy and SEM (scanning electron microscopy) characterisation techniques.

Experimental

The materials to be studied, were polyamide 6.6 knitwear without any special treatment and cotton. Commercial soybean lecithin containing 22% phosphatidylcholine, 20% phosphatidylethanolamine, 14% phosphatidylinositol and 10% phytoglycolipids was supplied by Stern (USA). Phosphatidylcholine (PC) and phosphatidylethanolamine (PE) 99% pure were purchased from Sigma. N-Rh-PE (*N*-lissamine rhodamine B sulphonyl phosphatidylethanolamine) was obtained from Molecular Probes (Netherlands). Soybean lecithin, PC, PE and fluorescent probe, N-Rh-PE were used without further purification.

Liposomes preparation

The various phospholipids were dissolved in diethyl ether at the desired molar ratio. The solution was evaporated to dryness under nitrogen. Vesicles were formed by hydrating the lipids with an aqueous solution containing 2 g/L ammonium sulphate (pH = 5.5) and sonication (Heat Systems-Ultrasonic Model W-225R) for 10 minutes at 20 °C and 60 W.

Dyeing conditions

Laboratory dyeing was done in an Ahiba Turbo Color dyeing machine with a dyebath ratio 50:1.

The following dyeings were carried out: 1) with soybean lecithin liposomes (2 g/L); 2) with PC liposomes (0.1 g/L); 3) with PE liposomes (0.1 g/L); 4) with N-Rh-PE liposomes (0.01 g/L) and 5) with soybean lecithin (2 g/L) and N-Rh-PE (0.01 g/L) mixed liposomes.

Dyeing of polyamide

To the aqueous solution with lecithin liposomes, 2 g/L of ammonium sulphate was added and the pH adjusted with acetic acid to reach pH five. Dyeing was started at 40 °C and the

Fig. 1 FTIR spectra of textile fibres: **A** polyamide; **B** cotton

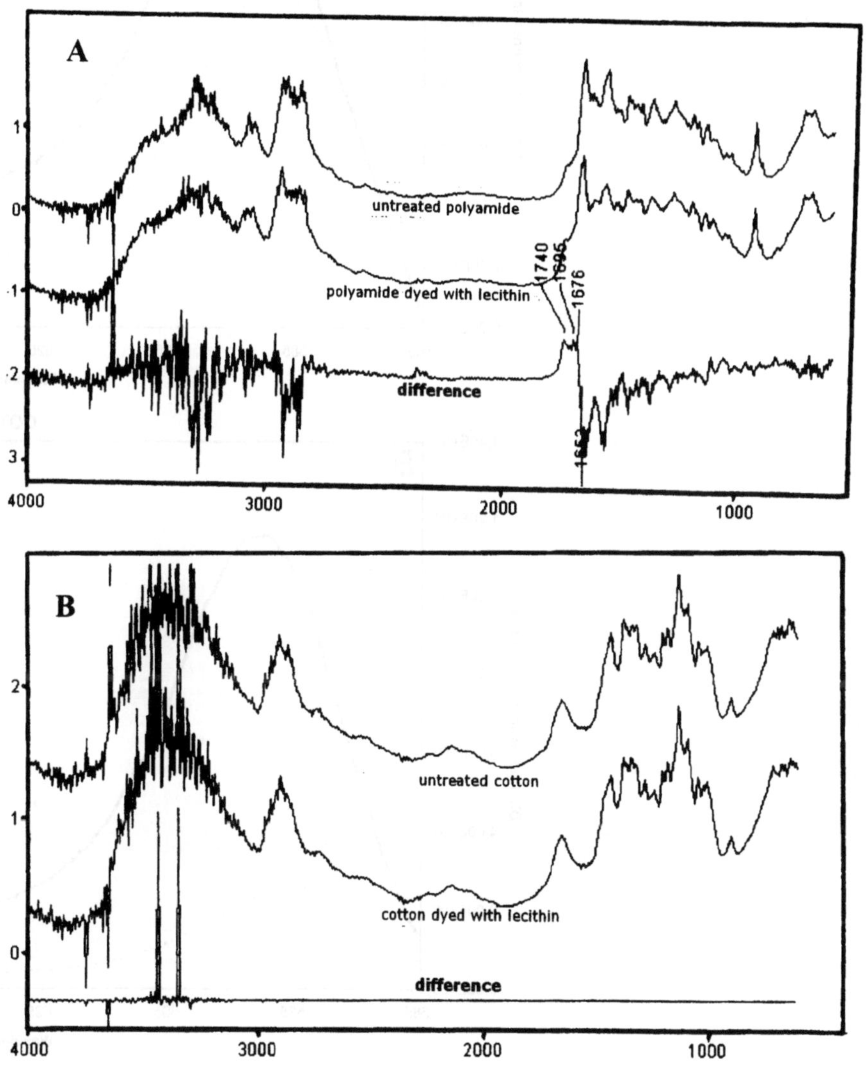

temperature was increased to 98 °C at a gradient of 10 °C/min, keeping the dyebath at this temperature during 30 minutes.

Dyeing of cotton

To the aqueous solution with lecithin liposomes, 50 g/L of sodium sulphate was added and the dyeing carried on at 45 °C during 30 minutes, then the pH was adjusted to 12 using 15 g/L of sodium carbonate and the dyeing process was kept during further 60 minutes at the same temperature.

Reflectance measurements

The diffuse reflectance of the samples was measured with a spectrophotometer Shimadzu UV-3101 PC.

Fluorescence measurements

The fluorescence of the samples dyed with N-Rh-PE was measured with a spectrofluorometer (Spex Fluorolog 212), using the rhodamine group emission by excitation at 463 nm.

FTIR

The Michelson Series FT-IR Spectroscopy from Bomem was used to obtain infrared spectra of the cotton and the polyamide samples that were untreated, i.e., without any dyeing process, and dyed with lecithin. The pieces of fabrics were cut, put on top of a sampling cup containing potassium bromide powder and scanned 100 times by diffuse reflectance spectroscopy.

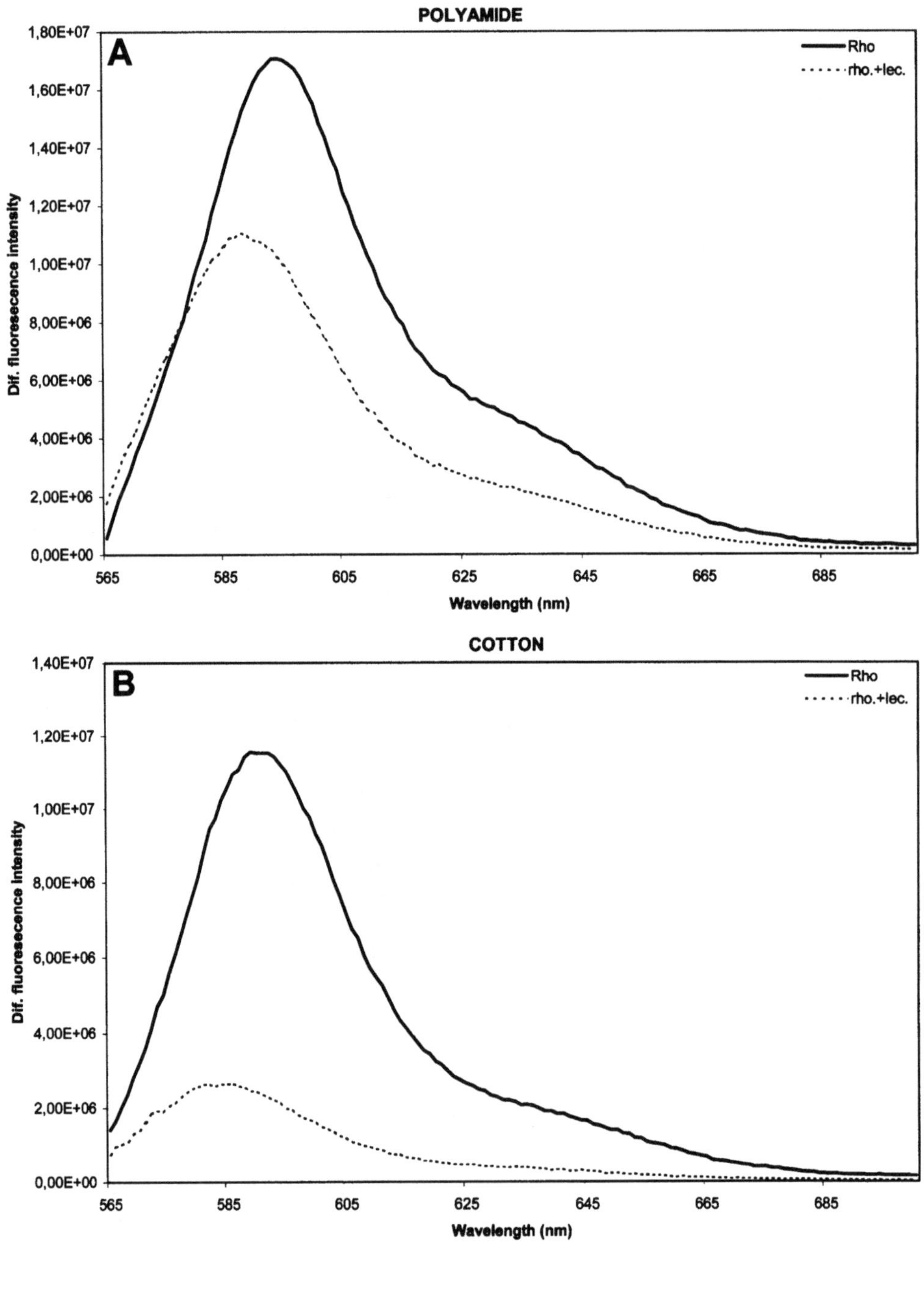

Fig. 2 Reflectance spectra of textile fibres: **A** polyamide; **B** cotton

SEM

The images of the fibres' surfaces and their cross-sections were obtained by a scanning electron microscope, Leica Cambridge S360.

Results and discussion

The FTIR spectra of polyamide and cotton before and after dyeing with lecithin are shown in Fig. 1. The resulting difference spectrum for polyamide (Fig. 1A) shows four peaks at around 1663, 1676, 1695 and 1740 cm^{-1} that correspond to the frequency region of carbonyl group present in lecithin showing that some or all of lecithin component is incorporated in the polyamide matrix. With the same technique the incorporation of lecithin in cotton is not so evident as in the case of polyamide.

In order to gain more insight about the interaction of lecithin with the fibres we tried other spectroscopic techniques. In previous work [5] we used N-Rh-PE, which is, a phospholipid labelled with a rhodamine group for studying surfactants-liposomes interactions. In this work we used the fluorescence of this labelled phospholipid to understand the interaction of lecithin with the fibres.

Fig. 2 shows the fluorescence spectra of N-Rh-PE in polyamide (A) and cotton (B) samples.

We observe that the rhodamine spectra are red shifted relative to, for example, an ethanolic environment ($\lambda_{max} = 581$ nm). In aromatic compounds the usually electronic transitions are $\pi\pi^*$ and $n\pi^*$[6]. The molecule in different electronic states interacts differently with environment originating from the dependences of the transition energy on the properties of the medium. $\pi\pi^*$ transition shows a red shift as the environment polarity increases. As the involved transition is a $\pi\pi^*$, we conclude that in both fibres rhodamine feels a polar media. Comparing both fibres we can conclude that in polyamide the sites where rhodamine is located, are more polar than in cotton as the emission maximum in polyamide are red shifted in relation to cotton. This can be understood if we consider the structures of the fluorescence probe [2] and the fibres [7].

During the dyeing process some polyamide sites are positively charged. As the probe has a negative charge, there is an electrostatic interaction (ionic binding) between the phospholipid fluorescent probe and polyamide. For cotton the ionic binding is not feasible and thus N-Rh-PE goes to the fibre via dispersive interactions. With this information we could expect that negatively charged PE would be lecithin's main binding component for polyamide and the zwitterionic PC should interact more efficiently with cotton. To see whether lecithin components can compete with N-Rh-PE for the fibres binding sites, we used mixed liposomes (dyeing 5). We observed a reduction of 35% of rhodamine emission in polyamide and of 71% in cotton. As in soybean lecithin PC and PE contents are approximately the same, it would seems that PC has a stronger interaction with cotton than N-Rh-PE and binds less efficiently with polyamide sites than N-Rh-PE. In both fibres the incorporation of both lecithin and N-Rh-PE induces a small blue shift in the rhodamine emission. This is a result of a different and less polar environment of the fluorescence probe. The possibility of some N-Rh-PE molecules being imbibed in a local "film" of lecithin present in the fibres could contribute to the observed blue shift (lecithin polar group has a polarity similar to ethanol).

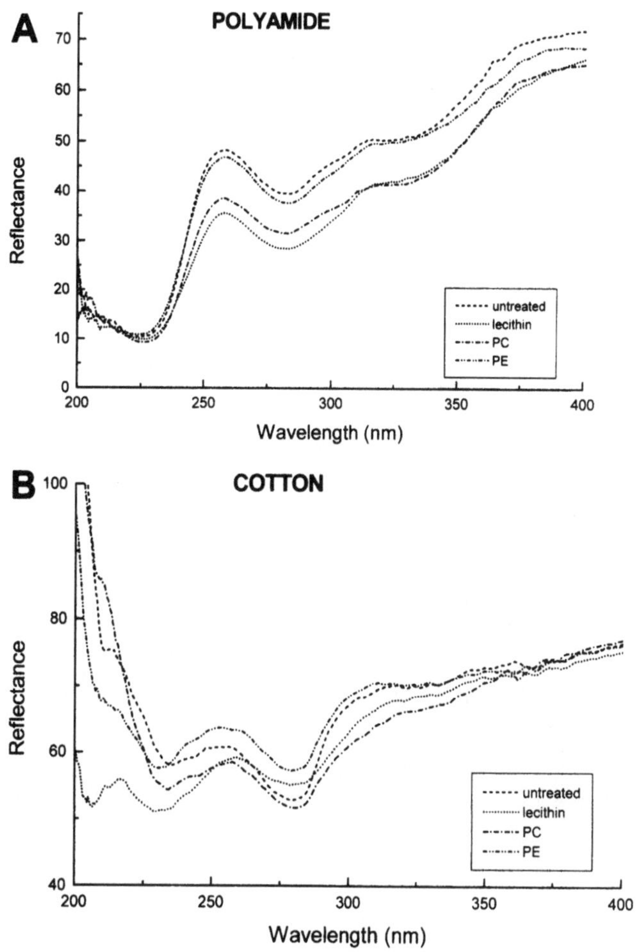

Fig. 3 Fluorescence spectra of N-Rh-PE of textile fibres: **A** polyamide; **B** cotton. The light emitted/dispersed by the fibres was subtracted from the fluorescence spectra of the dyed fibres

In order to clarify which lecithin components are incorporated in the fibre we obtained the reflectance spectrum (specular and diffuse) of the fibres untreated and dyed with soybean lecithin, PC and PE (Fig. 3A and 3B). The phospholipids usually absorbed in the region 220–300 nm. In polyamide we conclude that the main

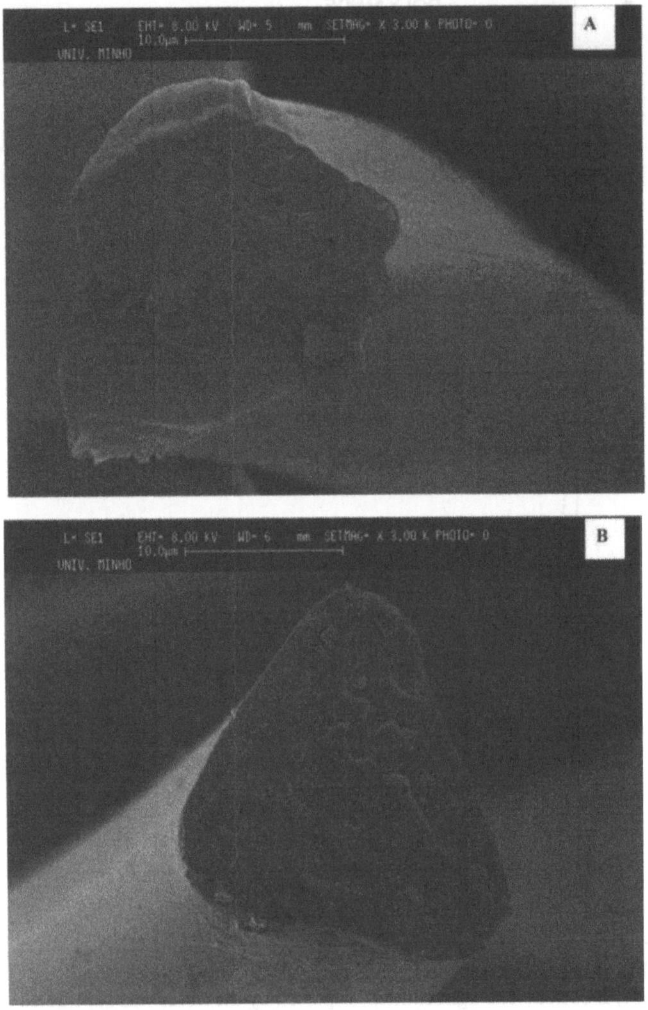

Fig. 4 Cross-section optical views of polyamide (× 3000): **A** untreated; **B** dyed with lecithin

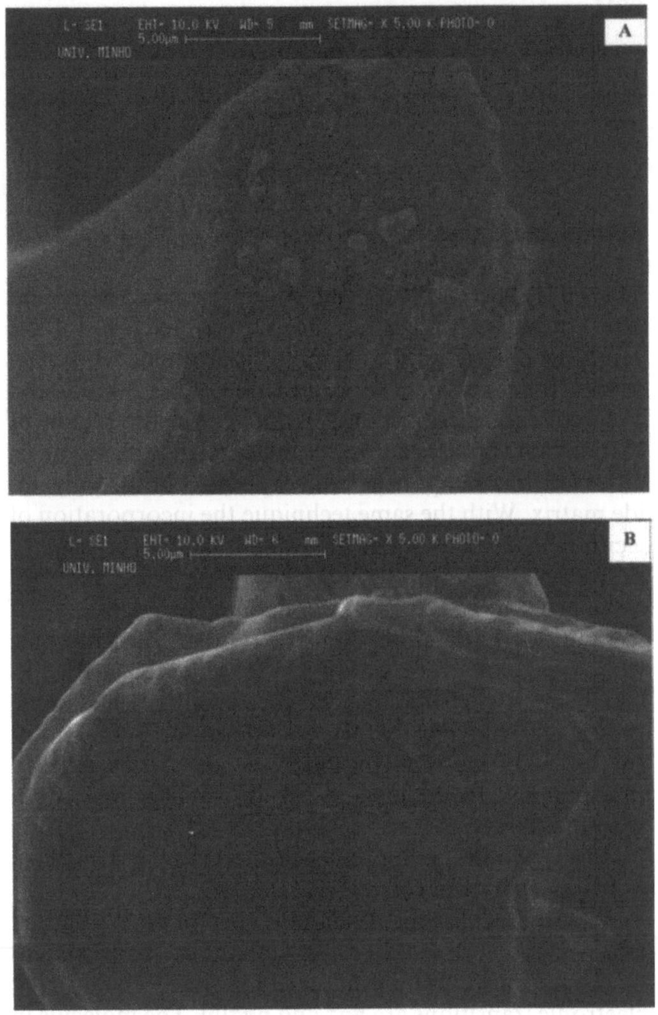

Fig. 5 Cross-section optical views of cotton (× 5000): **A** untreated; **B** dyed with lecithin

lecithin component that is incorporated in this fibre is PC. PE is incorporated only slightly and probably on the positive binding sites due to its anionic characteristics. In cotton the variations in the reflectance spectra are more complicated. Cotton acts as a mirror in the spectral region of 200–210 nm. We can see that PE has no effect in this region whereas PC decreases the mirror effect and when lecithin is present it is practically lost. This is mainly a surface effect and thus we can conclude that some other lecithin component interacts strongly with cotton surface, probably the phosphatidylinositol [8], which has a headgroup similar to glycose, a monomer of the cellulose polymer, present in the structure of the cotton. The region of phospholipid absorbance is not so indicative which is a confirmation of the difficulty in assigning phospholipid characterisation in the FTIR experiment. Nevertheless, we can see a little absorption due to lecithin and PC.

In order to understand what kind of interaction the lecithin has with the fibres, we decide to use the scanning electron microscope to analyse the surface and the interior of the two different fibres, untreated and treated with lecithin.

Fig. 4 shows the cross section of polyamide untreated (A) and dyed with lecithin (B).

Fig. 5 shows the cross section of cotton untreated (A) and dyed with lecithin (B).

The three-dimensional images clearly show surface features, such as the presence of surface or cross-section modifications. This confirms the reflectance results and shows that the interaction of the lecithin with the cotton is more at the surface through a coating layer. The interaction with the polyamide is more in the interior, because we can observe changes in the morphology of the cross-section of the fibre, which are not so evident in cotton.

Conclusions

In conclusion, we can say that lecithin has more affinity for polyamide than cotton. The interactions between polyamide and lipids occur especially in the interior of the fibre while in cotton those interactions are located at the surface, as we can observe in SEM images. In polyamide the lecithin component responsible for those interactions is essentially phosphatidylcholine as we can see in reflectance spectra, and it seems that phosphatidylinositol, with a chemical structure similar to cellulose, interacts more strongly with cotton.

Acknowledgements We wish to thanks to Fundação para a Ciência e Tecnologia (Portugal) for the scholarship (BD/ 17186/ 98) of A.L.F.B.

References

1. Gomes JINR, Baptista AL (2001) Textile Res J 71(2):153–156
2. Gomes JINR, Baptista ALF (1998) Proceedings of the 7th International Conference on Organic Dyes and Pigments "Colorchem' 98". Check Republic, p 11
3. Maza A, Coderch L, Manich A, Bosh P, Parra J (1996) In: Lasic D, Barenholz Y (eds), Handbook of nonmedical applications of liposomes, vol 4, chap 12. CRC Press, Boca Raton, pp 165–182
4. Kim I, Kono K, Takagishi T (1996) Textile Res J 66(12):763–770
5. Baptista ALF, Coutinho PJG, Real Oliveira MECD, Gomes JINR (2000) J Liposome Res 10(4):419–429
6. Valeur B (2001) Molecular fluorescence – principles and applications, chap 3. Wiley-VCH, Weinheim, pp 34–70
7. Gohl EPG, Vilenky LD (1983) Textile science. Longman, Cheshire, UK
8. New RRC (1989) Liposomes. IRL Press, Oxford

Progr Colloid Polym Sci (2004) 123:94–97
DOI 10.1007/b11649

E. Hatzara
E. Karatza
S. Avramiotis
A. Xenakis

Spectroscopic mobility probing studies of lecithin organogels

E. Hatzara · E. Karatza
S. Avramiotis · A. Xenakis (✉)
Industrial Enzymology Unit,
Institute of Biological Research
and Biotechnology, The National
Hellenic Research Foundation,
48, Vassileos Constantinou Ave.,
11635, Athens, Greece
e-mail: arisx@eie.gr
Tel.: +30210-7273762
Fax: +30210-7273758

Abstract The rigidity of lecithin film into an organogel has been studied via electron paramagnetic resonance and fluorescence spectroscopy. Well-known spin probes such as doxylated derivatives are used in order to monitor the rigidity and the polarity of the lecithin bilayer in different microdomains of the organogel. Stearic acids are located in the lecithin film in different depths while the alkane and the ester are located in the continuous organic phase. The corresponding 5-DSA spectrum in a lecithin/isooctane gel system shows more important immobilisation than the spectra of 12-DSA and 16-DSA in the same lecithin system. These differences of the immobilisation of the three probes are attributed to the different position of the nitroxide ring of the fatty acid aliphatic chain. In the case of 5-DSA probe the nitroxide is located closer to the polar head of the amphiphilic fatty acid and consequently closer to the lecithin polar head. The spectra of 10-DN and 12-DMSE show that these probes are located in the continuous organic phase and that they are not immobilised. The same behaviour is observed in lecithin/isopropyl palmitate or isopropyl myristate gel systems with a lower immobilisation. Fluorescence quenching experiments showed that the probes used are located as in a homogeneous media in a lecithin/isopropyl palmitate/system and come in contact instantly.

Keywords Lecithin · Organogels · EPR spectroscopy · Fluorescence

Introduction

Lecithin is a natural phospholipid, it can be used as a surfactant in biological applications such as pharmaceutical and cosmetics formulations and is preferable to bio-incompatible surfactants [1, 2]. When surfactants are dispersed in non-polar solvents, water-in-oil microemulsions or reverse micelles can form spontaneously [3]. Lecithin reverse micelles in many non-polar solvents possess the unusual property of forming cylindrical tubes when small amounts of water are added to the microemulsion [4–7]. The viscosity of reverse micellar solutions of lecithin in a number of different organic solvents increases dramatically upon the addition of very small amounts of water and the system often changes to an organogel [8]. The formation of such organogels possessing a viscoelastic behaviour has been observed in different solvents.

The structure of these organogels is not yet totally clarified [9]. Electron paramagnetic resonance (EPR) and fluorescence spectroscopy are spectroscopic methods that can be applied to investigate the mobility of spin probes in microemulsions [10, 11] and provide information on the structure of these organogels. In the present study lecithin organogels formed with various organic solvents such as isooctane, isopropyl palmitate and isopropyl myristate were studied.

Most specifically, for EPR studies have been used probes such as 5-doxylstearic acid (5-DSA), 12-doxylstearic acid (12-DSA), 16-doxylstearic acid (16-DSA),

12-doxylstearic methyl ester (12-DMSE) and 10-doxylnonadecane (10-DN) that differ as regards the position or the depth that the spin probe is located within the packed lecithin molecules [12]. Three different systems: solvent/spin probe, solvent/lecithin/spin probe (*binary system*) and solvent/lecithin/spin probe/water (*gel system*) were studied. Also, the effect of various parameters such as pH, lecithin concentration and organic solvents on the spectral characteristics has been examined. In order to monitor the dynamics of a spin probe the rotational correlation time parameter τ_R was calculated.

In the case of fluorescence spectroscopy we have used as probes $Ru(bpy)_3Cl_2$ as fluorophor and $K_3Fe(CN)_6$ as a quencher in order to examine the behaviour of aqueous microdomains where the probes are located [13, 14].

Experimental

Materials

Soybean lecithin (*Epikuron 200*) containing 96% phosphatidylcholine was purchased from Lucas Meyer. Isooctane was purchased from Merck; isopropyl palmitate and isopropyl myristate were from Fluka. The spin-labelled doxylated derivatives 5-doxylstearic acid [5-(1-oxyl-2,2-dimethyloxazolidine)stearic acid; 5-DSA], 12-doxylstearic acid [12-DSA], 16-doxylstearic acid [16-DSA], 12-doxylstearic methyl ester [12-DMSE] and 10-doxylnonadecane [10-DN] were obtained from Sigma. The fluorescent probe tris (2,2-bipyridine)ruthenium dichloride hexahydrate, $Ru(bpy)_3Cl_2$, was from GFS Chemicals, and the quencher potassium ferricyanide, $K_3Fe(CN)_6$ from Merck. High-purity water was obtained by a Millipore Milli Q Plus water purification system.

Preparation of the samples

The first system *solvent/doxylated derivatives* were prepared as follows: each spin probe was added in each solvent to give a final concentration 10^{-2} mM. The binary system was prepared as follows: in 5%w/w lecithin in isooctane or in 10%w/w lecithin in isopropyl palmitate or isopropyl myristate each spin probe was added to give a final concentration 10^{-2} mM. The gel system was prepared like the binary system in which small amounts of water were added depending on the experiment to give a final w_o ($w_o = [H_2O]/[\text{surfactant}]$) ranging between 0.5 and 5. In the case of fluorescence spectroscopy studies the fluorophor $Ru(bpy)_3Cl_2$ and the quencher $K_3Fe(CN)_6$ were added in a gel system to a final concentration 10^{-2} mM.

EPR measurements

EPR spectra were recorded at room temperature using a Bruker ER 200 D spectrometer. Spectra were accumulated and treated using the DAT-200 software for PC (University of Lubeck). The samples were contained in an E-248 cell. Typical settings were: centre field, 3471 G; scan width, 100 G; time constant, 0.5 s; scan time, 100 ms; microwave power, 7.5 mW; microwave frequency, 9.76 GHz; modulation amplitude, 1 G.

Fluorescence measurements

Fluorescence studies were recorded at room temperature using a Perkin-Elmer 650-40 Fluorescence Spectrophotometer. Fluorescence intensities were measured in the absence and presence of various concentrations of quencher. These measurements have been done in a lecithin/isopropyl ester system due to the disability of fluorophor probes to be dissolved in the isooctane system.

Results and discussion

EPR studies

Solvents/spin probes

In order to monitor the dynamics of a spin probe the rotational correlation time parameter τ_R was calculated [12]. The values of τ_R in the isooctane system are higher than those observed in esters systems due to the increased viscosity of the latter. The values of α_N in the isooctane system, a parameter that reflects the polarity of the solvents, are lower than those of the isopropyl palmitate system. However, the higher values are observed in the isopropyl myristate system. This is more obvious in the case of 5-DSA, 12-DSA and 16-DSA. That means that the position of the paramagnetic ring moiety effects the rotational correlation time of doxylated derivatives.

Binary systems – gel systems

In binary systems (solvent/lecithin/probe) doxylated derivatives are more immobilised due to the existence of lecithin. The level of immobilisation is higher in binary systems than in gels (solvents/lecithin/probe/water) because these are homogenous and so there are more interactions between spin probe and lecithin molecules. All DSA probes (5-DSA, 12-DSA, 16-DSA) are located in the lecithin film and their mobility, as expressed by the rotational correlation time parameter τ_R, reflect the rigidity of the lecithin film in different depths, while 12-DSME and 10-DN are located in the continuous organic phase as indicated by their mobility (Fig. 1). Another parameter studied was the effect of lecithin concentration. Despite the increase of macroviscosity of a lecithin gel in high concentrations, the lecithin concentration did not significantly alter the spectral characteristics.

Studying the effect of pH, we observed that the more the pH value increases the higher the ratio of molecules located in the lecithin film becomes (data not shown). This is more obvious in the case of 5-DSA in the isooctane system.

Fluorescence studies

To extend our EPR findings, we have employed the static fluorescence quenching method using $Ru(bpy)_3^{2+}$ as the fluorophor and $Fe(CN)_6^{3-}$ as the quencher. The quenching process occurs exclusively in the aqueous domains,

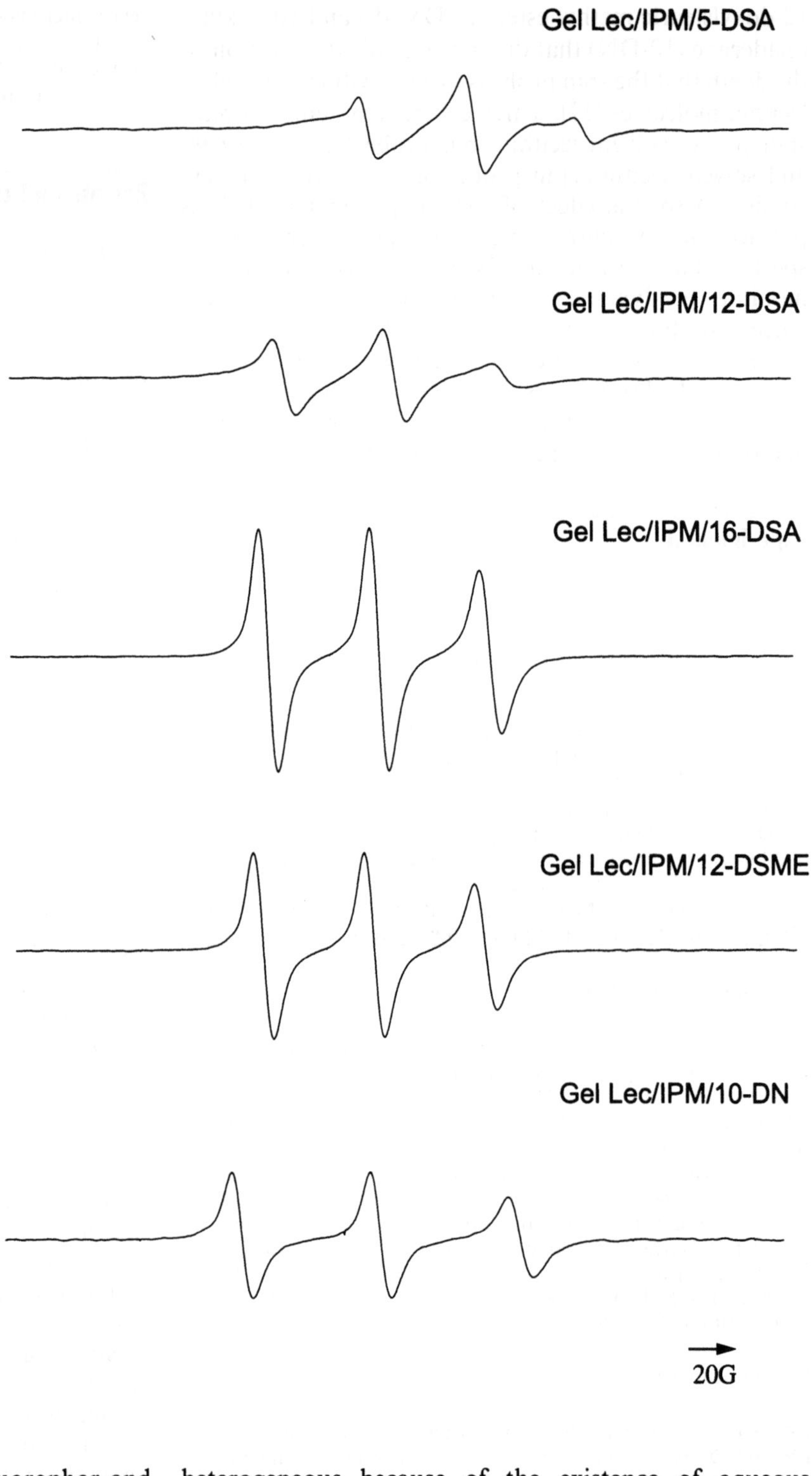

Fig. 1 EPR spectra of doxylated derivatives in lecithin/IPM/water gels

due to the strong hydrophilicity of the fluorophor and the quencher [14]. Plots of F_o/F vs. [Q] have shown a linearity required by the Stern-Volmer relationship in homogeneous media, a similar behaviour as in microemulsions [15] (F_0, F are the fluorescence intensities in the absence and presence of various concentrations of quencher ranged between 10^{-2} and 10^{-1} mM). Although the system of lecithin organogel is known that is heterogeneous because of the existence of aqueous microdomains where these fluorophor probes are located, the results show a fluorescence quenching behaviour corresponding to a homogeneous media (Fig. 2; F_o/F vs. concentration of quencher). This observation suggest that the investigated systems are water-connected as it was recently reported by Shchipunov et al. [16].

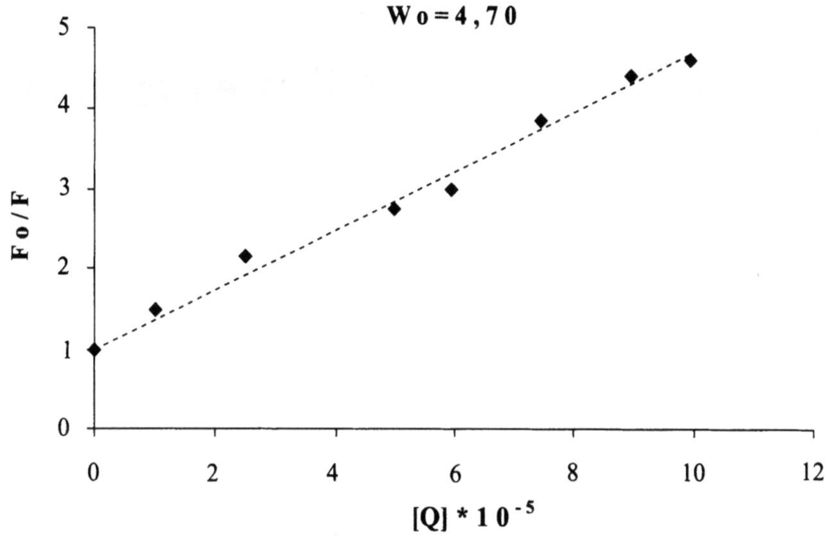

Fig. 2 Fluorescence quenching between $Ru(bpy)_3Cl_2$ a probe that was used as fluorophor and $K_3Fe(CN)_6$ a probe that was used as quencher. F_0, F are the fluorescence intensities in the absence and presence of various concentrations of quencher [Q]

Conclusions

EPR spectroscopy can be applied to study the structure of lecithin organogels. Spin probes, such as doxylated derivatives can be located in different areas of lecithin gels according to their polarity. The rigidity of the lecithin film varies between binary systems and gels because binary systems are homogeneous while in gels the presence of water effects the interactions between molecules. A change of the pH value effects the location of the doxylated stearic acids in the lecithin film. Fluorescence spectroscopy shows that fluorescence quenching is restricted in local areas and intermicellar interactions cannot be observed within the lifetime of probes.

Acknowledgements The Greek General Secretariat of Research and Technology has financed this work, within the frame of the program ΕΠΕΔ 99-ΕΔ-78.

References

1. Willard DM, Riter RE, Levinger NE (1998) J Am Chem Soc 120:4151
2. Zoumpanioti M, Karavas E, Skopelitis C, Stamatis H, Xenakis A (2003) Progr Colloid Polymer Sci 123
3. Danielsson I, Lindman B (1982) Colloids Surf 3:391
4. Schurtenberger P, Scartazzini R, Magid LG, Leser ME, Luisi PL (1990) J Phys Chem 94:3695
5. Schurtenberger P, Magid LG, Lindner P (1991) J Phys Chem 95:4173
6. Shchipunov Y, Hoffman H (1999) Langmuir 15:7108
7. Shchipunov Y, Dürrschmidt T, Hoffman H (2000) Langmuir 16:297
8. Luisi PL, Scartazzini R, Haering G, Schurtenberger P (1990) Colloid Polymer Sci 268:356
9. Shchipunov Y (2001) Colloids Surf A 183–185:541
10. Xenakis A, Cazianis CT, Malliaris A (1992) Colloids Surf 62:315
11. Avramiotis S, Stamatis H, Kolisis FN, Lianos P, Xenakis A (1996) Langmuir 12:6320
12. Avramiotis S, Cazianis CT, Xenakis A (1999) Langmuir 15:2375
13. Lakowicz JR (1983) Principles of fluorescence spectroscopy. Plenum Press, New York
14. Malliaris A (1987) Adv Colloid Interfaces 27:153
15. Avramiotis S, Xenakis A, Lianos P (1996) Progr Colloid Polymer Sci 100:286
16. Shchipunov YA, Mezzasalma SA, Koper GJM, Hoffman H (2001) J Phys Chem B 105:10484

Progr Colloid Polym Sci (2004) 123: 98–103
DOI 10.1007/b11650

Jürgen Hauck
Klaus Mika

Self-assembly of homogeneous systems

J. Hauck (✉) · K. Mika
Institut für Festkörperforschung,
Forschungszentrum Jülich, 52425 Jülich,
Germany
e-mail: j.hauck@fz-juelich.de
Tel: +49-2461-614237
Fax: +49-2461-612280

Abstract The circle packings including non-periodic packings can be analysed for physical and thermodynamic stability for a triangular net with $T_4=3$–6 neighbours at distance $d = \sqrt{7}$. Two-dimensional structures A_xB_y are characterised by the numbers of T_4 neighbours and the ratio r = y/x of vacant/occupied positions of the triangular net. A small number of structures with maximum density $\rho=6.35/(r+1)$ approach natural patterns of sunflower seeds, blackberries, bacteria, colloids, thistles or cactus spines. Two-dimensional patterns can be approached by combinations of the same structural units.

Keywords Circle packings · Patterns · Colloids · Lipid monolayers · Natural products

Introduction

Some clusters of the circle packings are similar to natural patterns of sunflower seeds, blackberries, cactus spines, thistles, bacteria [1] or colloids in confinements [2]. Also mixtures of liquids or gases and moving chemicals or grains can show a closely related ordering like surfactant patterns on water surfaces [3], convection patterns, cellular patterns in a shallow cylindrical flame, chemical Turing patterns or wave patterns of granular layers [4].

This investigation presents a system, where distorted squares with $\alpha\approx82°$ can be combined with triangles or hexagons to two-dimensional patterns. These patterns are related to circle packings with different contact numbers T_i to neighbouring circles A and different ratios r = y/x of occupied (●) and vacant (○) positions of the triangular net (T_1 T_2 T_3 T_4; r). The same type of ordering (0 0 0 5; 6.5a circle packing) with $T_4=5$ contacts of each circle can be observed for colloids, bacteria or the centres of DPPE (1,2-dipalmitoylphosphatidylethanolamine) stars [3] (Fig. 1, last row). The dashed lines correspond to varied densities of two-dimensional patterns. The combinations of triangles and distorted squares are outlined within the smallest periodic area (the unit cell).

Circle packings

A circle packing is a two-dimensional set of non-intersecting circles with mutual contacts between the circles [5]. It is homogeneous, if all circles have the same neighbourhood of other circles. A heterogeneous circle packing consists of at least two subsets of circles. The radii of symmetrically distinct circles can be either equal or different. There exists an infinite number of homogeneous and heterogeneous circle packings, but only 11 types of homogeneous circle packings. Two circle packings belong to the same type, if the number of contacts to other circles is preserved by distortion. It is stable, if no circle can be moved without moving neighbouring circles at the same time. A minimum of three contacts per circle is required. Not all contacts must fall in one semicircle [5].

The different types of circle packings can be approximated in a triangular net without occupation of nearest, next-nearest and third neighbour positions (T_1–$T_3=0$). Each circle has $T_4=12$ fourth-nearest neighbour positions at distance $d = \sqrt{7}$, which can be reached by a left or right turn after $l=2$ straight positions on a hexagonal net (Fig. 1, first row). The steps are similar to a knight's move in chess. A maximum of $T_4=6$ positions can be

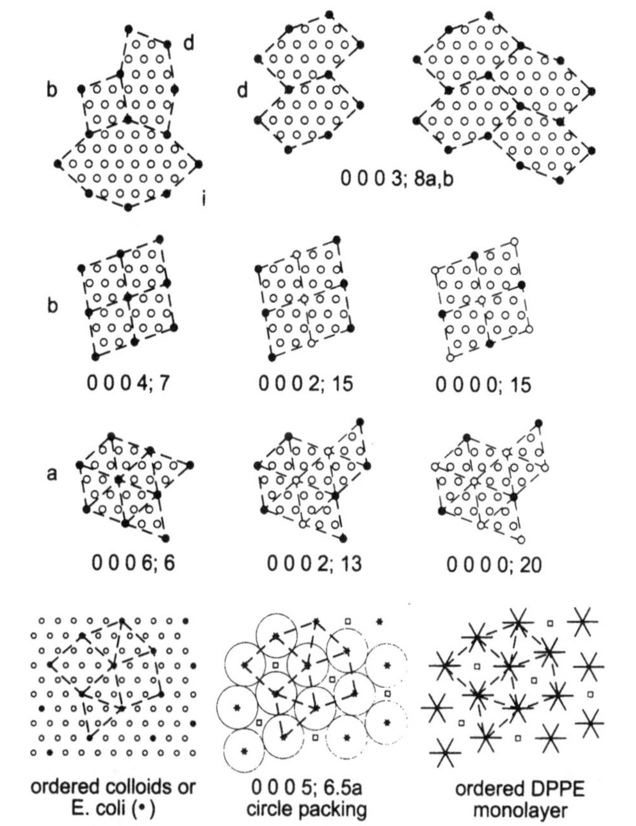

Fig. 1 Combination of structural units a, b, ... with composition A_xB_y (A = ●, B = ○). The structures are characterised by the self-coordination numbers T_i (T_1–T_3 = 0) and r = y/x values

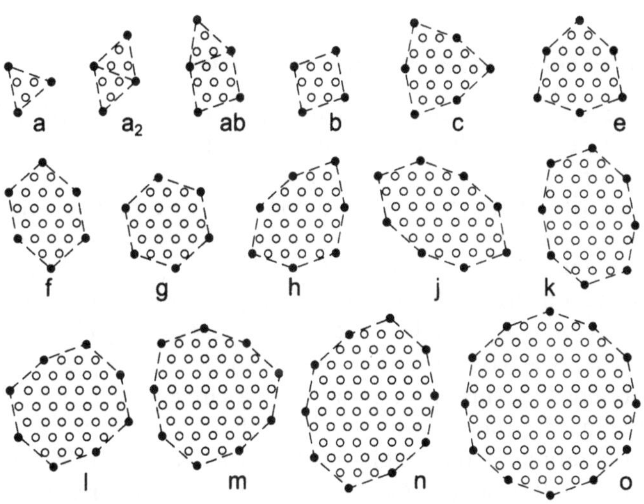

Fig. 2 Structural units a–o of the triangular net with x A = ● and y B = ○ positions

occupied for non-intersecting circles with diameter $d = \sqrt{7}$. The angle α at the central circle to the centres of two neighbouring circles amounts to 60° (Fig. 2). This angle can increase to $\alpha <$ 180° for stable circle packings with $\alpha_i \approx 82°$, 98°, 120°, 142° and 158°. The triangle with $\alpha = 60°$ and y = 3 vacant sites contains the smallest loop of a knight's move on the triangular net (structural unit a in Fig. 2). Two different rhombi a_2 with $\alpha_i = 60°$, 120° and y = 6 or b with $\alpha_i \approx 82°$, 98° and y = 7 vacant sites are obtained by 4 steps. The combination of structural units a and b to a b similar to a_2 combination corresponds to 5 steps s. The number of steps s is increased to $s = 6$ (structural units c–g), 7 (h), 8 (i–l), 9 (m), 10 (n) and 12 (o). The maximum angles $\alpha_i \approx 142°$ and 158° alternate in structural unit o. Closed loops are only possible for pairs of α_i with $\alpha_1 + \alpha_2 = 120°$, 180°, 240° or 300° for $s = 3$–12 steps and an additional $\alpha_i = 60°$ ($s = 3, 5$) or 120° ($s = 7, 9$) for odd s values (Table 1). The angles α_i satisfy for each loop $\Sigma\ \alpha_i = 180s - 360$. The different types of circle packings are obtained by combinations of structural units (Figs. 1 and 2). They are characterised in the present investigation by the numbers T_4 of contacts to neighbouring circles ($T_1 - T_3 = 0$ as outlined before) and the ratio r = y/x of y vacant sites and x occupied sites of the triangular net. The T_4 and r values of each circle packing correspond to a point (●) of the structure map with T_4 and r values as parameters (Fig. 3). The r values are related to the density ρ of the circle packing [5] by $\rho(r+1) = 3.5\pi/\sqrt{3} \approx 6.35$. The maximum and minimum values of the present circle packings amount to $\rho = 0.907$ for 0 0 0 6; 6 and $\rho = 0.393$ for 0 0 0 3; 15.17. A lower value would be obtained, if the structural units o could be assembled two-dimensionally with less dense units than the triangle a. The r = y/x values of structural units a–o can be obtained from the numbers y = 3–85 of vacant positions (B = ○) and the total amount of circles (A = ●) within the unit (Table 1). The densities of the structural units decrease from 0.907 (a) to 0.353 (o) (Table 1). The values of homogeneous circle packings are identical for combinations of a and g and deviate by ≈1% for other units [5].

The structural units a, b, d, f and g can be combined two-dimensionally for complete coverage of the plane. Other circle packings are obtained by combinations of two or more structural units (Table 1). The structural units a, b and g, for example, can be combined to 0 0 0 5; 6.5a,b (Fig. 1), 0 0 0 4; 7.7, 0 0 0 4; 8.33 and 0 0 0 5; 7.17 (Table 1) which are inside the dashed triangle of the structure map (Fig. 3). There are two relations for the numbers of circles x_i, apices of structural units s_i and contacts between circles T_i:

$$\Sigma x_i = \Sigma s_i / T_4 \tag{1}$$

$$x_i = 0.5 s_i - 1 \tag{2}$$

The unit cell of the 0 0 0 5; 6.5a structure as an example (Fig. 1) contains two distorted squares ($s = 4$), four triangles ($s = 3$) and $x = 4$ circles (20/5 = 4). The second equation gives the fraction of circles ($x = 0.5$ or 1) inside the triangle or distorted square. The combinations

Table 1 Structural units with x A and y B positions, sequences of α_i values, T_4; r = y/x values for circle packings, densities ρ, open structures (*) with $T_4 < 3$ and a minimum of A′ (○) positions

Struct. units	x	y	Sequences of α_i in °	T_4; r	ρ
a	0.5	3	$(60)_3$	6; 6	0.907
a_4b_2	4	26		5; 6.5a	0.846
a_2 b	2	13		5; 6.5b	0.846
b	1	7	$(82,98)_2$	4; 7	0.794
a_8 g	6	43		5; 7.17	0.777
a_2 b_3 g	6	46		4; 7.7	0.732
c	2	16	$(82,158)_3$		0.705
d	2	16	$(82,158,120)_2$	3; 8	0.705
a_2 g	3	25		4; 8.33	0.680
g c_2	6	51		3; 8.5	0.668
e	2	18	$(98,142)_3$		0.635
f	2	18	$(98,142,120)_2$	3; 9	0.635
g	2	19	$(120)_6$	3;9.5	0.605
h	2.5	24	(98,142,158,82,158, 120,142)		0.599
i	3	30	$(82,158,142,158)_2$		0.577
b k	4	41		3; 10.25	0.564
b l	4	42		3; 10.5	0.552
j	3	32	$(98,142,158,142)_2$		0.544
k	3	34	$(142,158,120,120)_2$		0.515
l	3	35	$(142,120,158,120)_2$		0.501
b_3 g_2 o	12	144		3; 12	0.488
m	3.5	45	$(120,142,158)_3$		0.458
a_8 *	2	26		2; 13	0.453
n	4	56	$(120,142,158,142, 158)_2$		0.423
b_2 *	1	15		2; 15	0.397
b_2 *	1	15		0; 15	0.397
a_2o	6	91		3; 15.17	0.393
o	5	85	$(142,158)_6$		0.353
a_6 *	1	20		0; 20	0.302

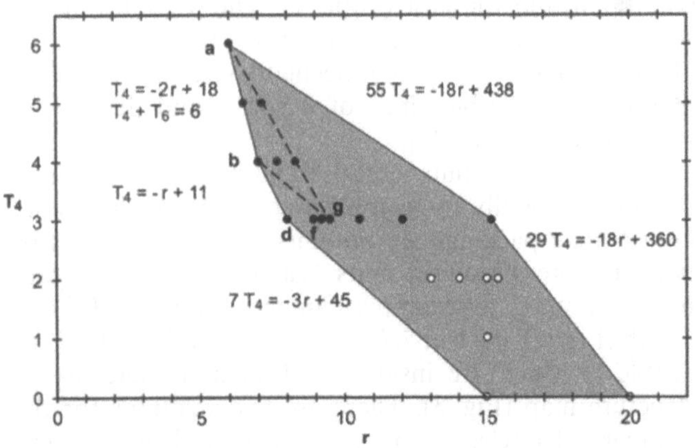

Fig. 3 Structure map of circle packings (●) with contact numbers T_4 and ratio r = y/x as parameters, different linear borders and structural units a–g (Fig. 2)

of structural units for certain T_4 values like $T_4 = 5$ are restricted by the sum of the five values $\alpha_1 + \alpha_2 + \alpha_3 + \alpha_4 + \alpha_5 = 360°$ (Fig. 1). The 0 0 0 5; 6.5a and b structures are obtained from different combinations of triangles and distorted squares.

Non-Periodic circle packings and clusters

The structural units a and b can also be assembled in non-periodic Fibonacci sequences with stepwise substitutions a→a b and b→a like a b b a b in Fig. 4. The majority of circles has $T_4 = 5$ contacts. One circle of the 5 × 5 sequence (* in Fig. 4) has $T_4 = 6$ (or $T_4 = 4$) contacts. The deviation of T_4 from $T_4 = 5$ and the deviation from the composition y/x = 6.5 depend on $\tau = (1+\sqrt{5})/2$ by $1/\tau^6$ and $1/(2\tau^6)$, respectively ($1/\tau^6 = 9 - 4\sqrt{5} \approx 0.0557$). Other non-periodic patterns can be obtained by clusters containing c = 1 circle with $T_4 = 6$ in the middle and six other circles with $T_4 = 5$ outside. The orientation of the neighbouring shells can be different like in 0 0 0 5.08; 6.46a–d (Fig. 5). These structures with the same distances between $T_4 = 6$ circles (⊙) can be combined in non-periodic patterns. The numbers of circles of the j-th shell increase by 6 in the $j+1$-th shell in 0 0 0 5.08; 6.46a. The same applies for $c = 2, 3, 4, \ldots, 12$ like $c = 3$ or 4 circles of structural unit a or b in the centre (Fig. 6). The clusters get unstable for $c > 12$. The total number x of a cluster increases by $x = 3j^2 + (3+c)j + c (c > 1)$ and $x = 3j^2 + 3j + 1$ $(c = 1)$.

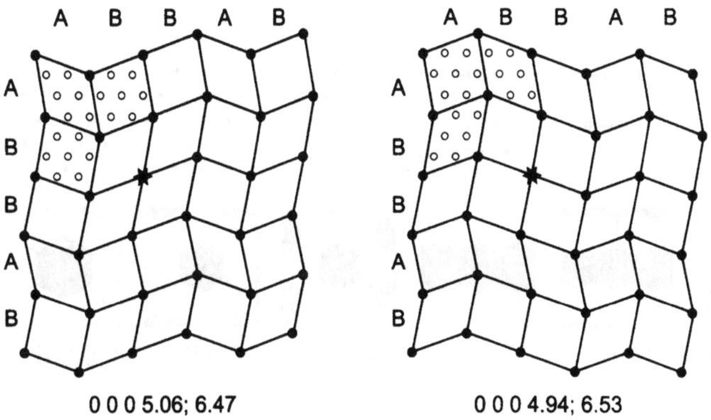

Fig. 4 Part of an aperiodic Fibonacci tiling of structural units a and b in the sequences a b b a b or b a a b a in two directions (5 × 5)

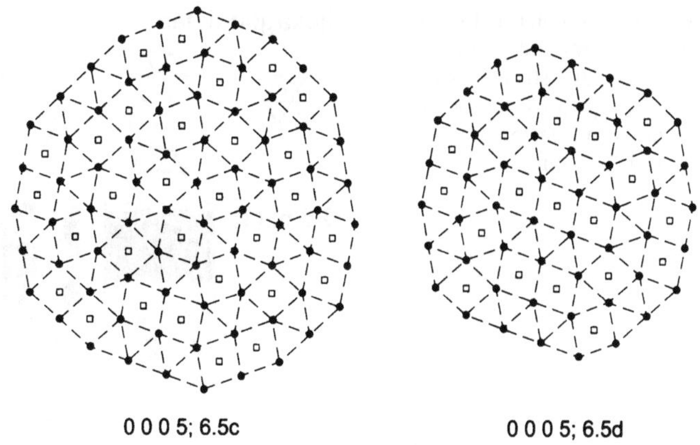

Fig. 6 Flower like patterns containing the structural unit a or b in the centre

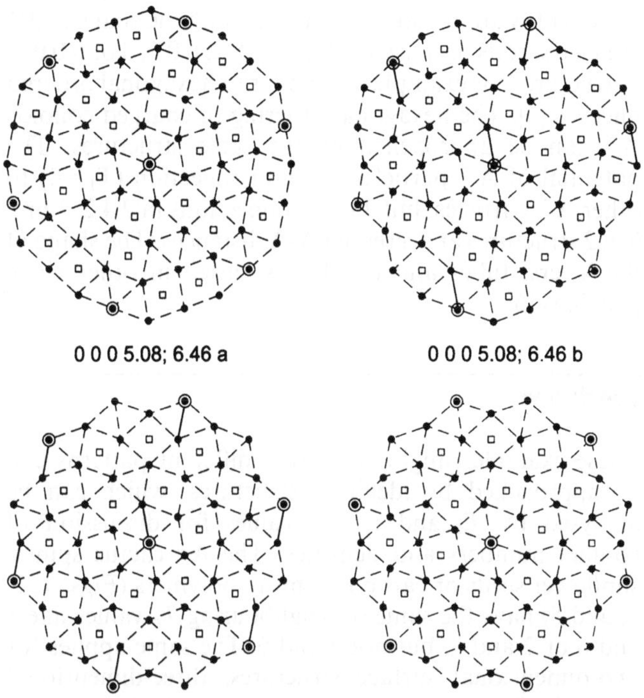

Fig. 5 Combinations of structural units a and b to flower like patterns containing circles with 5 (●) or 6 (⊙) neighbours at dashed lines (some solid lines indicating different directions)

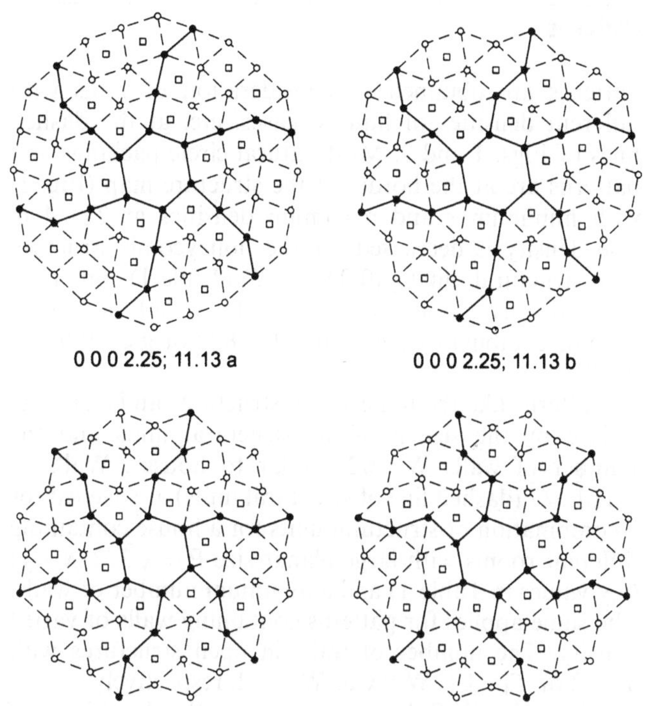

Fig. 7 Patterns similar to Fig. 5 for star polymer-like structures containing A = carbon atoms (●) with three or two A neighbours

Polymers

Part of the A positions can be occupied by A′ to approach two-dimensional star polymers like structures with A = carbon and A′ = hydrogen atoms as example (Fig. 7). The T_4 values are decreased further to $T_4 = 2$ in polymers or $T_4 = 0$ (Fig. 1). The two-dimensional polymers are assembled from different structural units like a and b (Fig. 1) with $T_4 = 2$ contacts to neighbouring A positions. The A′ positions (○) are vacant or occupied by H atoms. The y values of Table 1 are increased for these structures.

The angle $\alpha \approx 82°$ of the distorted square is increased to 92°, 98°, 102°, ... as the number l of straight steps increases to $l = 3$, 4, 5, etc. The angle $\alpha = 120°$ of the 0 0 0 6; 6 structure is approached for $l \to \infty$ in

$$-(l^2 + l + 1)2\cos\alpha = l^2 - 2l - 2. \qquad (3)$$

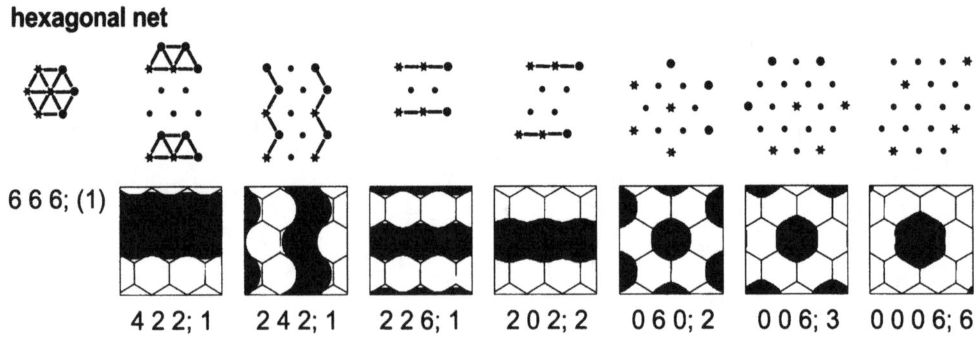

Fig. 8 Hexagonal net of A particles (●) and all homogeneous A_x A'_y structures with identical positive (*black*) or negative (*white*) potential of A and A′ atoms

The angle $\alpha = 90°$ is obtained for $l = 1 + \sqrt{3}$ with homogeneous circle packings (Table 1, except 0 0 0 3; 10.5). The number of different angles is reduced to four (60°, 90°, 120°, 150°).

Patterns

Patterns are obtained by consideration of walls with different densities of liquids, gases, etc. at the dashed lines of Figs. 1 and 2. Most natural circle packings and patterns are at the border of the structure map (Fig. 3) with minimum r and maximum densities at $T_4 = 4$–6. The density is decreased for the honeycomb structure with structural units g (0 0 0 3; 9.5) (Table 1) compared to the distorted honeycombs (d, 0 0 0 3; 8a,b or f, 0 0 0 3; 9) or combinations g c_2 (0 0 0 3; 8.5) or g e_2 (0 0 0 3; 9.17).

Patterns like the honeycomb structure can be characterised by the size P of the structural units and the number of walls $W = s/2$ similar to houses. Walls of length $\sqrt{7}$ (dashed line of structural units) are needed for a combination of structural units for a house containing different rooms with large planar size $P = \sqrt{3}/2\,(x + y)$ (x,y values in Table 1) and a minimum number of walls. The same applies for patterns containing walls of varied density. The number of walls in open structures with $T_4 = 2$ or $T_4 = 1$ is $W = x$ or $W = x/2$, respectively.

The ratios W/P decrease similar as the densities ρ of circles with diameter $\sqrt{7}$ (Table 1). The honeycomb structure (0 0 0 3; 9.5) of convection patterns and chemical Turing patterns [4] can change to waves (0 0 0 2; 15) or distorted honeycombs (0 0 0 3; 8) with decreased or increased density (Table 1).

Surfactants and colloids

The surfactant structures are approached by the zero-potential surfaces of A and A′ atoms with positive and negative charges, respectively. The 6; 6 structure for example (first structure of Table 1) corresponds to a hexagonal net of A and A′ atoms with a maximum of $T_1 = T_2 = T_3 = 6$ nearest, next-nearest and third neighbours. The numbers are reduced to 4 2 2, 2 4 2 or 2 2 6 for the ratio $A'/A = 1$ and other values at $A'/A = 2$, 3 or 6 (Fig. 8). A and A′ atoms are in the centre of the hexagons (Dirichlet domains). The positive potential (black) of A atoms can be compared with the shape of 2D lamellars, rods or micelles. The 0 6 0; 2, 0 0 6, 3 or 0 0 0 6; 6 structures for example are similar to hexagonal ordered micelles [4]. The hexagonal ordering of charged colloids A^+ can be approached by the same structures. The potential of A^+ particles has no connection ($T_1 = 0$) to other A^+ particles, but $T_1 = 2$ connections in 2 4 2; 1 or 2 0 2; 2 structures at decreased A-A distances. The shape of the layers, tubes and micelles is varied in other circle packings [6].

Conclusion

These examples might show that different patterns can be approached by idealised structures, which can be analysed by T_4 and $r = y/x$ values for discussion of contact numbers and densities. The A = cactus spines, sunflower seeds or carbon atoms in polymers (Figs. 1, 2, 4, and 5) have the same or neighbouring T_4 values like 5 and 6 or 2 and 3, but not 2 and 6. The same applies for two-dimensional surface structures, three-dimensional ordered alloys or magnetic ordering [7]. A relatively small number of structures is at the corners or edges of the structure maps (Fig. 3) compared to the large numbers of other structures. Maximum densities are obtained by combination of distorted squares and triangles ($4 \leq T_4 \leq 6$) or the distorted hexagons of structural unit d ($3 \leq T_4 \leq 4$). The combination of structural units a and o (a_2 o in the 0 0 0 3;15.17 structure, Table 1) has the lowest density of a circle packing. Very complex patterns containing hexagonal, dodecagonal and other voids are obtained for other T_4 values. Similar patterns were observed for colloids at increased charge numbers [8]. The honeycomb structure has the lowest density for combinations of single structural unit g. The large number of possible structures was outlined for different computer simulation methods of long polymer chains [9]. Clusters of circles like those shown in Fig. 5 and 6 were obtained by variation of interaction parameters in a special billiards simulation method after about 10^6 collisions [10]. The structures at the left border of the structure map (Fig. 3) are probably approached by variation of the interaction parameters.

References

1. Ben-Jacob E, Cohen I, Shochet O, Aranson I, Levine H, Tsimring L (1995) Nature 373:566
2. Bubeck R, Neser S, Bechinger C, Leiderer P (1998) Progr Colloid Polym Sci 110:41
3. Möhwald H, Dahmen U, Meijere K de, Brezesinski G (1998) Progr Colloid Polym Sci 109:3
4. Ball P (1999) The self-made tapestry: pattern formation in nature. Oxford University Press, Oxford
5. Koch E, Fischer W (1992) Sphere packings and packings of ellipsoids. In: Wilson AJC (ed), Intern tables of crystallogr, vol C. Kluver, Dordrecht, pp 654–659
6. Hauck J, Mika K (2002) Z Phys Chem 216:1281
7. Hauck J, Mika K (2000) Progr Solid State Chem 28:1
8. Ito K, Yoshida H, Ise N (1994) Science 263:66
9. Binder K (1992) In: Bicerano I (ed.) Computational modeling of polymers. Marcel Dekker, New York, p 221
10. Lubachevsky BD, Graham RL (1997) Discrete Comput Geom 18:179

Progr Colloid Polym Sci (2004) 123: 104–109
DOI 10.1007/b11651

Aonghus Lawlor
Gavin D. McCullagh
Emanuela Zaccarelli
Giuseppe Foffi
Kenneth A. Dawson

Interactions in systems with short-range attractions and applications to protein crystallisation

A. Lawlor (✉) · G.D. McCullagh
E. Zaccarelli · G. Foffi · K.A. Dawson
Irish Centre for Colloid Science and Biomaterials, Department of Chemistry, University College Dublin, Belfield, Dublin 4, Ireland
e-mail: aonghus@fiachra.ucd.ie

Abstract The problem of phase behaviour of solutions of globular proteins is approached by means of a Hard Core Yukawa fluid model with short-ranged attractions. We have determined the phase behaviour of this model system for different well widths, using a variety of high quality methods. The essential phase behaviour of systems with short-ranged attractions is reproduced. The typical phase behaviour of solutions of globular proteins is well represented by the Hard Core Yukawa fluid with short-ranged attractions. The formation of amorphous precipitates in protein crystal growth experiments is identified on our model phase diagram with the formation of an attractive glass. As the range of attraction is decreased, the formation of an attractive glass state dominates the phase behaviour in the meta-stable regime above the critical point. We show how the addition of a long-ranged attraction, in the Two-Yukawa model, has the effect of eliminating the formation of attractive glass, while preserving the equilibrium features of the short-ranged attractive system. This opens the possibility of using attractive long-range interactions to avoid the formation of an attractive glass state.

Introduction

In recent years, rapid developments in molecular biology now mean that one can synthesise any protein relatively cheaply and easily. However, although the sequence of such a protein – natural or synthetic – can be determined (or in the case of synthesis pre-determined), obtaining the structure of its native or folded state is a far less trivial matter. It has been shown by various techniques including NMR studies that the crystallised protein has in general the same structure as the solute protein [1].

At present the most common technique for this investigation is X-ray crystallography of the folded protein. At present, the structures of less than 10% of known proteins have been determined. The bottleneck in this process is the difficulty of producing good quality crystals of proteins in their native state. We attempt to model globular proteins, apparently the simplest case.

One of the most difficult aspects of this field is the many varied experimental observations associated with attempts, successful or otherwise, to crystallise proteins. These include the presence of a "cloud-point" which is meta-stable with respect to the equilibrium solubility curve [2], Hofmeister's empirical series to determine which metal ion to use in 'salting out' crystals [3, 4], enhancement of crystallisation by addition of different molecular weight polymers [5–7], a strong tendency to form amorphous gel-like solid structures, often at surprisingly low density, which can inhibit [8] or prevent crystal growth [2].

A system consisting of relatively large solute particles dissolved in a solvent behaves analogously to a monatomic substance (e.g., a noble gas). There are distinctive phases of very low concentration (gas) and very high concentration (liquid) with fluid-like structure, as well as a solid regular array (crystalline) phase. Liquid-gas coexistence and a critical point as well as solid-liquid coexistence are observed. The simplest such system is the so-called "hard-sphere" system, consisting of rigid particles which only interact with each other on contact. Here, only two phases are observed: crystal and fluid. Only one distinct fluid phase exists. This reflects the driving force for liquid-gas phase separation. If one considers a similar system with an added attraction at high temperature, the same homogeneous fluid exists. As one lowers the temperature however, the attractive force becomes more important. A situation then occurs where the lowest free energy for the system may be the gas phase driven by optimisation of the entropy (for low densities), exclusively the liquid phase driven by optimisation of enthalpy of attractions (for high densities), or a two phase liquid-gas coexistence that optimises the free energy by separately maximising the entropy (gas) and minimising the enthalpy (liquid) contributions to the free energy. The critical point is the first sign of incipient coexistence, a single point on the temperature-density phase diagram where the liquid and gas phases have identical chemical potential and pressure.

In the case of and proteins, X-ray scattering experiments have shown that the range of the attraction is small relative to the size of the particle [for lysozyme on the order of 8% [9] (the diameter of lysozyme was fixed at 32.4 Å and the width of the attractive part of the potential was found to be 3 Å)]. Such short-ranged attractive systems, which we will discuss here, display markedly different characteristics to the usual gas (dilute-solution), liquid (concentrated solution) and crystal systems.

With this in mind, we propose a simple model of hard spheres interacting through a short-ranged Yukawa potential, which can reproduce the main features of typical protein phase diagrams [10]. We and others have recently shown that one of the most prominent features of short-ranged attractive systems, independent of the details of the potential, is attractive glass formation [11–14]. Glasses (solids with a liquid-like structure) have traditionally formed at high density driven purely by close packed repulsions. This state we call the "repulsive glass". However, when the inter-particle attractive potential is narrow (as it is in our model) a new type of "attractive glass" may form at lower density. The lifetime of the attractive glass is typically such that crystal growth may not occur over experimental time-scales.

Naturally, the attractive glass only forms at low temperatures, where the thermal energy of the particles is small compared to the attractive well. One point is that, in addition, such a glass forms more easily when the range of the potential narrows. However, proteins denature at even moderately high temperature, so a pertinent question is whether one can reach a high enough temperature to avoid the glass without denaturing the protein.

We first discuss the methods we have used to compute the phase diagrams. We then calculate the equilibrium phase diagrams for different well widths and overlay the glass lines, showing at what temperature and density the glass forms. We aim to identify the different widths of attraction for which protein crystallisation may be feasible.

Methods

The Hard Core Yukawa fluid is described by the following inter-particle potential

$$v(r) = \begin{cases} \infty & r < \sigma \\ -A_0 \frac{e^{-b(r-\sigma)}}{r/\sigma} & r \geq \sigma \end{cases} \quad (1)$$

The parameter b determines the range of the interaction – larger b values correspond to shorter ranges of attraction. The parameter A_0 defines the energy scale. In this paper we set the particle diameter $\sigma = 1$ and $A_0 = 1$, so that the range b is in units of hard core diameter and the temperature is in units of energy.

We computed the phase diagram for the Hard Core Yukawa fluid using a variety of methods. We use a self consistent approximation described below to calculate the properties of the liquid, gas and fluid phases, and perturbation theory for the crystal phase, locating the phase boundaries by imposing the standard conditions on the chemical potential μ and pressure P,

$$\mu_{solid} = \mu_{fluid} \quad (2)$$

$$P_{solid} = P_{fluid} \quad (3)$$

The glass lines, as calculated from Mode Coupling Theory, are overlaid on the phase diagram. We now briefly describe the methods of the Self Consistent approximation, perturbation theory and Mode Coupling Theory.

Self Consistent Ornstein-Zernike Approximation (SCOZA)

The Ornstein-Zernike (OZ) equation for the pair correlation function $h(r)$ is

$$h(r) = c(r) + \rho \int d\mathbf{r}' c(|\mathbf{r} - \mathbf{r}'|) h(|\mathbf{r}'|) \quad (4)$$

where $g(r) = h(r) + 1$ is the radial distribution function and $c(r)$ the direct correlation function [15].

The Self Consistent Ornstein-Zernike Approximation (SCOZA) is designed to provide a closure to the OZ equation. Various approximations exist (e.g., mean spherical approximation, Percus-Yevick approximation, hyper-netted chain approximation, etc.) which relate the direct correlation function $c(r)$ to the potential with one or more state dependent parameters that can be adjusted to satisfy various exact thermodynamic relations [16].

The Self Consistent approximation alluded to here uses a Yukawa function to account for the contribution to the direct correlation function $c(r)$ due to the hard-core repulsion. Two thermodynamic conditions are imposed: one requires that the compressibility and the virial route to the thermodynamics lead to the same system properties – namely the Carnahan-Starling equation of state for the hard-sphere fluid; the other condition requires that the energy and compressibility routes yield the same result. These two conditions provide a means to calculate the thermodynamics in a self-consistent manner. Using the Yukawa interaction potential we can establish relations that allow us to compute all the thermodynamic properties of the system [16, 17].

The application of SCOZA to an attractive hard-core Yukawa fluid provides a semi-analytic calculation of the thermodynamic properties of the fluid, liquid and gas states that arise in systems interacting through short-range forces [16]. SCOZA has been applied to Yukawa systems with relatively large values of the range of the potential, with a satisfactory reproduction of the liquid-vapour binodal curves and a good description of the critical point region. The accuracy of SCOZA is maintained at high density and low temperature, in contrast to other methods. In fact, the agreement of SCOZA for narrower wells with recent Gibbs Ensemble Monte-Carlo simulations [18] are very good [19], giving us confidence in the accuracy of the phase diagrams calculated by the SCOZA method for the well widths that we present.

Perturbation Theory

Perturbation Theory is employed to calculate the phase behaviour of the crystalline phase following the lead of Gast et al. [20, 21]. At the simplest level this theory makes the assumption that one can divide the free energy of the system up into the contribution caused by the hard core part of the potential and a perturbation composed of an attractive tail as follows,

$$v(r) = v_0(r) + v_{att}(r)a \qquad (5)$$

The hard core being the greatest contribution to the potential is the zero order or "reference" free energy. The attractive tail is the perturbation. This amounts to a series of corrections to the reference where each subsequent term in the series is hopefully smaller than the previous one. Unfortunately even to calculate the second order term in this expansion is extremely demanding. However, Barker and Henderson [22] have developed a very accurate approximation to the second order term which we have used:

$$\frac{\beta F}{N} = \frac{\beta F_0}{N} + \frac{\beta\rho}{2}\int v_{att}(r)_0(r)d\mathbf{r} - \frac{\beta\rho}{4}\left(\frac{\partial\rho}{\partial P}\right)_0\int v_{att}^2(r)g_0(r)d\mathbf{r} \qquad (6)$$

The free energy of the reference system is calculated by thermodynamic integration in the packing fraction using an equation of state for the hard-sphere FCC solid proposed by Hall [23].

The perturbation terms were evaluated using an analytic form of the pair distribution function $g_0(r)$ for the hard-sphere FCC solid proposed by Kincaid and Weiss [24]. Once we have the Helmholtz free energy we can calculate the Gibbs free energy and hence the pressure.

For extremely narrow well widths a solid-solid transition occurs analogous to the liquid-gas binodal. The corresponding solid-solid critical point compares with quantitative accuracy to Monte-Carlo simulations of the Hard Core Yukawa fluid [25, 26]. From this evidence we are confident of this method's accuracy.

Mode Coupling Theory (MCT)

MCT describes the transition of super-cooled liquids to a non-ergodic state [27]. The transition of the super-cooled liquid to the glass state represents a critical slowing down of the particle motions, leading to structural arrest. A characteristic property of the arrested state is that it has the static structure of a liquid. Apart from the parameters describing the microscopic motion, the static structure factor $S(q)$ is the only input to MCT, which aims to give a complete description of the dynamic properties of the system.

MCT has been successfully applied to the study of certain aspects of the arrest transitions of colloidal particles [28, 29], and details of time correlation functions are well reproduced [30]. MCT is expected to provide a good description of the principal phenomena of glass formation in the regions of the phase diagram we are most interested in.

To achieve a high quality description of the arrest transition a good quality static structure factor is required. The SCOZA method we have used provides us with many thermodynamic properties of the system. We use the direct correlation function $c(r)$ as produced by SCOZA to construct the structure factor $S(q)$. The structure factor calculated in this way is expected to be highly accurate. With this structure factor input we solve the MCT equations to determine the final dynamic state of the system.

Results

Yukawa $b = 5.0$

The Yukawa $b = 5.0$, corresponds to an attractive range of about 12% of the particle radius, if we consider the range at half the potential depth (Fig. 1). The phase diagram displays all the features of a conventional Van der Waals type phase diagram. Three distinct phases exist, solid, liquid and gas. At high temperatures only a single fluid and the usual solid phase exists. As the temperature is lowered past the critical point C_p, two distinct fluids become apparent, one more (liquid) and one less (gas) dense. These two phases coexist at moderate densities. As the temperature lowers further a unique triple point T_p is reached where all three phases solid, liquid and gas coexist.

With the usual equilibrium lines, we have shown also the MCT glass transition line, beyond which the system may no longer reach equilibrium. For this range of attraction, we see only the presence of a repulsive glass. The formation of this high density repulsive glass is almost entirely temperature independent. For this well width there is almost no attractive glass formation.

Yukawa $b = 30.0$

The $b = 30.0$ well corresponds to an attractive range of about 3% of the particle radius. The first thing we observe is that the critical point has become meta-stable with respect to the solubility curve. In such a situation no triple point can exist as only two equilibrium phases are ever possible. This phase diagram, typical of short-ranged attractions, is similar to that observed for the proteins γ_{II}-crystallin [31] and lysozyme [2]. We have labelled three zones *I*, *II* and *III* in Fig. 2 in accordance with the ideas of Muschol and Rosenberger [2], and we believe the microscopic picture offered underlies the corresponding zones in their schematic phase diagram. Note that zone *I* is the regime where one can obtain good crystals.

On decreasing the temperature, the narrower range of attraction causes a low density, attractive glass to form. The attractive glass line lies at a higher temperature than the liquid-gas coexistence for a wide range of densities.

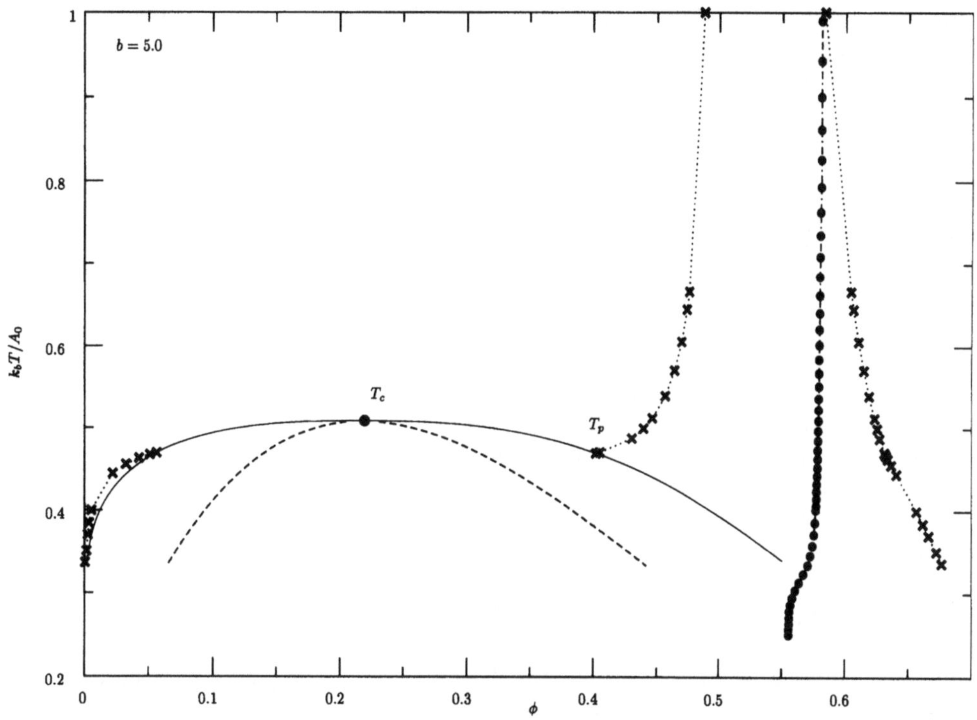

Fig. 1 Yukawa $b = 5.0$. The × denotes the fluid-solid coexistence and • denotes the glass line

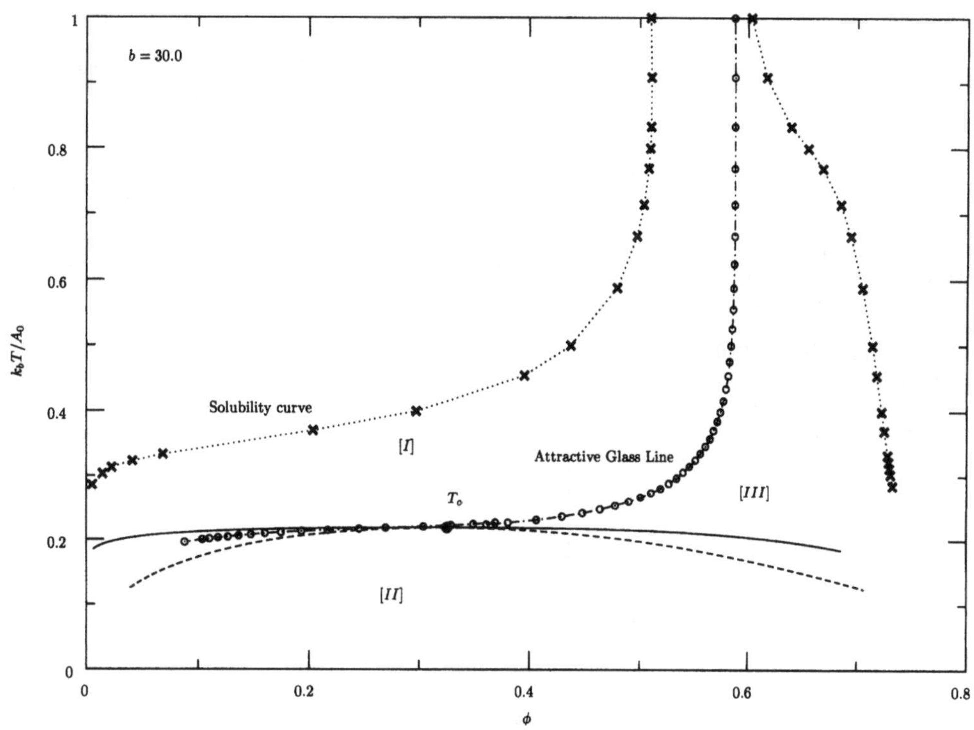

Fig. 2 Yukawa $b = 30.0$. The label are the same as Fig. 1. Zone *I* is the region where one can obtain good crystals. Zone *II* is the meta-stable region of fluid-fluid phase separation and zone *III* corresponds to the region of attractive glass formation. The attractive glass line has been rescaled in density from 52% so that it agrees with the actual density of the hard sphere arrest transition at 58%

Changing the range of attraction clearly alters both the equilibrium phase behaviour and the tendency to glass formation. To appreciate the significance of this on the phase behaviour, consider quenching down below the solubility curve towards the critical temperature in Fig. 2. It is clear that the range of temperatures in which equilibrium can be achieved is now limited by the presence of the attractive glass. This situation becomes more extreme as the well is narrowed further [19] with the attractive glass dominating at still lower densities. If the quench goes below this glass line, we expect to see the formation of various types of gel. These non-equilibrium structures may be long-lived and can thus prohibit or delay crystal formation beyond experimental time-scales. This sort of behaviour is commonly observed in protein crystal growth experiments where amorphous precipitates interfere with crystallisation.

Two-Yukawa $b_1=30.0$, $b_2=5.0$

As stated previously, the protein phase diagram is extremely sensitive to the addition of monovalent salts. The Hofmeister series describes the relative efficiency of different ions on protein crystal growth. Protein crystallisation can also be enhanced by the addition of various polymers such as polyethylene glycol. The addition of the polymer is associated with depletion interactions. Both of these cases show that modifying the range and nature of the interactions lead to important changes in the phase behaviour. Without investigating the origin of these interactions, we chose to study a modification of our model, by including a longer ranged attraction to our short-ranged Yukawa Hard Core fluid [32].

Starting from our previous short-ranged $b_1=30$ well we added the longer-ranged $b_2=5$ well

$$v(r)=\begin{cases}\infty & r<\sigma\\ -\frac{1}{A_1+A_2}\left[A_1\frac{e^{-b_1(r-\sigma)}}{r/\sigma}+A_2\frac{e^{-b_2(r-\sigma)}}{r/\sigma}\right] & r\geq\sigma\end{cases}\qquad(7)$$

constraining the overall strength of attractions to be the same as the One-Yukawa model $A_1+A_2=1$.

In Fig. 3 we see that the phase behaviour displays interesting features of both short- and long-ranged attractive systems. Like the $b=30$ case the critical point is meta-stable with respect to the solubility curve. Apart from a minor shift in the temperature and density, the equilibrium phase behaviour of the Two-Yukawa system is essentially the same as the One-Yukawa system.

More importantly, the attractive glass line, which dominates above the critical temperature of the $b=30$ model, is no longer present and we see only the repulsive glass line at high density. The effect of adding a long-ranged attraction is to diminish the role of glass formation in the phase behaviour. The role of long-range attractions in lifting the meta-stable critical point out of the region where glass formation dominates has been investigated before [32]. This is most interesting. It implies that zone *I* in Fig. 3 can be considerably extended by moving the glass curve.

Conclusions

We have presented the phase diagrams and glass lines of a model hard-core Yukawa fluid for various ranges of attraction, which we have computed using a variety of high quality methods. The resulting phase diagrams are consistent with simulated phase diagrams of the Hard Core Yukawa fluid [33] and bear striking resemblance to the reported phase diagrams of globular proteins [2, 31].

We find that when the range of attraction becomes smaller in comparison with the particle size, the attractive glass dominates an increasingly larger region of the phase diagram above the metastable critical point. The

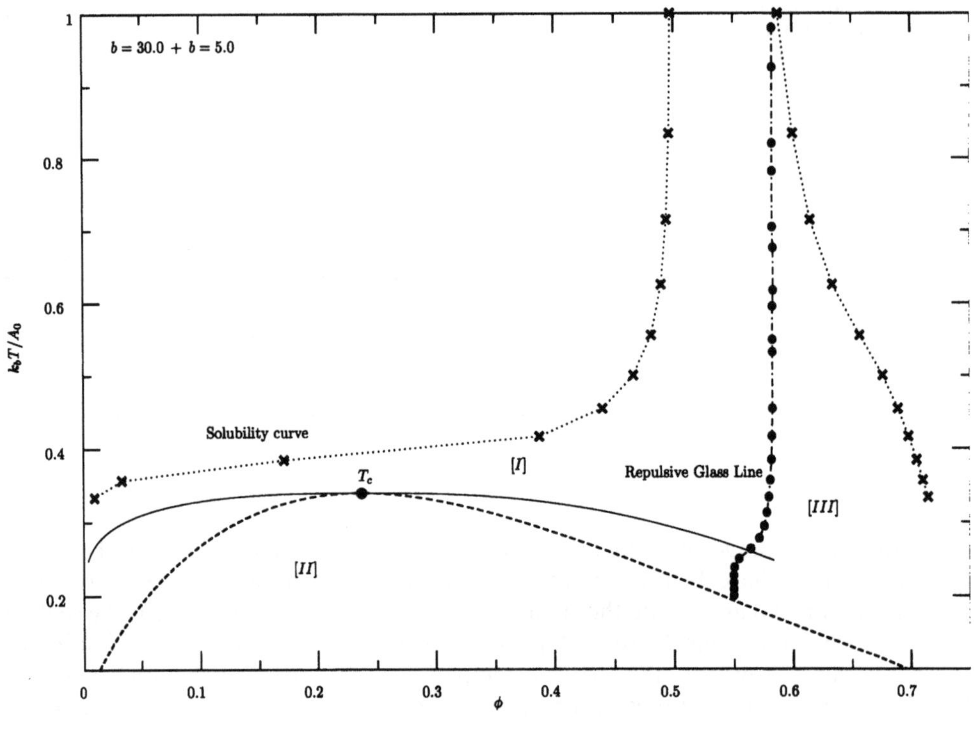

Fig. 3 Yukawa $b_1=30.0+b_2=5.0$. The labels and zones are as in Fig. 2. The important difference from the $b=30.0$ model is zone *I* has been considerably extended at the expense of the attractive glass (zone *III*)

formation of the attractive glass is responsible for many of the problems associated with crystallising such systems (proteins, colloids). By decreasing the width of the well we find that the region of the phase diagram where crystals can grow is steadily reduced in favour of the formation of a glassy state. This behaviour is seen in real protein crystallisation experiments where amorphous precipitates interfere with crystal growth.

We have developed highly accurate methods for calculating the phase diagram of short-ranged attractive systems, allowing us to identify the regions of the phase diagram where the formation of the attractive glass can be avoided. It should now be possible to identify the optimal conditions for crystal growth in these systems.

The Two-Yukawa model can combine multiple interactions of varying strengths, from which we can accurately calculate phase behaviour (coexistence and glass lines).

We have studied the Two-Yukawa model as simple extension to account for long-ranged interactions. In the Two-Yukawa model we find phase behaviour similar to that of the short-ranged attractive system. However, attractive glass formation does not occur. At high densities a temperature independent repulsive glass does form, but the absence of the attractive glass at lower densities suggests a role for long-ranged attractions in enhancing crystal growth by avoiding the formation of glass state. The purpose of this work is to accurately model the phase behaviour of short-ranged attractive systems, and to extend this model to investigate the role of longer-ranged interactions, which might represent for example a depletion potential. Work is in progress to study the Two-Yukawa model in more detail.

Acknowledgements This work is supported by COST P1.

References

1. Fersht A (1999) Structure and mechanism in protein science. Freeman
2. Muschol M, Rosenberger F (1997) J Chem Phys 107:1953
3. Ducruix A, , Guilloteau J, Ries-Kautt M, Tardieu A (1996) J Crystal Growth 168:28
4. Piazza R, Pierno M (2000) J Phys Condens Matter 12:443
5. Kulkarni A, Chatterjee A, Schweizer K, Zukoski C (1999) Phys Rev Lett 83:4554
6. Degiorgio V, Piazza R, Pietro GD (1996) Prog Colloid Polym Sci 100:210
7. Broide M, Tominc T, Saxowsky M (1996) Phys Rev E 53:6325
8. Illet S, Orrock A, Poon W, Pusey P (1995) Phys Rev E 51:1344
9. Tardieu A et al. (1999) J Crystal Growth 196:193
10. Rosenbaum D, Zamora PC, Zukoski CF (1996) Phys Rev Lett 76:150
11. Dawson K et al. (2001) Phys Rev E 63:011401
12. Foffi G et al. (2000) J Stat Phys 100:363
13. Zaccarelli E et al. (2001) Phys Rev E 63:031501
14. Bergenholtz J, Fuchs M (1999) Phys Rev E 59:5706
15. Hansen J, McDonald (1986) Theory of simple liquids. Academic Press, New York
16. Caccamo C et al. (1999) Phys Rev E 60:5533
17. Caccamo C (1996) Phys Rep 274:1
18. Shukla KP (2000) Phys Rev E 112:10358
19. Foffi G et al. (2001)
20. Gast A, Hall C, Russel W (1983) J Colloid Interface Sci 96:251
21. Gast A, Russell W, Hall C (1986) J Colloid Int Sci 109:161
22. Barker J, Henderson D (1967) J Chem Phys 47:2856
23. Hall K (1972) J Chem Phys 57:2252
24. Kincaid J, Weis J (1977) Mol Phys 34:931
25. Bolhuis P, Hagen M, Frenkel D (1994) Phys Rev E 50:4880
26. Bolhuis P, Frenkel D (1994) Phys Rev Lett 72:2211
27. Bengtzelius U, Götze W, Sjölander A (1984) J Phys C 17:5915
28. Pusey P (1991) Liquids, freezing and the glass transition. North Holland, Amsterdam, p 763
29. Megen W van, Underwood S (1993) Phys Rev Lett 70:2766
30. Götze W (1999) J Phys Condens Matter 11:A1
31. Asherie N, Lomakin A, Benedek GB (1996) Phys Rev Lett 77:4832
32. Noro MG, Kern N, Frenkel D (1999) Europhys Lett 48:332
33. Hagen M, Frenkel D (1994) J Chem Phys 101:4093

Progr Colloid Polym Sci (2004) 123:110–113
DOI 10.1007/b11652

M. Boström
D.R.M. Williams
B.W. Ninham

Specific ion effects: why colloid science has failed to contribute to biology

M. Boström (✉) · D.R.M. Williams
B.W. Ninham
Department of Applied Mathematics,
Research School of Physical Sciences and Engineering, Institute of Advanced Studies,
Australian National University, Canberra,
Australia 0200

Abstract Surface tension of salt solutions and double layer forces between charged interfaces show marked specific ion effects not accommodated by the classical double layer theory for charged interfaces. We will demonstrate how this can be partly remedied when the ionic dispersion potential that acts between ion and interface is included on the same level as the electrostatic potential. We will focus on the important influence of these dispersion forces on ionic distributions near air-water and mica-water interfaces. Dispersion potentials that accommodate the experimental surface tension of KBr also gives the physical reason why Pashley et al. had to postulate a 90% binding of bromide ions to the charged mica surfaces to explain their measured force.

Keywords Specific ion effects · Ionic dispersion potentials · Hofmeister series · DLVO · Surface tension

Introduction

The polarisability of ions in water is different from an equivalent volume of water. Due to the excess polarisability ions will experience dispersion potentials towards or away from an interface.

The purpose of the present work is to demonstrate how these previously ignored dispersion forces offer an obvious way to understand the specific ion effects that haunts biology. We argue that the neglect of these dispersion forces have too long been standing in the way of real progress towards a predictive theory for biocolloid and colloid science.

We will consider two important examples, surface tension of salt-solutions and double layer interaction between mica surfaces across electrolytes. The surface tension of ionic solutions show marked specific ion effects [1] not explained by the existing theories [2]. We recently showed how these effects could be accommodated when we included the ionic dispersion force between ions and the air-water interface [3]. As an example we show in Fig. 1 the calculated surface tension change with added salt for a few different dispersion potentials. The solid line representing our model values for potassium ions may be compared with the experimental value for *KBr*, which is approximately 1.36 10^{-3} J m^3 mol^{-1} litre. Clearly, ion-specific surface tensions can be accommodated. For *KCl* a higher value is found whereas for CH_3COOK the surface tension change is much smaller. The origin of these differences is the difference in ionic dispersion potentials. Bromide ions, with high excess polarisability, are driven away from the interface by repulsive dispersion potentials. Since cations (here K^+) usually experience a different and smaller dispersion potential a double layer is created. This gives rise to a self-consistent potential and also, if two interfaces are close together, a double layer force. The celebrated DLVO theory for the double layer interaction fails in exactly the same way to accommodate Hofmeister effects that are legion in biology. Dubois et al. [4] used the osmotic measurements in an experiment that varied concentration as well as nature of counterion. At 0.1 M there is a decrease of a factor 5 of the pressure when chloride counterions are replaced with bromide for a spacing of 40 Å. At lower concentration, i.e., high dilution, the

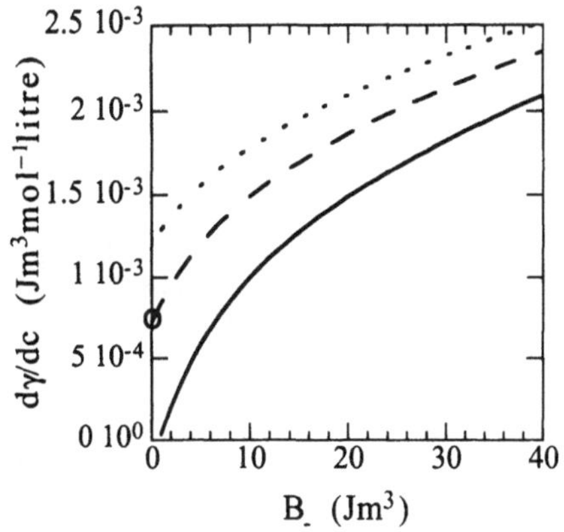

Fig. 1 $d\gamma/dc$ in a 1 M electrolyte as a function of the anion dispersion coefficient B_{-}. Counted from above in the figure, the three different curves correspond to $B_{+} = -5, 0, 5 \times 10^{-50}$ J m^3, counted from below. The solid line corresponds to our model value for potassium. The value obtained without any dispersion interactions (the Onsager-Samaris limit) has been marked with a circle

two equations of state merge for both counterions. In fact, forces can vary in magnitude by more than a factor of 50 by simply changing the counterion from, e.g., bromide to acetate [5]. The problem lies in the separation of forces between particles into double layer and van der Waals forces [6]. We have recently shown that when ionic dispersion potentials are included in the formalism on the same level as electrostatic forces, a new theory emerges which can accommodate specific ion effects [7]. The ionic dispersion potentials have important influence on ion distributions and hence on the double layer force. How important the effect of dispersion potentials are depends on several factors, e.g., on salt concentration and on magnitude and sign of dispersion potentials and surface charge.

At the high salt concentrations relevant for biological systems, say above 0.1 M, the DLVO theory has lost all pretence of predictability. Here electrostatics becomes strongly screened and the ionic dispersion potentials dominate. Indeed, we recently showed that this force might even become attractive at high salt concentrations and low surface potentials (i.e., low surface charge) [7]. Ion specific, or Hofmeister, effects in biological systems [8] have been found in studies of protein stability [9], ion transport [10], cutting efficiency of DNA [11], and thermal stability of nucleic acid helices [12]. Although detailed investigation of ion-DNA interactions may require a microscopic theory [13] we believe, and have recently shown [14], that the much simpler macroscopic model can contribute to our understanding of the physics of specific ion-DNA interactions when dispersion potentials are included.

We will outline the theory in Sect. 2. In Sect. 3 we show that the same excess polarisabilities that can accommodate surface tension of KBr also offers the obvious physical reason why Pashley et al. [5] had to postulate that roughly 90% of the bromide ions were bound to the surface in order to explain the measured force between charged mica surfaces across a 6×10^{-4} M KBr solution. The highly polarisable bromide ions experience an attractive dispersion force, in addition to the electrostatic force, which drives them towards the mica surface. There is no need to postulate that the bromide ions are "bound" to the surface. This highlights the common origin of the ion-specific surface tension of KBr and the very specific observation found in this force measurement. Finally, in Sect. 4 we summarise our results.

Theory

We find the self-consistent potential (ϕ) and ion concentrations ($c_{\pm}$) for 1:1 electrolytes near planar surfaces from a macroscopic mean-field theory by numerically solving the non-linear Poisson-Boltzmann (P-B) equation:

$$\frac{d^2\phi}{dx^2} = -e(c_{+} - c_{-})/\varepsilon_w\varepsilon_0 \tag{1}$$

where the concentrations for the ions are given by

$$c_{\pm} = c_0 \exp\left[-\beta\left(\pm e\phi + U_{image} + U^{\pm}_{dispersion}\right)\right]. \tag{2}$$

Here e is the proton charge, ε_w is the dielectric constant of water (more exactly that of the solution), $\beta = 1/(k_BT)$, k_B is the Boltzmann constant, and T is the temperature. It is vital that image and dispersion potentials that act on ions are included in the formalism on the same level as the electrostatic potential. The ionic dispersion potential and image potential between two interfaces a distance 2 L apart can be written as:

$$U^{\pm}_{dispersion} \approx B_{\pm}\left[x^{-3} + (2\,L - x)^{-3}\right], \tag{3}$$

and

$$U_{image}(\kappa, x, 2\,L) \approx \frac{e^2(S_1 + S_2)}{16\pi\varepsilon_w\varepsilon_0}, \tag{4}$$

where:

$$S_1 = -\ln\left[1 - \Delta^2 \exp(-4\kappa\,L)\right]/L, \tag{5}$$

$$S_2 = \Delta^{-1}\sum_{p=1}^{\infty}\left[\frac{e^{-2\kappa[2\,L(p-1)+x]}}{2\,L(p-1)+x} + \frac{e^{-2\kappa(2\,Lp-x)}}{2\,Lp-x}\right], \tag{6}$$

Here $\Delta \equiv (\varepsilon_w - \varepsilon_{a/m})/(\varepsilon_w + \varepsilon_{a/m}) \approx 1$, with $\varepsilon_{a/m}$ being the dielectric constant of air/mica and $B_{\pm}$ the dispersion coefficient. The inverse Debye length is $\kappa_D = \sqrt{(2\beta e^2 c_0)/(\varepsilon_0\varepsilon_w)}$. To solve the P-B equation we use the two boundary conditions that follows from global charge neutrality: no electric field in the midpoint between two surfaces and the electric field at the closest point of approach (one ion radius) given by $-\sigma/\varepsilon_w\varepsilon_0$. Here we take the ion radius to be the same for the bromide and potassium ions to focus on the effects of dispersion potentials ($r_{ion} = 2$ Å). At an air-water interface we assume that there is no surface charge (σ), at the mica-water interface we use $\sigma = 0.27$ Cm^{-2} for comparison with Ref. [5].

When retardation is neglected we can calculate the dispersion coefficients ($B_{\pm}$) as a sum over imaginary frequencies ($i\omega_n = i2\pi k_BTn\hbar$, where $\hbar$ is Plancks constant) [6]

$$B_\pm = \sum_{n=0}^{\infty} \frac{(2-\delta_{n,0})\alpha^\pm(i\omega_n)\left[\varepsilon_w(i\omega_n)+\varepsilon_{a/m}(i\omega_n)\right]}{4\beta\varepsilon_w(i\omega_n)\left[\varepsilon_w(i\omega_n)+\varepsilon_{a/m}(i\omega_n)\right]} \,. \tag{7}$$

We recently showed that the surface tension of different salts could be accommodated when ionic dispersion potentials were accounted for. For bromide and potassium we assumed that the values were $B_{Br^-} = 21 \times 10^{-50}$ N m^3 and $B_{K^+} = -5 \times 10^{-50}$ N m^3. Of course, the result was at high concentrations to some extent additive for the dispersion contributions, and it is not straightforward to obtain an unique set of dispersion coefficients. However, we keep these values for the purpose of demonstration.

The increment of the refractive index of water when a salt solution is added is different for different salt solutions. The refractive index of pure water is $n_w = 1.3333$, when 5 g/L KBr is added the refractive index increase is $\delta n = 6 \times 10^{-4}$. We know the mass of KBr (119.01 u) and can calculate the added concentration (c_{ion}). The sum of static excess polarisabilities for bromide and potassium can then be estimated from the following approximation

$$\varepsilon(0) \approx \varepsilon_w(0) + 4\pi c_{ion}(\alpha_+ + \alpha_-) \approx (n+\delta n)^2 \,. \tag{8}$$

We find that the sum of static excess polarisability is ≈5 Å^{-3}. We will for the sake of simplicity and demonstration here take the electron affinities (ionic resonance frequencies) to be the same for these ions (ω_0). We model the excess polarisability as

$$\alpha_\pm(i\omega_n) = \alpha_\pm(0)/\left(1+\omega_n^2/\omega_0^2\right) \,.$$

Obviously, within this approximation, and using the B values given above, the separation of the sum becomes trivial. We find that $\alpha_- = 6.6$ Å^{-3} and $\alpha_+ = -1.6$ Å^{-3}. Using the experimental dielectric function for water [15] and solving the expression for $B_\pm$ numerically at an air-water interface we find that $\omega_0 = 2.23 \times 10^{16}$ rad/s. When the air-water interface is replaced with a mica-water interface using the model dielectric function for mica given in [16], and the model excess polarisability, we find that $B^{mica}_{BR^-} = -50 \times 10^{-50}$ N m^3 and $B^{mica}_{K^+} = 12 \times 10^{-50}$ N m^3. These values only give an estimate, but a reasonable one as long as the electron affinities are not too different for bromide and potassium. One should also point out at this stage that the mica surfaces were covered with a 2.7 nm thick bilayer. An investigation beyond this demonstration obviously demands that the dielectric properties of these layers are included when the ionic dispersion potentials are evaluated. There are a large number of uncertainties in this demonstration, but all are quantities that should be readily accessible from experiments. We are now ready to discuss how ionic dispersion potentials influence the ion distribution of potassium and bromide ions near air-water and mica-water interfaces.

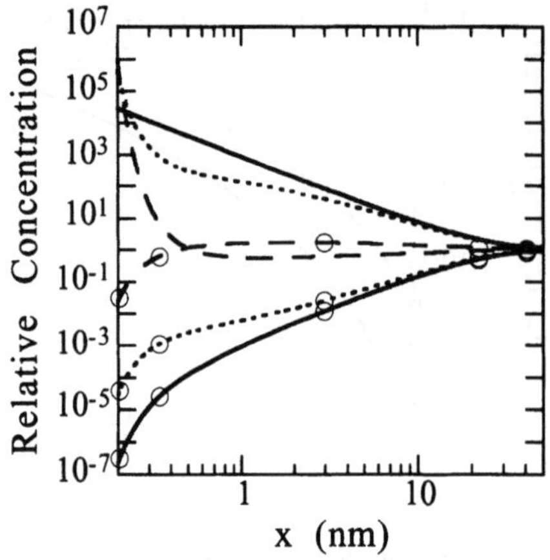

Fig. 2 Ion distribution (c/c_0) of a 6×10^{-4} M KBr salt solution between charged mica surfaces ($\sigma = 0.27$ Cm^{-2}) 100 nm apart. Images potentials are included and $B_+ = 12$. We compare different ionic dispersion potentials acting on the counterions: 0 (*solid line*); −25 (*dotted line*); and −50 (*dashed line*). All dispersion coefficients are given in units of 10^{-50} J m^3. The co-ion distributions have been marked with circles

Ion distributions near air-water and mica-water interfaces

We will here demonstrate by numerically solving the ion-specific Poisson-Boltzmann equation given in the previous section that ionic dispersion potentials have very important influence on ionic distributions near air-water and mica-water interfaces. We argue this is one of the main physical reasons behind the many ion specific Hofmeister effects found in biology and colloid science.

The effect of different ionic dispersion forces on ion distributions near an air-water is that although the system is overall charge neutral, it is no longer charge neutral in every point: a double layer is formed near the interface. To accommodate the experimental surface tension of KBr we used the known fact that bromide ions are highly polarisable, more so than an equivalent volume of water or potassium ions. The bromide ions are pushed away from the interface by repulsive dispersion forces. Near a mica-water interface we find that the dispersion force of bromide ions becomes attractive. This explains why bromide as a counterion accumulates close to a charged interface to a much higher degree than say the less polarisable acetate ions. In Fig. 2 we show the ionic distributions between two charged mica interfaces 100 nm apart. Three different cases are considered: the bromide ions experiences no dispersion potentials; they experience half the dispersion potential given in the previous section; and finally they experience the full dispersion potential. We include the effect of images and the dispersion force that act on the potassium ions in all three cases. There are important effects of including the dispersion potentials that act on the counterions. The ratio of counterions ions within a distance less than the ion size is in both cases above 90%. Pashley et al. investigated this system and found that they could explain the measured force if they assumed that roughly 90% of the bromide were bound to the interface. As we have demonstrated this binding is most likely due to adsorption excess caused by attractive ionic dispersion potentials acting on the bromide ions. In Fig. 3 we show the electric field for the same system. Outside the first ion layer(s) the electric fields behave as if the actual surface charge was reduced when the attractive ionic dispersion interaction that acts on the counterions is turned on. For $B_- = -25 \times 10^{-50}$ J m^3 it behaves as if the surface charge was roughly 10% of its actual value. When

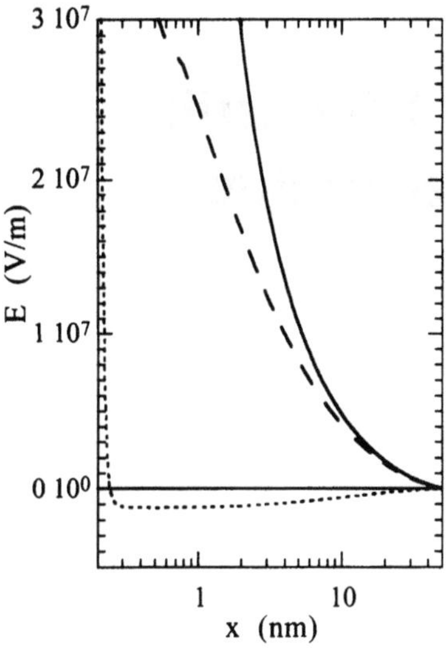

Fig. 3 The electric field ($E=-d\phi/dx$) for the systems described in Fig. 2

$B_{-}=-50\times10^{-50}$ J m^3 it is even more extreme, the system behaves as if there was a surface charge with a magnitude that is less than 1% of the actual value and with the opposite sign. The result should in both cases be reduced electrostatic double layer forces. This offers an explanation why the measured force between charged mica surfaces across 0.6 and 2 mM KBr salt solution could be accommodated when the effective surface charge were taken to be roughly 10% of the surface charge of the DHDAA bilayers.

Conclusion

When reasonable values are used for the excess polarisability of bromide and potassium the ion specific surface tension KBr can be accommodated, and an explanation given why bromide ions accumulate to a much higher degree near a charged mica surface than less polarisable ions such as acetate. This offers the obvious explanation why Pashley et al. had to postulate that 90% of the bromide ions were bound to the surface, while acetate ions were assumed completely dissociated [5]. The effect of ionic dispersion potentials furthermore become more and more important with increasing concentration. This is in accordance with the known fact that specific ion effects usually are observed at high concentrations. The traditional DLVO theory is in serious trouble when it comes to explain specific ion effects and we have shown a way out of this. Of course, detailed comparison with experiments can only be done if different ions sizes are accounted for. It will also be important to consider dissolved gas [17] and the way dispersion forces are handled in the surface region [18]. Furthermore, one can not avoid the fact that some of the many additional non-DLVO forces that have been invoked through the years may be real. It is however important that the double layer force is evaluated correctly before additional non-DLVO forces are introduced.

Acknowledgement M.B. would like to acknowledge financial support from STINT, the Swedish Foundation for International Cooperation in Research and Higher Education.

References

1. Weissenborn PK, Pugh RJ (1996) J Colloid Interface Sci 184:550
2. Onsager L, Samaras NT (1934) J Chem Phys 2:528
3. Boström M, Williams DRM, Ninham BW (2001) Langmuir 17:4475
4. Dubois M, Zemb T, Fuller N, Rand RP, Parseigan VA (1998) J Chem Phys 108:7855
5. Pashley RM, McGuiggan PM, Ninham BW, Brady J, Evans DF (1996) J Phys Chem 90:1637
6. Ninham BW, Yaminsky V (1997) Langmuir 13:2097
7. Boström M, Williams DRM, Ninham BW (2001) Phys Rev Lett 87:168103
8. Collins KD, Washabaugh MW (1985) Q Rev Biophys 18:323 (and references therein)
9. Baldwin RL (1996) Biophys J 71:2056
10. Grigorjev PA, Bezrukov SM (1994) Biophys J 67:2265
11. Kim H-K, Tuite E, Nordén B, Ninham BW (2001) Eur J Phys E 4:411
12. Tomac S, Sarkar M, Ratilainen T, Wittung P, Nielsen P, Nordén B, Gräslund A (1996) J Am Chem Soc 118:5544
13. Lyubartsev AP, Laaksonen A (1999) J Chem Phys 111:11207
14. Boström M, Williams DRM, Ninham BW (2002) J Phys Chem B (submitted)
15. Anonymous (1971) CRC Handbook of Chemistry and Physics, 52nd edn. Chemical Rubber Co, Cleveland, USA
16. Richmond P, Ninham BW (1972) J Colloid Int Sci 40:406
17. Alfredsson M, Ninham BW, Wall S (2000) Langmuir 16:10087
18. Marrink S-J, Marcelja S (2001) Langmuir 17:7929

Progr Colloid Polym Sci (2004) 123: 114–118
DOI 10.1007/b11653

A. Martín-Molina
M. Quesada-Pérez
F. Galisteo-González
R. Hidalgo-Álvarez

Charge inversion of latex particles in the presence of electrolyte

A. Martín-Molina
F. Galisteo-González
R. Hidalgo-Álvarez (✉)
Grupo de Física de Fluidos y Biocoloides
Departamento de Física Aplicada
Facultad de Ciencias
Universidad de Granada
Granada 18071, Spain
e-mail: rhidalgo@ugr.es

M. Quesada-Pérez
Departamento de Física
Universidad de Jaén
Escuela Universitaria Politécnica de Linares, 23700 Linares, Jaén, Spain

Abstract Recent studies have demonstrated the huge interest in the inversion of charge in systems such as solid particles, DNA molecules, charged membranes, etc. in the presence of electrolyte. Electrophoresis measurements show the charge of colloidal particles can be inverted when an electric field is applied. Some researchers suggest this fact could be due to the size of the counterions as well as electrostatic correlations between them. These effects become more important for asymmetrical electrolytes and high particle charges. In this work, this phenomenon is discussed for the case of polystyrene latex particles when several electrolytes are used. Up to now a considerable theoretical effort has been devoted to this topic. Nevertheless, there exists a lack of experimental work particularly at very high salt concentrations. We intend to address this matter by carrying out mobility measurements for ionic strengths as large as 2 M by using symmetric electrolytes. These experimental results allow us to test some theoretical models.

Keywords Colloids · Electric double layer · Electrophoretic mobility · Overcharging

Introduction

As many colloidal dispersions are stabilised electrostatically, the structure of the electrolyte around the charged particles, termed the electric double layer (EDL), plays an important role in colloid science. Several models of double layer have been developed such as the Gouy-Chapman (GC) model and the primitive model (PM). The former is based on the Poisson-Boltzmann (PB) theory and assumes that the ions in solution are point charges. This assumption means ions are able to approach right up to the surface of a particle considered as an infinite and flat impenetrable interface. The PB theory, for its part, is a mean-field approach in which correlations between ions are partially neglected. In contrast, the PM model treats the ions as small charged hard spheres and consequently, the ion-ion correlations are taken into account. Nevertheless, both theories have in common some approximations such as assuming the solvent is a uniform medium with properties that are independent of the distance and the surface charge in the interface is uniformly smeared out over the surface.

Theoretical works show the results obtained by the GC model match well with the PM model ones when the electrolyte concentration and the surface charge are sufficiently low. On the contrary, there exist marked differences in other cases where the role of the ion-ion correlations becomes important. Under these conditions, unexpected changes in the sign of the electrostatic potential and surface charge were reported and interpreted in terms of an *overcharging* or *inversion charge* phenomenon. Furthermore, unlike the GC model, integral equations theory based on the PM model such a *the hypernetted chain/mean-spherical approximation* (HNC/MSA), predicts that the electrostatic potential is non-monotonic (as a function of the distance from the surface and the surface charge density).

Although there is a plethora of theoretical works about overcharging, there exists a considerable lack of experimental evidence. Up to now, nevertheless, the reversal of mobility (for spherical particles) has been observed *almost exclusively* for solutions with trivalent counterions [1–4]. In such cases, the inversion has been attributed to the specific counterion adsorption.

The aim of this work is to revise some of the mentioned theoretical treatments and show to what extent the overcharging effects predicted by them are confirmed by electrophoretic mobility experiments carried out with different latex particles and electrolytes.

Theory

As mentioned before, ion-ion correlations are taken into account in the PM context, and particularly, integral equation theories can be used. For instance, in 1982 Lozada-Cassou et al. applied the so-called HNC/MSA approximation to the *planar* EDL. The electrolyte was supposed to be a fluid of hard spheres with charge $z_i e$ (e is the elementary charge) and radius a (identical for coions and counterions) immersed in a dielectric continuum whose permitivity was $\varepsilon_r\varepsilon_0$ (ε_0 is the vacuum permitivity). Furthermore, the HNC integral equations for the *wall-ion* distribution function of species i, $g_i(x)$ turns out to be [5–7]

$$g_i(x) = \exp\left[-\beta z_i e\psi(x) + 2\pi\sum_j \rho_j \int_{-\infty}^{\infty} h_j(x')C^0_{ij}(x,x')dx'\right] \tag{1}$$

where $\beta = 1/k_B T$ (T is the absolute temperature), $h_i(x) = g_i(x) - 1$ is the *wall-ion total* correlation function, $\psi(x)$ is the electrostatic potential at a distance x from the wall, ρ_i is the bulk density of i-ions and $C^0_{ij}(x,x')$ are *integrals* over the *direct* correlation functions of the bulk species. These functions can be calculated using the MSA approximation, which has the advantage of being analytical [8] and leads to accurate results. The electrostatic potential can be related to the correlation functions through

$$\psi(x) = \frac{e}{\varepsilon_r\varepsilon_0}\sum_j z_j\rho_j \int_x^{\infty} (x-t)h_j(t)dt \tag{2}$$

It should be emphasised that the $C^0_{ij}(x,x')$ functions include ion-ion correlations (due to size and charge) and, in fact, they depend on certain parameters characterising the electrolyte solution, such as its salt molar concentration (c_{salt}), and the size and charge of ions (see the closed expressions for them in [8] and [9]).

As we noticed in the introduction, one of the most important consequence of the ionic correlation effects is the overcharging. This phenomenon may cause an inversion in the sign of the electrostatic potential when is calculated at a determined distance from the surface. This is a key point in our discussion since our experimental technique is electrophoresis and, consequently, we are interested in the ζ-potential. This quantity is defined at the shear plane (SP). Hereafter, however, the approximation $\zeta \approx \psi_d = \psi(a)$ will be applied, where ψ_d is the potential at the closest approach of the hydrated ions to the wall (known as diffuse potential).

Concerning the planar approximation, Lozada-Cassou and co-workers concluded that there were not significant differences in the ψ_d-values obtained for larger particles and a planar wall [10]. Accordingly, we have assumed our colloidal particles as planes in order to facilitate the conversion of the ζ-potential into electrophoretic mobility (μ_e) by using the Helmholtz-Smoluchowski equation

$$\mu_e = \varepsilon_0\varepsilon_r\zeta/\eta \tag{3}$$

This result is valid at the limit, $\kappa A \to \infty$, (where A is the particle radius) which is in agreement with the size of our colloidal particles and the electrolyte concentrations considered here. In addition, Eq. (3) is *formally* identical to that obtained from the PB approach because the *entire* mean electrostatic potential profile at equilibrium is not required in this limiting case. Therefore, ion-ion correlations are actually included in calculating ζ.

Material and methods

Two polystyrene latexes were used, a sulphonated one, SN10, and a carboxylated latex, CC2. The former was prepared by a two stage *shot growth* emulsion polymerisation process in absence of emulsifiers and subsequently, a styrene/sodium styrenesulphonate copolymer was obtained. The whole synthesis method is described in [3]. The latter was synthesised following another free emulsifier polymerisation in a two-step process. In the first step, a latex analogous to the previous one was used as a core. Thereafter a shell of styrene and acrylic acid was put onto the core, using potassium persulphate as initiator. The particles sizes, obtained by photon correlation spectroscopy (PCS), were 196 ± 3 nm and 180 ± 6 nm for the SN10 and CC2 latexes, respectively. The polydispersity indexes, which tend to zero for monodisperse samples, were about 0.05 for SN10 and 0.16 for CC2. Conductimetric and potentiometric titrations were used to determine their surface charge density. In the case of the sulphonated latex, a surface charge density of (11.5 ± 1.7) $\mu C/cm^2$, which did not depend on pH, was found. Unlike the previous case, the surface charge of the latex CC2 is controlled by pH. In particular, the electrophoretic mobility experiments were performed at pH of 5.8 where $\sigma_0 \approx -40$ $\mu C/cm^2$ for this latex.

A new instrument, known as *Brookhaven ZetaPALS*, based on the principles of phase analysis light scattering (PALS) is used to obtain electrophoretic mobilities (μ_e). The set-up is especially useful at high ionic strengths, where mobilities are usually low. In this sense, the PALS configuration has been shown to be able to measure μ_e at least two orders of magnitudes lower than traditional light scattering methods based on the shifted frequency spectrum (spectral analysis) [11].

Finally, electrophoretic mobility measurements were performed at 25 °C and the electrolytes used were LiCl and $MgSO_4$.

Results and discussion

In relation to the monovalent salt, the electrophoretic mobility measured for both latexes versus electrolyte concentration are shown in Fig. 1 and Fig. 2. As can be seen, the magnitude of μ_e decreases with increasing c_{salt}, but is always negative (mobility reversal is not observed). In these figures, besides the experimental data, the HNC/MSA approach and the GC model predictions are plotted. In the latter case, the mobility values were obtained from Eq. (3) with the approximation $\zeta \approx \psi_d$ and using the GC relationship for symmetrical electrolytes ($z_1 = -z_2 = z$)

$$\psi_d = \frac{2kT}{ez} a\sinh\left(\frac{ze\sigma_0}{2\varepsilon_0\varepsilon_r\kappa kT}\right) \tag{4}$$

which assumes a diffuse double layer just beyond the outer Helmholtz plane (OHP), located at the closest approach of the hydrated ions, $x = a$.

In order to calculate the HNC/MSA predictions for μ_e the hydrated ion radius has to be specified as input parameter in the theory. In particular, we have chosen $a = 0.36$ nm for latex SN10 (Fig. 1) and $a = 0.28$ nm for latex CC2 (Fig. 2), which are practically identical to that reported for Li^+ by certain authors [12]. It should also be mentioned that this quantity depends on how it is measured (as pointed out in [12]). In addition, some authors have reported that the hydrated ion size could be reduced in presence of the highly charged surfaces [13]. As can be observed in these first figures, although the HNC/MSA μ_e-values also exceed (in magnitude) those obtained from measurements, the differences between theory and experiment become smaller than in the GC analysis.

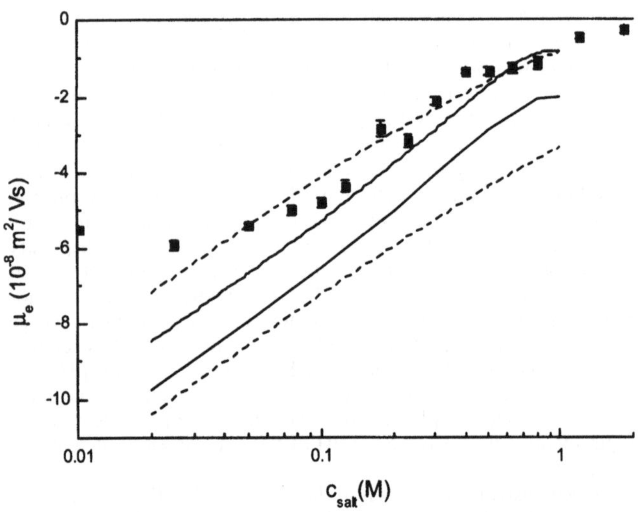

Fig. 1 Electrophoretic mobility vs. LiCl molar concentration for latex SN10. The results corresponding to experiments (*squares*) as well as the HNC/MSA (*lower solid line*) and GC (*lower dashed line*) approaches are shown. The predictions considering a distance δ between the SP and the OHP are also plotted. The values $\delta = 0.12$ nm and $\delta = 0.40$ have been used for the HNC/MSA (*upper solid line*) and GC (*upper dashed line*) calculations, respectively. The value $a = 0.36$ nm has been used for the Li^+ radius

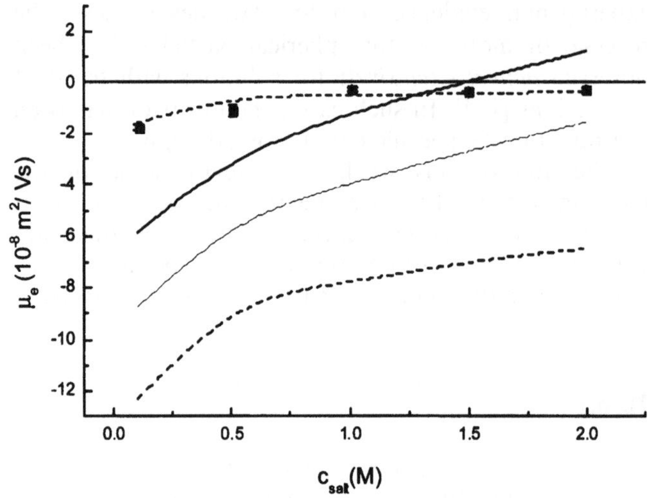

Fig. 2 Electrophoretic mobility vs. LiCl molar concentration for latex CC2. The results corresponding to experiments (*squares*) as well as the HNC/MSA (*lower solid line*) and GC (*lower dashed line*) approaches are shown. The predictions considering a distance δ between the SP and the OHP are also plotted. The values $\delta = 0.10$ nm and $\delta = 0.70$ have been used for the HNC/MSA (*upper solid line*) and GC (*upper dashed line*) calculations, respectively. The value $a = 0.28$ nm has been used for the Li^+ radius

For much more charged latex particles in a solution of monovalent ions (Fig. 2), the electrophoretic mobility seem to be insensitive to this increase in surface charge density. The differences between the GC predictions and the experimental values are, therefore, greater for SN10.

Whichever the case, such discrepancies can be reduced assuming a certain distance, δ, between the OHP and the SP. Fig. 1 shows that good agreement can be found for the first latex taking $\delta = 0.40$ nm and $\delta = 0.12$, respectively, in the GC and HNC/MSA calculations. With regard to the CC2 latex, it takes the values $\delta = 0.70$ nm for the GC model and $\delta = 0.10$ for the HNC/MSA approach. In view of the fact that the distance between the OHP and the SP may be up to 2–3 water molecule diameters away from the surface [14, 15], more realistic results are obtained in the second approximation than the GC theory. Moreover, the distance of the SP from the surface, $a + \delta$, is almost identical for SN10 and CC2 (0.48 and 0.38 nm). In other words, its location (relative to the particle surface) would barely depend on the latex.

Electrophoretic mobility versus $MgSO_4$ concentration is shown in Fig. 3 (latex SN10) and Fig. 4 (latex CC2). Again, significant discrepancies between both approaches are found. In the GC case, the distance δ turns out to be $\delta = 0.25$ nm for both latexes. Concerning with the computation of the HNC/MSA results, the hydrated ion radii given in [12] cannot be used any more since it is

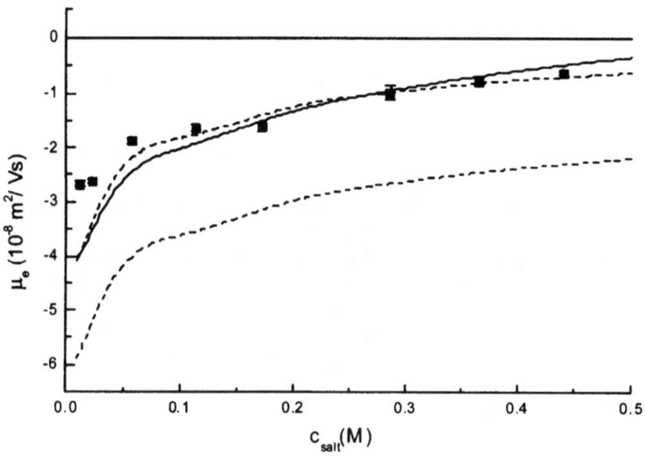

Fig. 3 Electrophoretic mobility vs. $MgSO_4$ molar concentration for latex SN10. The results corresponding to experiments (*squares*) as well as the HNC/MSA (*solid line*) and GC (*lower dashed line*) approaches are shown. The predictions considering a distance $\delta = 0.25$ nm between the SP and the OHP for the GC calculations (*upper dashed line*) are also plotted. The value $a = 0.30$ nm has been used for the Mg^{++} radius

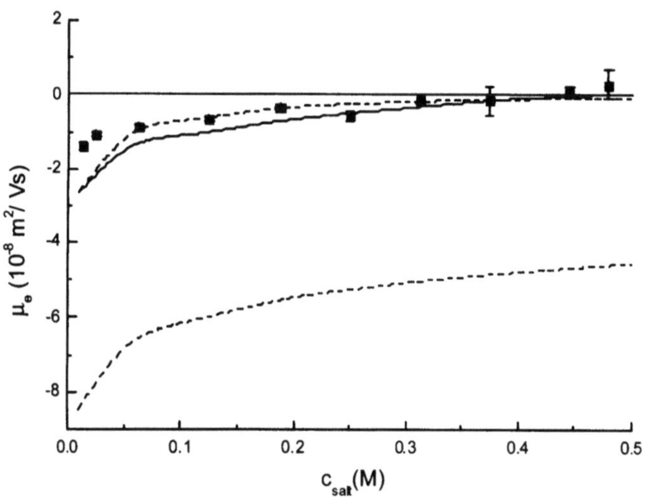

Fig. 4 Electrophoretic mobility vs. $MgSO_4$ molar concentration for latex CC2. The results corresponding to experiments (*squares*) as well as the HNC/MSA (*solid line*) and GC (*lower dashed line*) approaches are shown. The predictions considering a distance $\delta = 0.80$ nm between the SP and the OHP for the GC calculations (*upper dashed line*) are also plotted. The value $a = 0.18$ nm has been used for the Mg^{++} radius

not reliable at high electrolyte concentrations and, if this fact were ignored, a reversed mobility under such conditions would be obtained. After several trials, it can be easily shown that the measurements are matched with moderate accuracy *without requiring additional adjustable parameters* using $a = 0.30$ nm and $a = 0.18$ nm for both SN10 and CC2 latexes, respectively. These values are somewhat smaller than the reference value (0.43 nm) but completely acceptable remembering that an increase in the surface charge produces a reduction in the size of the hydrated ion size [13]. Furthermore, similar values were measured by using neutron diffraction experiments [16].

Conclusions

We would like to complete this work summing up results and discussing several conclusions. Regarding monovalent ion solutions, neither the HNC/MSA approach nor the GC model can justify the mobility measurements found for these charged latexes considering exclusively the surface charge density determined by titration, although it should be stressed that the HNC/MSA predictions are better than the GC ones. The discrepancies could be reduced assuming a certain distance between the OHP and the SP.

In divalent electrolytes, the role of ion correlations is more relevant. This conclusion is supported by the fact that the HNC/MSA can fit acceptably mobility measurements without requiring additional effects or parameters.

But our main concern was to look into whether overcharging is revealed by mobility reversal. This phenomenon is predicted by using completely hydrated ionic sizes in the HNC/MSA approach. However, all our experimental results show that mobility reversal does not take place. In relation to this, it should be kept in mind that large surface charge densities could reduce the size of the hydrated ions. Consequently, the overcharging would not be so intense as expected, and the mobility reversal might fade away. In spite of this, the electrostatic potential could remain reverse far from the particle surface which might be revealed experimentally in other ways.

Acknowledgements We would like to thank Dr. D. Bastos for providing us one of the latexes used in this work and Prof. M. Lozada-Cassou for valuable discussions on theoretical aspects. The authors also acknowledge the financial support from 'Ministerio de Ciencia y Tecnología, Plan Nacional de Investigación, Desarrollo e Innovación Tecnológica (I+D+I)'. Particularly, M.Q.P. expresses his gratitude to project MAT2000-1550-C03-03, whereas R.H.A. is grateful to project MAT.

References

1. Ottewill RH, Shaw JN (1968) J Colloid Interface Sci 26:110
2. Elimelech M, O'Melia CR (1990) Colloids Surf 44:165
3. Bastos D, Nieves FJ de las (1993) Colloid Polym Sci 271:870
4. Galisteo F, Nieves FJ de las, Cabrerizo M, Hidalgo-Álvarez R (1990) Progr Colloid Polym Sci 82:313
5. Henderson D, Blum L, Smith WR (1979) Chem Phys Lett 63:381
6. Henderson D, Blum L (1981) J Electroanal Chem 111:217
7. Carnie SL, Chan DYC, Mitchell DJ, Ninham BW (1981) J Chem Phys 96:1472
8. Wertheim M (1963) Phys Rev Lett 10:321
9. Lozada-Cassou M, Saavedra-Barrera R, Henderson D (1982) J Chem Phys 77:5150
10. González-Tovar E, Lozada-Cassou M, (1989) J Phys Chem 89:3761
11. McNeil-Watson F, Tscharnuter W, Miller J (1998) Colloids Surf A 140:53
12. Israelachvili J (1992) Intermolecular and surface forces. Academic Press, London
13. Grygiel W, Starzak M (1995) J Lumin 63:47
14. Chan DYC, Horn RG (1985) J Chem Phys 83:5311
15. Hunter RJ (1981) Zeta potential in colloid science. Principles and applications. Academic Press, London
16. Howell I, Neilson GW (1996) J Phys Condens Matter 8:4455

Progr Colloid Polym Sci (2004) 123: 119–122
DOI 10.1007/b11654

A. Moncho-Jordá
M. Quesada-Pérez
F. Martínez-López
R. Hidalgo-Álvarez

Structure and interaction forces in colloidal monolayers

A. Moncho-Jordá · F. Martínez-López
R. Hidalgo-Álvarez (✉)
Grupo de Física de Fluidos y Biocoloides, Departamento de Física Aplicada, Facultad de Ciencias, Universidad de Granada, Granada 18071, Spain
e-mail: rhidalgo@ugr.es

M. Quesada-Pérez
Departamento de Física, Universidad de Jaén, Escuela Universitaria Politécnica de Linares, 23700 Linares, Jaén, Spain

Abstract The interaction potential between colloidal particles is the key to understand and control the aggregation processes in which they are involved. However, interaction forces are also responsible for spatial ordering. This kind of phenomena has been used to obtain valuable information on the interaction potential in three-dimensional dispersions. In this work, this possibility is investigated for colloidal monolayers spread at the water-air interface. The radial distribution function has been determined for an assembly of negatively charged polystyrene particles. The salt concentration in the subphase was 1 mM. First, a recent model was applied in the attempt of justifying the experimental result. Then, a simple inversion scheme based on the HNC closure has been used. Apart from the expected repulsive forces, the existence of a long-range attractive interaction is suggested in the latter case.

Keywords Colloidal monolayers · Interaction potential · Inverse problem · Radial distribution function

Introduction

Aggregation of colloidal particles also takes place at the liquid-air interface. Remarkable examples are the manufacture of emulsion polymers in stirred-tank reactors and separation processes such as froth extraction. In order to understand and control such processes (of scientific, technological and industrial interest), the pair interaction potential, $u(r)$, plays an essential role. But interaction forces also bring out structural ordering in colloidal dispersions, which can be easily observed for concentrated and/or interacting systems.

Regarding three-dimensional dispersions of charged particles, it has been known that, at very low ionic strength, they exhibit spatial ordering over distances considerably greater than the particle diameter as a result of long-ranged electrostatic forces. The liquid-like structures formed under such conditions have been revealed by different techniques (e.g., light scattering, turbidimetry) and are used to extract information about the interaction potential [1]. Concerning two-dimensional systems, there exist several studies that deal with the spatial structure of colloidal monolayers trapped at diverse interfaces [2–8]. Particularly, valuable information about the interaction potential has been obtained from the liquid-like structures formed on a liquid subphase *without additional electrolyte* applying an inversion method [9]. This paper is devoted to the case of colloidal monolayers formed with a subphase whose ionic strength is moderate (1 mM). First, a brief theoretical background is given. Then the experimental technique for determining the radial distribution function, $g(r)$, is described and some details about the latex particles used in this work are given. Finally, the results will be presented and discussed.

Theoretical background

It is well known that the structure of stable colloidal dispersions (especially at high particle concentrations and/or strongly interacting systems) can be predicted with the help of liquid-state theories treating colloids as macroscopic analogues of molecular (monoatomic) fluids. Given an interaction potential, the correlation function $h(r) \equiv g(r) - 1$ and the direct correlation function, $c(r)$, can be calculated solving the set formed by the Ornstein-Zernike equation [10]

$$h(r) = c(r) + \rho \int h(r')c(|\vec{r} - \vec{r}'|)d^3r' \quad (1)$$

and a closure relationship linking $h(r)$, $c(r)$ and $u(r)$ (ρ is the particle concentration). The latter can be formally expressed by

$$\beta u(r) = h(r) - c(r) - \ln[h(r) + 1] + B(r) \quad (2)$$

where $\beta = 1/k_B T$ (T is the absolute temperature) and $B(r)$ is known as *the bridge function*. Since $B(r)$ is analytically intractable, several approximations for Eq. (2) have been proposed. For instance, the hypernetted-chain (HNC) closure takes as a fact that $B(r) = 0$. At any rate, it is possible to fit experimental data if a functional form for $u(r)$ is assumed to be valid, which is called as *forward* method.

However, there exists an alternative method for obtaining information on $u(r)$ that does not require any assumption (a priori) about its functional form. This way, which is termed the *inverse* problem, is briefly outlined as follows. Having measured $g(r)$ from experiments, $c(r)$ can be calculated from Eq. (1) and, subsequently, $u(r)$ is easily extracted from Eq. (2). This *single-step* procedure was applied within the HNC approximation for three-dimensional colloidal dispersions [1]. A long-ranged potential with a minimum at a *large* distance was reported. This result, which contrasts strongly with the Derjaguin-Landau-Verwey-Overbeek (DLVO) theory, must be carefully analyzed. Inversion methods require quite accurate structure factors, particularly at low values of the scattering vector modulus, since $g(r)$ is calculated from the structure factor. However, such accurate data are not easily available. On the contrary, the radial distribution function is directly determined from experiments for colloidal monolayers.

Experimental

The colloidal particles of our experiments were polystyrene latex beads, prepared from styrene by free-emulsifier emulsion polymerisation with potassium peroxidisulphate as initiator and $KHCO_3$ a buffer. The polymerisation was carried out in a reactor fitted with a reflux condenser, stainless steel stirrer and nitrogen inlet tube. The size distribution was determined by transmission electron microscopy. The average particle diameter (d) turned out to be (600 ± 25) nm. The sample polydispersity can be considered negligible because the size distribution was extremely narrow (see Fig. 1). The polydispersity index (defined as the weight average diameter to number average diameter ratio) is 1.004. The latex dispersion was cleaned first by serum replacement and second using a mixed-bed ion exchange resin. Finally, the surface charge density

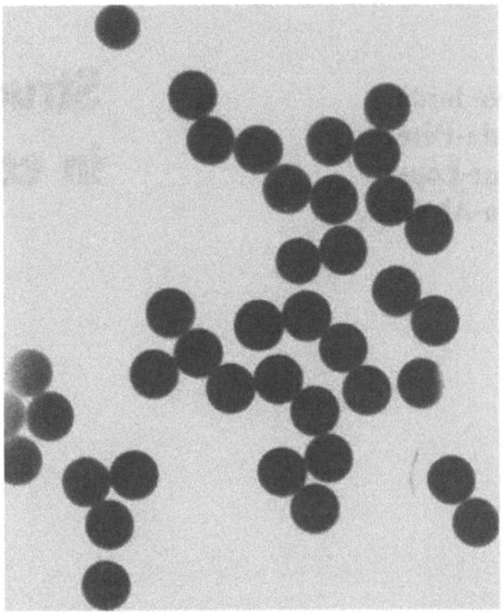

Fig. 1 Micrograph of latex AMJ2

was determined by conductiometric titration. A value of $-(0.053 \pm 0.006)$ C/m^2 was obtained.

Concerning the two-dimensional structure formation, an aqueous solution of KBr 1 mM was prepared (with distilled water) and introduced into a Teflon cell. Then, latex particles were deposited at the air/water interface using a microsyringe. The interface has to be as planar as possible in order to prevent the particle emigration as a consequence of gravity. The immersion of latex beads into water was avoided by dispersing the particles using methanol as spreading agent, according to procedures described elsewhere [3, 4]. For this purpose, colloidal suspensions were prepared in a methanol solution and sonicated for five minutes to assure a very monodisperse sample. It should be stressed that the use of methanol in the particle deposition is necessary to obtain a uniform monolayer. After methanol evaporation, the cell is covered with a thin glass plate to prevent the contamination of the colloidal monolayer as well as convective fluxes produced by the air motion. Furthermore, possible vibrations were avoided by placing the system formed by the cell, microscope and camera on an antivibratory table.

Images of the interface were recorded by a digital camera (with 1280 × 1024 pixels) incorporated to a phase contrast microscope. The magnification of the objective was fixed so that the pixel diameter was close to the particle size. The images were first acquired by a frame grabber and then transformed to binary images and analysed to determine the radial distribution function, $g(r)$. Several experiments of spreading were carried out and some images were recorded for each one. If the surface particle concentration was low enough, an almost homogeneous monolayer was obtained. A representative experiment, corresponding to $\rho = 5.5 \cdot 10^{10}$ particles × m^2, was chosen for analysis. Changes were not observed in the particle distribution after deposition.

The radial distribution function was determined from images using the relation

$$g(r) = \Delta N/(\rho \Delta S_r) \quad (5)$$

In this equation, ΔN is the number of particles whose distance from a given one is in the range $[r - \Delta r/2, r + \Delta r/2]$ and $\Delta S_r = 2\pi r \Delta r$ is the area of the circular shell of radius r and thickness Δr around the tagged particle. The result is averaged over all the particles in the system. In computer simulations, the radial distribution function is usually calculated assuming periodic boundary conditions. This method allows us to obtain the distribution function for $r < L/2$, where L is the width of the simulation box. Nevertheless, in real experiments we have a finite box and, consequently, these conditions produce particle correlations without physical meaning near of the edges. A first solution for this problem is to consider only the shells that do not cut the boundary, but this procedure leads to a poor statistic for large distances since only tagged particles at the centre of the box contribute at such distances. Hence, an alternative method was used. Accordingly, ΔS_r is the surface of the r-shell *inside* the box. In other words, if S_{out} is the outer area, this quantity is given by

$$\Delta S_r = 2\pi r \Delta r - S_{out} \tag{6}$$

This procedure improves the calculation of contributions to $g(r)$ for large distances considerably.

Result and discussion

The radial distribution function determined by means of the procedure described in the preceding section is plotted in Fig. 2. As can be seen, this function resembles a lot the one corresponding to molecular monoatomic liquids, but its main peak is located at a larger distance (in comparison with the particle diameter). At this point, our aim is to obtain information about the interaction potential from these structural data applying a formalism based on a two-dimensional OZ equation

$$h(r) = c(r) + \rho \int h(r')c(|\vec{r} - \vec{r}'|)d^2r' \tag{7}$$

(in which ρ is now the number of particles per unit of surface) and the HNC closure. As our experiment was carried out at moderate ionic strength (1 mM), the forward method is feasible since there exist theoretical expressions for $u(r)$ that work under these conditions. Particularly, Martínez-López et al. have proposed a

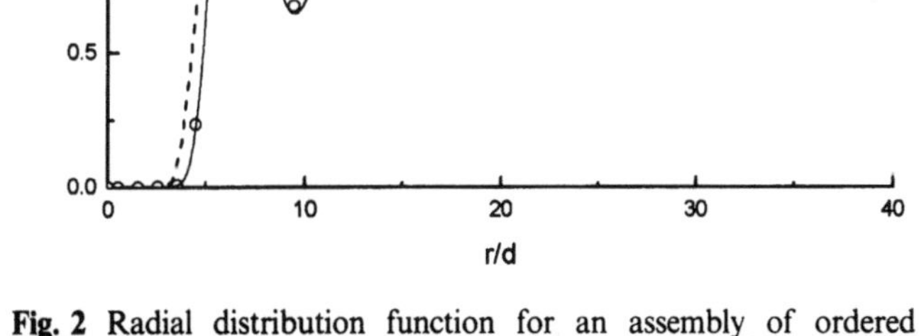

Fig. 2 Radial distribution function for an assembly of ordered colloidal particles whose concentration is $5.5 \cdot 10^{10}$ particle/m^2, obtained experimentally (*open circles*) and from several MC simulations: from the potential obtained through inversion and the HNC closure (*solid line*); from the potential calculated through the model described in Ref. [11] (*dashed line*).

model for colloidal particles trapped at the air-liquid interface [11]. This model includes interactions that have no analogue in the DLVO theory for three-dimensional systems. A relevant example for our purpose is the dipolar interaction since, at 1 mM, this contribution dominates considerably at large distances. In these kinds of two-dimensional systems, the dipoles originate from the chemical groups on the particle surface and the cloud of counterions around the wetted part of the particle. The interested reader can find a detailed discussion on this matter in Ref. [11]. Here, our main concern is to find out if this model accounts for our experimental data. In order to do that, first, we calculate $u(r)$ for our colloidal particles assuming that the number of formed dipoles from surface groups is as large as possible. Additional parameters, such as the ζ-potential, the contact angle or two Hamaker constants, are also required. However, the dipole–dipole interaction, which is the main contribution at large distance (as noted earlier), depends essentially on the number of dipoles. Then, a theoretical $g(r)$ was obtained with the help of Monte Carlo (MC) simulations using 400 particles, 15,000 configurations per particle to equilibrate the systems and 20,000 in the statistical analysis. In these simulations we have assumed monodisperse particles although the degree of order may be influenced by the size polydispersity. However, this parameter is quite small in our experiment, so this effect is suppose to be negligible. The resulting radial distribution function is also shown in Fig. 2. As can be concluded, the range of this potential is not long enough to justify the spatial ordering found in our experiments. At this point, some questions come out. For instance, one could think that estimating the maximum number of dipoles is not an easy task and the actual value is considerably larger. The possibility of purely Coulombic electrostatic forces (due to some surface groups of the emergent parts) might also be considered. In fact, studies on compression of monolayers suggest that this kind of long-ranged repulsion is responsible for the highly ordered structures observed at the octane/water interface [8, 12].

At any rate, a previous study has shown that the HNC within the inverse formalism can give valuable information about the interaction providing that the analysed systems are not highly correlated, which is clearly revealed by the measured radial distribution function. The inversion of structural data has been carried out for a colloidal monolayer without additional electrolyte in the subphase [1]. In view of this, we decided to apply a single-step inversion to our dilute two-dimensional assembly. The interaction potential obtained within this approximation is plotted in Fig. 3. Its most significant feature is a long-range repulsive core for $r < 5d$. In order to test this result, another MC simulation using this potential as input was run. For $r < 3.45d$, the resulting potential is not reliable because of $g(r)$ being extremely small. Therefore, $u(r)$ was estimated by extrapolation. The radial distribution function so obtained is also plotted in Fig. 2. As can be easily observed, the agreement between this radial distribution function and the experimental one is outstanding, which supports our confidence on the interaction potential calculated within the HNC approximation. Whichever the case, the origin of this repulsive core, which presumably accounts for the great stability found for colloidal particles at the liquid-air interface, remains unclear. Obviously, at moderate ionic strengths the electrostatic forces between the immersed parts of particles are not long-ranged. The dipolar interaction could contribute but does not explain these experimental results by itself, as discussed earlier.

Finally, we will focus on the minimum observed (in Fig. 3) at large distances. Its depth is rather small (about the thermal energy, $k_B T$), but a previous study have proved that it must be considered. Consequently, a long-range attraction compensating the repulsive interaction should be required. Very recently, Stamou et al. have proposed a mechanism for long-range attractive interactions that is based on a irregularly shaped meniscus [13]. Some estimates made by these authors suggest, however, that the minimum would be located at a distance of about twice the particle diameter (2 μm, in round numbers), whereas the minimum that we have found is further. Certain experiments carried out with spheres confined between two parallel glass plates should also be kept in mind. If the distance separating the plates is short enough, a long-range attraction that cannot be explained by the DLVO theory has been reported [14]. The minimum found in colloidal monolayers might therefore be caused by confinement. This is still an open question.

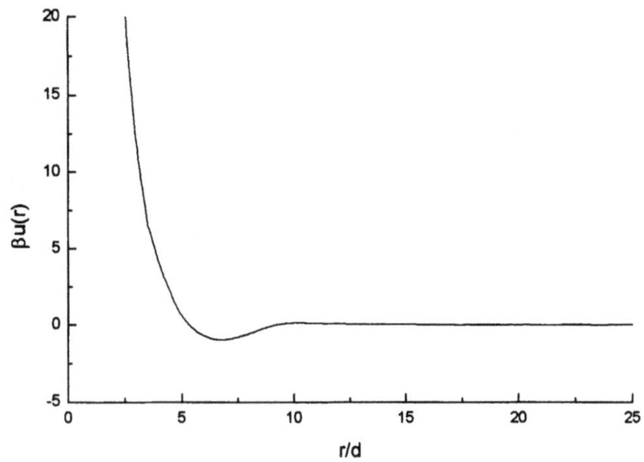

Fig. 3 Interaction potential obtained through an inversion scheme within the HNC closure

Acknowledgements This work is supported by 'Ministerio de Ciencia y Tecnología, Plan Nacional de Investigación, Desarrollo e Innovación Tecnológica (I + D + I)', project MAT2003-08356-C04-01.

References

1. Quesada-Pérez M, Callejas-Fernández J, Hidalgo-Álvarez R (2002) Adv. Colloid Interface Sci 5:295
2. Pieranski P (1980) Phys Rev Lett 45:569
3. Robinson DJ, Earnshaw JC (1992) Phys Rev A 46:2045
4. Robinsosn DJ, Earnshaw JC (1993) Langmuir 9:1436
5. Zahn K, Méndez JM, Maret G (1997) Phys Rev Lett 79:175
6. Ghezzi F, Earnshaw JC (1997) J Phys Condens Matter 9:L517
7. Bubeck R, Neser S, Bechinger C, Leiderer P (1998) Prog Colloid Polym Sci 110:41
8. Averyard R, Clint JH, Nees D, Paunov VN (2000) Langmuir 16:1969
9. Quesada-Pérez M, Moncho-Jordá A, Martínez-López F, Hidalgo-Álvarez, J Chem Phys (2001) J Chem Phys 115:10879
10. Hansen JP, McDonald IR (1986) Theory of simple liquids, 2nd edn. Academic Press, London
11. Martínez-López F, Cabrerizo-Vílchez MA, Hidalgo-Álvarez R (2000) J Colloid Interface Sci 232:303
12. Sun J, Stirner T (2001) Langmuir 17:3103
13. Stamou D, Duschl C, Johannsmann D (2000) Phys Rev E 62:5263
14. Crocker JC, Grier DG (1996) Phys Rev Lett 77:1897

Progr Colloid Polym Sci (2004) 123: 123–126
DOI 10.1007/b11744

N.M. Kovalchuk
D. Vollhardt

Direct numerical simulation of the mechanism of surface tension auto-oscillation

D. Vollhardt (✉)
Max-Planck-Institute of Colloids and Interfaces, 14424 Potsdam/Golm, Germany

N.M. Kovalchuk
Institute for Problems of Material Science, 03142 Kiev, Ukraine

Abstract A theoretical investigation of the surface tension auto-oscillations is performed by direct numerical simulation of the behaviour of a system where a surfactant droplet dissolves under the free water surface. The chosen model allows for the geometry of the measuring cell. The Marangoni effect is accepted as responsible for giving rise to the instability. The simulation shows that the system behaviour to prior the first oscillation does not depend on the radial dimension of the system. The development of a second and following oscillations is only possible in systems with limited dimensions. The calculated dependence of the surface tension on time is in good agreement with the experimental results.

Keywords Surface tension auto-oscillations · Instability · Numerical simulation · Marangoni effect

Introduction

Dissolution of a surfactant droplet situated on the tip of a capillary under the free liquid surface can be the cause of a periodic change in surface tension denoted as auto-oscillations [1]. Auto-oscillations begin after a certain induction time, wherein the surface tension remains nearly constant, and can continue over several hours. They are characterised by a rapid decrease of surface tension followed by its gradual increase. The appearance of the auto-oscillations is related to instability phenomena that can be observed in systems far from equilibrium where energy or mass transfers take place [3, 4].

A comprehensive analysis of the auto-oscillations mechanism was started in Ref. [5]. There was considered a model system representing a semi-infinite liquid volume consisting of a spherical surfactant droplet under a free surface. The numerical simulation allows the tracing of the system behaviour during the induction period as well as development of the first oscillation. The rise of the instability reveals itself in a sharp decrease of the surface tension. It shows also that after some time the instability faded due to mixing of surfactant in the bulk near the surface. However, the model chosen in Ref. [5] cannot describe the return of the system in the equilibrium state and the development of subsequent oscillations. In the present paper, we consider a model more adequate to experimental conditions, namely with a limited liquid volume. Here, we focus on the question whether the simulation of this system provides a series of auto-oscillations. The first results of the numerical simulation of the behaviour of such a system are presented below.

Experimental background

Surface tension auto-oscillations were observed in experiments with diethyl phthalate [1], aliphatic alcohols [2], and fatty acids used as surface-active substances. Their characteristics depend on the substance used. For example, the period of the auto-oscillations for middle-chain alcohol changes from nearly 40 s for pentanol to more than 10 min for octanol.

The experimental set-up and method used are described elsewhere [1, 2]. An example of the experimental results is presented in Fig. 1, which shows that there is an induction period at the beginning of the process when the surface tension remains nearly constant. After that, long-time regular oscillations of the surface tension start. Sometimes the oscillations could reveal a more complicated form (Fig. 2). It is interesting to note that a small amount of the surfactant dissolved in water does not prevent the

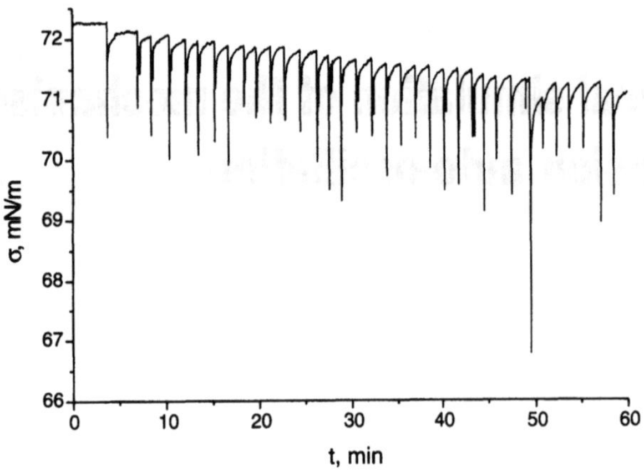

Fig. 1 Surface tension auto-oscillations in the hexanol-water system

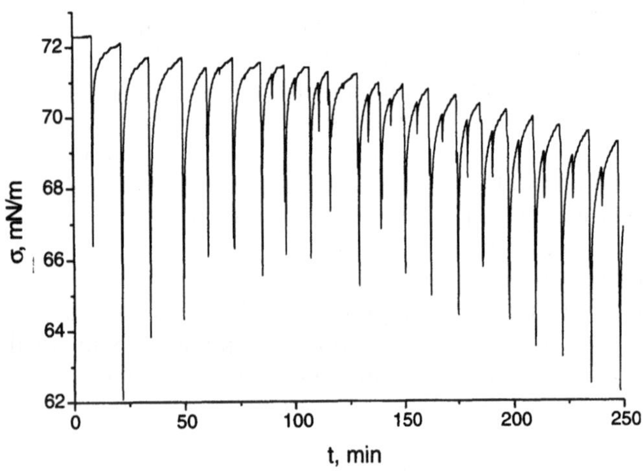

Fig. 2 Surface tension auto-oscillations in the octanoic acid-water system

appearance of auto-oscillations. In this case, the gradual decrease of the surface tension owing to adsorption of the dissolved surfactant is superimposed with the ordinary shape of the auto-oscillations (Fig. 3).

Mathematical formulation

A cylindrical vessel with a rigid wall and bottom filled with a Newtonian incompressible viscous liquid in contact with a passive gas phase is taken as a model system. A cylindrical capillary is placed in the liquid so that the capillary axis coincides with the vessel axis. A droplet of a soluble surfactant is on the capillary tip.

The behaviour of the system under consideration is described by the set of equations written in cylindrical co-ordinate system including the transport equation for the vorticity Eq. (1), the Poisson equation for the stream function Eq. (2) and the convective diffusion equation, Eq. (3):

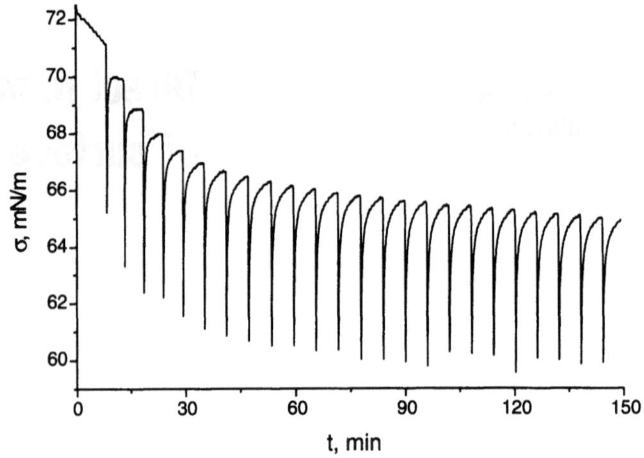

Fig. 3 Surface tension auto-oscillations in the octanol-water system with contaminations

$$\frac{\partial\omega}{\partial t}+\frac{\partial(v_r\omega)}{\partial r}+\frac{\partial(v_z\omega)}{\partial z}=\nu\left(\frac{\partial^2\omega}{\partial r^2}+\frac{\partial^2\omega}{\partial z^2}+\frac{1}{r}\frac{\partial\omega}{\partial r}-\frac{\omega}{r^2}\right) \quad (1)$$

$$\frac{\partial^2\Psi}{\partial r^2}+\frac{\partial^2\Psi}{\partial z^2}-\frac{1}{r}\frac{\partial\Psi}{\partial r}=\omega r \quad (2)$$

$$\frac{\partial c}{\partial t}+\frac{\partial(v_r c)}{\partial r}+\frac{\partial(v_z c)}{\partial z}+\frac{v_r c}{r}=D\left(\frac{\partial^2 c}{\partial r^2}+\frac{\partial^2 c}{\partial z^2}+\frac{1}{r}\frac{\partial c}{\partial r}\right) \quad (3)$$

where t is time, r is the radial co-ordinate, z is the normal to the surface co-ordinate, ν is the cinematic viscosity of the liquid, c is the surfactant concentration, D is the volume diffusion coefficient, Ψ is the stream function, $v_r=\frac{1}{r}\frac{\partial\Psi}{\partial z}$ is the velocity component in radial direction, $v_z=-\frac{1}{r}\frac{\partial\Psi}{\partial r}$ is the velocity component in normal to the surface direction, and $\omega=\frac{\partial v_r}{\partial z}-\frac{\partial v_z}{\partial r}$ is the vorticity.

The set of Eqs. (1)–(3) was solved numerically by using appropriate initial and boundary conditions. At the initial time the liquid is supposed motionless, the surfactant concentration on the droplet surface is equal to its solubility in water and it is zero elsewhere. The no-slip condition is used for the wall and bottom of the vessel and for the droplet surface. The free liquid surface is assumed to be non-deformable. Viscosity of gas, intrinsic surface viscosity and evaporation of the surfactant are neglected. The Marangoni force is accepted as responsible for the development of instability. It is taken into account in the tangential stress balance. The Langmuir isotherm and the Szyszkowsky-Langmuir equation for the surface tension are used for the calculations.

The results presented below were obtained for the following values of the variables: solubility of the surfactant in water $c_0=5.8\cdot10^{-5}$ mol/cm^3; volume diffusion coefficient $D=7.8\cdot10^{-6}$ cm^2/s; parameters of the Langmuir isotherm $\Gamma_m=6.2\cdot10^{-10}$ mol/cm^2, $K_L=$

$2.5 \cdot 10^5$ cm^3/mol; vessel radius $R = 30$ mm; vessel high $H = 20$ mm; capillary radius $r_c = 1$ mm; capillary immersion depth $h_c = 10$ mm; droplet radius $r_d = 1$ mm. The c_0, D, Γ_m and K_L values are close to those for hexanol.

Simulation was performed on the regular grid 20 × 30 by using the finite difference method. Eq. (2) and Eqs. (1) and (3) were solved by successive approaches and by explicit up-wind scheme, respectively.

Results of the numerical simulations

The numerical simulation allows us to get the values of concentration and velocity in each grid node at any time. It is thus possible to watch the evolution of the system with time. The comparison of the experimental results (Fig. 1) with surface tension vs. time dependency obtained from the numerical simulation (Fig. 4) shows that the chosen model describes adequately well the main regularities of the system behaviour. At the same time both the calculated induction time and the oscillation period are larger than experimentally observed. The amplitude of the oscillations is smaller than that of the experiments. The most probable cause for the discrepancy between calculated and observed induction times and oscillations periods is the neglection of the buoyancy effect in the accepted model. As shown in Ref. [6], for hexanol the buoyancy should be taken into account. The rather large vessel radius accepted for the calculations could be the cause for the small amplitude values of the oscillations obtained by the calculations. The further theoretical investigations should clear up the effect of the surfactant properties and geometrical factors on the auto-oscillations characteristics. The results of calculations now performed can be used for the explanation of the auto-oscillation mechanism.

The comparison of the calculations performed here for a cylindrical vessel with those unlimited in the radial direction of the liquid layer [5] shows that the behaviour of the system during the induction period and at the beginning development of the instability does not depend on the radial dimensions and coincides for both cases. The induction period is characterised mainly by the diffusion mass transfer. The dissolution of surfactant from the drop leads to an increase of its concentration in the bulk and at the air-water interface. According to the system geometry, the concentration distribution is inhomogeneous in the bulk and at the surface. Both the tangential and the normal to the interface concentration gradients increase over the time. Due to the Marangoni effect the increase of the tangential concentration gradient leads to an increase of the surface velocity, thus also of the bulk velocity and therefore, to an increase of the convective mass transfer. Convection brings more concentrated solution to the surface resulting in a fast increase of the normal concentration gradient and to the development of instability.

Initially, instability arises in the vicinity of the capillary where concentration and concentration gradient have a maximum, when the normal concentration gradient near the surface overcomes a threshold value. The instability propagates then to more distant regions. Its development is characterised by a rapid increase of the surface concentration (Fig. 5) and the velocity (Fig. 6) as well as by an abrupt decrease of the surface tension (Fig. 4).

According to the velocity distribution in radial direction (Fig. 6), surface dilatation takes place in the region near the capillary, and surface compression near the container wall. This causes the appearance of a reverse surface concentration gradient near the wall (Fig. 5). At the same time, the diffusion flux to the surface near the capillary decreases due to convective mixing of the solution. The surface concentration

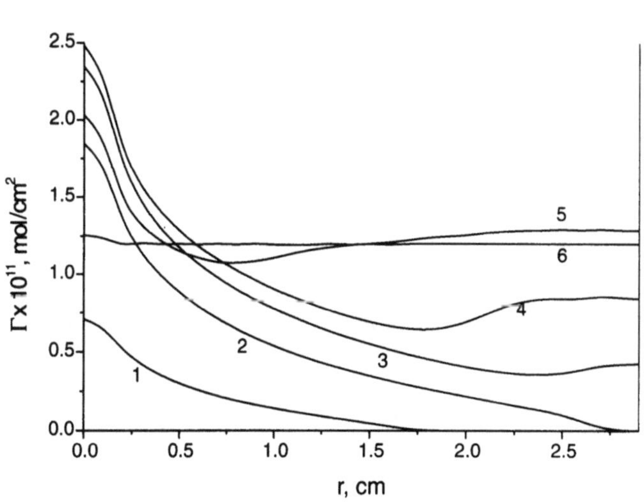

Fig. 4 Surface tension vs. time (numerical simulation)

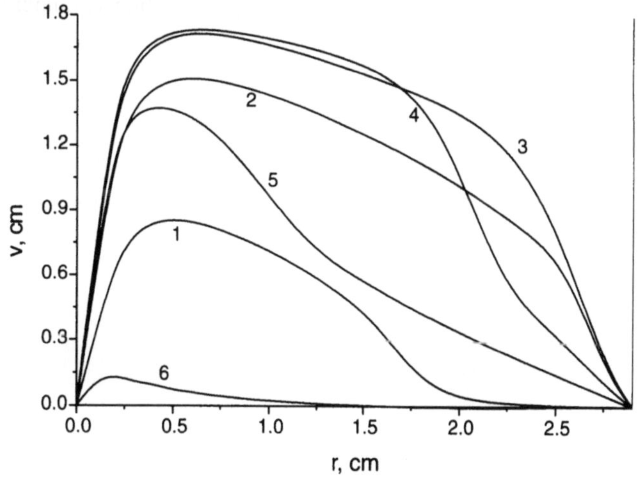

Fig. 5 Radial distribution of the surface concentration (numerical simulation). 1–25 min. 30 s, 2–25 min. 32 s. 3–25 min. 33 s, 4–25 min. 34 s, 5–25 min. 36 s, 6–26 min

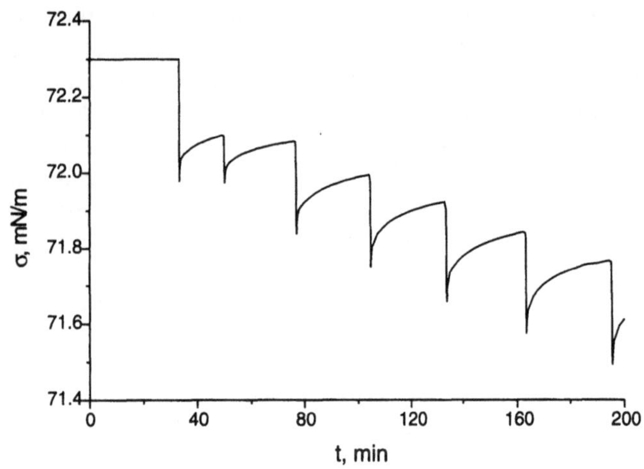

Fig. 6 Radial distribution of the surface velocity (numerical simulation). 1–25 min. 30 s, 2–25 min. 32 s. 3–25 min. 33 s, 4–25 min. 34 s, 5–25 min. 36 s, 6–26 min

decreases in the vicinity of the capillary. The changes in the concentration distribution lead to a decrease of the surface velocity and consequently, of the bulk velocity. The system returns gradually to the slow diffusion stage which is characterised by a gradual increase of the surface tension due to desorption of the solute from the surface into the adjacent dilute solution (Fig. 6). Afterwards, all repeats again (Fig. 4).

Conclusions

The performed numerical simulation shows that the behaviour of the system having a surfactant droplet under the free water surface during the slow diffusion stage and at the beginning of the development of instability is independent of the radial dimensions. The appearance of the first oscillation is determined by a concentration gradient near the surface in the vicinity of the capillary. The behaviour of the system during return to stable state and the possibility of the development of second and following oscillations is determined by the presence of the vessel wall. The surface compression near the wall leads to the appearance of a reverse surface concentration gradient causing a fast decrease of the velocity on the surface and in the bulk and the return of the system in a state with predominance of the diffusion mass transfer. This is a pre condition for the appearance of the following oscillations.

The comparison of the calculated surface tension vs. time dependency with the experimental results shows that the chosen model describes well enough the main regularities of the system behaviour.

Acknowledgements N.M.K. thanks gratefully the Max-Planck Gesellschaft for financial support.

References

1. Kovalchuk VI, Kamusewitz H, Vollhardt D, Kovalchuk NM (1999) Phys Rev E 60: 2029
2. Kovalchuk NM, Vollhardt D (2000) J Phys Chem B 104: 7987
3. Zieper J, Oertel H (eds) (1982) Convective transport and instability phenomena. Braun, Karlsruhe
4. Koschmieder EL (1993) Benard cells and Taylor vortices. Univ Press Cambridge, Cambridge
5. Kovalchuk NM, Kovalchuk VI, Vollhardt D (2001) Phys Rev E 63: 031604
6. Vollhardt D, Kovalchuk NM (2003) Prog Colloid Polym Sci (submitted)

Progr Colloid Polym Sci (2004) 123: 127–130
DOI 10.1007/b11745

V. Hrust
V. Tomišić
N. Kallay

Characterisation of aqueous solutions of ionic surface active agents by conductometry

V. Hrust · V. Tomišić · N. Kallay
Laboratory of Physical Chemistry,
Faculty of Science, University of Zagreb,
P.O. Box 163, 10001 Zagreb, Croatia

Abstract Conductometry data were analysed to determine some equilibrium properties of sodium decylsulphonate aqueous solution. Temperature dependence of conductivity above solubility and below Krafft temperature enabled a determination of the enthalpy of solid phase dissolution. At 15 °C $\Delta_r H = 8.0$ kJ mol^{-1} which agrees well with calorimetry (8.3 kJ mol^{-1}). Increase of enthalpy with temperature is due to the heat capacity change, being 850 J K^{-1} mol^{-1}. Conductivity data above the Krafft temperature enabled estimation of the degree of association of counterions with charged heads of surfactant chains comprising micelles, as well as the corresponding equilibrium constant. The latter slightly decreases with surfactant concentration which is due to decrease in electrostatic potential affecting the state of associated counterions.

Keywords Anionic surfactants · Sodium decylsulphonate · Conductometry · Calorimetry · Counterion association

Introduction

Conductometry is a useful method for the determination of equilibrium properties of ionic surface active agents (SAA) in aqueous systems since the measurements are simple and accurate. It has been commonly used as a method for determination of the critical micellisation concentration (cmc) and the Krafft temperature [1–6]. In this work conductometry was used for the determination of enthalpy of dissolution in the region below the Krafft temperature and above solubility, as well as for the estimation of the equilibrium constant and degree of counterion association above the Krafft temperature and above cmc [7]. An anionic surface active agent, sodium decylsulphonate (NaDS), having a Krafft temperature at 22 °C [3], was used as a model system.

Experimental

The conductivity of aqueous solutions of sodium decylsulphonate was measured by means of a Metrohm AG 712 conductometer in the thermostatted cell. NaDS was obtained from Y. Moroi (Kyushu University, Fukuoka, Japan) as a pure salt and was not further purified.

Calorimetric measurements were carried out in a reaction isoperibolic calorimeter, described elsewhere [8].

Theoretical

Dissolution

Above solubility and below the Krafft temperature the solid phase of NaDS is in equilibrium with the ions:

$$\mathrm{MA(s)} \rightarrow \mathrm{M^+(aq)} + \mathrm{A^-(aq)} \tag{1}$$

The corresponding equilibrium constant is defined using the activity coefficient y as:

$$K_s = y^2 (c(\mathrm{MA})/c^\circ)^2; \quad c^\circ = 1 \text{ mol dm}^{-3} \tag{2}$$

$$c(\mathrm{MA}) = c(\mathrm{M^+}) = c(\mathrm{A^-}) \approx K_s^{1/2} c^\circ; \quad y \approx 1 \tag{3}$$

The temperature dependence of K_s is given by:

$$\ln K_s = -\Delta_r H^\circ / RT + \Delta_r S^\circ / R \tag{4}$$

where $\Delta_r H^\circ$ and $\Delta_r S^\circ$ are the enthalpy change and the entropy change of dissolution, respectively. The variation of these quantities with temperature can be expressed as:

$$\Delta_r H^\circ(T) = \Delta_r H^\circ(T_{ref}) + \Delta_r C_p(T - T_{ref}) \quad (5)$$

$$\Delta_r S^\circ(T) = \Delta_r S^\circ(T_{ref}) + \Delta_r C_p \ln(T/T_{ref}) \quad (6)$$

where T_{ref} is an arbitrary chosen reference temperature.

Below the Krafft temperature, the concentration of dissolved SAA is given by the ratio of conductivity and molar conductivity

$$c(\mathrm{MA}) = \kappa/\Lambda \quad (7)$$

According to Walden's rule, the following relation can be written

$$\Lambda(T) = \Lambda(T_{ref})\eta(T_{ref})/\eta(T) \quad (8)$$

where η denotes viscosity of the medium.

The combination of Eqs. (3), (4) and (7) gives:

$$\ln(\kappa/\Lambda c^\circ) = -\Delta_r H^\circ(T)/2RT + \Delta_r S^\circ(T)/2R \quad (9)$$

and by introducing Eqs. (5), (6), (8) one can derive the expression:

$$\begin{aligned} Y &= \ln(\kappa(T)\eta(T)/\Lambda_\infty \eta(T_{ref})c^\circ) \\ &\quad - (\Delta_r C_p/2R)(T_{ref}/T + \ln(T/T_{ref})) \\ &= (\Delta_r S^\circ(T_{ref}) - \Delta_r C_p)/2R - (\Delta_r H^\circ(T_{ref})/2R)(1/T) \end{aligned} \quad (10)$$

The above equation predicts a linear relation of Y vs. $1/T$ which enables the determination of the enthalpy change of dissolution from the conductivity data.

Conductivity of the micellar system

In the presence of micelles, counterions are partially associated with surfactant ionic heads in the micelle [7]:

$$\mathrm{M^+(aq) + A^-(mic) \rightarrow MA(mic)} \quad (11)$$

The equilibrium constant of the counterion association is defined as:

$$\begin{aligned} K &= [\mathrm{MA(mic)}]/(c(\mathrm{M^+})[\mathrm{A^-(mic)}]) \\ &= (N - z)/(zc(\mathrm{M^+})) \end{aligned} \quad (12)$$

where N is the micelle aggregation number and z denotes the absolute value of the micelle charge number.

Conductivity above cmc can be expressed as a sum of contributions of the ions and of the charged micelles:

$$\kappa = \kappa(\mathrm{M^+}) + \kappa(\mathrm{A^-}) + \kappa(\mathrm{mic}) \quad (13)$$

Molar conductivity of the spherical micelles of radius r in the medium of viscosity η is given by:

$$\Lambda(\mathrm{mic}) = z^2e^2L/6\pi\eta r = z^2 f \quad (14)$$

Assuming that $c(\mathrm{A^-})$ is approximately equal to cmc, denoted as c(mic), it follows:

$$\begin{aligned} \kappa &= \Lambda(\mathrm{MA})c(\mathrm{mic}) \\ &\quad + [(c - c(\mathrm{mic}))z/N][\lambda(\mathrm{M^+}) + Nf(z/N)] \end{aligned} \quad (15)$$

where c denotes total SAA concentration, while relative micellar charge number z/N is:

$$\begin{aligned} z/N &= \{[(Kc(\mathrm{mic}) + 1)^2 + 4K(c - c(\mathrm{mic}))]^{1/2} \\ &\quad - Kc(\mathrm{mic}) - 1\}/[2K(c - c(\mathrm{mic}))] \end{aligned} \quad (16)$$

The counterion association equilibrium constant and the relative micellar charge number can be calculated for each total concentration of SAA by solving Eqs. (15) and (16).

Results and discussion

Conductivity of an aqueous solution of sodium decylsulphonate at various temperatures is shown in Fig. 1. Two different regions can be seen. At lower temperatures the solid phase is in equilibrium with the ions while above 25 °C the micelles are formed and the conductivity again increases with temperature. Eq. (10) was applied to analyse the behaviour of the system in details. The linear relationship between Y and inverse temperature exists in the temperature region below 14 °C (Fig. 2). The value of $\Delta_r H^\circ$ calculated from the slope of the straight line corresponding to the reference temperature of 15 °C, was found to be 8.0 kJ mol^{-1}. In calculations the value of the heat capacity change of $\Delta_r C_p$ = 850 J mol^{-1} K^{-1} was used. The enthalpy of dissolution was also determined independently by reaction calorimetry. The temperature dependence of enthalpy was established earlier; it was concluded that heat capacity change is due to the water structuring around the hydrophobic tails [9–12]. The heat capacity change was determined according to Eq. (5) from the slope of the linear function of $\Delta_r H$ vs. T (Fig. 3) as $\Delta_r C_p$ = 850 J mol^{-1} K^{-1}. At 15 °C the calorimetry yielded $\Delta_r H$ = 8.3 kJ mol^{-1} which is close to the value determined by conductometry.

In Fig. 2 three temperature regions could be distinguished. The linearity below 14 °C agrees with Eq. (10) which means that in this region free ions are in equilibrium with the solid phase. It can be also concluded that above 25 °C free ions are in equilibrium with micelles. In order to analyse the system in the region between 14 and 25 °C the solubility of the surfactant

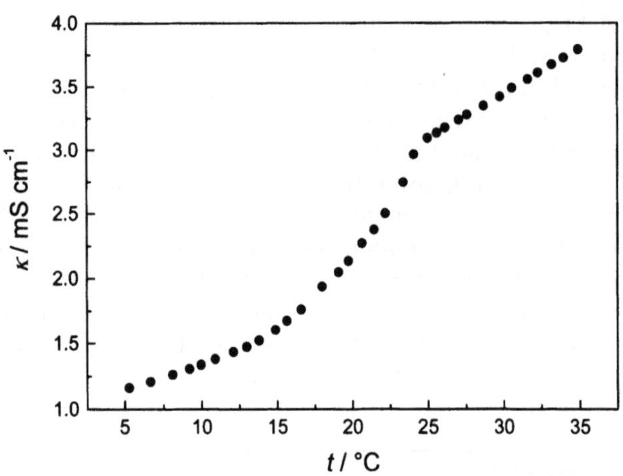

Fig. 1 Conductivity of NaDS aqueous system as a function of temperature at surfactant content n(NaDS)/V = 0.064 mol dm^{-3} > c_{sat}, c_{mic}

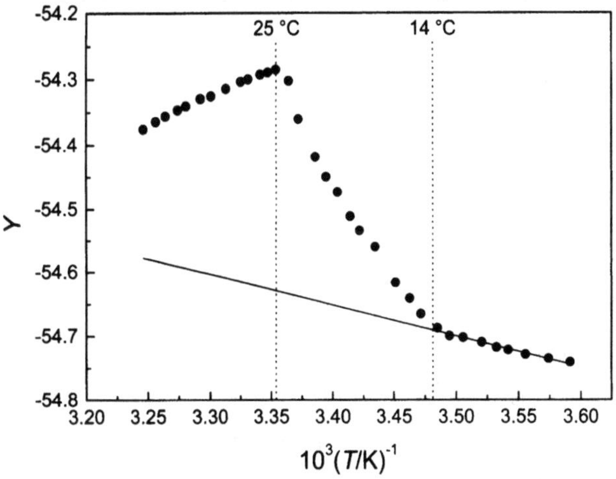

Fig. 2 Interpretation of the conductivity data of NaDS aqueous system presented in Fig. 1 according to Eq. (10)

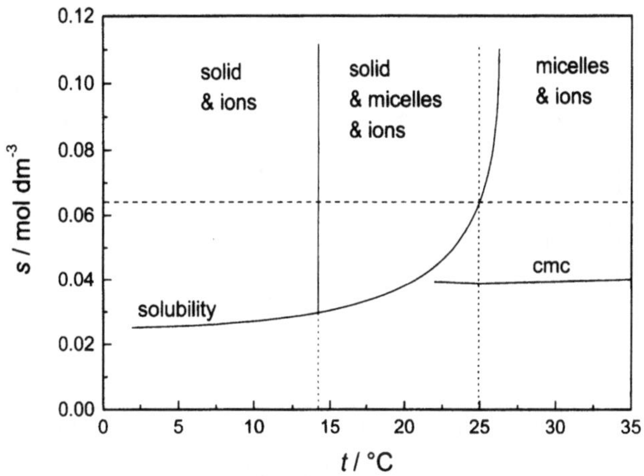

Fig. 4 Schematic representation of the temperature dependence of solubility and cmc according to Ref. [1]

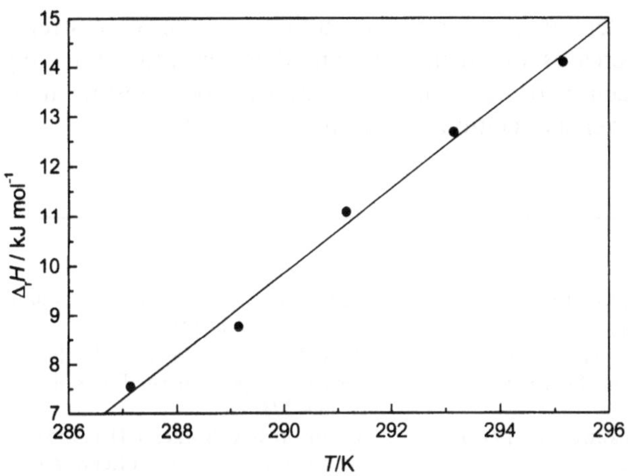

Fig. 3 Dissolution enthalpy of NaDS determined by calorimetry as a function of temperature

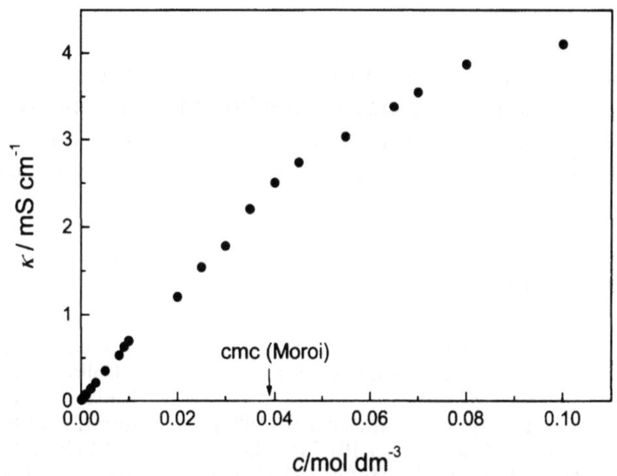

Fig. 5 Conductivity of aqueous solutions of NaDS as a function of concentration at 25 °C

should be considered. Figure 4 is schematic representation showing the temperature dependence of solubility and cmc [1]. It is clear that the system is not ideal since the solubility is not infinite above the Krafft temperature but rather markedly increases. (Note that ideal behaviour may be expected in the case of non-ionic surfactants and large micelles.) In our conductometric experiment the amount of NaDS per volume was 0.064 mol dm^{-3} so that complete dissolution may be expected at 25 °C which is higher than the Krafft temperature. Therefore, one may conclude that between 14 and 25 °C the micelles coexist with the solid phase and that below 14 °C no micelles exist.

The conductivity results at 25 °C are presented in Fig. 5. Data above cmc were analysed according to Eqs. (15) and (16). The value of $\lambda(Na^+) = 50.1$ S cm^2 mol^{-1} was taken from the literature [13], while the molar conductivity of the surfactant anion (18.8 S cm^2 mol^{-1}) was obtained from the slope of κ vs. c function below cmc. Critical micellisation concentration ($3.88 \cdot 10^{-2}$ mol dm^{-3}) as determined by Moroi et al. [1] was used. In order to examine the influence of the micelle radius and its aggregation number on the calculated equilibrium constant and degree of counterion (Na^+) dissociation, z/N, two values of the size parameter Nf were assumed, i.e. 200 and 500 S cm^2 mol^{-1}. As can be seen in Fig. 6, the value of counterion association equilibrium constant slightly decreases with increasing total concentration, while z/N remains approximately the same in the concentration range studied (Fig. 7). The patterns of K_s and z/N dependence on c are similar for both Nf values assumed. The Nf value of 200 S cm^2 mol^{-1} may be taken as more realistic because it corresponds approximately to aggregation number of 50, and hydrodynamic radius of 2 nm. If so, the degree of counterion

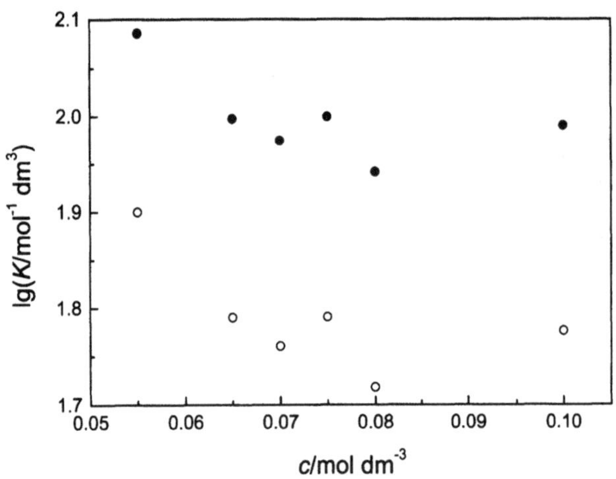

Fig. 6 Equilibrium constant of counterion association as a function of total concentration of NaDS at 25 °C as calculated by Eqs. (15) and (16) using the conductivity data presented in Fig. 5. Two different values of fN in S cm^2 mol^{-1} were assumed: ○ 200 and • 500

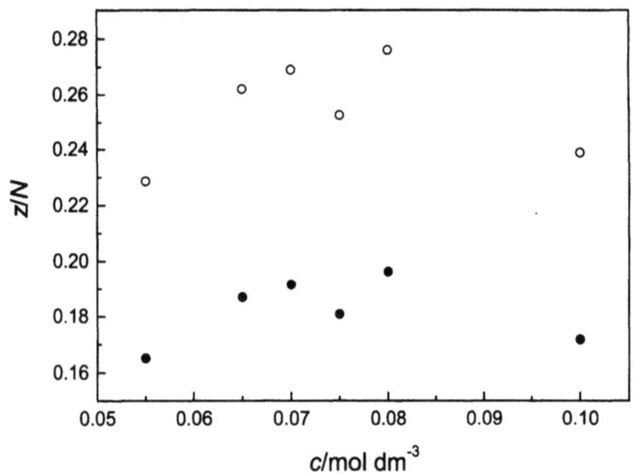

Fig. 7 Relative charge number of micelles as a function of total concentration of NaDS at 25 °C as calculated by Eq. (15) using the conductivity data presented in Fig. 5. Two different values of fN in S cm^2 mol^{-1} were assumed: ○ 200 and • 500

dissociation from micelles is about 25%. The decrease of counterion association equilibrium constant with surfactant concentration could be explained considering the effect of ionic strength on the electrostatic potential at the micellar surface. As the ionic strength increases the surface potential is reduced, and consequently decrease in the attractive force results in the lower association equilibrium constant.

References

1. Moroi Y, Sugii R, Akine C, Matuura R (1985) J Colloid Interface Sci 108:180–188
2. Moroi Y, Ikeda N, Matuura R (1984) J Colloid Interface Sci 101:285–288
3. Saito M, Moroi Y, Matuura R (1982) J Colloid Interface Sci 88:578–583
4. Saito M, Moroi Y, Matuura R (1980) J Colloid Interface Sci 76:256–258
5. Hato M, Shinoda K (1973) Bulletin Chem Sci Japan 46:3889–3890
6. Mukhayer GI, Davis SS (1976) J Colloid Interface Sci 56:350–359
7. Kallay N, Tomašić V, Žalac S, Chittofrati A (1994) Colloid Polym Sci 272:1576–1581
8. Simeon V, Ivičić N, Tkalčec M (1972) Z Phys Chem 78:1–12
9. Nemethy G, Scheraga HA (1962) J Chem Phys 36:3382–3400
10. Kallay N, Tomašić V, Hrust V, Dugandžić V, Tomić M (1989) J Surface Sci Technol 5:255–266
11. Kallay N, Hrust V, Moroi Y (1990) Colloids Surf 47:125–133
12. Hrust V, Branisavljević T, Kallay N (1998) J Dispersion Sci Technol 19:369–378
13. Robinson RA, Stokes RH (1955) Electrolyte solutions. Butterworths, London

Progr Colloid Polym Sci (2004) 123: 131–135
DOI 10.1007/b111746

Elisa González-Romero
Begoña Fernández-Calvar
Carlos Bravo-Díaz

Electrochemical determination of the stability constant of an aryl radical with β-cyclodextrin

E. González-Romero
B. Fernández-Calvar
Universidad de Vigo,
Facultad de ciencias,
Dpto. química analítica y alimentaria,
36200 Vigo, Pontevedra, Spain

C. Bravo-Díaz (✉)
Universidad de Vigo,
Facultad de ciencias,
Dpto. química física, 36200 Vigo,
Pontevedra, Spain
e-mail: cbravo@uvigo.es

Abstract Addition of β-cyclodextrin (β-CD) to an aqueous acid solution of *p*-nitrobenzenediazonium (PNBD) tetrafluoroborate causes a substantial decrease in peak currents and shifts in opposite directions the apparent peak potential values, E_{app}, of the -N_2^+ and -NO_2 groups. β-CD hinders the reduction of the nitro group since E_{app}(-NO_2) is shifted towards more negative values, in contrast with the E_{app} shift observed for the one-electron reduction of the -N_2^+ group to form the arenediazenyl radical, $ArN_2^{\cdot}$, which is shifted towards more positive values and thus favouring the reduction of PNBD. Either the decrease in peak currents or E_{app} shifts are interpreted in terms of the formation of an inclusion complex between β-CD and PNBD, with the -NO_2 group inserted into the CD cavity. Such specific spatial configuration contrasts with that observed when crown ethers and calix-[n]-arenes are employed as host molecules. Values for the association constants of the parent arenediazonium ions and of the electrochemically-generated arenediazenyl radicals with β-CD were estimated from the electrochemical measurements.

Keywords Arenediazonium ions · Aryl radicals · Cyclodextrins · Polarography

Introduction

Probably one of the most interesting aspects of cyclodextrin chemistry is their ability for recognising guest molecules mainly by their sizes, making cyclodextrins, CDs, to be one of the most widely studied host compounds among a variety of them including crown ethers and calix-n-arenes. The CDs have very interesting catalytic properties, and they have been used as models of hydrolytic enzymes [1–3]. Important applications have also been found in electroorganic synthesis, either when CDs are added to solution or bound to the electrode surface, causing substantial beneficial changes in the selectivity of the reactions [4]. A commonly proposed mechanism for the oxidation-reduction process of a substrate in the presence of host molecules, Scheme 1, involves electron transfer to/from the electrode from/to the guest molecule in the inclusion complex forming a reactive radical-intermediate (e.g., a radical or anion radical) that may be held within the CD cavity [5]. Certainly, the reactivity of this intermediate will differ from that of the uncomplexed intermediate and of the parent substrate, leading to a substantial change in the selectivity of the reaction, which is frequently observed in the presence of CDs.

The carcinogenic properties of *N*-nitrosamines and other *N*-nitroso compounds have been known for over 30 years [6, 7]. *N*-Nitrosamide type compounds are known to be direct mutagenic agents since they require no biological activation but the *N*-nitrosamine analogues must be metabolised in order to elicit their mutagenic or carcinogenic properties [6, 7]. The mode of carcinogenic biochemical activation is not known for all nitrosamines but it is believed that many of them are activated through the process of α-hydroxylation, resulting in an unstable α-hydroxynitrosamine which decomposes readily to yield

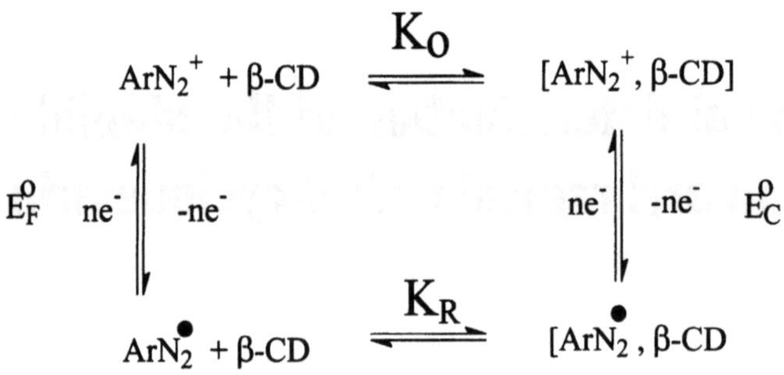

Scheme 1 Simplified square mechanism illustrating the electrochemical reaction of ArN_2^+ in the presence of *β*-CD involving *n* electrons. E_F and E_C represent the formal potentials of the "free" and "complexed" ArN_2^+

diazonium ions, which in turn are aggressive alkylating agents due to their ability to generate aryl radicals [8, 9]. On the other hand, arenediazonium compounds, ArN_2^+, are strong oxidising agents that undergo homolytic fragmentation to produce aryl radicals upon reacting with certain electron donors [10, 11] and there is good evidence [12–14] that the ability of arenediazonium ions to generate radicals may be responsible, to some extent, for the mutagenic and carcinogenic properties of aromatic diazonium compounds [8, 9, 12–14, 15, 16].

From a mechanistic stand point of view, the most straightforward way of promoting electron transfer to an arenediazonium ion is at the surface of an electrode [11, 17–19]. The basic aspects of the electrochemical reduction process of arenediazonium ions in aqueous and in some binary systems are known [18–23]. Polarograms of aqueous acid solutions of a variety of arenediazonium ions show that two polarographic waves can be observed in potential regions of +0.02 V to −0.1 V (vs. Ag/AgCl), a potential where few organic compounds are electroactive [24, 25], and around −0.5 V to −0.70 V. Microcoulometry [18] at the DME shows that the two waves correspond to the uptake of one electron yielding a diazenyl radical, and four (overall) electrons, respectively, yielding phenylhydrazine [17]. Thus, when an arenediazonium ion acquires an electron it forms an aryldiazenyl radical, $ArN_2^{\cdot}$, a rather labile specie that in turn yields an aryl radical, $Ar^{\cdot}$ and dinitrogen.

Dediazoniation reactions have been extensively studied under a variety of conditions but almost nothing is known about the details of electron-transfer reactions of arenediazonium ions in the presence of host molecules like CDs [10, 11, 26], so we set out to investigate a suitable model system by employing *p*-nitrobenzenediazonium ion and *β*-CD. Previous dediazoniation work in the presence of CDs suggest that arenediazonium ions form inclusion complexes with the $-N_2^+$ group located in the vicinity of the secondary hydroxy groups of the CDs [26, 27], in contrast with the spatial configuration observed when employing other host macrocycles like crown ethers [11, 28–30].

Experimental

Differential pulse polarographic (DPP) measurements were obtained with a Metrohm E506 Polarecord, in conjunction with the following cell system. A 663 VA-Stand (Metrohm) equipped with water jacketed voltammetric cell was used. The multimode-working electrode was used in the DME mode. The three-electrode system was attached to a glassy carbon rod (2 x 65 mm) auxiliary electrode and Ag/AgCl (3 M KCl) reference electrode. All potentials given in this work are relative to the mentioned Ag/AgCl electrode. Instrumental conditions were: scan rate $= 7.5$ mV s^{-1}, drop time $t_g = 0.4$ s, pulse amplitude $\Delta E = -80$ mV, initial potential $E_i = +0.2$ V. Solutions used in polarographic measurements were bubbled with N_2 gas (99.999%) for at least 10 min and kept under a nitrogen atmosphere during the electrochemical runs.

Materials

Reagents were of maximum purity available and were used without further purification. The reagents used in the preparation of the diazonium salt (as tetrafluoroborate) and in the preparation of the Britton-Robinson (BR) buffer, *p*-nitrophenol, PNBOH, nitrobenzene, PNBH, and *β*-cyclodextrin, *β*-CD, were purchased from Aldrich or Fluka. Other materials employed were from Riedel de Hen. All solutions were prepared by using Milli-Q grade water. PNBD was prepared under non-aqueous conditions as described in previous work [31, 32] and was stored in the dark at low temperature to minimise its spontaneous decomposition. The UV/VIS spectrum of an aqueous acid (2.0×10^{-3} M HCl) PNBD solution and the ^{1}H-NMR spectra of PNBD in CD_3CN at 25 °C (not shown) are in agreement with literature data [32]. PNBD stock solutions were prepared by dissolving the diazonium salt in aqueous HCl, to minimise diazoacetate formation [33], to give final concentrations of [PNBD]~2×10^{-4} M and they were freshly prepared and used immediately to minimise their decomposition.

Results and discussion

In the absence of *β*-CD, the polarogram of an aqueous acid solution of PNBD (not shown) confirms the two polarographic waves observed in previous dediazoniation work [26, 34] at a given pH, but the morphology of the polarogram is substantially modified in the presence of *β*-CD, Fig. 1. E_{app} values are significantly shifted and the corresponding peak currents decrease. In addition, a new polarographic peak, marked with an arrow in Fig. 1, not observed in the absence of CD, is detected at $E_p \sim -0.40$ mV. Preliminary experiments indicated that i_p values are time dependent, confirming that there exists a reaction between PNBD and *β*-CD.

With increasing *β*-CD, peak potential values, E_{app}, for the $-N_2^+$ group are shifted towards more positive values and peak currents, i_p drops significantly, Fig. 2. The observed i_p decrease is consistent with the expected formation of inclusion complexes because the diffusion coefficients, which are related with peak currents through the Ilkovic [25] equation, of the complexed substrates, D_C, are smaller than those of the "free" substrates, D_F, since molecular sizes of guest molecules are usually much smaller that those of CDs. Alternatively, the observed E_p

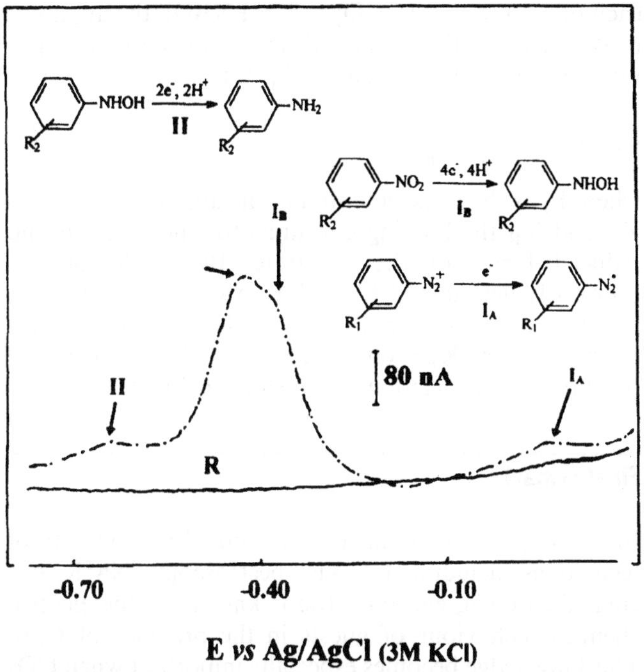

Fig. 1 Polarogram of a β-CD/PNBD aqueous acid solution. [β-CD] = 40 [PNBD] = 3.95×10^{-5} M, T = 4.2 °C. The chemical processes associated to each polarographic peak are also indicated according to literature data

shift indicates that β-CD favours the reduction of the diazonium group, favouring an homolytic dediazoniation mechanism. In contrast, E_{app} values for the -NO_2 group are shifted towards more negative values, suggesting a more difficult electron transfer associated to the reduction of the nitro group and thus consistent with theformation of PNBD·β-CD inclusion complexes with the -NO_2 group inserted into the CD cavity.

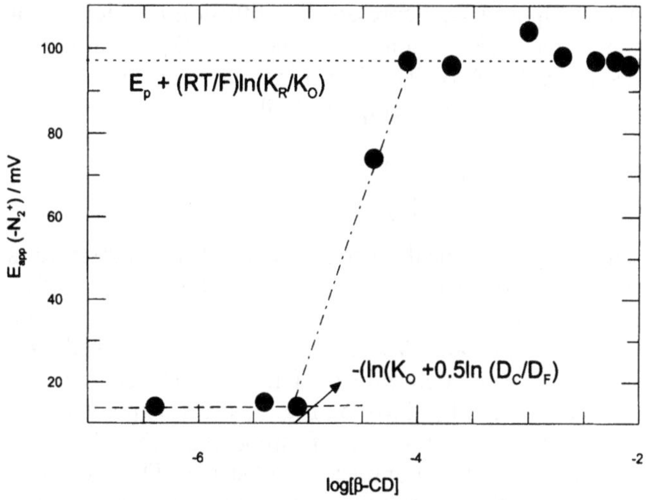

Fig. 3 Variation of E_{app} (peak I_A in Fig. 1) with [β-CD]. The inflection point takes place when [PNBD] = [β-CD] indicating a 1:1 inclusion complex. [PNBD] = 2.0×10^{-4} M, T = 25 °C

Polarographic titration of PNBD with β-CD, Fig. 3, shows the variation of $E_{ap}(I_A)$ with β-CD. The inflection point of the variation of either i_p (not shown) or shifts in E_{app} takes place when [β-CD] = [PNBD], indicating that a 1:1 inclusion complex is being formed. Quantitative analysis of the peak current dependence with [β-CD] provides an easy way to determine the corresponding association constants assuming that complexation reactions remain at equilibrium everywhere in the diffusion layer [5, 35]. Under these conditions, the apparent diffusion coefficient, D_{app}, can be obtained by means of Eq. (1):

$$D_{app}[S_T] = D_C[S_C] + D_F[S_F] \qquad (1)$$

Fig. 2 Variation of the peak current, i_p (○) and E_{app} (●) of peak I_A (see Fig. 1 for peak identification) upon increasing [β-CD]. [PNBD] = 4×10^{-5} M, pH = 4.5 (BR buffer), T = 30 °C

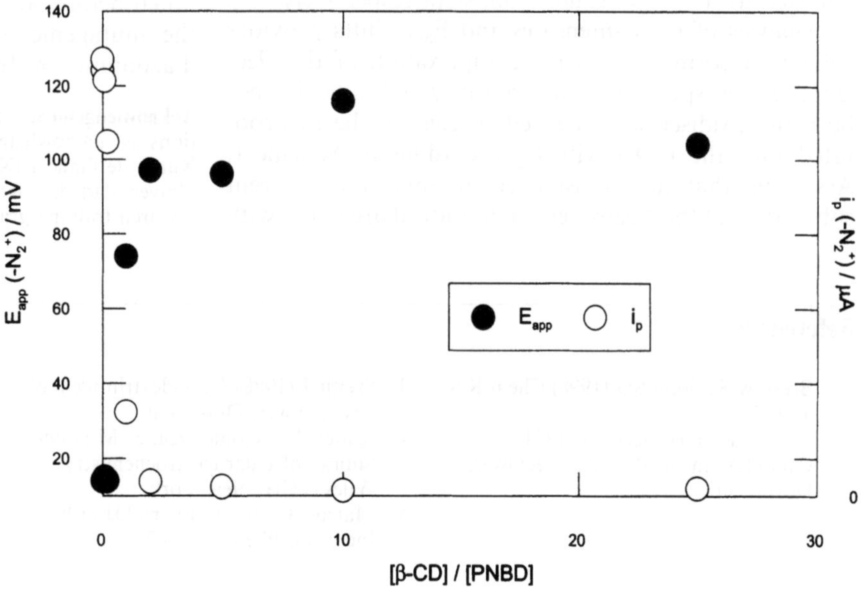

where D_F and D_C represent the diffusion coefficients of the free (S_F) and complexed (S_C) substrate ($[S_T]=[S_C]+[S_F]$). To determine the association constant, Eq. (2) can be employed [5]

$$\frac{i_{pc}^2}{i_{p0}^2}=\frac{1}{K_O[CD]}\left(1-\frac{i_{pc}^2}{i_{p0}^2}\right)+\frac{D_C}{D_F} \quad (2)$$

where i_{pc} is the peak currents when all the substrate is incorporated into the CD, i.e., when $[\beta\text{-CD}]>>>[S_T]$, i_{p0} is the peak current in the absence of CD and K_O the inclusion constant. Values of $K_O=2282\ M^{-1}$ and $D_C/D_F=0.03$ were determined. The ratio D_C/D_F can be calculated from E_P shifts [5], Fig. 3, and as expected, $D_C/D_F<<1$ in all cases, confirming that the diffusion coefficient of the complexed substrate, D_C, is much lower than that of the free substrate. The obtained K_O value for PNBD is much higher that that expected for the inclusion of a positively charged molecule, which is usually lower than that for the neutral specie. The stability of a CD-complexes is improved upon increasing the hydrophilic character of the guest molecules: electron-releasing groups in a given guest improve the stability of the complexes; meanwhile electron-withdrawing substituents hinder complex formation [2, 36]. In addition, the charge of the guest may also be crucial [36]: anionic compounds bind more tightly than neutral ones and much more than cationic ones. Thus the stability constant for the formation of the PNBD inclusion complex should be lower that those of PNBOH, PNBH or even the parent *p*-nitroaniline. This high value has been explained [26] by assuming two rapid preequilibrium steps prior to the slow step of the reaction, the formation of the inclusion complex followed by the formation of a highly unstable transient Z-diazoether which further decomposes yielding the arenediazenyl radical. Values of $K_1\sim 20\ M^{-1}$ and $K_2\sim 115\ M^{-1}$ were reported for those preequilibrium steps respectively [26].

Analysis of peak intensities and E_{app} shifts provides valuable information on the complexation of the electrogenerated specie, i.e., the arenediazenyl radical, since both the oxidised and reduced species can be incorporated into the CD cavity [5] according to Scheme 1. Assuming that the diffusion coefficients of the parent substrate and the electrogenerated radical are equal with each other and that $[\beta\text{-CD}]>>>[\text{PNBD}]$, the apparent peak potential, E_{app}, when all the substrate is incorporated into the CD is given by Eq. (3)

$$E_{app}=E_F+\frac{RT}{nF}\ln\frac{K_R}{K_O} \quad (3)$$

where E_F is the peak potential in the absence of CD and K_O and K_R the binding constants for the oxidised and reduced forms of the substrate. By employing the previously determined $K_O=K_1K_2$ value, $2282\ M^{-1}$, the stability constant for the $ArN_2^{\cdot}$-β-CD complex was estimated to be $K_R\approx 760\ M^{-1}$, much higher than that of the parent arenediazonium ion ($K_1\sim 20\ M^{-1}$).

Final remarks

Inclusion phenomena have been studied by a variety of techniques including UV-Vis spectroscopy, NMR, circular dichroism, etc. but little is known on the electrochemical behaviour of guests in the presence of CDs. This knowledge becomes especially important when CDs act as catalysts or inhibitors in electron transfer reactions in biological systems, where CDs are widely employed either as drug or food carriers. In this work, we have shown a relatively easy way to determine the stability constants of the substrates and the electrochemically-generated radicals.

Indeed, when a molecule undergoes an electron-transfer process, both the oxidised and reduced species can coexist in solution and, because of the relatively hydrophobic nature of the CD cavity, the reaction-intermediate radicals may be stabilised and such an interaction may play an important role in the behaviour of the guests that needs to be taken into consideration in various applications of CDs. Indeed, this is the case for the electrochemically generated aryl cations, which as noted before, are believed to be partially responsible for the mutagenic and carcinogenic properties of arenediazonium ions [9, 16, 23].

Acknowledgements Financial support from the following institutions is acknowledged: MCYT of Spain (BQU2000-0239-C02), Xunta de Galicia (XUGA 38301A92 and XUGA 38305A94) and Universidad de Vigo. B.F.-C. thanks Universidad de Vigo for a research training grant.

References

1. Breslow R, Song SD (1998) Chem Rev 98:1997
2. Frömming KH, Szejtli J (1994) Cyclodextrins in Pharmacy. Kluwer, Dordrecht
3. Szejtli J (1988) Cyclodextrin technology. Kluwer, Dordrecht
4. Kaifer AE, Gómez-Kaifer M (1999) Supramolecular electrochemistry. Wiley-VCH, New York
5. Matsue T, Osa T, Evans DJ (1984). Inclusion Phenom 2:547
6. Loeppky RN, Michejda CJ (1994) Nitrosamines and related N-nitroso compounds. In: Chemistry and biochemistry, vol 553. American Chemical Society, Washington

7. Preussmann R, Stewart BW (1984) N-nitroso carcinogens, vol 182. American Chemical Society, Washington
8. Reszka KJ, Chignell CFJ (1993) J Am Chem Soc 115:7752–7760
9. Reszka KJ, Chignell CFJ (1995) Chem-Biol Interact 96:223
10. Galli C (1988) Chem Rev 88:765
11. Zollinger H (1994) Diazo chemistry I. Aromatic and Heteroaromatic Compounds. VCH, Weinheim
12. Augusto O (1993) Free Rad Biol Med 15:329
13. Kato T, Kojima K, Hiramoto K, Kigugawa K (1992) Mutat Res 268:105
14. Ohshima H, Friesen M, Malaveille C, Brouet I, Hautefeulli A, Bartsch H (1989) Food Chem Toxicol 27:193
15. Riordan J, Vallee BL (1972). Methods Enzymol 25:521–523
16. Quintero B, Morales JJ, Quiros M, Martinez-Puentedura M, Cabeza MC (2000) Free Radic Biol Med 29:464
17. Ben-Efrain DA (1978) Detection and determination of diazo and diazonium groups. In: Patai S (ed), The chemistry of diazonium and diazo groups, vol 1. Wiley, New York
18. Fry AJ (1978). Electrochemistry of the diazo and diazonium groups. In: Patai S (ed) The chemistry of diazo and diazonium groups. Wiley, New York
19. Viertler H, Pardini VL, Vargas RR (1994) The electrochemistry of triple bond. In: Patai S, Viertler H, Pardini VL, Vargas RR (eds), The chemistry of triple-bonded functional groups, supplement C. Wiley, New York
20. Bravo-Diaz C, Gonzalez-Romero E (1999) Anal Chim Acta 385:373
21. Bravo-Díaz C, Romero-Nieto ME, Gonzalez-Romero E (2000) Langmuir 16:42
22. Pazo-Llorente R, Bravo-Díaz C, González-Romero E, Fresenius J (2001) Anal Chem 369:582
23. Costas-Costas U, Gonzalez-Romero E, Bravo Díaz C (2001). Helv Chim Acta 84:632
24. Zuman P (1969) Physical organic polarography. In: Zuman P, Perrin CL (eds), Organic polarography. Wiley, New York
25. Sawyer DT, Sobkowiak A, Roberts JL (1995) Electrochemistry for chemists, 2nd end. Wiley, New York
26. González-Romero E, Malvido-Hermelo B, Bravo-Díaz C (2002) Langmuir 18:46
27. Bravo-Díaz C, Sarabia-Rodriguez MJ, Barreiro-Sio P, Gonzalez-Romero E (1999) Langmuir 15:2823
28. Bartsch RA (1983) Complexation of aryl diazonium ions by polyethers. In: Saul Patai (ed), The chemistry of triple-bonded functional groups, vol 2. Wiley, New York
29. Kuokkanen TJ (1997) Phys Org Chem 10:67
30. Kuokkanen T, Palokangas J, Talvensaari MJ (2000) Phys Org Chem 13:452
31. Garcia-Meijide MC, Bravo-Diaz C, Romsted LS (1998) Int J Chem Kinet 30:31
32. Bravo-Diaz C, Romsted LS, Harbowy M, Romero-Nieto ME, Gonzalez-Romero EJ (1999) Phys Org Chem 12:130
33. Zollinger H, Wittwer C (1952) Helv Chim Acta 35:1209
34. Romero-Nieto ME, Bravo-Diaz C, Gonzalez-Romero E (2000) Int J Chem Kin 32:419
35. Osa T, Matsue T, Fujihra M (1977) Heterocycles 6:1833
36. Szente L (1996) Preparation of cyclodextrins complexes. Szente L (ed.), Elsevier, Amsterdam

Progr Colloid Polym Sci (2004) 123: 136–140
DOI 10.1007/b11747

L. Lehmann
E. Kudryashov
V. Buckin

Ultrasonic monitoring of the gelatinisation of starch

L. Lehmann · E. Kudryashov
V. Buckin (✉)
Department of Chemistry,
University College Dublin,
Belfield, Dublin 4, Ireland
e-mail: vitaly.buckin@ucd.ie
Tel.: +353-1-7162436
Fax: +353-1-7162127

Abstract High-resolution ultrasonic spectroscopy was applied to analysis of starch gelatinisation. Temperature profiles of ultrasonic attenuation and velocity were obtained in three different starch suspension samples (corn, rice and wheat) at 25% w/w starch-water ratio using the ascending temperature ramp regime (2 °C/min) in the range 37 to 95 °C. Both ultrasonic velocity and attenuation decrease upon the gelatinisation process. Three regions in the temperature profile of the ultrasonic velocity and attenuation were distinguished during formation of starch gel and attributed to the successive steps involved in the gelatinisation process. The effects of concentration and chemical composition of starch on ultrasonic parameters were observed and interpreted in terms of microrheological parameters of the starch suspensions and gels. Temperature profiles of both ultrasonic velocity and attenuation were also found to be frequency dependent and this was related to the microstructure of starch.

Keywords Ultrasonic spectroscopy · Starch · Gelatinisation · Ultrasonic velocity · Ultrasonic

Introduction

Starches are widely used in food and other industries to achieve the desired viscosity of various products. This involves gelatinisation, which is formation of gel by heating of starch-water mixtures. The main steps of this process include swelling and progressive hydration of the starch granule when heated in excess of water. After further swelling, amylose molecules are solubilised and released into the starch-water suspension. Eventually a three-dimensional gel network of amylose is formed, reinforced by strong interactions between the swollen starch particles. Starch origin, composition, heating procedure and other factors affect the properties of the gel formed. Understanding of the details of gelatinisation requires analytical techniques capable to provide microstructural information in gelation process in starches. Numerous methods have been employed for analysis of the starch gelatinisation. The most popular of them are rheological measurements [1–3] and differential scanning calorimetry (DSC) [3–7]. However interpretation of results is often complicated as it was found that the rheological data depend on specific instrumentation [8] and the absence of stirring in DSC technique results in the measurements in inhomogeneous suspension due to the rapid sedimentation of starch granules in cold water [9]. Other techniques, which were employed to characterise the gelatinisation in starches, include Fourier transform infrared spectrometry (FT-IR) [10], NMR spectroscopy [11], thermomechanical analysis (TMA) [12], electrical conductivity measurements [13] and monitoring of the loss of birefringence [14]. These techniques have also some limitations such as complicated post-experimental analyses or microscopic interpretation of the results. Apart from very expensive NMR, they cannot provide structural information on the bulk sample. Therefore there is a need for a new experimental technique, which avoids these limitations in analysis of the starch gelatinisation.

In the present paper we describe the application of novel high-resolution ultrasonic spectroscopy for analysis of gelation process in starches. This method is based on the measurements of parameters of acoustical (ultrasonic) waves propagating through the analysed sample. It is capable to perform measurements in optically opaque systems. Scattering of ultrasonic waves on micro non-homogeneities allows analysis of the microstructure of the sample. Measurements of ultrasonic velocity provide information on microelasticity and hydration of starch components. Overall, ultrasonic spectroscopy presents the advantage of providing access to both structural and thermodynamic information.

High-resolution ultrasonic spectroscopy has successfully been used previously to monitor gelatinisation processes in food colloids (e.g., yoghurt) [15–19]. It has the advantages of being non-destructive, non-invasive, fast and highly sensitive. As ultrasonic waves are the waves of high frequency mechanical deformations, by its nature this is a high frequency rheological technique and the results can be interpreted in terms of rheological characteristics. Létang and co-workers [20] used ultrasonic measurements for analysis of wheat-flour-water doughs, which, at microscopic level, are equivalent to concentrated suspensions of starch granules in a gluten matrix. Other work was carried out by Povey and Rosenthal [21] on the ultrasonic detection of the degradation of starch by α-amylase. No study concerning the actual starch gelatinisation process by ultrasonic measurements has been reported.

Recently a new generation of ultrasonic devices, high-resolution ultrasonic spectrometers, were introduced to the market [22]. It was demonstrated that these instruments are a high-resolution tool for the study of heat induced phase transitions and provide high frequency viscoelastic properties of liquids and gels. By working in the megahertz frequency range, it is possible to access processes and interactions with characteristic times of 10^{-7} s and shorter and analyse their contribution to the viscoelasticity of gels, thus allowing a better understanding of the molecular mechanism involved in the gelatinisation process. In the present work this new technique was employed for characterisation of starch gelatinisation.

Experimental

Materials

Starch

Three different starch samples were analysed: corn starch (S-4126), rice starch (S-7260) and wheat starch (S-5127). All of the products were purchased from Sigma (St. Louis, MO.). The 25% (w/w) starch-water mixtures were prepared by mixing the appropriate amount of starch powder with deionised water (Mili-Q water) and hydration was allowed for over six hours. Before the loading into ultrasonic cell, the mixture was shaken vigorously until optically homogenous.

Methods

The HR-US 101 ultrasonic spectrometer from Ultrasonic Scientific Ltd. was used for measurements of ultrasonic velocity and attenuation in the starch-water suspensions. This spectrometer provides resolution in ultrasonic velocity of 10^{-5}% and attenuation of 0.2%. All measurements were done differentially with two identical cells of 1 cm^3. The measuring cell was filled with the starch-water solution. The reference cell was filled with deionised water. The samples in both cells were kept under continuous stirring during measurements, which was efficient to prevent the sedimentation of the starch particles. Three frequencies were monitored: 4.5 MHz, 7.6 and 11.0 MHz. All measurements were done in the temperature ramp regime in the temperature range from 37 °C to 95 °C with the temperature ramp rate 2 °C/min. The temperature in the cell was monitored by the incorporated thermometer provided with the HR-US 101 spectrometer.

Results and discussion

Ultrasonic velocity, u, and attenuation, α, are determined by the rheological characteristics of the medium. In a homogenous medium the following Eq. (1) is valid [23]:

$$u \cong ((K' + 4/3G')/\rho)^{1/2}$$

where G' is the shear modulus and K' the volume modulus of the medium, ρ is the density of the medium and f the frequency of the ultrasonic wave. Similarly, ultrasonic attenuation can be related to the shear and volume moduli. Prior the gelation the shear storage modulus, G', is negligibly small and therefore the volume storage modulus, K', is the major contributor to ultrasonic velocity in the suspension. Monitoring the ultrasonic velocity and attenuation of starch-water mixtures during gelatinisation allows the characterisation of the viscoelastic properties of the starch granules and gel.

Fig. 1 illustrates the evolution of the ultrasonic velocity during the heat-induced gelatinisation of the different starch-water mixtures. For comparative purposes, the baseline of the gelatinised starch samples was subtracted from the ultrasonic velocity temperature profiles of the different starches studied and the data were adjusted so that the value of the ultrasonic velocity is zero at 90 °C. The baseline used in the calculation is shown in the insert. Fig. 2 illustrates the changes in ultrasonic attenuation temperature profile. The data were plotted using the ultrasonic attenuation at 90 °C as reference values.

The results of the ultrasonic measurements for 25% (w/w) starch-water mixtures are summarised in Table 1. The gelatinisation can be characterised by an onset temperature (T_O), a transition temperature (T_T) and a

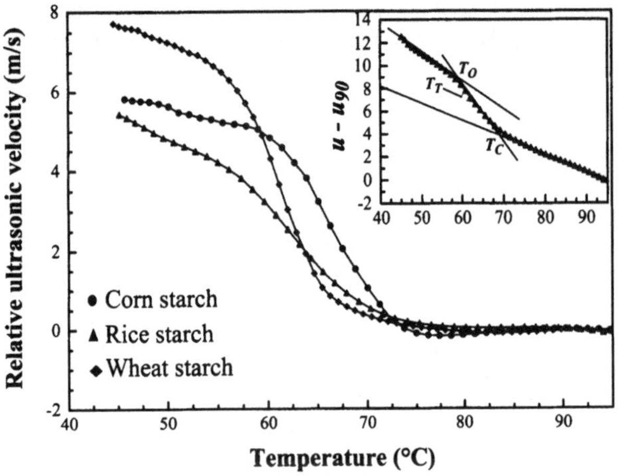

Fig. 1 Temperature profile of ultrasonic velocity in 25% (w/w) starch-water mixtures at 7.6 MHz throughout a temperature ramp of 2 °C/min. For comparison purposes, the baseline for gelatinised starch was subtracted from the ultrasonic velocity temperature profiles of the different starches studied. The inserted picture shows u-u_{90} (the ultrasonic velocity adjusted so that its value is zero at 90 °C) and the used baseline; the relative ultrasonic velocity plotted on the picture is the adjusted data (u-u_{90}) after subtraction of the baseline

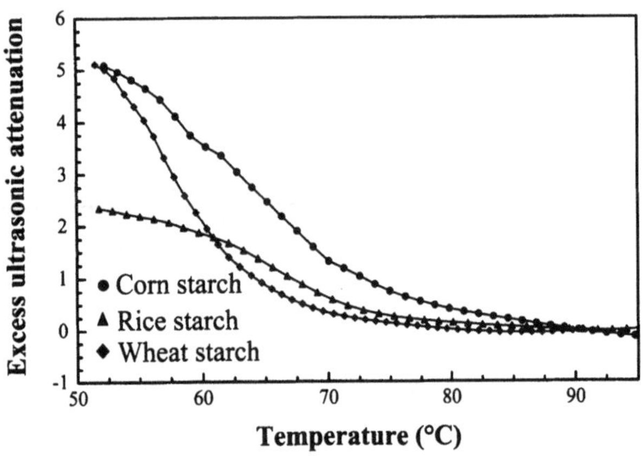

Fig. 2 Excess ultrasonic attenuation of 25% (w/w) starch-water mixtures at 7.6 MHz throughout a temperature ramp of 2 °C/min. For comparative purposes, the value of the attenuation of the starch samples at 90 °C was subtracted to all the temperature profiles of the ultrasonic attenuation

Table 1 Starch gelatinisation temperatures as determined by ultrasonic measurements

	T_O (°C)	T_T (°C)	T_C (°C)	ΔT (°C)
Corn S-4126	61.0	67.9	73.7	12.7
Rice S-7260	56.6	65.8	70.8	14.2
Wheat S-5127	55.7	62.0	67.0	11.3

T_O is the gelatinisation onset temperature, T_T the transition temperature and T_C the conclusion temperature. ΔT is the temperature interval over which the transition occurs ($\Delta T = T_C - T_O$)

conclusion temperature (T_C). T_T is the inflection point in the velocity temperature profile, while T_O and T_C are determined as the intersections of the linear fit of the velocity decrease during the transition with the baseline, respectively, before and after the transition. ΔT is the temperature interval over which the transition occurs ($\Delta T = T_C - T_O$). Three stages can be distinguished in both velocity and attenuation temperature profiles for all the starch samples.

Pre-transition stage

At the first stage, ($T < T_O$), a decrease in the ultrasonic velocity upon heating is observed. For quantitative analysis of the temperature dependence of ultrasonic velocity in starches at this stage the concentration increment of ultrasonic velocity of starch, A, was calculated as

$$A = (u - u_o)/(u_o c)$$

where u is the ultrasonic velocity in starch, u_o is the ultrasonic velocity in water and c is the starch concentration in g/cm^3. The slope of the concentration increment of ultrasonic velocity, $\delta A/\delta T$ (T is the temperature), within this stage, is similar for all types of starch with an average value of -1.2×10^{-3} cm^3 g^{-1} K^{-1}. To analyse the nature of the temperature profile of ultrasonic velocity in the starch samples, the ultrasonic velocity profile of the gelatinised starch sample in down scan regime was measured in the temperature interval 95 to 30 °C. We obtained a monotonous increase of ultrasonic velocity with a decrease of temperature. The mean temperature slope of the concentration increment of ultrasonic velocity in these samples (under cooling) was -1.2×10^{-3} cm^3 g^{-1} K^{-1} at temperatures of 40 to 50 °C, which is the same as the slope for the heated starch samples. For comparison, the corresponding value for solutions of glucose (monomer unit of the starch) of similar (as for starches) concentration were calculated using the ultrasonic velocity temperature profiles in aqueous solutions of glucose obtained earlier. These values are about -1.7×10^{-3} cm^3 g^{-1} K^{-1} within the temperature range 30 to 50 °C [24] and about -1.9×10^{-3} cm^3 g^{-1} K^{-1} within the temperature range 40 to 50 °C [25], which is close but slightly less than the value obtained for our starches. This indicates that hydration of the starch is one of the main factors responsible for the decrease in the ultrasonic velocity at the first stage. Indeed, hydration of starch changes the temperature dependence of the compressibility and density of water in the hydration shell of atomic groups of polysaccharides compared to the bulk water, thus changing the temperature dependence of ultrasonic velocity. (See reference [26] for discussion of relationship between hydration phenomenon and the temperature slope of ultrasonic velocity increment.) Other factors such as swelling and leaching of amylose from the starch granules into solution occurring at this stage [27] could also contribute to the change of ultrasonic velocity.

Transition stage

At the second stage ($T_O < T < T_C$) a dramatic decrease in the ultrasonic velocity and attenuation is observed, that shows a significant structural change within the starch granules. As described by French [28], the starch granule is made of alternately amorphous and semicrystalline concentric rings called growth rings. The sub-structure of the semicrystalline rings consists of stacks of alternating crystalline and amorphous lamellae. Robin and co-worker [29] described the semicrystalline ring as a radial arrangement of clusters of amylopectine: the amorphous lamella is a region rich in amylopectin branch point whereas the crystalline lamella contains short chain segments of amylopectin forming double helices. Globally the starch granule can be modelled as a finite number of alternative crystalline and amorphous lamellae embedded in a background region, which corresponds to the amorphous growth ring. According to the model developed by Jenkins [27], the swelling takes place in the amorphous background region, which causes the disruption of the crystalline lamellae. At this stage the crystalline regions of the starch granules melt, that should result in the increase of the compressibility of the granules thus the decrease in the ultrasonic velocity. Additional contribution to the observed decrease in ultrasonic velocity could come from the change in the scattering of ultrasonic waves due to increase of the size of the granules and a change of their intrinsic physical parameters responsible for the scattering effects. The observed drop of ultrasonic attenuation at this stage can be attributed to the decrease in the scattering of ultrasonic waves as result of the structural reorganisation. A quantitative analysis of the contribution of these processes to the ultrasonic parameters will be a subject of further study.

Another observation raised from our measurement is the frequency dependence of the amplitude of the changes in ultrasonic velocity and attenuation. This reflects differences in particle size distribution at pregelation stage and frequency dependence of the loss and the storage moduli of the gel. Interpretation of these data will require a quantitative analysis in terms of theory of scattering of ultrasonic waves and will be a subject on another publication.

Post-transition stage

The third stage, ($T > T_C$), shows a decrease of the relative ultrasonic velocity but at a smaller rate than in the first region. This corresponds to the end of the gelatinisation process and is simply due to the temperature dependence of the ultrasonic velocity in the gelled sample.

Effects of composition and concentration

Table 1 and Figs. 1 and 2 show that the temperatures at which gelatinisation occurs, the broadness of the transition and the amplitude of the changes in velocity and attenuation depend on the origin of the starch. The biological origin of the starch determines its chemical composition (i.e., its amylose content). The supplier (Sigma) indicates an amylose content of approximately 27% for the corn starch. Wheat starch typically has an amylose content of around 25% and common rice starch 20%. Another difference due to the various origins of the starches is the size of the starch granules. Common corn starch has irregular shaped granules ranging between 5 microns and 20 microns. Rice starch has small irregularly shaped granule whose sizes range from 3 to 8 microns whereas wheat has round shaped granules ranging form 22 to 36 microns in diameter. The difference in sizes and shapes of the starch granules explains the different attenuations recorded for the various samples and the different temperatures of gelatinisation can be related to their chemical compositions that affect the granular organisation and its inherent crystallinity. The amplitude of the change in attenuation for the rice starch sample is about 50% of the change for the other samples while the amplitude of the change in velocity for rice starch is much closer to that observed for the other samples. This is due to the fact that the amplitude of attenuation is mainly affected by the scattering of ultrasonic waves, which is determined by the size of the particles, whereas ultrasonic velocity is mainly determined by the micro-elasticity of the starch. Quantitative analyses of the velocity and attenuation temperature profiles can allow access to both the structural and thermodynamic/microrheological characteristics of the system.

To characterise the effects of concentration on the gelatinisation process, one of the starch sample (corn starch) was studied at another, low concentration, 2% w/w. A comparison of the results obtained at the two concentrations for this sample is given in Table 2. We can first notice that the transition temperature interval,

Table 2 Effect of starch concentration on the gelatinisation temperatures of corn starch obtained by ultrasonic measurements

Concentration	T_O (°C)	T_T (°C)	T_C (°C)	ΔT (°C)
25%	62.5	71.8	78.3	15.8
2%	64.6	69.8	73.3	8.7

Legends were used same as in Table 1

ΔT, is narrower for the 2% mixture than for the 25% mixture. In addition, the transition temperature is lower at lower concentration. This phenomenon has previously been observed in DSC studies; Biliaderis and co-workers found that the melting temperature of starch appears to be proportional to the inverse of the volume fraction of water [4]. The authors explained this as follows: when a large amount of water is present, the melting of the starch is facilitated by extensive hydration and swelling of the amorphous regions whereas in more concentrated starch solutions, the destabilising effect of the amorphous regions decreases and only partial melting of the crystallites is initially achieved. Subsequent redistribution of water around the unmelted crystallites will assist their melting upon further heating, which explains why ΔT is broader in more concentrated starch solutions.

Conclusion

High-resolution ultrasonic spectroscopy is an effective tool for the analysis of gelatinisation of the starch. It allows monitoring of the hydration and microstructural transformations in the gelatinisation process. Further work will be carried out to evaluate contributions of various processes to ultrasonic temperature profile of starches and analyse the details of the gelatinisation processes.

Acknowledgments This work was supported by the grant ST/2000/046 of Enterprise Ireland. We are grateful to Ultrasonic Scientific Ltd. for providing instrumentation in this work and consultancy on the optimal measuring regimes. We would like to acknowledge Alan Grogan for providing results of ultrasonic measurements in glucose solutions.

References

1. Hsu S, Lu S, Huang C (2000) J Food Sci 65(2):215–220
2. Eerlingen RC, Jacobs H, Block K, Delcour JA (1997) Carbohydr Res 297:347–356
3. Morikawa K, Nishinari K (2000) Carbohydr Polym 43:241–247
4. Biliaderis CG, Maurice TJ, Vose JR (1980) J Food Sci 45:1669–1674
5. Hwang CH, Heldman DR, Chao RR, Taylor TA (1999) J Food Sci 64(1):141–144
6. Takaya T, Sano C, Nishinari K (2000) Carbohydr Polym 41(1):97–100
7. Maaurf AG, Che Man YB, Asbi BA, Junainah AH, Kennedy JF (2001) Carbohydr Polym 45(4):335–345
8. Dolan KD, Steffe JF (1990) J. Texture Studies 21:265–294
9. Schriebaum F, Taufel K, Ullman J (1962) Stärke 14:161–164
10. Iizuka K, Aishima T (1999) J Food Sci 64(4):653–658
11. Tang HR, Brun A, Hills B (2001) Carbohydr Polym 46(1):7–18
12. Yuryev VP, Nemirovskaya IE, Maslova TD (1995) Carbohydr Polym 26(1):43–46
13. Karapantsios TD, Sakonidou EP, Raphaelides SN (2000) J Food Sci 65(1):144–150
14. Liu H, Lelievre J, Ayoung-Chee W (1991) Carbohydr Res 210:79–87
15. Buckin V, Smyth C (1999) Sem Food Anal 4:89–105
16. Kudryashov E, Smyth C, Duffy G, Buckin V (2000) Progr Coll Polym Sci 115:287–294
17. Smyth C, Dawson KA, Buckin VA (1999) Progr Coll Polym Sci 112(39):37–43
18. Smyth C, Kudryashov ED, Buckin V (2001) Coll Surf 183–185:517–526
19. Buckin V, Kudryashov E (2001) Adv Coll Interf Sci 89–90:401–422
20. Létang C, Piau M, Verdier C, Lefebvre L (2001)Ultrasonics 39:133–141
21. Povey MJW, Rosenthal AJ (1984) J Food Techn 19:115–119
22. Buckin V, O'Driscoll B (2002) Lab Plus Int 16:17–21
23. Herzfeld KF, Litovitz TA (1959) Absorption and dispersion of ultrasonic waves. Academic Press, New York
24. Grogan A (2001) MSc. thesis, University College Dublin
25. Pryor AW, Roscoe R (1954) Proc Phys Soc B 67:70–81
26. Buckin VA (1988) Biophys Chem 29:283–292
27. Jenkins PJ, Donald AM (1998) Carbohydr Res 308:133–147
28. French D (1984) In: Whistler RL, Bemi JN (eds), Starch: chemistry and technology. Academic Press, London, p 183
29. Robin JP, Mercier C, Charbonniere R, Guilbot A (1974) Cereal Chem 51:389–406

Progr Colloid Polym Sci (2004) 123: 141–146
DOI 10.1007/b11748

F. Scheffold
S. Romer
F. Cardinaux
H. Bissig
A. Stradner
L.F. Rojas-Ochoa
V. Trappe
C. Urban
S.E. Skipetrov
L. Cipelletti
P. Schurtenberger

New trends in optical microrheology of complex fluids and gels

F. Scheffold (✉) · S. Romer
F. Cardinaux · H. Bissig · A. Stradner
L.F. Rojas-Ochoa · V. Trappe
C. Urban · P. Schurtenberger
Department of Physics,
University of Fribourg,
1700 Fribourg, Switzerland
e-mail: Frank.Scheffold@unifr.ch

C. Urban
LS Instruments, c/o Department of Physics,
University of Fribourg,
1700 Fribourg, Switzerland

S.E. Skipetrov
Department of Physics,
Moscow State University,
119899 Moscow, Russia

L. Cipelletti
GDPC, Université Montpellier II,
34095 Montpellier Cedex 05, France

Abstract We have studied various complex systems from particle and biopolymer gels to concentrated surfactant solutions using classical rheometry and optical microrheology. Optical microrheology uses dynamic light scattering, usually in the multiple scattering regime, to obtain information about the microscopic dynamic properties of complex media. This can be done either by direct investigation or by addition of tracer particles to otherwise transparent systems. Based on the local dynamics the macroscopic viscoelastic properties are predicted. We have implemented several new approaches to extend the range of application for optical microrheology: Taking advantage of the recently developed "two-cell technique" we will show how dynamic multiple light scattering (Diffusing Wave Spectroscopy) can be used to investigate the properties of fluid and solid-like media. Furthermore we have significantly extended the range of accessible correlation times to 10^{-8}–10^{4} s using a CCD based multispeckle analysis scheme. Our experiments cover such different materials as polystyrene latex dispersions and gels, ceramic green bodies, casein micellar gels (yogurt) and giant micelle solutions. Excellent quantitative agreement is found when comparing the results obtained from DWS to classical rheological measurements. However, compared to classical rheology, we were able to significantly increase the range of accessible frequencies using optical microrheology, thereby opening up a wealth of new possibilities for the study of these fascinating materials.

Keywords Microrheology · Diffusing wave spectroscopy · Colloids · Biopolymers · Micelles

Introduction

In recent years significant progress has been made in the development of modern optical techniques to study and characterise the rheological properties of complex fluids [1–9]. While these techniques have been mostly restricted to fundamental research they now become increasingly available to both industrial and applied researchers [6–11].

The underlying idea of optical microrheology is to study the thermal response of small (colloidal) particles embedded in the system under study. In this case the particle can either be artificially introduced, which is then called "tracer-microrheology", or can be part of the system itself, e.g., like in the case of ceramic green bodies. By analysing the thermal motion of the particle it is possible to obtain quantitative information about the loss and storage moduli, $G'(\omega)$ and $G''(\omega)$ over an extended range of frequencies [1–3].

One of the most popular techniques to study the thermal motion of the particles is diffusing wave spectroscopy (DWS) which is an extension of standard photon correlation spectroscopy (PCS) to turbid media. Here the analysis of (multiply) scattered laser light is used to determine the time evolution of the probe particles mean square displacement [12–14]. DWS allows access to a broad range of time scales which results in the above mentioned large frequency range covered by DWS-based optical microrheology.

The aim of this article is twofold. First we want to show how modern optical techniques can extend and improve classical rheology both in the sense of frequency range and applicability. In particular we will introduce two novel approaches to extend standard optical microrheology, normally restricted to liquid samples, to viscoelastic solids [6, 7, 11]. Secondly we discuss the application of optical microrheology to different applied systems. We will show that it is perfectly suited for a fast and non-invasive determination of the key rheological properties as given by the frequency dependence of $G'(\omega)$ and $G''(\omega)$.

Experimental set-up

Diffusing Wave Spectroscopy (DWS)

Dynamic light scattering (DLS), or photon correlation spectroscopy (PCS), analyses the fluctuations of the light intensity scattered from a system under study. The light fluctuates due to the local motion of the scatterers. While in conventional light scattering experiments the sample has to be almost transparent (and hence often highly diluted), diffusing wave spectroscopy (DWS) extends "conventional" dynamic light scattering (DLS) to media with strong multiple scattering, treating the transport of light as a diffusion process [12–14]. Analogous to DLS it is possible to express the measured intensity autocorrelation function $g_2(\tau) - 1 = \langle I(t)I(t+\tau)\rangle / \langle I\rangle^2 - 1$ in terms of the mean square displacement of the scattering particle

$$g_2(\tau) - 1 = \left[\int_0^\infty ds\, P(s) \exp\left(-(s/l*)k^2\langle \Delta r^2(\tau)\rangle\right) \right]^2 \quad (1)$$

with $k = 2\pi n/\lambda$ being the wave number of light in a medium with refractive index n. $P(s)$ is the distribution of photon trajectories of length s in the sample and it can be calculated within the diffusion model taking into account the experimental geometry. The transport mean free path l^* characterises the typical step length of the photon random walk, given by the individual particles scattering properties and particle concentration. l^* can be determined independently and enters the analysis as a constant parameter [13]. From Eq. (1) it is possible to numerically calculate the particle mean square displacement $<\Delta r^2(\tau)>$ from the measured autocorrelation function $g_2(t)$.

Two-cell technique

To overcome the problem of non-ergodicity in dynamic light scattering [15], arising from the constraint particle motion in a solid like material, we have recently developed a non-invasive efficient new method [6, 7]. We prepare a sandwich consisting of two independent glass cells [thickness $L_2 = L_1 = 1$ mm] where the first cell contains the sample to be investigated, which can be either a stable ergodic or a gelling non-ergodic sample. The second cell, which serves to properly average the signal of the first cell only, contains an ergodic system with very slow internal dynamics and moderate turbidity (Fig. 1). The correlation function $g_2(\tau)-1$ of the two-cell set-up can be expressed by a product of the correlation functions of the two independent cells

$$g_2(\tau) - 1 = [g_2(\tau, L_1) - 1] \cdot [g_2(\tau, L_2) - 1], \quad (2)$$

which we call the "multiplication rule" [7]. This relation holds if the first layer has a high optical density $L_1/l_1^* >> 1$ while the second layer shows only moderate multiple scattering $L_2/l_2^* \tilde{\ } 2–3$ [7]. From this it is possible to determine directly the contribution of the first cell $g_2(\tau, L_1)-1$ by dividing the signal of the two cell sandwich, $g_2(\tau)-1$, with the separately measured autocorrelation function $g_2(\tau, L_2)-1$ of the ergodic system in the second cell

$$g_2(\tau, L_1) - 1 = [g_2(\tau) - 1]/[g_2(\tau, L_2) - 1] \quad (3)$$

Multispeckle DWS

Another very useful extension of standard DWS is the use of a CCD camera to follow temporal fluctuations of the scattered light (see Fig. 2). Instead of analysing the fluctuations of intensity at a single spatial position (one speckle spot) we now analyse a large area of the intensity pattern of the scattered light (hence multispeckle) using a

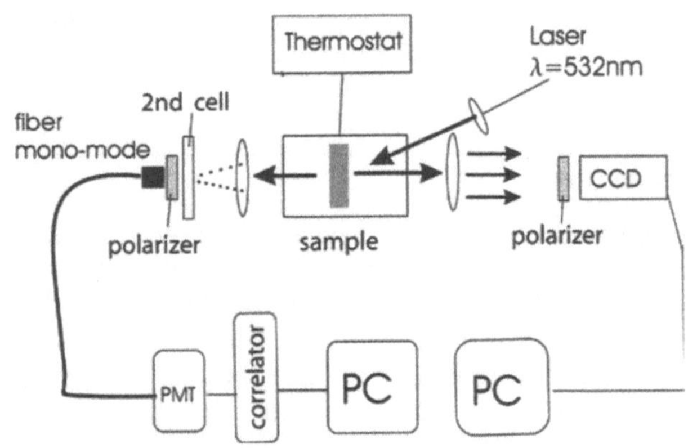

Fig. 1 Schematic plot of the two-cell set-up. The first cell is filled with a liquid or solid-like sample and the second cell is filled with an ergodic system (e.g., polystyrene latex spheres in glycerol)

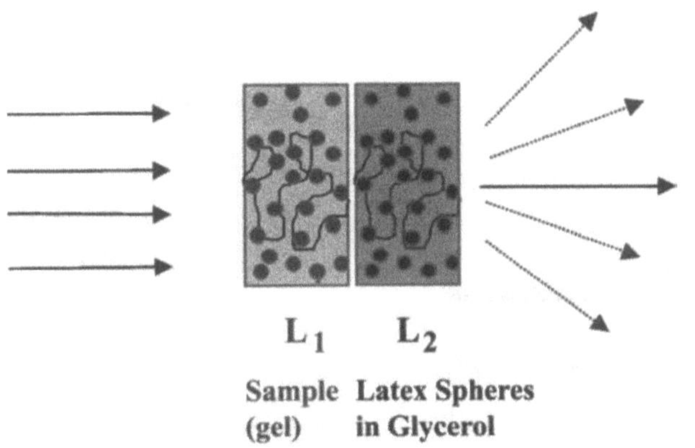

Fig. 2 DWS set-up: an intense laser beam (Verdi from Coherent) is scattered from a turbid sample contained in a temperature controlled water bath. The scattered light is detected in transmission or backscattering with a mono mode fibber or a CCD camera and subsequently analysed digitally (correlator and PC)

CCD camera [4, 11, 17]. The main advantage of this set-up in DWS based microrheology is the significantly improved data acquisition time, since a large number of DWS-scattering experiments is actually performed simultaneously. In standard DWS (or DLS) measurements the data acquisition time has to be several orders of magnitude larger than the typical relaxation time of the correlation function $g_2(\tau)-1$, a restriction that does not apply to multispeckle-DWS. Furthermore, since different configurations of the sample are probed simultaneously the measured correlation function never suffers the problems of non-ergodicity described above.

The main drawback of camera based DWS is the currently still much limited time resolution of CCD cameras. Typically correlation times τ down to approx. 10 ms can be accessed (as compared to 10 ns with a standard photo-multiplier – digital correlator set-up), which is hence not sufficient for most of the fast relaxation processes usually encountered in DWS. However a combination of the novel two-cell technique and multi-speckle DWS turns out to be a perfect combination to overcome most of the commonly encountered experimental limitations. Using both techniques enables us to cover a range of 10 ns to at least 10,000 seconds, hence more than twelve orders of magnitude in correlation time [11].

Microrheology

Colloidal gels

Aggregation and gelation in complex fluids has been for a long time a field of intense research where both fundamental as well as applied questions are equally important [5–10, 18, 19]. Applications of gels and sol-gel processing include such different areas as ceramics processing, cosmetics and consumer products, food technology, to name only a few. Gels are formed by chemical or physical reactions of small sub-units (molecules, polymers or colloids) which can be either reversible or irreversible. The macroscopic features that bring together such different materials are based on the microstructural properties of all gels, which can be described as random networks built up by aggregation of the individual sub-units. Starting from a solution of the sub-units the systems is destabilised, which leads to aggregation, cluster formation and gelation. At the gel point a liquid-solid transition is observed which can be characterised by the appearance of a storage modulus in rheological measurements.

We have recently reported the first study of the sol-gel transition in a concentrated colloidal suspensions based on DWS [6]. In our systems a concentrated suspension of monodisperse polystyrene latex spheres is destabilised by increasing the solvent ionic-strength with a catalytic reaction. Thereby the electrostatic repulsion of the double layer is reduced and the particles aggregate due to van-der-Waals attraction. Fig. 3 shows the measured autocorrelation function as

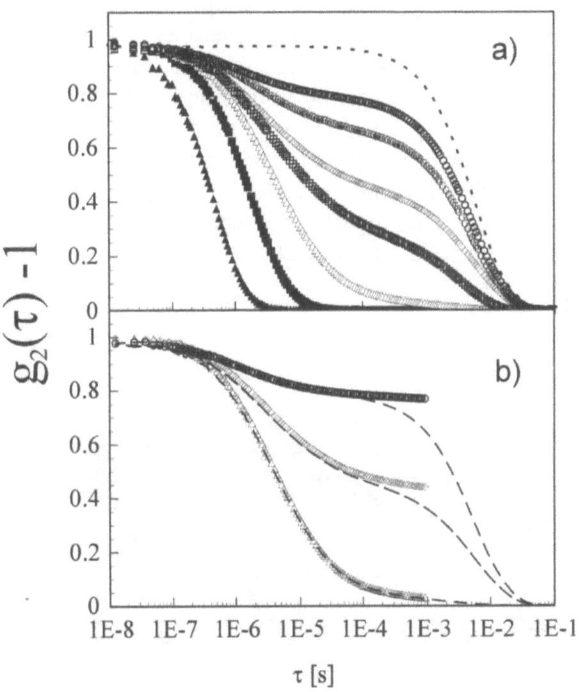

Fig. 3 Sol-gel transition of a concentrated colloidal suspension measured over 10 days with a two-cell set-up (Fig. 2). (**a**) The destabilised system shows a transition from a liquid state, characterised by an almost exponential decay of the correlation function, to a solid state (*open symbols*) after about 80 min. In the gel-state a continuously increasing plateau builds up in the correlation function, $g_2(\tau)-1$, characteristic for the finite storage modulus of a solid-like system. Dashed line: separately measured autocorrelation function $g_2(\tau,L_2)-1$ of the ergodic system in the second cell. (**b**) Correlation function of the gel, $g_2(\tau,L_1)-1$, determined from Eq. (3). The dashed lines show uncorrected data, $g_2(\tau)-1$

a function of time ($t=0$, 18, 108, 256, 734, 4683, 14,400 mm). At early stages, clusters form due to particle aggregation and the decay of the correlation function shifts to higher correlation times due to the slower motion of the clusters. Gelation occurs when a single cluster fills the entire sample volume. After the sol-gel transition we observe that the correlation function $g_2(\tau)-1$ does not decay to zero but remains finite (Fig. 3). At the gel time the short time behaviour changes qualitatively from diffusion to a sub-diffusive motion well described by a power law $<\Delta r^2(\tau)> \sim \tau^p$ (Fig. 4). We find, within our time resolution, that the exponent for diffusion $p=1$ drops rapidly at the gel point and takes a value of $p \sim 0.7$ for all $t > t_{gel}$ (see also Fig. 5). Once the gel spans over the whole sample the signal is dominated by a broad distribution of elastic gel modes. In the gel-state the average mean square displacement is well described for all τ by a stretched exponential

$$\langle \Delta r^2(\tau) \rangle = \delta^2 \left[1 - e^{-(\tau/\tau_c)^p}\right], \quad (4)$$

leading to a plateau at long times (with $p \sim 0.7$).

It is also possible to directly link the results from DWS to the macroscopic storage modulus taking advantage of a recent model developed by Krall and Weitz [18]. They found that for the case of fractal (dilute) gels the storage modulus is given by $G' = 6\pi\eta/\tau_c$ [see also Eq. (4)]. Recently we could demonstrate that indeed this macroscopic storage modulus G′ deduced from the DWS measurements is in good agreement with classical rheological measurements over a large range of particle concentrations up to 30% volume fraction [8].

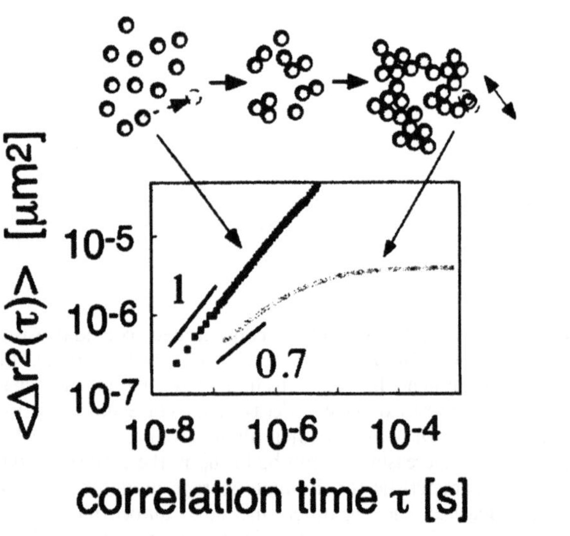

Fig. 4 Particle mean square displacement of a colloidal system from sol to gel determined from Eq. (1)

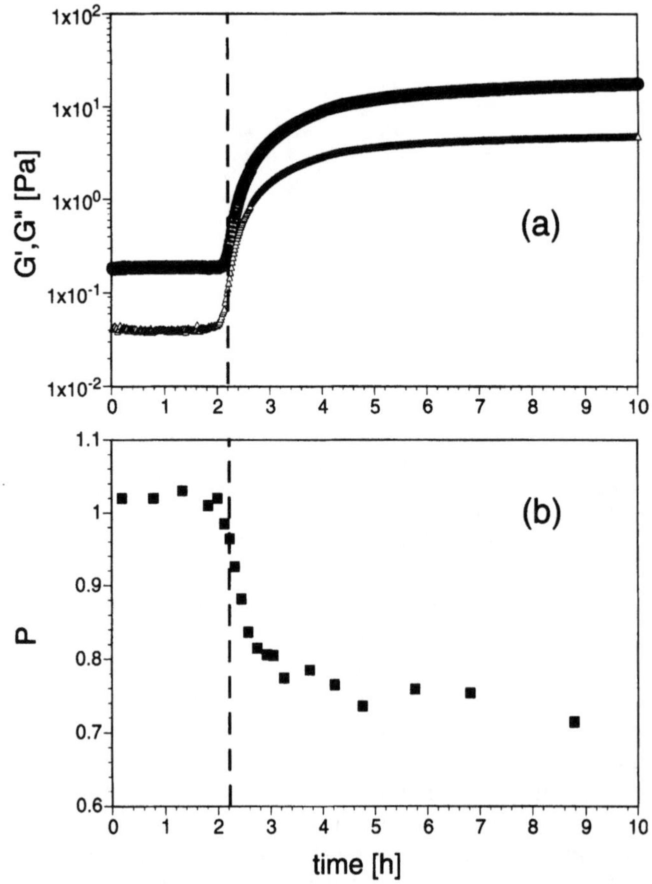

Fig. 5 Comparison of results from classical rheology and DWS during the yogurt making process in fat-free milk. (**a**) Time evolution of the storage modulus G′(t) (*big grey circles*) and the loss modulus (*small black triangles*) obtained from an oscillating rheological measurement (1 Hz, 1% amplitude). (**b**) Time evolution of the exponent p obtained from DWS

Ceramic slurries and green bodies

Ceramic green bodies can be formed by casting colloidal suspension of high solid loading and then coagulating them in situ with an enzyme catalysed hydrolysis reaction [19]. This forming process is called direct coagulation casting (DCC). The process can be carried out along two different destabilisation routes: a) increasing the ionic strength or b) shifting the pH of the solvent (for details see refs. [5, 6, 19]). Surprisingly it has been found that the mechanical properties of the wet green bodies produced from these two systems appear to be completely different. We have studied the time evolution of the correlation function of destabilised alumina suspension using two-cell DWS. The higher mechanical strength of the ionic strength destabilised system can be clearly identified in the DWS experiments (data not shown). By furthermore analysing the optical density (i.e., 1/l*) of the system we could show that the pH destabilised-system is much more homogeneous

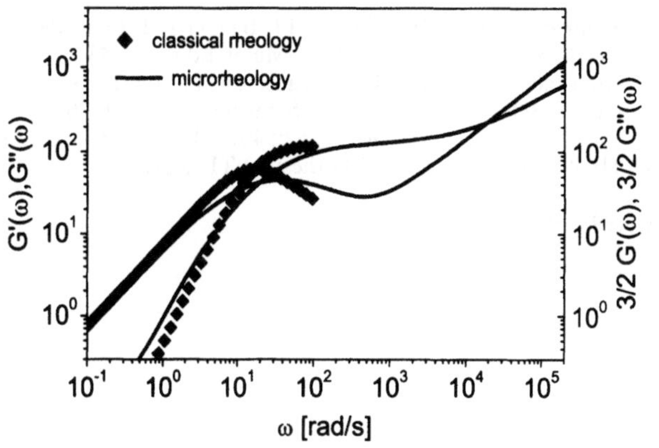

Fig. 6 Frequency dependent elastic moduli obtained from classical rheology (*open symbols*) and DWS based microrheology (*solid line*, right axis). The microrheology results have been multiplied by a factor 3/2 to allow a quantitative comparison of both experiments

which apparently gives rise to the observed different mechanical properties [5].

Biopolymers

Another interesting example is the gelation of casein micellar system in the yoghurt and cheese making process. Fig. 5 shows how DWS can be used to monitor the rheological properties of such a process. We find that the enormous increase in G′ and the drop of the exponent coincide (dashed line) demonstrating the link between microscopic particle dynamics and macroscopic sol-gel transition and viscoelastic properties [9, 10].

Concentrated surfactant solutions

The previous examples addressed systems where the scattering particles are part of the system under study. Now we want to show some very recent measurements where we have introduced tracer particles (polystyrene spheres, diameter 720 nm) in an otherwise transparent matrix consisting of a concentrated surfactant solution. Under the chosen conditions these surfactants form giant, polymer like, micelles, which results in a surprisingly strong viscoelastic liquid [20]. In this case we can take advantage of the formalisms derived for tracer-microrheology which allows to link directly the particle mean square displacement, as obtained from DWS, to the macroscopic viscoelastic moduli $G'(\omega)$, $G''(\omega)$ [1–3]. Fig. 6 shows the excellent agreement between classical rheology and DWS based microrheology, with a dramatically increased frequency range for the latter technique [11]. For a quantitative comparison a scaling factor of 3/2 has been introduced. The origin of this factor is not well understood yet but, as we think, it likely reflects the coupling of the tracer sphere to the medium not well described by current theoretical approaches (see also [3, 11]).

Conclusions

In conclusion, we have shown that DWS based microrheology is a versatile tool to study and characterise the rheological properties of complex fluids and gels. With the advent of new techniques such as the two-cell method and the use of CCD cameras for slow relaxation processes it is now possible to study an increasingly large number of liquid and solid-like complex media. This is expected to have an equally strong impact on both applied and fundamental research. In particular the fast and non-invasive way the information can be obtained from laser light scattering should make these optical techniques an ideal method for industrial applications, e.g., for system characterisation and process monitoring.

References

1. Mason TG, Weitz DA (1995) Phys Rev Lett 74:1250; Mason TG, Gang HU, Weitz DA (1997) Opt Soc Am A 14(1):139
2. Gittes F, Schnurr B, Olmsted PD, MacKintosh FC, Schmidt CF (1997) Phys Rev Lett 79:3286
3. Levine AJ, Lubensky TC (2000) Phys Rev Lett 85:1774
4. Knaebel A, Bellour M, Munch J-P, Viasnoff V, Lequeux F, Harden JL (2000) Europhys Lett 52:73–79
5. Wyss H, Romer S, Scheffold F, Schurtenberger P, Gauckler LJ (2001) J Coll Interface Sci 241:89–97
6. Romer S, Scheffold F, Schurtenberger P (2000) Phys Rev Lett 85:4980
7. Scheffold F, Skipetrov SE, Romer S, Schurtenberger P (2001) Phys Rev E 63:61404
8. Romer S, Bissig H, Trappe V, Scheffold F, Schurtenberger P (in preparation)
9. Schurtenberger P, Stradner A, Romer S, Urban C, Scheffold F (2001) CHIMIA 55:155–159
10. Romer S, Urban C, Stradner A, Kruif CG de, Schurtenberger P (2003) Food Hydrocoll (submitted)
11. Cardinaux F, Cipelleti L, Scheffold F, Schurtenberger P (2003) (submitted)
12. Maret G, Wolf PE (1987) Z Phys B 65:409
13. Pine DJ, Weitz DA, Chaikin PM, Herbolzheimer E (1988) Phys Rev Lett 60:1134
14. Weitz DA, Pine DJ (1993) In: Brown W (ed.) Dynamic light scattering. Oxford University Press, New York

15. Xue JZ, Pine DJ, Milner ST, Wu XL, Chaikin PM (1992) Phys Rev A 46:6550; Pusey PN, Megen W van (1989) Physica A 157:705
16. Romer S, Scheffold F, Schurtenberger P (2000) Suisse Institute of Intellectual Property, patent application filed 27th February 2000 under number 200 0335/00
17. Cipelletti L, Weitz DA (1999) Rev Sci Instrum 70:3214
18. Krall AH, Weitz DA (1998) Phys Rev Lett 80:778
19. Graule TJ, Baader FH, Gauckler LJ (1994) J Mater Educ 16:243
20. Cannavacciuolo L, Sommer C, Pedersen JS, Schurtenberger P (2000) Phys Rev E 62:5409–5419; Sommer C (2001) PhD thesis, ETH Zürich

Progr Colloid Polym Sci (2004) 123: 147–151
DOI 10.1007/b11749

Wuge H. Briscoe
Roger G. Horn

Electrical double layer interactions in a non-polar liquid measured with a modified surface force apparatus

Presented at XV European Colloid and Interface Society Conference

W.H. Briscoe · R.G. Horn (✉)
Ian Wark Research Institute,
ARC Special Research Centre for Particle and Material Interfaces,
University of South Australia,
Mawson Lakes, SA 5095, Australia
e-mail: roger.horn@unisa.edu.au

Present address: W.H. Briscoe
Physical and Theoretical Chemistry Laboratory, University of Oxford,
South Parks Road, Oxford,
OX1 3QZ, United Kingdom

Abstract Surface force measurements have been made between mica surfaces immersed in a non-polar liquid, decane, in the presence of an amphiphilic electrolyte, sodium di-2-ethylhexyl sulphosuccinate (AOT), using a modified Surface Force Apparatus (SFA). The modifications, notably the employment of a pair of sensitive force measuring springs and the implementation of a magnetic drive mechanism, have appreciably improved the sensitivity of SFA, which enables the detection of a weak long-range repulsion in the non-polar system. The origin of the repulsion is ascribed to electrical double layer interactions, and it may be satisfactorily described by a counterion-only double layer theory. This comprises the first report of electrical double layer interactions in a non-polar liquid.

Keywords Surface force · Non-polar · Electrical double layer interactions · Counterion-only

Introduction

When a solid is immersed in a polar liquid, such as water ($\varepsilon_r \sim 80$), it will most likely develop a surface charge via various charging mechanisms such as the preferential adsorption of ions from the solution or the ionisation of surface groups [1, 2]. What happens when a solid is brought into contact with a non-polar liquid? This is a question frequently encountered in many circumstances of practical importance, such as petroleum processing [3], various imaging process technologies [4], and finely-controlled synthesis of particulate materials [5], to name a few. Concurrently, academic interest is stimulated by fundamental questions regarding the applicability of DLVO theory in the non-polar medium and various related issues [6–8]. A non-polar liquid is distinguished from a polar liquid by its very low dielectric constant, e.g., around 2 to 4, although there is no exact boundary between non-polar and low polar liquids. As a result, the Coulomb attraction between the cation and anion in an electrolyte is strengthened by a factor of 20 to 40 in comparison to that in water. Consequently, the dissociation of electrolyte, and in turn the ionic concentration, is minimal, and the validity of the charging mechanisms, which are operative in the aqueous medium, becomes questionable. It is for this reason that it has been somewhat controversial to ask whether the solid can acquire a surface charge and if the electrical double layer interaction plays a role in colloidal stability in the non-polar liquid [6, 9].

However, the difficulty of electrolyte dissociation in the non-polar liquid can to a certain extent be alleviated if some large structures such as inverse micelles or crown ethers are present [2, 3, 10, 11]. Indeed, ample evidence has emerged from both academic investigations and industrial applications that solids can become charged in contact with non-polar liquids in the presence of amphiphilic electrolytes [3, 7, 8, 10, 12–16]. If so, an electrical double layer should develop at the solid-liquid interface. Hitherto, the possible presence of such an electrical double layer has been probed by electrokinetic studies [3, 17–23]. In principle, the direct measurement of an electrical double layer repulsion would also provide evidence of surface charging but is yet to be reported. If present, such repulsion would be very long-ranged and weak [24], which imposes a substantial difficulty on its

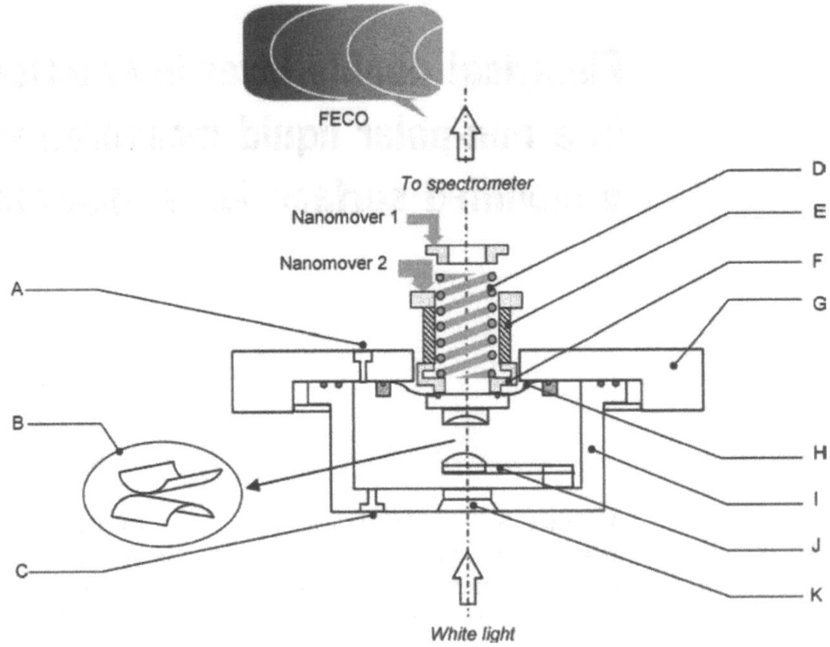

Fig. 1 Schematic representation of Mk Π SFA. *A* Air inlet/outlet; *B* orthogonally crossed cylinder geometry; *C* liquid inlet/outlet; *D* soft helical spring; *E* piezoelectric tube; *F* stiff spring; *G* top plate; *H* Teflon diaphragm; *I* liquid chamber; *J* double cantilever spring; *K* silica window. The measurement is taken between two molecularly smooth substrates, usually mica, back-silvered and glued onto two orthogonal cylindrical silica discs with silvered sides down. The bottom disc is mounted on a pair of calibrated double cantilever springs. The top surface is positioned either directly by a Nanomover microstepping motor or indirectly by a second Nanomover through a differential spring mechanism. More precise positioning is achieved by the expansion/contraction of the piezoelectric tube attached to the top surface holder. The surface separation is monitored by the technique of the fringes of equal chromatic order (FECO).

detection. It turns out none of the force measuring techniques available with their current implementations [25–29] can achieve the detection of such a weak long-range interaction. In this study, modifications of an SFA have been carried out to enhance its capacity, with which an attempt has been successfully made to measure an electrical double layer interaction in a non-polar liquid.

Experimental

Mark Π surface force apparatus

Before proceeding to describe the modifications, a brief description of a version of the SFA which we call the "Mark Π" is presented here as such a description has not previously appeared in the literature. The Mk Π SFA is a redesigned version of the original Mk 2 SFA, whose operating principle remains the same as that of Mk 2, but there are a few distinct features. The most notable one is that the set-up for the optics is fixed while the position of the apparatus itself is adjustable in all three directions by means of a translation stage onto which the apparatus mounts. This has significantly simplified the optics alignment and fringe finding procedure. The chamber is cylindrical (with a volume of ~75 mL), with its axis coincident with the optical path, which has reduced the difficulty in machining appreciably. A further feature that differs from Mk 2 is that it is the top surface that is positioned with the differential spring mechanism which is implemented by two computer-controlled Nanomovers (Melles Griot, France), a precision stepping motor with 50 nm step-size resolution. Such a configuration avoids the mechanical drift sometimes augmented by the dovetail slide mechanism in the original SFA design [30], and has also afforded the versatility of replacing the bottom surface with a mercury drop in a separate effort to modify Mk Π [31], which allows the interactions involving a deformable surface to be investigated. A schematic representation of the Mk Π SFA is illustrated in Fig. 1, and a description of the key parts is given in the caption. Not shown in the Figure is the temperature control unit that the apparatus is enclosed in, which employs a Proportional Integral Differential (PID) controller[1] that enables an optimal temperature stability of ±0.02 K.

Consideration and implementation of modifications

In an SFA measurement, the surface force $F(D)$ at a particular surface separation D is measured by gauging it with the deflection of a paired cantilever spring of spring constant K

$$F(D) = K[\Delta x(D) - \Delta x(\infty)] = K\Delta d(D), \tag{1}$$

where $\Delta x(D)$ and $\Delta x(\infty)$ are the spring deflections at separation D and infinity, and $\Delta d(D)$ the deviation of the spring deflection from that if the surface force is absent. In practice, "infinity" is chosen to be a large enough separation such that the surface interaction can be regarded as zero, and thus $\Delta d(D)$ is obtained by subtracting a zero force baseline extrapolated from this large separation. When the interaction is long-ranged, the zero force baseline has to be established when the surfaces are further apart, e.g., a couple of μm, and many of the implemented surface positioning mechanisms would exhibit undesirable non-linearity over a long travel distance. The consequently prolonged measurement time further introduces

[1] The PID controller was constructed by David Antelmi in our laboratory.

vulnerability to the thermal drift caused by ambient temperature fluctuations, and the system also becomes more susceptible to ambient vibrations. In a recent investigation [32], the possible sources for the thermal and mechanical drifts have been carefully considered and identified, which leads to suggestions on some intricate aspects of mechanical design and temperature control. These considerations and suggestions are pertinent to the particular set-up in that study. However rigorous the implementations of the temperature and vibration isolation control are, the temperature fluctuations and vibrations will always be present and often manifest their effects in an unpredictable manner; and whatever their sources are, the effects of the drifts may be grouped together and expressed in terms of an additional deflection $\Delta\tilde{a}$ of the spring in addition to that due to the surface force

$$K\Delta\tilde{d}(D) = K[\Delta d(D) + \Delta\tilde{a}] = F(D) + K\Delta\tilde{a}, \qquad (2)$$

where $\Delta\tilde{d}$ is the overall deflection of the spring observed at a separation D. These uncertainties may be tolerated if the force is strong, as $\Delta\tilde{a}$ is insignificant in comparison to $\Delta d(D)$. In the case of a non-polar liquid, however, the magnitude of the electrical double layer interaction is exceedingly small, and easily masked by such uncertainties [24].

Examination of Eq. (2) leads to the proposal of a two-pronged approach to alleviate the effects of these uncertainties. Firstly, the error in force is proportional to the spring constant K, hence can be reduced if weaker or more sensitive force measuring springs are used. This is intuitive yet effective, and the consequent benefit turns out to be twofold, since the amplitude of the deflection of the spring Δd is also amplified as a result. The most straightforward option is to use thinner springs, but the practicality of machining and handling limits how thin they can be. The springs employed here have a nominal thickness of 0.05 mm, and made with the technique of spark corrosion from a stainless steel sheet. In the SFA the springs also play the role of suspension system for one surface, and a consequence of reducing K is that the springs become too weak to support the weight of the silica disc and the disc holder in the original horizontal configuration. However, this problem is easily solved by tilting the apparatus to a vertical configuration as schematically illustrated in Fig. 2. The second prong of the approach focuses on the improvement of the linearity of the positioning mechanism. SFA set-ups in different laboratories have adopted various drive mechanisms to implement the differential spring mechanism. Hitherto, the most successful attempt has been made by Stewart and Christenson [33], which varies the surface separation by the coupling between a small magnet and a pair of electromagnetic coils. A similar attempt will be pursued here, and a calibration procedure has also been developed for the magnetic drive and the soft springs and described elsewhere [34, 35]. It is worth noting that the employment of electromagnetic coupling as positioning mechanisms or actuators has long existed in the field of microbalance technology [36, 37].

Model experimental system

The model system in which the measurements are conducted is a 2.2 mmol·L^{-1} solution of sodium di-2-ethylhexyl sulphosuccinate (AOT) in decane, under the ambient moisture condition, i.e., the non-polar solution in the SFA chamber is exposed to the ambient humidity (57%) via a filter with 0.45 μm pore size (Pall Gelman). AOT (99%) is purchased from Sigma, and used without further purification. Decane (99+%), obtained from Aldrich, is distilled under nitrogen in the presence of molecular sieves (BDH, type 4 Å, 1/16th inch pellets) immediately before use, and the distillation is carried out in a laminar flow hood (LFH). Distilled decane is directly collected into a Pyrex collecting bottle fitted with an O-ring (Teflon-coated Viton) sealed Teflon plug, through a plumbing path consisting of Kel-F tubing and Omnifit valves. All subsequent transfer of decane and injection into the SFA chamber is undertaken with gas tight Hamilton syringes, where all the necessary connections in the plumbing path are implemented with Omnifit valves and tubing. The measurements are performed at 26 ± 0.02 °C.

Results and discussion

Figure 3 shows the force-distance profile measured between mica surfaces immersed in decane with 2.2 mM AOT undertaken under the ambient moisture condition (filled symbols). For comparison, the electrical double layer interaction measured in pure water is also

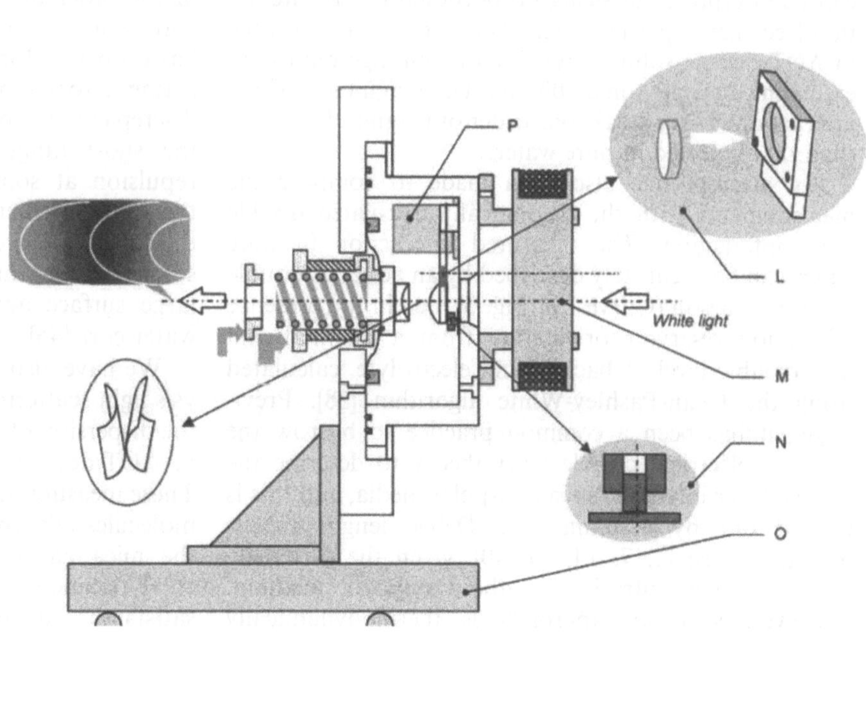

Fig. 2 Schematic representation of the modified Mk II SFA. *L* Small magnet with PCTFE encapsulate (Teflon lid not shown); *M* electromagnetic coil; *N* adjustable stage for holding the surfaces when the apparatus is mounted horizontally for initial alignment; *O* rigid base with kinematic mounts; *P* drying agent chamber

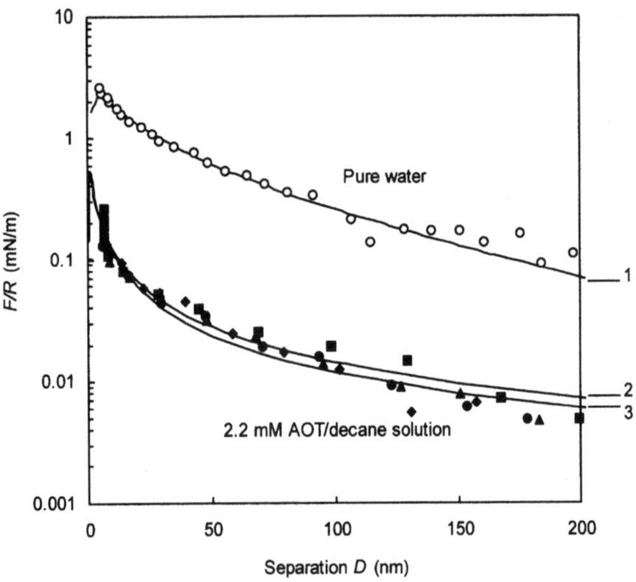

Fig. 3 Force-distance profile of the electrical double layer interaction measured between mica surfaces immersed in pure water (*open circles*), and in decane with 2.2 mM AOT after the solution has been exposed to ambient moisture (*filled symbols*, with different symbols representing different measurements). Curve 1 is calculated from the aqueous double layer theory using the Chan-Pashley-White algorithm [38], assuming a constant surface potential of 155 mV and 1:1 electrolyte concentration of 1.5×10^{-5} M. Curves 2 and 3 are the theoretical predictions from a counterion-only double layer theory [24], assuming an absolute constant surface charge density of 1×10^{-3} C/m^2 and constant potential of 100 mV respectively. The Hamaker constants used to calculate the van der Waals attraction are 2.0×10^{-20} J in water [43] and 4×10^{-22} J in decane respectively

presented in the figure (open circles). The surface force is plotted on a log scale, and F/R, i.e., force divided by the local radius of the curvature of the interacting surfaces, can be interpreted as being proportional to the interaction free energy per unit area. The force-distance profile in AOT/decane solution reveals a long-range repulsion, extending beyond some 200 nm. The magnitude of the interaction is very weak, one order of magnitude weaker than that detected in pure water.

An attempt has also been made to compare the measurements with the theoretically calculated double layer interaction. The measured interaction in pure water can be accurately described by an aqueous double layer interaction (Curve 1 in Fig. 3), i.e., in the presence of an ion reservoir originating from a minimal and unavoidable level of background electrolyte, calculated using the Chan-Pashley-White algorithm [38]. Previously, it has been a common practice to borrow the aqueous electrical double layer theory to describe the double layer interactions in non-polar media, and this is carried out by assuming the Debye length κ^{-1} is infinitely large [2, 7, 14, 39, 40], given the extremely low ionic concentration in the non-polar medium. However, such an approach is thermodynamically inappropriate, which has been established in a recent theoretical study [24], since setting κ^{-1} to infinity removes the thermodynamic foundation of constant chemical potential that the aqueous double layer is based upon [41]. Instead, it is appropriate to invoke a counterion-only theory. Curves 2 and 3 in Fig. 1 are calculated using the counterion-only theory developed in Ref. [24], respectively assuming an absolute constant surface charge density of 1×10^{-3} C/m^2 and constant surface potential of 100 mV. The counterion-only theory predicts that the magnitude of the interaction at such charge density or potential would already approach the upper limit of the double layer repulsion in a non-polar liquid. That is, it represents the strongest detectable double layer interaction. Thus, the fitting values of the surface charge density and the surface potential should not be taken as strictly accurate. However, such a surface charge density is some 2 orders of magnitude lower than that typically acquired in aqueous media, which is indeed expected in the case of a non-polar medium. The scatter in the data prevents us from committing to distinguishing between different boundary conditions, but it is fair to comment that the agreement between the counterion-only theory and the experimental data is satisfactory at large separations.

At short range, the theory predicts that van der Waals attraction overtakes the double layer repulsion. In order to calculate the van der Waals component in the total interaction energy, the Hamaker constant has to be estimated, which may be carried out by modelling the surfaces with AOT aggregates as liquid hydrocarbons interacting across decane, ignoring underlying mica. Then a straightforward calculation from the Lifshitz theory [42] or a simple estimate from the expression supplied by Israelachvili [43] yields a Hamaker constant in the order of 10^{-22} J. A value of 4×10^{-22} J is used here, which is comparable to the value that Christenson has estimated for surfactant-coated mica surfaces interacting across a hydrocarbon medium [44]. However, a discrepancy is observed in the measured interactions at the short range, as the surfaces encounter a steep repulsion at some 6.7 ± 0.2 nm apart, ascribable to the steric interaction due to the adsorption of AOT on the surfaces. Given that the molecular length of AOT is some 1.2 nm, this indicates the possible formation of large surface aggregates, most likely incorporating a water core [45].

We have also performed complementary phase analysis light scattering (PALS) and FTIR measurements on the dispersion of the plate-like dry-ground mica particles in AOT/decane solutions, as described elsewhere [34]. These measurements indicate two things: first, that AOT molecules adsorb onto mica surfaces, and second, that the mica particles become negatively charged in the AOT/decane solution. Such evidence, together with the satisfactory agreement between the observed interaction and the counterion-only theory (Fig. 3), enables us to ascribe the detected weak long-range repulsion to an electrical double layer interaction. The interaction has been measured under the ambient moisture condition, and the presence of trace amounts of water is expected to play an important role in the generation of the electrical double layer at the mica-decane interface, in turn having significant effects on the double layer interactions. The extension of these measurements to other experimental conditions to investigate such effects of water is reported elsewhere [35].

Acknowledgements We are indebted to many helpful discussions from Phil Attard. Image analysis software developed by Jason Connor has been used for data analyses. PALS and FTIR measurements were performed by Roland Keir and Kathryn Hanton. Financial support from the Australian Research Council is acknowledged.

References

1. Hunter RJ (1986) Foundations of colloid science. Clarendon Press, Oxford
2. Lyklema J (1968) Adv Colloid Interface Sci 2:65
3. Klinkenberg A, Minne JL van der (1958) Electrostatics in the petroleum industry. The prevention of explosion hazards. Elsevier, Amsterdam
4. Pearlstine K, Page L, El-Sayed L (1991) J Imaging Sci Technol 35:55
5. Arriagada FJ, Osseo-Asare K (1995) J Colloid Interface Sci 170:8
6. Osmond DWJ (1966) Disc Faraday Soc 42:247
7. Parfitt GD, Peacock J (1978) Stability of colloidal dispersions in nonaqueous media. In: Matijevic E (ed), Surface and colloid science, vol 10. Plenum Press, New York, pp 163–226
8. Kitahara A (1984) Nonaqueous systems. In: Kitahara A, Watanabe A (eds), Electrical phenomena at interfaces. Fundamentals, measurements, and applications, vol 15. Marcel Dekker, New York, pp 119–143
9. Albers W, Overbeek JTG (1959) J Colloid Sci 14:510
10. Hoeven PC van der, Lyklema J (1992) Adv Colloid Interface Sci 42:205
11. Morrison ID (1993) Colloids Surf A 71:1
12. Keir RI, Suparno, Thomas JC (2002) Langmuir 18:1463
13. Fowkes FM (1985) Interface acid-base/charge transfer properties. In: Andrade JD (ed), Surface and interfacial aspects of biomedical polymers, vol 1. Plenum Press, New York, pp 337–372
14. Jenkins P, Ralston J, Thomas JC, Nicholls SL, Staples PE (1999) J Colloid Interface Sci 211:11
15. Miller JF, Schatzel K, Vincent B (1991) J Colloid Interface Sci 143(2):532
16. McKay RB (1987) Pigment dispersion in apolar media. In: Eicke H-F, Parfitt GD (eds), Interfacial phenomena apolar media, vol 21. Marcel Dekker, New York Basel, pp 361–404
17. Novotny V, Hair ML (1979) J Colloid Interface Sci 71(2):273
18. Kitahara A, Karasawa S, Yamada H (1967) J Colloid Interface Sci 25:490
19. Jayaram S, Cross JD, Weckman EJ (1995) J Electrostatics 34:1
20. Kitahara A, Satoh T, Kawasaki S, Kon-No K (1982) J Colloid Interface Sci 86(1):105
21. Labib ME, Williams R (1984) J Colloid Interface Sci 97(2):356
22. Delgado A, Gonzalez-Caballero F, Bruque JM (1986) Colloid Polymer Sci 264:435
23. Kornbrekke RE, Morrison ID, Oja T (1992) Langmuir 8:1211
24. Briscoe WH, Attard P (2002) J Chem Phys 117:5452
25. Ducker WA, Senden TJ, Pashley RM (1991) Nature 353:239
26. Grier DG (1997) Curr Opin Colloid Interface Sci 2:264
27. Israelachvili JN, Adams GE (1976) Nature 262:774
28. Prieve DC (1999) Adv Colloid Interface Sci 82:93
29. Parker JL (1992) Langmuir 8:551
30. Luesse C, Alsten JV, Carson G, Granick S (1986) Rev Sci Instr 59(5):811
31. Connor JN, Horn RG (2001) Langmuir 17:7194
32. Heuberger M, Zach M, Spencer ND (2000) Rev Sci Instr 71(12):4502
33. Stewart AM, Christenson HK (1990) Meas Sci Technol 1:1301
34. Briscoe WH (2001) PhD thesis, The University of South Australia
35. Briscoe WH, Horn RG (2002) Langmuir 18:3945
36. Gast T (1963) Vacuum Microbalance Techniques 3:45
37. Beams JW, Hulburt CW, Lotz WE Jr, Montague RMJr (1955) Rev Sci Instr 26(12):1181
38. Chan DYC, Pashley RM, White LR (1980) J Colloid Interface Sci 77:283
39. Koelmans H, Overbeek TG (1954) Disc Faraday Soc 18:52
40. Kitahara A (1973) Prog Organic Coatings 2:81
41. Attard P (2002) Thermodynamics and statistical mechanics: equilibrium by entropy maximisation. Academic Press, London
42. Hough DB, White LR (1980) Adv Colloid Interface Sci 14:3
43. Israelachvili JN (1991) Intermolecular and surface forces. Academic Press, London
44. Christenson HK (1986) J Phys Chem 90:4
45. De TK, Maitra A (1995) Adv Colloid Interface Sci 59:95

Progr Colloid Polym Sci (2004) 123: 152–155
DOI 10.1007/b11750

P. Dynarowicz-Łątka
J. Miñones Jr
K. Kita
P. Milart

The utility of Brewster angle microscopy in evaluating the origin of the plateau in surface pressure/area isotherms of aromatic carboxylic acids

P. Dynarowicz-Łątka (✉) · K. Kita
P. Milart
Jagiellonian University,
Faculty of Chemistry, Ingardena 3,
30-060 Kraków, Poland
e-mail: ucdynaro@cyf-kr.edu.pl

J. Miñones Jr
Universidad de Santiago de Compostela,
Facultad de Farmacia,
Departamento de Química Física,
Campus Sur s/n,
15 706 Santiago de Compostela, Spain

Abstract Our previous investigations were focused on characterising the surface behaviour of a group of polyphenyl carboxylic acids, namely 5′-phenyl-1,1′:3′,1″-terphenyl-4-carboxylic acid (PTCA) and its derivatives. Their pressure/area (π/A) isotherms show a broad plateau, which spans over a region corresponding to a decrease in molecular area by a factor of ca. 2. Upon analysing the surface pressure and surface potential isotherms, in addition to BAM images and surface potentials for deposited LB films, we could not derive unambiguously the cause for the plateau, however, the hypothesis of molecular tilting seemed to be more probable than a bilayer formation. We have employed herein a more sensitive technique, namely quantitative Brewster angle reflection measurements to support one of these hypotheses. Our results indicate that the formation of non-monomolecular structures is responsible for the plateau in the π/A isotherms.

Keywords Langmuir monolayers · Aromatic carboxylic acids · Brewster angle microscopy · Air/water interface

Introduction

Our previous investigations aimed at characterising a family of polyphenyl carboxylic acids [1–5], which are aromatic analogues of the well-investigated aliphatic polymethylene (aliphatic) amphiphiles, extensively explored in Langmuir monolayers. An interesting feature in surface pressure/area (π-A) isotherms of the basic compound, namely 5′-phenyl-1,1′:3′,1″-terphenyl-4-carboxylic acid (abbreviated as PTCA) and its derivatives, is a plateau with ca. twofold decrease in area, after which the surface pressure increases again sharply upon compression. To understand the origin of this plateau region, firstly a set of control monolayer experiments were performed. It was observed that changing the experimental conditions (such as, for example, spreading volume, compression speed, or water for an acidic subphase) did not influence significantly the π/A characteristics for any compound, i.e., the plateau still remained in the pattern of the isotherms [4], indicating that the plateau region is not due to experimental artefacts deriving from non-equilibrium compression of the monolayer or the loss of monolayer material from the surface. However, the plateau pressure was found to be affected by the subphase temperature, i.e., it increased with decreasing temperature [3, 4]. Such a behaviour is inverse to that of plateau's originating from phase coexistence, observed with Langmuir monolayers formed by model amphiphiles (e.g., n-pentadecanoic acid) [6, 7]. However, plateau regions attributed either to the collapse of a monolayer into a multilayer state [8] or due to orientation changes upon compression [9–11] exhibit an analogous trend with temperature as observed for PTCA and its derivatives.

In order to get further insight into the phenomena occurring at the plateau region, the electric surface potential (ΔV) was measured and effective dipole moments ($\mu_\perp$) were calculated according to the Helmholtz equation [12]. Fig. 1 presents the effective dipole moment-area dependence for p-tolyl-PTCA which is

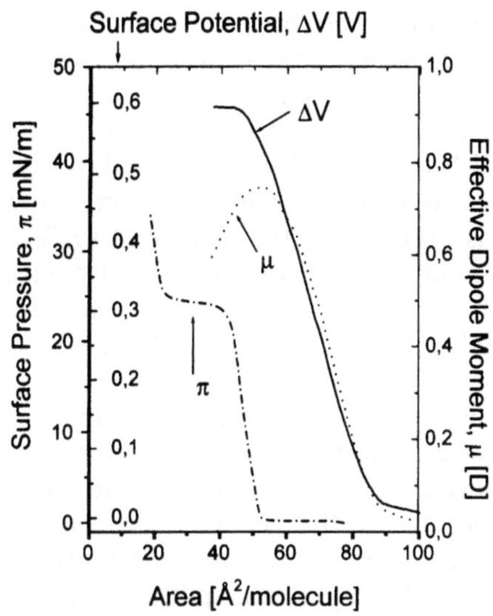

Fig. 1 Surface pressure (*dash-dotted line*), surface potential (*solid line*) and effective dipole moment (*dashed line*) – area isotherms of *p*-tolyl-PTCA spread on a 10^{-3} M HCl aq. subphase. Compression speed: 7.5×10^{17} Å^2/min, T = 20 °C

representative for all the monolayers formed by 4′-PTCA derivatives with hydrophobic substituents. The surface pressure and electric surface potential-area isotherms are reproduced in the same graph for comparison.

Both surface potential and effective dipole moment increase upon monolayer compression, and at the area per molecule corresponding to the beginning of the plateau, both parameters reach their maximum value (corresponding to the most upright orientation of film molecules), and afterwards they decrease. Such a decrease in effective dipole moment may again be due either to the bilayer formation (keeping in mind the twofold decrease in molecular area within the plateau region) or gradual inclination of film molecules; the former hypothesis was suggested to be unlikely based on the analysis of Brewster angle microscopy images (indicating monolayer homogeneity up to the middle of the plateau region [3, 13]) and also taking into consideration the high reproducibility of isotherms [4] as well as analysing the surface potential values for deposited LB films [5]. PTCA was thought to tilt over upon compression of the film from an upright into an interdigitated, inclined orientation [3–5]. The formation of non-monomolecular structures was considered possible only after the mid-point of the plateau, where stripes were observed to form in BAM images.

However, a problem with this interpretation arose from the quantification of the molecular area, *A*. Persistence of a monolayer up to the most condensed region, where the measured film area corresponds to $A \sim 20$ Å^2 per molecule under the assumption of a monomolecular surface film, could be immediately ruled out owing to the physical dimensions of the aromatic ring

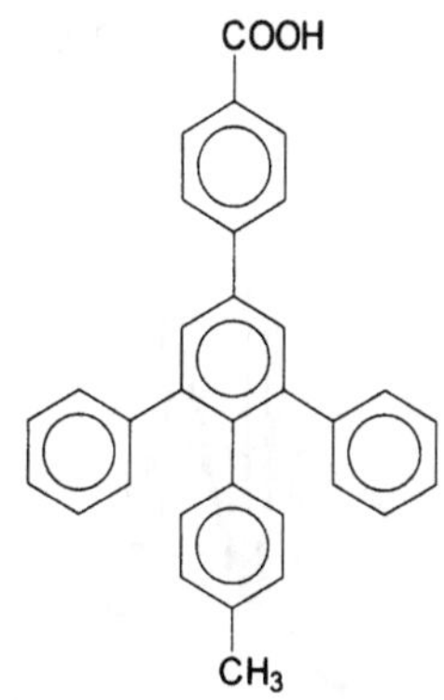

Scheme 1 Chemical structure of the investigated *p*-tolyl-PTCA

systems. But even at the beginning of the plateau, which starts at $A \sim 40$ Å^2, this area was close to a minimum value that might conceivably accommodate the bulky aromatic rings.

We have therefore revisited the problem and have broadened the spectrum of applied experimental techniques, aiming at a better quantification of surface film thickness, by using quantitative Brewster angle reflection measurements. For this study we selected *p*-tolyl-PTCA (Scheme 1) because of the relatively high stability of its monolayers.

Experimental

Langmuir monolayers

4′-(4-Methylphenyl)-5′-phenyl-1,1′:3′,1″-terphenyl-4-carboxylic acid, in short *p*-tolyl-PTCA, was synthesised as described elsewhere [2, 13]. Ultrapure water was used as subphase for Langmuir monolayers, which were prepared by spreading an aliquot (~150 μL) of a chloroform (Merck, p.a.) solution (concentration ~ 0.5 mg/mL) onto the air/water interface. Purified water was supplied by a Milli-RO/Milli-Q system (Millipore; resistivity 18 MΩcm, pH 5.6–5.8). *π*-*A* isotherms were measured in a computer-controlled LB film balance (KSV Instruments, Helsinki, Finland, model 5,000). After spreading, the solutions were left for 5 min to ensure complete solvent evaporation before compression was started. All measurements were performed at room temperature (ca. 20 °C).

Brewster angle microscopy (BAM)

A BAM 2 plus (NFT, Göttingen, Germany) Brewster angle microscope mounted on a Nima (UK) trough, equipped with a 30 mW laser emitting p-polarised light at a wavelength, λ 690 nm, was used to visualise the lateral structure of the monolayers as the light beam is reflected off the air-water interface at the Brewster angle. The shutter speed used was 1/1,000. The full dependence of the reflected intensity as a function of instrumental settings, film thickness and optical constants has been given by Berreman [14]. Since we are interested in the transformation of a monolayer into more complex surface film structures it is sufficient to quantify relative thickness changes. Hence a simplified approach to data evaluation was taken [15] which uses the fact that the intensity *I*

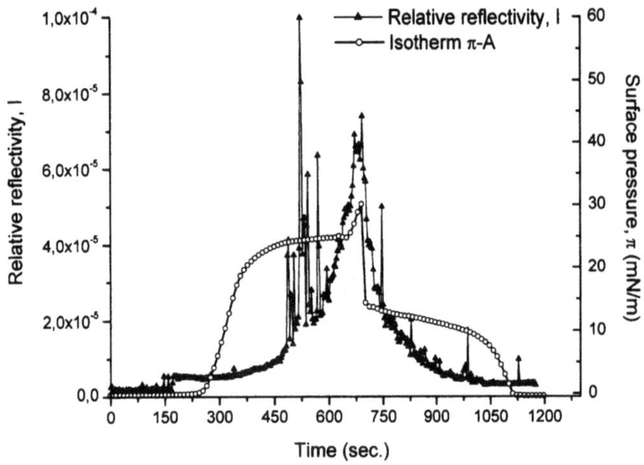

Fig. 2 Time evolution of surface pressure, π, and relative reflectivity, I, during a compression-expansion cycle for p-tolyl-PTCA monolayer spread on water subphase at 20 °C

reflected from an optically isotropic molecular film at the Brewster angle of the substrate depends on the film thickness, d, as [16]

$$I = C\, d^2 \tag{1}$$

where C is an instrument constant. For a quantification, the grey level output of the camera was converted into absolute intensity values after a calibration that was performed as described by Rodríguez Patino and co-workers [15].

Results and discussion

As it has already been mentioned above, the BAM images (published previously in [13]) showed stripes appearing after the mid-plateau region on compression, pointing to non-monomolecular structures, but there were no relevant features in the beginning of the plateau. However, when the relative reflectivity is computed, it is clear that the average thickness of the monolayer increases right from the start of the plateau. Fig. 2 shows that as the surface pressure plateau begins, an increase in reflectivity is observed, which becomes even steeper in the mid-plateau region. Also, spikes due to large oscillations in the measured reflectivity also appear after the mid-plateau region, consistent with the formation of collapsed structures already indicated in the BAM images.

While the increase in relative reflectivity from the beginning of compression until the start of plateau is negligible, a four-fold increase in relative reflectivity (from $1.2{\cdot}10^{-5}$ to $4.9{\cdot}10^{-5}$ (arb. units)) is observed throughout the plateau region. According to Eq. (1), this corresponds to the increase in monolayer thickness by a factor of 2. It should be stressed that until the mid-plateau, the monolayer becomes only 1.2 times thicker than at the beginning of the transition. This explains our previous observation of characteristic stripes in the BAM images, which became visible only after the mid-plateau region, had been reached. Upon monolayer decompression the relative reflectivity attains its original value, and hysteresis is observed as the plateau in surface pressure occurs at a lower pressure. This is also visualised in Fig. 2. When the film is left to relax for 20 or more minutes, subsequent compression-decompression cycles lead to essentially the same results.

The conclusion above was confirmed in several instances where the monolayer was compressed up to the beginning of the plateau, mid-plateau region and full compression up to the second steep rise in surface pressure. In all cases, the relative reflectivity increased at the beginning of the plateau, spikes appear in the mid-plateau region and the return to the original reflectivity upon decompression follows the surface pressure isotherm. This unambiguously indicates that changes in the film structure occur at the beginning of the plateau, and not in the middle, as it was previously concluded basing solely on BAM images [3, 13].

By analogy to the models from refs. [17–19] (wherein a 3-fold decrease in molecular area within a plateau was attributed to a trilayer formation), the decrease in area per molecule by a factor of 2 as well as the two-fold increase in monolayer thickness as inferred from the relative reflectivity data may be attributed to a transition of a monolayer into a bilayer. It is, however, difficult to suggest an adequate, energetically favourable model. The formation of a bilayer would involve either the formation of a second layer of molecules with polar groups exposed to the air or immersion of one layer of molecules into the water subphase. Such molecular arrangements are both unlikely from the thermodynamic point of view. In fact, a 2-fold decrease in area/increase in film thickness should be considered as an average value and may result from the formation of multilayer (e.g., tri-layer) patches coexisting with a monolayer. This model seems to be most plausible in view all the discussed results.

In conclusion, we have proven that the plateau in the surface pressure isotherms of *p*-tolyl-PTCA is associated with the formation of multilayer structures, rather than due to molecular reorientation, as it was previously believed [4, 5]. This non-monomolecular arrangement of molecules in the plateau region was unambiguously supported by the relative reflectivity data. What still remains indeed surprising is that the formation of the non-monomolecular structures is reversible, despite the evidence for the appearance of collapsed domains after the mid-plateau region.

References

1. Czapkiewicz J, Dynarowicz P, Milart P (1996) Langmuir 12:4966
2. Czapkiewicz J, Dynarowicz-Łątka P, Janicka G, Milart P (1998) Colloids Surf A 135:149
3. Dynarowicz-Łątka P, Dhanabalan A, Oliveira ON Jr (1999) J Phys Chem B 103:5992
4. Dynarowicz-Łątka P, Dhanabalan A, Cavalli A, Oliveira ON Jr (2000) J Phys Chem B 104:1701
5. Dynarowicz-Łątka P, Dhanabalan A, Oliveira ON Jr (2000) Langmuir 16:4245
6. Pallas NR, Pethica BA (1995) Langmuir 1:509
7. Marcelja S (1974) Biochem Biophys Acta 367:165
8. Seitz M, Struth B, Preece JA, Plesnivy T, Brezesinski G, Ringsdorf H (1996) Thin Solid Films 284/5:304
9. Kellner BMJ, Cadenhead DA (1978) J Coll Interface Sci. 63:452
10. Vogel V, Möbius D (1985) Thin Solid Films 132:205
11. Vila Romeu N, Miñones Trillo J, Conde O, Casas M, Iribarnegaray E (1997) Langmuir 13:71
12. Davies JT, Rideal EK (1963) Interfacial phenomena. Academic Press, New York
13. Dynarowicz-Łątka P, Czapkiewicz J, Kita K, Milart P, Brocawik E (1999) Progr Colloid Polym Sci 112:15
14. Berreman D (1972) J Opt Soc Am 62:502
15. Rodríguez Patino JM, Sánchez CC, Rodríguez Niño MR (1999) Langmuir 15:2484
16. Azzam RMA, Bashara NM (1992) Ellipsometry and polarized light. North Holland, Amsterdam
17. Xue J, Jung CS, Kim MW (1992) Phys Rev Lett 69:474
18. Schröter JA, Plehnert R, Tschierske C, Katholy S, Janietz D, Penacorada F, Brehmer L (1997) Langmuir 13:796
19. Seitz M, Struth B, Preece JA, Plesnivy T, Brezesinski G, Ringsdorf H (1996) Thin Solid Films 284/5:304

Progr Colloid Polym Sci (2004) 123: 156–159
DOI 10.1007/b11751

M. Brunner
C. Bechinger

Colloidal systems in intense, two-dimensional laser fields

M. Brunner · C. Bechinger (✉)
Physics Department,
University of Konstanz,
78457 Konstanz, Germany
e-mail: Clemens.Bechinger@uni-konstanz.de

Abstract The properties of two-dimensional (2D) colloidal systems in the presence of a strong laser field are studied. By interfering three laser beams we obtain a 2D laser field, which acts as an artificial substrate potential to the particles. In contrast to topographical substrate potentials, the use of a light potential allows changes in the potential strength and the periodicity of the substrate potential continuously. We report first results on charge stabilised polystyrene particles on a triangular light potential.

Keywords Colloids · Two-dimensional phase transitions · Optical tweezers

Introduction

Melting and freezing are among the most common yet fascinating physical phenomena encountered in everyday life. While in three-dimensional systems, there is an abrupt change in the system density when crossing the melting point, two-dimensional (2D) systems show a different scenario. Due to thermal fluctuations, in 2D systems no longer a first order transition but a two-stage, continuous transition is observed which is driven by an unbinding of topological defects (dislocations and disclinations) [1–3]. Besides theoretical interest in the physics during 2D melting, the underlying physics is also important from a technological point of view, as can be seen from novel 2D semi-conducting or magnetic devices.

While most of the experimental work considered only the interactions within 2D objects [4, 5], real 2D systems do not exist as self-contained objects but are typically confined to solid (crystalline) or liquid substrates. Accordingly, it is the interplay between the particle-particle interactions within the 2D system and the particle-substrate interaction, which finally determine the phase behaviour in such a situation.

In the present study, we used colloidal suspensions, which have been established as ideal model systems for fundamental studies of the phase behaviour of 2D systems [4, 5]. As substrate potentials we used an optical interference pattern which is formed by overlapping several laser beams. In contrast to former experiments, where periodic, one-dimensional substrate potentials were considered [6, 7], here we interfere three laser beams which results in 2D substrate potentials. Since the geometry and the substrate potential strength can be easily varied (in contrast to atomic systems), this allows systematic measurements on the phase behaviour of 2D systems on periodic substrate potentials, e.g., crystalline surfaces.

Experimental

As colloidal system we used aqueous suspensions of sulphate-terminated polystyrene (PS) particles of 1.5 μm radius with an average polydispersity below 4% (Interfacial Dynamics Corporation). The suspended particles are negatively charged and interact by means of a partially screened electrostatic repulsion which can be described by [8, 9]

$$\Psi(r) = \frac{(Z^* e)^2}{4\pi\varepsilon\varepsilon_0} \left(\frac{\exp(\kappa R)}{1+\kappa R}\right)^2 \frac{\exp(-\kappa r)}{r} \quad (1)$$

Here, Z^*e is the renormalised charge of the particles which has been roughly determined to be $Z^* \approx 20{,}000$ [10], ε is the dielectric constant of water, κ is the inverse Debye screening length and r is the distance between particle centres. The experiments were performed in a closed circuit which is composed of the sample cell, a vessel of ion exchange resin and an electrical conductivity probe to control the ionic strength of the suspension [11]. A peristaltic pump is used to pump the highly deionised suspension through the circuit. The Debye screening length was calculated from the interparticle distribution of the colloidal spheres to be in the order of $\kappa^{-1} \approx 350$ nm. As sample cell we used a silica glass cuvette with a spacing of 1 mm. We first deionised the circuit almost completely as confirmed by the value of the ionic conductivity $\sigma = 0.07$ μS/cm. Then the sample cell was disconnected from the circuit to allow stable conditions during several hours.

The laser potential is created by three laser beams of a linearly polarised, frequency doubled Nd:YVO_4 laser (Coherent VERDI $\lambda = 532$ nm, $P_{max} = 5$ W) which overlap in the sample plane where they create an interference pattern (Fig. 1). Due to the polarisability of the PS spheres, which is larger than that of the solvent, the particles are attracted into the regions of highest light intensity. Accordingly, the interference pattern provides a periodic two-dimensional light potential for the particles. The periodicity and geometry of the potential can be adjusted by variation of the crossing angle. Due to the almost vertical incidence of the laser beams onto the sample the particles are additionally exposed to a vertical light force which pushes them towards the negatively-charged bottom silica plate of our cell. This vertical force is estimated to be in the range of pN, i.e., three orders of magnitude larger than the gravitational force on the particles. Accordingly vertical particle fluctuations are considerably reduced to a value below 50 nm and the system is effectively confined into two dimensions.

The sample illuminated with white light through a dichroic mirror from the top, to image the particles with a microscope objective (Zeiss, Achroplan 20×, 0.4) onto a CCD camera chip

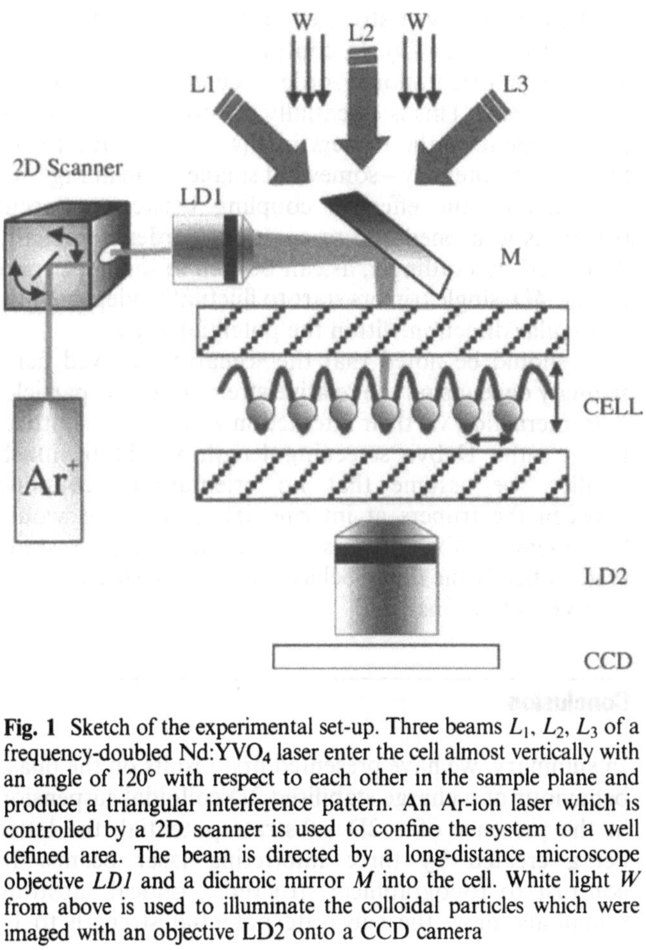

Fig. 1 Sketch of the experimental set-up. Three beams L_1, L_2, L_3 of a frequency-doubled Nd:YVO_4 laser enter the cell almost vertically with an angle of 120° with respect to each other in the sample plane and produce a triangular interference pattern. An Ar-ion laser which is controlled by a 2D scanner is used to confine the system to a well defined area. The beam is directed by a long-distance microscope objective *LD1* and a dichroic mirror *M* into the cell. White light *W* from above is used to illuminate the colloidal particles which were imaged with an objective LD2 onto a CCD camera

(Teli, CS 8310) which was connected to a computer. The intense Nd:YVO_4 laser light was blocked below the sample cell with an optical filter to protect the CCD camera from the exposure to the high laser intensity. The particle centre positions were analysed on-line with an imaging processing software (Visiometrics, IPS) and stored on the hard drive for further analysis.

In order to experimentally determine the depth of the laser potential as a function of the intensity of the interference fringes, we exposed highly diluted colloidal suspensions to the above mentioned optical interference pattern. From the measured probability distribution of the particles we obtained the laser potential acting on the particles by employing the Boltzmann statistics. The shape of the laser potential in the central region of the overlapping laser beams is found to be

$$\Phi(x) = -V_0\left(3 + \cos(2\pi/d^x) + 4\cos(\pi/d^x)\cos\left(\sqrt{3}\pi/d^y\right)\right) \quad (2)$$

with V_0 being the potential amplitude and d the period of the fringe spacing. The amplitude V_0 as obtained from our experiments increases linearly as a function of the laser intensity I. According to Loudiyi et al. the colloid-light interaction V_0 can be calculated by integration over the potential contributions for each infinitesimal volume element of the colloidal sphere [12].

$$V_0 = -6n_w^2\left(\frac{2n^2-1}{2n^2+2}\right)\left(\frac{j_1(2\pi/d\sigma_0)}{4\pi/d\sigma_0}\right)\frac{WP\sigma_0^3}{cr^2} \quad (3)$$

Here, P is the laser power, c the velocity of light in vacuum, n the ratio of the refraction indices of polystyrene n_P and water n_W, σ_0 the colloidal particle radius, j_1 the first order spherical Bessel function and r the waist radius of the Gaussian laser beam, respectively.

In addition to the interference pattern, an Ar^+-laser beam with 200 mW (Coherent INNOVA 90, $\lambda = 488$ nm) whose position was controlled by a galvanostatic 2D mirror unit (Scanlab, SCANgine 10) was focused into the sample cell. This allowed us to draw a rectangular box around the system, which acts as a lateral confinement and provides well defined conditions (e.g., constant particle number) during the measurements. Since the repetition rate of this box was more than 50 Hz, the confinement can be regarded as static, due to the slow relaxation times in colloidal systems. Behind the scanning unit, the laser beam passed a long-distance microscope objective (Nikon, CF Plan 20×, 0.35) and was reflected from a dichroic mirror into the sample cell. Special care was taken, in order to guarantee, that the focus of the scanned laser tweezers was in the sample plane, i.e., the focus-plane of the three interfering laser beams.

Because the scanning unit is controlled by a computer, the size of the rectangular bounding box can easily be changed. Accordingly, if no particles overcome the bounding box, this allows adjusting the particle density in the observed system.

Results

Fig. 2A shows a calculation of the triangular intensity distribution which is obtained when three beams of equal intensity and polarisation interfere under an angle of 60 degrees. The corresponding experimental intensity distribution is shown in Fig. 2B and shows good agreement. Depending on the relative angles between the three beams, also different geometries and lattice constants can be obtained. When a colloidal suspension is exposed to such an interference pattern, optical gradient forces drive the particles into the triangular

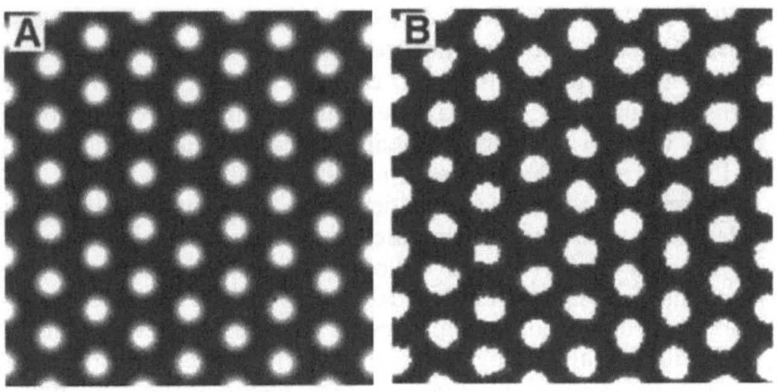

Fig. 2 (A) Calculated intensity distribution of three interfering laser beams of equal intensity and an intersection angle of 60°. (**B**) Experimentally observed laser intensity inside the sample cell

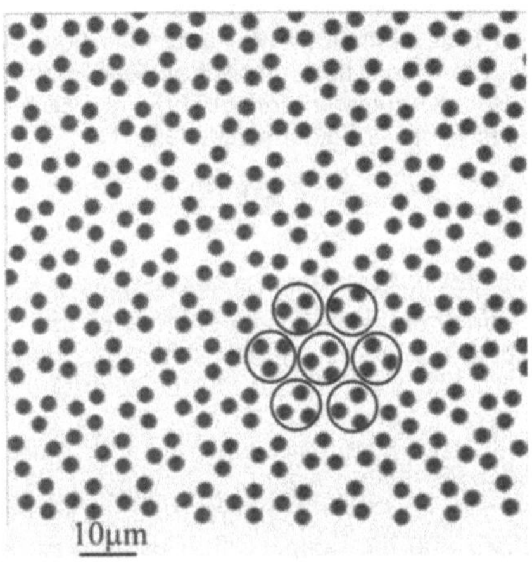

Fig. 3 Snapshot of a 2D colloidal system in the presence of a triangular light lattice (the latter being schematically indicated by black circles) where three particles (trimers) each are confined within one laser potential well

arranged regions of highest intensity. In addition to those light forces, the particles interact by means of the screened electrostatic repulsion as described in Eq. (1).

We have investigated different particle number densities, with the number density to be commensurate with the underlying light lattice, i.e., the number of particles to be an integer multiple m of the light lattice sites. Fig. 3 shows a snapshot for $m=3$. As can be seen, clusters of three particles (trimers) were confined to the triagonal laser lattice. As can be seen in the upper right corners, the centre of mass of the trimers is identical to the corresponding triangular geometry of the light lattice. This clearly demonstrates, that the optical interference pattern acts as a modulated surface potential for the colloidal spheres. In order to investigate the effect of potential depth of the interference pattern onto the structure of the colloidal system, we varied the light intensity of the interfering laser beams and determined the averaged particle density. This has been done by analysing 1000 pictures with 2 s time interval. The results are shown for the trimers in Fig. 4.

The particle density has been chosen to be somewhat below the critical density required for spontaneous crystallisation when no interference pattern was applied. Accordingly, the system forms in the absence of the laser field an isotropic, 2D liquid with no orientational and positional order (Fig. 4A). When the laser field is adjusted to a value $V_0=6\ k_BT$ the particle distribution is considerably affected as can be seen in Fig. 4B. In some regions of the samples trimers were localised already at the laser lattice sites defined by the interference pattern (indicated schematically as open circles in Fig. 4). However, due to thermal fluctuations, particles can still exchange between those laser potential wells. Accordingly, no positional order is observed at these laser intensities. Increasing the light intensity to a value of $V_0=17\ k_BT$, the trimers become more localised at the corresponding triangular lattice sites of the underlying laser field. Due to the interplay of the repulsive interparticle forces and the laser potential, now an ordered phase with both long-range positional and orientational order is obtained (Fig. 4C). In contrast to Fig. 4b, where inter-well diffusion is very pronounced, here only point defects were observed which can diffuse through the lattice without destroying long-range positional order.

When the laser potential is increased to $V_0=27\ k_BT$, the trimers are even stronger localised to the triagonal laser lattice (Fig. 4D). In contrast to Fig. 4C, however, long-range orientational order becomes partially destroyed again. This is essentially due to the fact that the lateral extent of the trimers in Fig. 4D is – due to the higher laser intensity – somewhat smaller than in Fig. 4C. Accordingly, the effective coupling between adjacent trimers is weakened and orientational order is partially destroyed. Accordingly, as can be seen in the upper part of Fig. 4D, single trimers start to fluctuate independently in angular direction within the potential wells.

It should be noted that the scenario observed here, strongly depends on the relative strengths of the particle-pair interaction vs. their interaction with the light lattice. If, e.g., the Debye screening length would be much smaller, we assume that no orientational coupling between the trimers at intermediate laser fields would be observed. Experiments are in progress, to study systematically the phase behaviour of 2D systems in such complex situations.

Conclusion

In summary, we have presented first results on the phase behaviour of a charge stabilised 2D colloidal suspension in the presence of a 2D substrate potential, the latter being created by three interfering laser beams. In contrast to topographical structures, optical substrate potentials allow the substrate potential strength to be

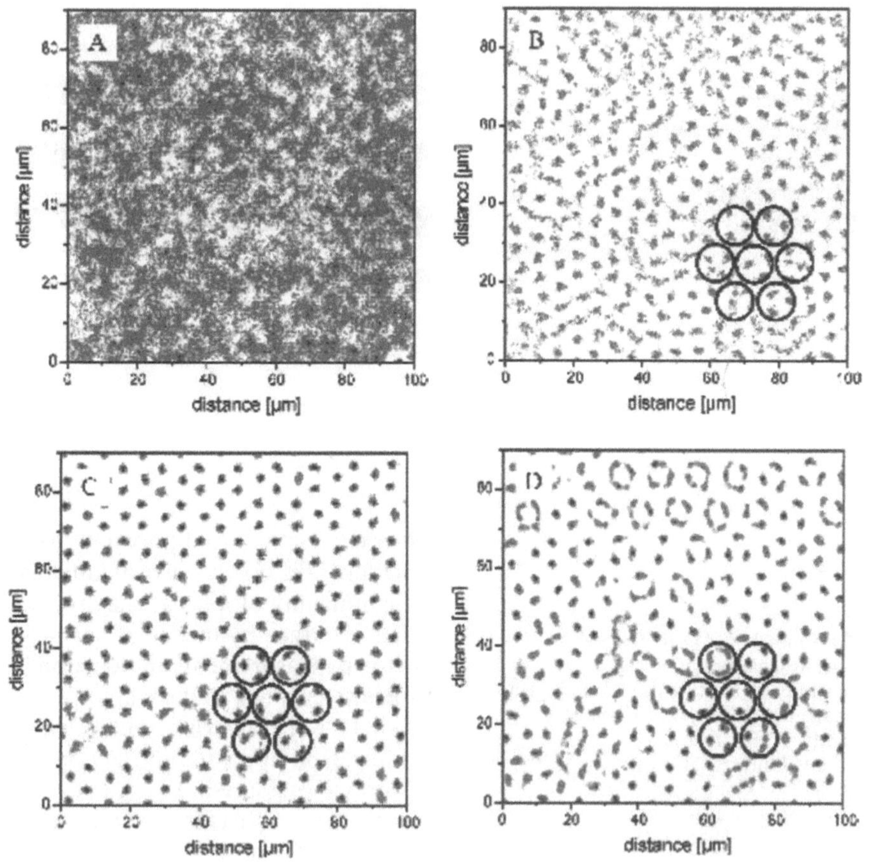

Fig. 4 Density plots at different laser potential strengths V_0. (**A**) In the absence of a laser potential the density distribution of the system equals the density of an isotropic liquid. (**B**) At a potential strength of about 6 k_BT the concentration of particles within the potential wells can be observed. (**C**) At 17 k_BT a true long-range crystalline order is present. (**D**) At higher intensities around 27 k_BT the orientational order is lost again

continuously adjusted. With increasing potential strength, we observe first the occurrence of positional and orientational order. Upon further increasing the potential depth, however, orientational order is lost again. We interpret this phenomenon by the interplay of the particle pair-interaction and their interaction with the optical substrate potential.

Acknowledgement We gratefully acknowledge financial support from the Deutsche Forschungsgemeinschaft (BE 1788).

References

1. Kosterlitz JM, Thouless DJ (1973) J Phys C 6:1181
2. Nelson DR, Halperin BI (1979) Phys Rev B 19:2457
3. Young AP (1979) Phys Rev B 19:1855
4. Murray CA, Winkle DHV (1987) Phys Rev Lett 58:1200
5. Zahn K, Lenke R, Maret G (1999) Phys Rev Lett 82:2721
6. Wei Q-H, Bechinger C, Rudhardt D and Leiderer P (1998) Phys Rev Lett 81:2606
7. Bechinger C, Brunner M, Leiderer P (2001) Phys Rev Lett 86:930
8. Derjaguin BV, Landau L (1941) Acta Physicochim URSS 14:633
9. Vervey EJW, Overbeek JTG (1948) Theory of the stability of lyophobic colloids. Elsevier, Amsterdam
10. Alexander S, Chaikin PM, Grant P, Morales GJ, Pincus P (1984) J Chem Phys 80:5776
11. Palberg T, Härtl W, Wittig U, Versmold H, Würth M, Simnacher E (1992) J Phys Chem 96:8180
12. Loudiyi K, Ackerson BJ (1992) Physica A 184:1

Progr Colloid Polym Sci (2004) 123 : 160–163
DOI 10.1007/b11752

J. Miñones Jr
P. Dynarowicz-Łątka
R. Seoane
E. Iribarnegaray
M. Casas

Brewster angle microscopy studies of the morphology in dipalmitoyl phosphatidyl glycerol monolayers spread on subphases of different pH

J. Miñones Jr (✉) · R. Seoane
E. Iribarnegaray · M. Casas
University of Santiago de Compostela,
Department of Physical Chemistry,
Faculty of Pharmacy,
Campus Universitario Sur s/n. 15706,
Santiago de Compostela, Spain
e-mail: jminones@usc.es
Tel.: +34-981-594629
Fax: +34-981-528011

P. Dynarowicz-Łątka
Jagiellonian University,
Faculty of Chemistry, Ingardena 3,
30-060 Kraków, Poland

Abstract Dipalmitoyl phosphatidyl glycerol (DPPG) was investigated as Langmuir monolayers at the air/water interface by means of surface pressure measurements in addition to Brewster angle microscopy (BAM). A characteristic phase transition region appeared in the course of surface pressure-area (π-A) isotherms for monolayers spread on alkaline water or on buffer subphases, while on neutral or acidic water the plateau region was found to disappear. This phase transition region was attributed to the ionisation of DPPG monolayer. Quantitative Brewster angle microscopy measurements reveal that at the transition the thickness of ionised DPPG monolayer increases by 4.2 Å as a result of conformational changes of the ionised polar groups, which tend to emerge from the bulk subphase up to the surface due to their dehydration upon compression.

Keywords Dipalmitoyl phosphatidyl glycerol monolayers · Langmuir films · Brewster angle microscopy

Introduction

In recent years, Brewster angle microscopy (BAM) [1, 2] has become one of the most frequently used techniques for in situ observation of monolayers. It does not require the presence of any additives like, for example, fluorescent probes, and therefore does not alter the physical state of the monolayer. With this technique we are presenting herein BAM images for monolayers of dipalmitoyl phosphatidyl glycerol (DPPG). This phospholipid shows, under particular pH and ionic strength conditions, a first-order phase transition, which manifests in BAM images as a structure containing small, circular and bright domains, similar to those at observed with BAM for other systems [3–5]. Moreover, the quantitative BAM measurements, based on the analysis of time evolution of the relative reflectivity (I) upon monolayer compression, allow for the determination of a change in the relative film thickness at the transition region due to the change in conformation of polar groups at the interface.

Experimental

Dipalmitoyl L-α-phosphatidyl-DL-glycerol (DPPG) (Sigma) was dissolved in chloroform/ethanol (4:1 v/v) mixture. The dropping solution was spread onto the air/water interface with a Microman Gilson microsyringe, precise to $\pm$ 0.2 μL.

Ultrapure water, used as subphase, was obtained from a Milli RO, Milli Q reverse osmosis system (Millipore Corp.). Its resistivity was 18 MΩcm. The temperature was maintained constant at 20.0 °C. The subphase pH was adjusted by addition of HCl or NaOH (p.a. grade). In some experiments, the Theorell-Stenhagen buffer solutions were used as a subphase.

The Langmuir-Blodgett KSV-5000 trough (Finland) was used to record the surface pressure/area (π-A) isotherms at a rate compression of 8.5 Å^2/molecule·minute. Each curve shown in this paper represents the average of four independent experimental π-A isotherms.

Morphological film studies as well as the determination of the relative film thickness were performed with a Brewster angle microscope, model BAM 2 (NFT, Germany). This apparatus is equipped with a 30 mW laser, emitting *p*-polarised light of 690 nm wavelength. The lateral resolution of the microscope was 2 μm. Images were digitised and processed in order to obtain the best quality BAM pictures. Different shutter speeds, corresponding to different time of exposition, were used (1/50 s–1/500 s).

To measure the relative film thickness, the camera calibration was necessary in order to determine the relationship between the intensity unit, so-called "grey level" (GL), and the relative reflectivity (I). It was done following the procedure described by Rodriguez Patino et al. [6, 7] for different shutter speeds.

Results and discussion

Figure 1 shows the π-A isotherms for DPPG monolayers spread on water of different pH. Monolayers deposited on subphases of pH 3 and 6 are of condensed type; their limiting areas are about 55 Å^2/molecule and collapse pressures between 57 and 62 mN/m were recorded. A pseudo-plateau region, corresponding to the liquid-expanded (LE)-liquid-condensed (LC) transition exists in the course of π-A curve for the monolayer spread on alkaline aqueous subphase (pH 10). BAM images recorded under these experimental conditions reveal the presence of small, bright and circular shaped domains at the beginning of the plateau (Fig. 2A). These domains increase in size upon monolayer compression, along the LE-LC region (Fig. 2B), and evolve into polygonal forms at the end of the transition (Fig. 2C). The shape of DPPG domains at the transition region is similar to that observed by different authors for other systems [3–5]. At the pressure of 20 mN/m, these domains occupy nearly all the available area of the surface. They are closely packed, without any separating gap, and exhibit optical anisotropy (images not shown here) which evidences for different orientation of the aliphatic tail groups. However, at higher surface pressures (above 45 mN/m), the images are homogeneous (Fig. 2D), without optical anisotropy, indicating vertical orientation of the non-polar groups.

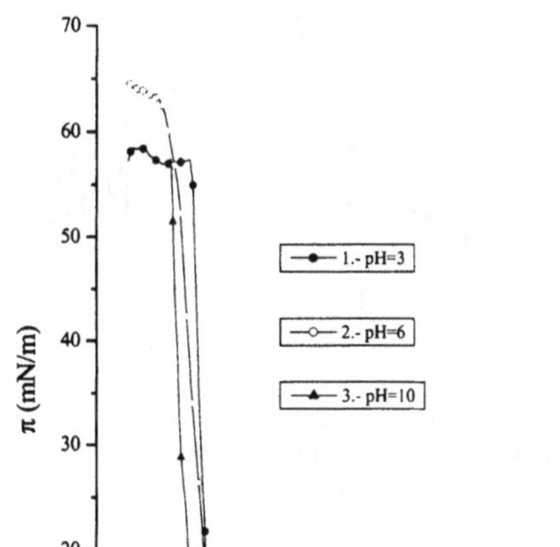

Fig. 1 Surface pressure-area (π-A) isotherms for DPPG monolayers spread on water

When the Theorell-Stenhagen buffer is used as a subphase, DPPG monolayers also show an LE-LC transition region at surface pressure, the value of which increases with increasing buffer pH (Fig. 3). Such a phase transition has already been observed by Koppenol et al. [8], however, the transition surface pressure value reported therein (~15 mN/m) was higher than observed in the present work. This discrepancy can be due to different experimental conditions applied by Koppenol et al., i.e., subphase composition (10 mM TRIS buffer and 15 mM NaCl) and higher temperature (25 °C). Similarly, Borissevitch et al. [9] reported the existence of a plateau in the course of the π-A isotherms for DPPG spread on phosphate buffer solution, which disappear when the monolayer was deposited on pure water subphase of pH 5.9. This phenomenon has also been observed in the present work, as evidenced in Fig. 1.

On an aqueous subphase of pH 10, the DPPG monolayer is highly ionised due to the fact that its pK value is 2.9 [10]. On the other hand, it is know that the addition of salts (buffer) into aqueous subphases causes the increase of the ionisation state of phospholipids [9, 11]. Accordingly, it can be supposed that the ionisation state of DPPG spread either on alkaline water or on buffer subphases of pH >6 is similar. Consequently, the plateau in the π-A curves (ascribed to the LE-LC transition) could be attributed to the high film ionisation, because when the monolayer is only slightly ionised (i.e., on water at pH<6), this plateau does not appear.

Fig. 4 shows the time evolution of the relative reflectivity (I-t curves) and the surface pressure (π-t curves) during a compression-expansion cycle for DPPG monolayer spread on alkaline aqueous subphase (pH 10). Upon compression, in the gaseous (or expanded, E) phase and in the liquid-expanded (LE) phase, the relative reflectivity increases continuously. In the LE-LC transition region, I increases significantly from 5.1×10^{-7} to 9.4×10^{-7}, which corresponds to an increase of 1.35-times in the film thickness. (The relation between I and the relative thickness, δ, of film regions is $I = k\,\delta^2$, where k is a constant) [12]. Finally, in the LC and solid (S) regions, the monolayer's relative thickness increases 1.17 times. It is also important to point out that, upon monolayer compression, there are visible few, low intensity reflectivity peaks, due to the fact that the domains are very small and of low reflectivity (Fig. 2).

From the relative reflectivity (I) results (shown in Fig. 4), using the relation $I = k\,\delta^2$, one can deduce that the relative thickness of the monolayer at the beginning of the liquid-condensed state is 16.1 Å, assuming the value of 19 Å for the film thickness in the solid phase (where the hydrocarbon tail groups of the film molecules are supposed to be vertically oriented in the *trans*

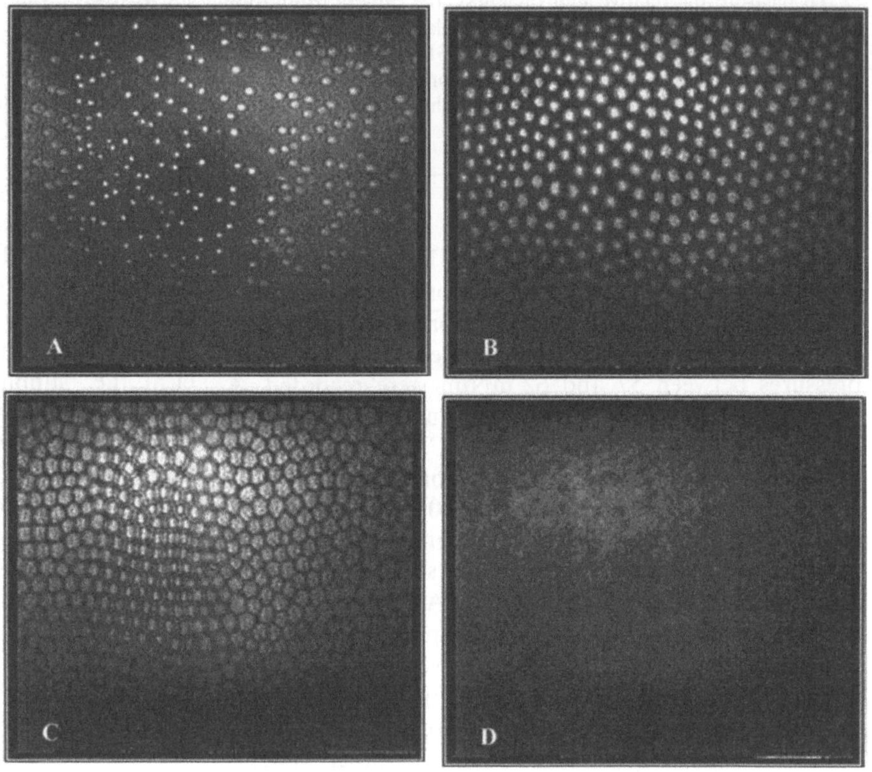

Fig. 2 BAM images taken at different stages of compression for DPPG monolayer spread on water, pH 10. (see explanation in the text)

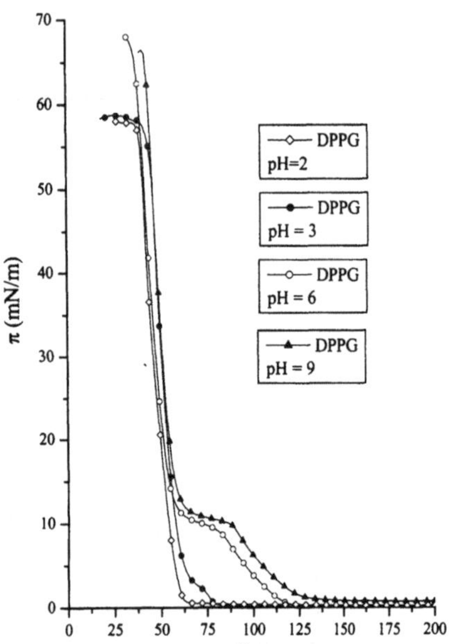

Fig. 3 Surface pressure-area (π-A) isotherms for DPPG monolayers spread on Theorell-Stenhagen buffer subphases

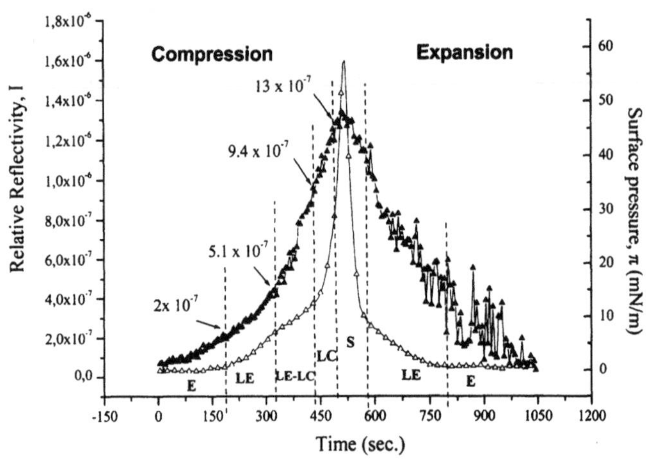

Fig. 4 Time evolution of surface pressure (Δ-Δ) and relative reflectivity (▲-▲) during a compression-expansion cycle for a DPPG monolayer spread on water, pH 10

conformation, such as it is deduced from images in Fig. 2D):

$$L_{tail} = N_{CH_2} x (1.53 \sin 56°)Å + 1.27 Å = 19 Å \quad (1)$$

Wherein $N_{CH_2} = 15$ (taking the length of the C-C bond as 1.53 Å and the angle between neighbouring carbon atoms as 112°) [13].

Since the hydrocarbon tail of DPPG is 2.54 Å longer than that of dimiristoyl phosphatidyl glycerol (DMPG), the value of 16.1 Å obtained here for DPPG monolayer is quite reasonable since for DMPG the thickness of the film in the condensed state was reported by Bayerl et al. [14] to be 14.2 ± 1 Å.

As mentioned above, the LE-LC transition was attributed to the ionisation of the molecule. So, the plateau in the π-A curves could be caused by conformational changes of the ionised polar groups. Accordingly, it can be postulated that in the liquid-expanded phase the hydrated polar groups are drawn into the subphase. Film compression along the transition region provokes their dehydration and subsequent change of conformation, surfacing the DPPG molecule and simultaneously increasing the monolayer thickness. The experimental results show that in the LE-LC transition region the monolayer thickness increases from 11.9 Å to 16.1 Å. This increase of 4.2 Å corresponds to the hydrocarbon tail length increased by ca. 3 carbon atoms as a result of the change in conformation experienced in the transition region by the ionised polar group, which tends to emerge from the bulk subphase up to the surface.

Acknowledgements The Xunta de Galicia (Project PGIDT99P-XI20302B) supported this work.

References

1. Hénon S, Meunier J (1991) Rev Sci Instrum 62:936
2. Höning D, Möbius D (1991) J Phys Chem 95:4590
3. Györvary E, Albers WM, Peltonen J (1999) Langmuir 15:2516
4. Hénom S, Meunier J (1992) Thin Solid Films 210/211:121
5. Prieto I, Martín Romero MT, Camacho L, Möbius D (1998) Langmuir 14:4175
6. Rodriguez Patino JM, Sánchez CC, Rodríguez Niño MR (1999) Langmuir 15:2484
7. Rodríguez Patino JM, Sánchez CC, Rodríguez Niño MR (1999) Food Hydrocolloids 13:401
8. Koppenol S, Yu H, Zografi G (1997) J Colloid Interface Sci 189:158
9. Borissevitch GP, Tabak M, Oliveira ON (1996) Biochim Biophys Acta 1278:12
10. Cevc G, Marsh D (1987) Phospholipid bilayers, physical principles and models. Wiley Interscience, New York
11. Abramson MB, Katzman R, Wilson CE, Gregor HP (1964) J Biol Chem 239:4066
12. Mul MNG de, Mann JA Jr (1998) Langmuir 14:2455
13. Small DM (1988) In: Hanahan DJ (ed.) The physical chemistry of lipids, 2nd edn. Plenum Press, New York London, p 22
14. Bayerl TM, Thomas RK, Penfold J, Rennie A, Sackmann E (1990) Biophys J 57:1095

Progr Colloid Polym Sci (2004) 123: 164–168
DOI 10.1007/b11754

L.L. Peel
J.R. Lu

The interaction of $C_{12}E_5$ with olive oil films studied by neutron reflection

L.L. Peel · J.R. Lu (✉)
Department of Physics, UMIST,
PO Box 88, Manchester, M60 1QD, UK
e-mail: J.Lu@umist.ac.uk
Tel.: +44-161-2003926;
Fax: +44-161-2004303

Abstract Neutron reflection has been used to explore the action of $C_{12}E_5$ on olive oil films coated on SiO_2. The measurements were made at the solid/solution interface at the CMC of $C_{12}E_5$ with the oil films fixed at 50 Å and 200 Å dry thickness. It was found that the total amount of $C_{12}E_5$ interacting with the oil was higher for the thicker film and that the ratio of absorption to surface adsorption was 3:2. The results indicate that $C_{12}E_5$ interacts with the oil film via adsorption and absorption.

Keywords Absorption · Adsorption · Surfactant · Surface cleaning · Neutron reflection

Introduction

Surfactants have been used for many years in various commercial formulations to clean oil stains. Many studies have been made to improve the performance of surfactants, but the molecular level of interactions has been little understood [1–5]. A major limitation has been that few existing techniques are sensitive to the molecular structure details. Neutron reflection is a newly developed technique capable of probing molecular structure profiles at different interfaces. Detailed information on the theory and procedures of neutron reflection experiments is available in the literature [6–9] and the particular application to the study of wet interfaces is reviewed [8]. In this work, neutron reflection was used to determine the molecular positions of $C_{12}E_5$ at the oil-water interface prior to any film destruction and hence learn about *how* a surfactant can clean oil from a surface. This work is in contrast to the previously reported studies where the emphasis has been on the final removal of an oil residue.

Experimental

Materials used

Silicon blocks of (111) orientation were used as substrates. The blocks were polished and prepared for coating following procedures given previously [10, 11]. This set of treatment produced an optically flat surface with a smooth native SiO_2 layer around 15–20 Å. The subsequent piranha treatment involving the immersion of the block to a solution of H_2SO_4 (600 mL, 98%) and H_2O_2 (50 mL, 30%) at 90 °C for 2 min helped to keep the surface hydrophilic. The blocks were then rinsed with ultrahigh quality (UHQ) water to remove the acid. The edible oil used was Extra Virgin Olive Oil (Tesco). To facilitate film coating and to mimic the formation of an oil stain, the oil was heated at 180 °C for 40 h. The protonated surfactant used was $C_{12}H_{25}(OC_2H_4)_5OH$, $hC_{12}hE_5$, (Fluka, 99%) and the chain deuterated surfactant $dC_{12}hE_5$ (98% D) was synthesised following a standard procedure [12]. All solutions were prepared in D_2O (Fluorochem, 99.9%) with the concentration fixed at the critical micelle concentration (CMC).

Film formation

A dip-coating rig set at a dip speed of 4 mm/sec was used to prepare the oil films on the silicon blocks. The general set up of the rig, together with more detailed information of dip-coating procedures has been presented previously [13, 14]. Solutions of 0.4% wt/vol and 2.0% wt/vol of the treated olive oil in hexane were used to coat films of approximately 50 Å and 200 Å dry thickness respectively. The films were annealed at 180 °C for 5 h under standard atmospheric pressure.

Data collection

The neutron reflection data was recorded using the CRISP reflectometer at the Rutherford Appleton Laboratory near Oxford. Specially designed Teflon troughs were clamped against the oil film surface of each block and the troughs could be filled and drained with pure water or surfactant solution as necessary without needing to dismantle the cell. The reflectivity profiles were measured at three glancing angles of incidence, 0.35°, 0.8° and 1.8° and the data were combined to give a total momentum transfer range of 0.012–0.3 Å^{-1}. Above about 0.2 Å^{-1}, the reflectivity showed no further changes with increasing momentum transfer, and the signal corresponded purely to that of the background. This enabled the reflectivity value observed at high momentum transfer to be used for background subtraction. A percentage transmission file was recorded by comparing the signal obtained when the beam was directed straight through the silicon block with that of straight through air and was used to normalise the observed reflectivity data to an absolute scale.

Results and discussion

Wet oil film structure

Neutron reflectivity profiles were first measured at the oil-water interfaces of the two samples using two different water contrasts, D_2O and CM4 (D_2O/H_2O mixture to give a scattering length density of 4×10^{-6} Å^{-2}). Information about the wet film structures was extracted using model fitting based on the optical matrix theory [15]. A model could contain any number of layers whereby the layer boundaries corresponded to a change in structure perpendicular to the interface. Each layer in the model had two variable parameters, thickness (τ_i) and scattering length density (ρ_i). The aim was to find the simplest model, using the least number of layers that would fit the data and show agreement from both contrasts.

The measured profiles were fitted to the highest momentum transfer (Q) value possible before the signal was lost due to the background scattering. For the thin film, a Q range of 0–0.15 Å^{-1} was used. However, for the thick film, the data became smeared above about 0.1 Å^{-1}, causing a reduction in the amplitude of oscillation. These thick film profiles could not be modelled above 0.1 Å^{-1} by adding more layers to the model but only by using interfacial roughness parameters. However, the roughness parameters affected the profile as a whole, not just the high Q region, so there was a risk that the fits would become meaningless. Instead, the profiles for the thick film were fitted within a Q range of 0–0.1 Å^{-1} and roughness parameters were not used. The data above this range was not neglected completely. It was still important that the oscillations of the fitted profile fell within the correct place, even if the amplitudes did not match.

The models and parameters used to achieve the fitted profiles for both films are summarised in Table 1. Since the individual scattering length densities of the oil and water were known, their volume fractions (ϕ) in each layer could be calculated using Eqs. (1) and (2), where ρ_{layer} is the fitted scattering length density.

Table 1 Wet film structures

Sample	Layer	τ/Å	ϕ_{oil}	ϕ_{water}
"50 Å"	0 (SiO_2)	13		
	1	53	0.739	0.261
	2	13	0.183	0.817
"200 Å"	0 (SiO_2)	12		
	1	221	0.739	0.261

$$\rho_{layer} = \rho_{oil}\phi_{oil} + \rho_{water}\phi_{water} \tag{1}$$

$$\phi_{oil} + \phi_{water} = 1 \tag{2}$$

The results in Table 1 show that both films contained a bulk oil region consisting of about 26% water. This water penetration was to be expected due to the attraction between the oil ester groups and the water molecules. However, most importantly, the films were found to remain stable in the water after the initial swelling. The wet "200 Å" film could be described by a single layer, whereas the wet "50 Å" film structure included a second thin water diffuse outer layer. This suggests that the water molecules were more evenly distributed throughout the thick film. A diffuse outer region either did not exist, or was either sufficiently thin or sufficiently diffuse that it had a negligible effect on the total reflectivity profile in contrast to the thicker and more oil dense region beneath.

Interaction of CMC $C_{12}E_5$

Since the oil films contain water molecules and oxygen atoms, surfactant molecules added in solution may not only *adsorb* on to the outer film surface, but they may also *absorb* into the oil film. These effects were first examined using the oil film of dry thickness 50 Å and the experimental profiles measured from this film in contact with CMC $C_{12}E_5$ are shown in Fig. 1.

For both contrasts, the results show a deviation from the profiles measured in pure D_2O by way of a shift in the troughs to lower Q. This increase in oscillation frequency implies an increase in thickness and indicates that surfactant molecules had positioned themselves at the interface.

By gradually adapting the wet film model, the experimental data was fitted to the simplest model possible. Besides the addition of an outer surfactant/water layer, it was found that the scattering length densities of both the inner and outer oil regions had to be changed indicating that the surfactant had penetrated the oil film as well as adsorbing on to the oil film surface. The volume fractions (ϕ) of the constituent molecules in each oil film layer were obtained using the following equations:

$$\rho_{pi} = \rho_{oil}\phi_{oili} + \rho_{D_2O}\phi_{D_2Oi} + \rho_{hC_{12}}\phi_{C_{12}i} + \rho_{E_5}\phi_{E_5i} \tag{3}$$

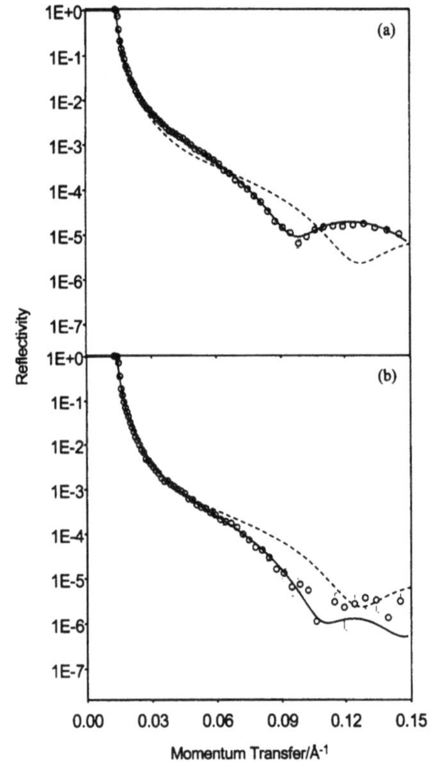

Fig. 1 Neutron reflectivity profiles (O) to show the interaction of protonated (**A**) and chain deuterated (**B**) CMC $C_{12}E_5$ at the "50 Å" olive oil film–water interface together with the fitted data (—). The profiles measured in the absence of surfactant (---) are also shown for comparison

$$\rho_{di} = \rho_{oil}\phi_{oili} + \rho_{D_2O}\phi_{D_2Oi} + \rho_{dC_{12}}\phi_{C_{12}i} + \rho_{E_5}\phi_{E_5i} \tag{4}$$

$$\phi_{oili} + \phi_{D_2Oi} + \phi_{C_{12}i} + \phi_{E_5i} = 1 \tag{5}$$

where ρ corresponds to scattering length density and ρ_{pi} and ρ_{di} are the scattering length densities of the ith layer as determined by model fitting the data obtained using $hC_{12}hE_5/D_2O$ and $dC_{12}hE_5/D_2O$ respectively. To determine the volume fractions of the molecules in the adsorbed layer, the same equations were used but the volume fraction of oil was set to zero.

It was found that the oil layers had not changed in thickness (τ_i) following the addition of surfactant and that the surfactant could be rinsed away leaving the film intact. Therefore, it could be assumed that the volume fractions of oil in layers 1 and 2 (ϕ_{oil1} and ϕ_{oil2}) could be fixed at the values obtained from the measurements taken in absence of surfactant. This gave three equations with three unknowns for each layer. However, the complete system had to contain whole surfactant molecules only. Therefore, by equating the ratio of the total volume of head to the total volume of tail with the molecular volume (V) of head to tail, the following relationship was obtained:

$$\frac{\phi_{E_51}\tau_1 + \phi_{E_52}\tau_2 + \phi_{E_53}\tau_3}{\phi_{C_{12}1}\tau_1 + \phi_{C_{12}2}\tau_2 + \phi_{C_{12}3}\tau_3} = \frac{V_{E_5}}{V_{C_{12}}} \tag{6}$$

The previous six equations were solved for a given set of data fit values using Gaussian Elimination. Eq. (6) was used to provide an error indication by comparing the calculated volume ratio with the known volume ratio. Data fit values could then be found such that this error was negligible and all the volume fractions were positive. The surface excess concentration, Γ_n, for surfactant head or tail with scattering length b_n and scattering length density ρ_n in layer i, could be calculated using Eqs. (7) and (8) where A_{ni} is the area per molecule of n in layer i.

$$\Gamma_{ni} = \frac{1}{A_{ni}N_A} \tag{7}$$

$$A_{ni} = \frac{b_n}{\phi_{ni}\rho_n\tau_i} \tag{8}$$

The results for the thin film are summarised in Table 2. Layer 1 corresponds to the interfacial region closest to the silicon oxide and layer 3 corresponds to the region closest to the bulk solution.

Comparison of the results with those for the film in pure D_2O (Table 1) shows the same thickness and oil volume fraction values for layers 1 and 2 confirming that the oil film structure had not changed. The volume fraction of water in these layers has reduced as a result of displacement by surfactant molecules. The outer adsorbed layer (layer 3) contained a higher volume fraction of E_5 head groups compared to the fraction of C_{12} tail groups. Consequently, the oil film layers (layers 1 and 2) contained higher volume fractions of tail groups in relation to head groups.

The calculated surface excess values (Table 2) again highlight the distribution of the surfactant head

Table 2 Distribution of CMC $C_{12}E_5$ at the "50 Å" olive oil film–water interface

Layer	τ/Å	ϕ_{oil}	ϕ_{D2O}	ϕ_{C12}	ϕ_{E5}	Γ_{C12}/mmol $m^{-2} \times 10^{-3}$	Γ_{E5}/mmol $m^{-2} \times 10^{-3}$
1	53	0.768	0.170	0.041	0.021	1.04	0.57
2	13	0.183	0.172	0.398	0.247	2.45	1.64
3	14		0.521	0.154	0.325	1.02	2.32
					Total $C_{12}E_5$:	4.52	

and tail groups across the different layers. The total amount of surfactant interacting with the oil was found to be 4.52 mmol m^{-2}, as determined from the average of the total head and total tail excess concentrations. This corresponds to an area per molecule of 36.7 $Å^2$ $molecule^{-1}$, which, as compared with an area per molecule of 48 $Å^2$ $molecule^{-1}$ for CMC $C_{12}E_5$ at the air-water interface [12], again suggests strong absorption.

From the results for the individual layers (Table 2), the percentage of the total amount of surfactant interacting that had absorbed into the oil film was calculated as 64%, with the remaining 36% having been adsorbed onto the outer film surface. The main driving force for the absorption process is based on the hydrophobic effect, which is believed to have occurred by two main mechanisms. Firstly, there is the affinity between the surfactant and oil alkyl chains, which is likely to have been the dominant mechanism in the more dense oil region (layer 1). Secondly, the surfactant molecules can reduce the level of unfavourable contacts between the oil and water molecules in the film, which is likely to have been the dominant mechanism in the more water diffuse oil region (layer 2). However, these interactions are thought to be relatively weak since the surfactant could be removed by rinsing the film. This reversibility, together with the fact that the surfactant displaced the water in the film and not the oil, as shown by comparison of Tables 1 and 2, suggests that the surfactant molecules were, in effect, able to flow within the water regions of the film.

From the mechanisms described above for the thin film, it was expected that the absorbed amount would increase for a thicker film. Therefore, the experiment was repeated using a 200 Å thick oil film. The measured neutron reflectivity profiles are shown in Fig. 2. Similar to the trend seen for the thin film, comparison of the profiles before and after surfactant addition shows an increase in oscillation frequency, which implies an increase in thickness and indicates the presence of surfactant at the interface.

By applying the same model fitting methods and equations as used for the thin film, the results indicated such a high level of absorbed surfactant that the volume fractions could not be determined properly due to a lack of "space" in the oil. There were two ways the available space for the surfactant could have increased. Either the total film thickness had increased, or the film had been partially solubilised by the surfactant causing oil to be lost into the bulk solution. However, the latter of these suggestions was dismissed due to evidence that the surfactant interaction could be reversed by rinsing the film and the film structure remained intact. Comparison of the reflectivity profiles measured from the film in pure D_2O, before surfactant addition and after rinsing away the surfactant, showed no difference.

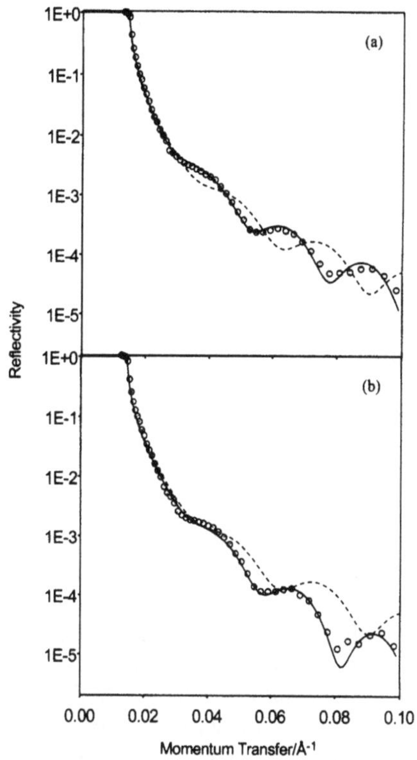

Fig. 2 Neutron reflectivity profiles (○) to show the interaction of protonated (**A**) and chain deuterated (**B**) CMC $C_{12}E_5$ at the "200 Å" olive oil film–water interface together with the fitted data (——). The profiles measured in the absence of surfactant (---) are also shown for comparison.

Therefore, a model was tried that allowed the film to increase in thickness.

Although the film thickness could increase, the amount of oil present in the layer could not change and had to be equal to the amount calculated before the surfactant was added:

$$\phi_{oil}\tau = \phi_{oil1}\tau_1 \tag{9}$$

where τ and ϕ_{oil} are the wet film thickness and oil volume fraction before surfactant addition and τ_1 and ϕ_{oil1} are the film thickness and oil volume fraction with surfactant present. This relationship was used to determine the volume fraction of oil in the oil/water/surfactant swollen layer (ϕ_{oil1}) before solving the remaining equations. All the data fit parameters could then be varied where possible, as before, until the ratio of surfactant head to tail was correct and all solutions were positive. The final results are summarised in Table 3.

The volume fractions of the layer constituents given in Table 3 show much higher fractions of surfactant in the outer adsorbed layer (layer 2) than in the oil film layer (layer 1). However, it must not be forgotten that the oil layer is much thicker than the adsorbed layer and in fact this layer contained the highest *amount* of surfactant

Table 3 Distribution of CMC $C_{12}E_5$ at the "200 Å" olive oil film–water interface

Layer	τ/Å	ϕ_{oil}	ϕ_{D2O}	ϕ_{C12}	ϕ_{E5}	Γ_{C12}/mmol $m^{-2} \times 10^{-3}$	Γ_{E5}/mmol $m^{-2} \times 10^{-3}$
1	234	0.725	0.163	0.069	0.043	7.62	5.19
2	24		0.286	0.265	0.449	3.02	5.50
					Total $C_{12}E_5$:	10.67	

molecules as is emphasised by comparing the surface excess concentrations for each layer.

It was calculated that 60% of the total amount of surfactant interacting had penetrated the film, which is similar to the result obtained for the thin film. However, the thick film also increased from 221 Å to 234 Å in thickness as a result of surfactant absorption. An explanation for this swelling can be provided by first considering the mechanism of absorption discussed previously where it was thought that the surfactant can flow within the oil structure by displacing the water. Theoretically, the limit of this process would be when all the water has been displaced. However, practically it is unlikely this point would ever be reached. The hydration of surfactant head groups would mean that the surfactant molecules always "carry" some water molecules with them. In either case, it is believed that when the surfactant concentration within the oil film is sufficiently high, the increase in pressure would cause the film structure to expand.

Despite the film expansion caused by surfactant absorption, the oil structure returned to its original state after the surfactant was rinsed away. This suggests that the oil had some elastomeric properties.

Conclusions

This work has demonstrated that appropriately designed neutron reflection experiments can unravel molecular details of surfactant distributions in oil films. More specifically, the extent of film swelling in water and how the addition of surfactant affected the oil/water structure could be determined. The detailed information of the surfactant distribution should be beneficial to understanding the process leading to oil removal by surfactants. The use of deuterium labelling to water and surfactant has been very effective in elucidating the structural details relating to the distributions of polymer film, water and surfactant across the interface. This means that although for each interfacial system a number of parameters were obtained from model fitting, the use of the necessary contrasts together with the physical constraints to the interfacial distributions of the three species make the model interpretation reasonably reliable.

Acknowledgements The authors thank the Engineering and Physical Research Council (EPSRC) for financial support. LLP also thanks Unilever Research for partial studentship.

References

1. Starkweather BA, Counce RM, Zhang X (1999) Separ Sci Technol 34: 1447
2. Scheuing DR (1991) ACS Symp Ser 447:251
3. Tamura T, Iihara T, Nishida S, Ohta S (1999) J Surfactants Deterg 2: 207
4. Beaudoin SP, Grant CS, Carbonell RG (1995) Ind Eng Chem Res 34: 3307
5. Weerawardena A, Boyd BJ, Drummond CJ, Furlong DN (2000) Colloid Surface A 169:317
6. Penfold J, Richardson RM, Zarbakhsh A, Webster JRP, Bucknall DG, Rennie AR, Jones RAL, Cosgrove T, Thomas RK, Higgins JS, Fletcher PDI, Dickinson E, Roser SJ, Mclure IA, Hillman AR, Richards RW, Staples EJ, Burgess AN, Simister EA, White JW (1997) J Chem Soc Faraday Trans 93: 3899
7. Bucknall DG (1999) Neutron reflection studies of polymers. In: Pethrick RA, Dawkins JV (eds), Modern techniques of polymer characterisation. Wiley, Chichester New York Weinheim Brisbane Singapore Toronto, pp 109–140
8. Lu JR, Thomas RK (1998) J Chem Soc Faraday Trans 94: 995
9. Hines JD, Fragneto G, Thomas RK, Garrett PR, Rennie GK, Rennie AR (1997) J Colloid Interface Sci 189: 259
10. Gilchrist VA, Lu JR, Staples E, Garrett P, Penfold J (1999) Langmuir 15: 250
11. Brzoska JB, Shahidzadeh N, Rondelez F (1992) Nature 360: 719
12. Lu JR, Li ZX, Thomas RK, Staples EJ, Tucker I, Penfold J (1993) J Phys Chem (1993) 97: 8012
13. Murphy EF, Keddie JL, Lu JR, Brewer J, Russell J (1999) Biomaterials 20:1501
14. Landau LD, Levich BG (1942) Acta Physiochim URSS 17: 42
15. Born M, Wolf E (1970) Principles of optics. Pergamon, Oxford

Progr Colloid Polym Sci (2004) 123: 169–173
DOI 10.1007/b11755

A. Terreros Gomez
B.J. Rubio Retama
B. Lopez Ruiz
P.A. Galera Gomez
C. Rueda Rodriguez
C. Arias Garcia
E. Lopez Cabarcos

Encapsulation of alkaline phosphatase in polyacrylamide microparticles using the concentrated emulsion polymerisation method

A. Terreros Gomez · B.J. Rubio Retama
P.A. Galera Gomez · C. Rueda Rodriguez
C. Arias Garcia · E. Lopez Cabarcos (✉)
Departamento Quimica-Fisica
Farmaceutica, Facultad de Farmacia,
Universidad Complutense de Madrid,
28040 Madrid, Spain
e-mail: cabarcos@farm.ucm.es
Tel.: +34-91-3941751
Fax: +34-91-3942032

B. Lopez Ruiz
Departamento Quimica Analitica,
Facultad de Farmacia,
Universidad Complutense de Madrid,
28040 Madrid, Spain

Abstract Polyacrylamide microparticles were prepared by the concentrated emulsion polymerisation technique. In distilled water the microparticles have spherical form and two size distributions with average diameters 1.5 μm and 5.7 μm. Alkaline phosphatase was encapsulated in the microparticles. The encapsulation of the enzyme slightly alters the size of the new microparticles that present average sizes of 1.4 μm and 6.2 μm. However, the activity of the enzyme is reduced to 12% as compared with the free enzyme.

Keywords Concentrated emulsion polymerisation · Polyacrylamide microparticles · Alkaline phosphatase encapsulation

Introduction

There are many potential applications of aqueous polymer systems in medicine, pharmacy, biotechnology, industry and environmental problems [1–3]. One possibility involves the immobilisation of enzymes into microscopic particles and gels [4]. A great many encapsulation techniques have been suggested and new ones are continually being developed [5–12]. In recent years the concentrated emulsion polymerisation method has been proposed and more recently it has been employed to encapsulate drugs [13, 14]. The main difference with conventional emulsion polymerisation consists in the large volume fraction of the dispersed phase, which in the concentrated emulsions is larger than 74% and sometimes reaches values as large as 99% [13–16]. In concentrated emulsions (also called gel-like emulsions) the dispersed phase form polyhedral cells, separated by a network of thin film of continuous phase. The polymerisation in these systems occurs in the monomer cells of the dispersed phase. In a first step, a stable colloidal dispersion of the monomer, cross-linking agent and the enzyme is prepared in aqueous solution. This colloidal dispersion is employed as the dispersed phase of a concentrated emulsion whose continuous phase contained the surfactant. The role of the surfactant is to stabilise the gel-like emulsion. Upon addition of the initiator the polymerisation takes place and the drug becomes encapsulated in a polymer microparticle, which has a similar size to the droplet of the precursor emulsion [17, 18].

The aims of this paper are twofold. First, to synthesise polyacrylamide (PAA) microparticles following the W/O concentrated emulsion polymerisation method [13, 16, 19, 20] and to encapsulate enzyme in the microparticles. For this study we have entrapped the alkaline phosphatase in poly(acrylamide-co-acrylic acid) microparticles. Second, to examine the changes in size and morphology of the microparticles due to the enzyme and to compare the activity of the encapsulated enzyme with that of the free enzyme.

Experimental

Materials

The monomer acrylamide (Aldrich), the initiator ammonium persulphate (Janssen Chimica), the cross-linker *N,N'*-methylenebisacrylamide (Janssen Chimica), the accelerator *N,N,N'*,N'-tetramethylethylenediamine (TMED, Bio Rad), the surfactant Span80 (Fluka) and the dodecane (Sigma) were used as received. The water was Milli-Q quality. The alkaline phosphatase (Boehringer) to encapsulate was dialysed against water to eliminate possible traces of $MgCl_2$ and $ZnCl_2$ which are potential inhibitors of radical polymerisation.

Methods

The shape and size of the microparticles was examined using optical and scanning electron microscopy. Scanning electron microscopy (SEM) studies were conduced on a JEOL model JSM-6400 microscope operating at 20 kV. The grid with the microparticles was dried and replicas were produced by shadowing with gold deposited with a Balzers Sputter Coater (SCD-004). The X-ray diffraction studies were performed using a Philips X'pert PW3050 diffractometer. Silicon powder was used to calibrate the sample to film distance. The granulometric analysis was performed with a Coulter Counter analyser. The enzyme activity was monitored on a Varian (Cary 300) UV-Visible spectrophotometer.

Results and discussion

Empty PAA microparticles

The synthesis of polyacrylamide microparticles using the concentrated emulsion pathway includes two steps: First, the preparation of a stable W/O concentrated emulsion. Second, the polymerisation of the gel-like emulsion. The concentrated emulsion was prepared according to procedures outlined in the literature [13–18]. The oil phase (14% v/v), composed of dodecane and the surfactant Span80, constitutes the low fraction continuous phase. The dispersion phase (86% v/v) is an aqueous solution of monomer, initiator and cross-linker. The detailed composition of the dispersed and the continuous phase is given in Table 1 under EPM. The concentrated emulsion was produced by dropwise addition, with a syringe, of the dispersed phase to the continuous oil phase. The system was homogenised under magnetic stirring and dry nitrogen was continuously bubbled through the emulsion to remove residual oxygen. This emulsion has a volume fraction $\phi = Vaq/(Voil + Vaq) = 0.86$ and its stability was followed by measuring the mean diameter of the droplets as a function of time. Coalescence leads to phase-separation after 16 days for the emulsions maintained at 25 °C. A more detailed study of the emulsion preparation and stability can be found in a previous publication [18].

To polymerise the concentrated emulsion a 200-mL glass reactor with magnetic stirring was employed and the polymerisation was started by adding the ammonium persulphate (11 mM) and TMED (32 mM) to the recently prepared gel like emulsion. During the polymerisation, the temperature inside the reactor rises from 22 °C to a maximum of 41 °C. The conversion of monomer into polymer was calculated by gravimetry as follows: 10 vials were filled with 250 μL of a recently prepared concentrated emulsion. The polymerisation was started in each vial at different time but was finished in all of them at the same time by adding freezing methanol. The microparticles were isolated by centrifugation at 2,000 rpm for 20 minutes, dried at 100 °C for 20 hours and then weighted. The conversion was estimated by subtracting the known weights of the microparticles from the total weight of monomer plus cross-linker agent in the sample. Fig. 1 shows conversion as a function of time. Polymerisation is fast and a conversion of over 98% is achieved in less than 50 minutes.

Table 1 Representative concentrated emulsion composition used in the preparation of empty polyacrylamide microparticles (column EPM) and microparticles with alkaline phosphatase encapsulated (column APM)

	% (v/v)	EPM	APM
Dispersed phase	86		
Bidestillated water		10 mL	
Acrylamide		3.5 M	3.5 M
Methylenebisacrylamide		86.3 mM	162 mM
Ammonium persulphatc		11.0 mM	11.0 mM
TMED		32.0 mM	32.0 mM
Buffer 7.2 pH			10 mL
Acrylic Acid			200 mM
Alkaline phosphatase			6 μg
Continuous phase	14		
Dodecane (% v/v)		10.5	10.5
Span 80 (% v/v)		3.5	3.5

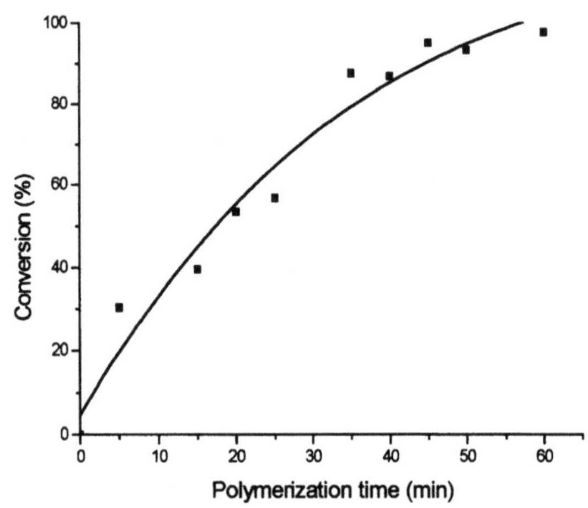

Fig. 1 Conversion vs. time curve of acrylamide W/O concentrated emulsion polymerisation at 25 °C

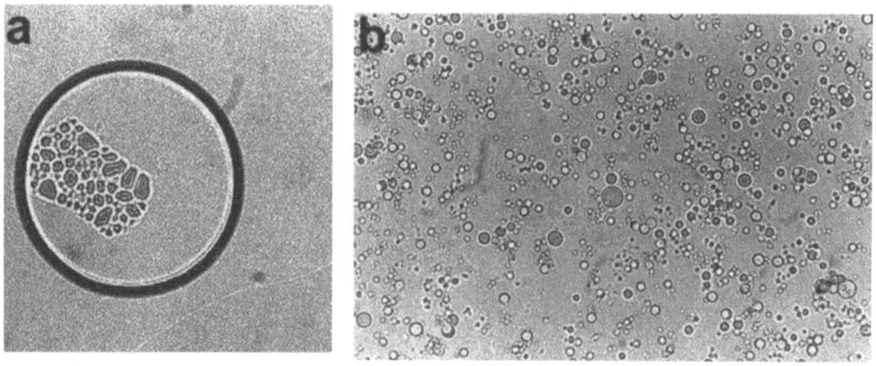

Fig. 2 A Optical microscopy image of the empty microparticles after the polimerisation in concentrated emulsion. The particles show a polyhedral shape similar to the polyhedral shape of the globules of the precursor concentrated emulsion. **B** The same particles dispersed in water after swelling

The reaction generates polyacrylamide microparticles with two size distributions; one with average diameter approximately the size of the precursor emulsion droplet and the other one, with smaller diameter, is associated to the size of the cells filling the space between the droplets. The cross-linking content, defined as the ratio between the cross linker *N,N'*-methylenebisacrylamide and the acrylamide monomer expressed in percentage, was 5.3%.

In Fig. 2A,B we compare optical microscopy images (magnification 120) of the empty polyacrylamide microparticles. The difference in appearance of these two pictures is caused by the fact that Fig. 2A depicted the original packing in polyhedral cells of the microparticles (here imprisoned in an air bubble) after the polymerisation in the concentrated emulsion, whereas Fig. 2B shows that the same particles dispersed in water have adopted the spherical shape because the swelling.

Fig. 3 presents scanning electron micrographs, taken with magnification 1100, of empty polyacrylamide microparticles. Their statistical diameter was measured with a Coulter Counter analyser and it was expressed in terms of the surface-volume diameter d_{av},

$$d_{av} = \frac{\sum nd^3}{\sum nd^2}$$

where n is the number of particles whose diameter is d. Two diameter distributions were obtained; one with $d_{av} = 5.7$ µm (standard deviation $\sigma = 2.6$) and the other one with $d_{av} = 1.5$ µm ($\sigma = 0.23$).

The microstructure of the microparticles was investigated by X-ray diffraction. The diffraction pattern, shown in Fig. 4, presents the broad halo with two maxima at approximately $2\theta = 21$ degrees and $2\theta = 35$ degrees characteristic of the amorphous cross-linked polyacrylamide polymer.

PAA microparticles with enzyme

Because its good stability alkaline phosphatase is a model enzyme to investigate enzyme immobilisation in microparticles. The encapsulation of alkaline phosphatase was carried out following the previous method slightly modified. To prevent enzyme denaturation, we have used as aqueous phase a tris(hydroxymethyl)aminomethane buffer solution of pH 7.2 in which the

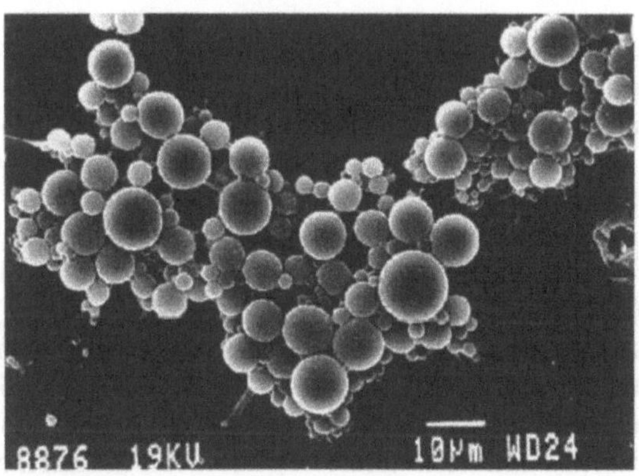

Fig. 3 Scanning electron micrographs of empty polyacrylamide microparticles (5.3% cross-linked with *N,N'*-methylenebisacrylamide) prepared by polymerisation in concentrated emulsion. The magnification is 1100

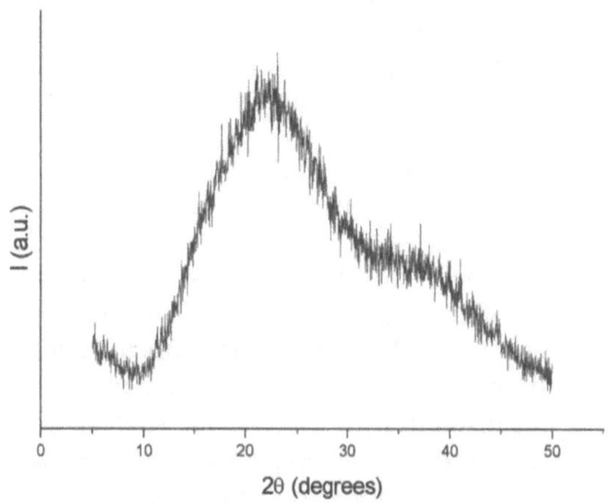

Fig. 4 Wide angle X-ray diffraction pattern of empty PAA microparticles showing the broad diffuse halo characteristic of the amorphous polymers

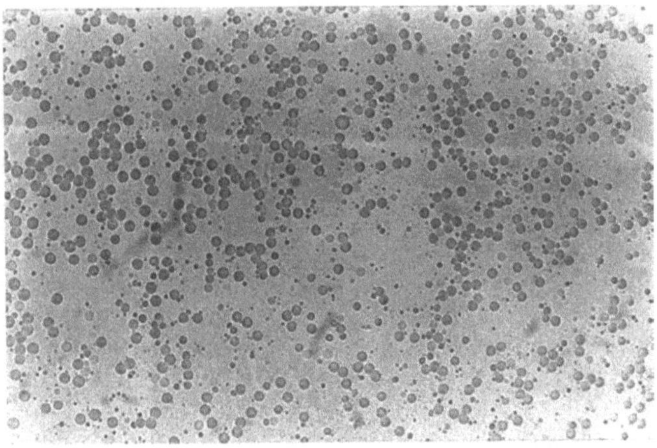

Fig. 5 Optical microscopy image of the microparticles with alkaline phosphatase encapsulated. using the concentrated emulsion polymerisation method. The particles show two size distributions with average diameters $d_{av} = 6.2$ µm and $d_{av} = 1.4$ µm

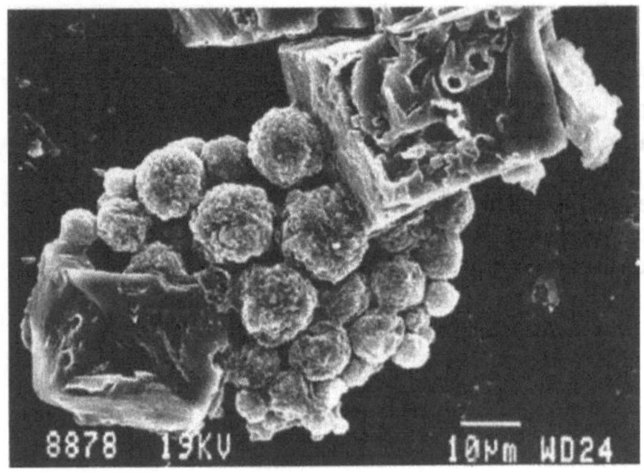

Fig. 6 Scanning electron micrographs of PAA microparticles with alkaline phosphatase encapsulated using the concentrated emulsion polymerisation method. The particles appeared between two crystals of tris(hydroxymethyl)aminomethane which was part of the buffer solution used to prevent enzyme denaturation

enzyme was solubilised. Acrylic acid was added to provide negative charges to the microparticles. The COOH groups are situated preferentially at the surface of the microparticles and they facilitate the anchorage of the microparticles for their possible use in clinical analysis. The full composition of the concentrated emulsion is given in Table 1 under APM. The addition of new compounds to the aqueous phase present the problem that sometimes the equilibrium hydrophilic-lipophilic can be broken and the gel-like emulsion does not form. As it was indicated above, the aqueous phase was added drop by drop under stirring with a magnetic stirrer and once the gel emulsion was prepared the polymerisation was performed. Since the temperature control is crucial to avoid the denaturation of the enzyme that we want to encapsulate, during the polymerisation the reactor was immersed in an ice-water bath to ensure that the temperature never surpassed 25 °C. After 45 minutes the polymerisation was completed and the resulting latex was precipitated with methanol and purified with acetone. Finally the microparticle suspension was used as such or freeze-dried.

An optical microscopy image of the two population of microparticles of polyacrylamide with the alkaline phosphatase encapsulated is shown in Fig. 5. The statistical diameter of the two populations measured with the Coulter Counter is $d_{av} = 6.2$ µm ($\sigma = 1.8$) and $d_{av} = 1.4$ µm ($\sigma = 0.2$). Both, the optical microscopy image and the Coulter Counter measurement indicate that the encapsulation of the enzyme slightly alters the size of the microparticles even though a tendency to increase the dimensions of the larger particles is observed. However, the scanning electron micrograph (taken with magnification 700) of these microparticles, depicted in Fig. 6, shows particles with wrinkled surfaces

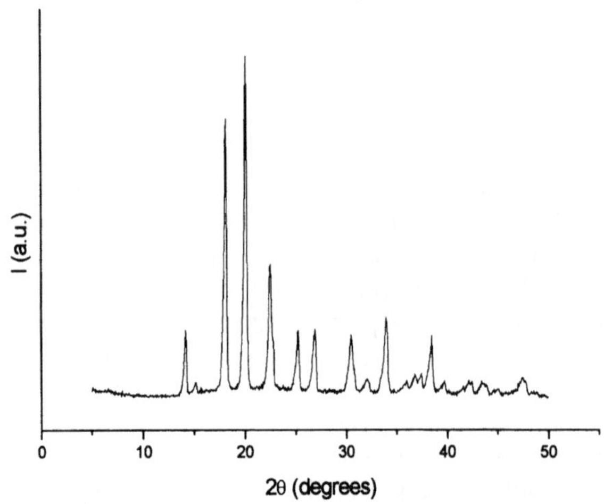

Fig. 7 Wide angle X-ray diffraction pattern of the microparticles with enzyme showed in Fig. 6

and diameters larger than 10 µm stuck to some blocks of crystalline material.

The X-ray diffraction pattern of this sample presents more than 30 reflections as can be seen in Fig. 7. These reflections can be indexed in an orthorhombic unit cell ($a = 8.791$ Å, $b = 8.860$ Å and $c = 7.808$ Å) which has been identified as coming from tris(hydroxymethyl)aminomethane crystals. Thus, the decrease of the intensity of the PAA amorphous halo in Fig. 7 seems to indicate that the size increase observed in the SEM micrographs, could be attributed to layers of tris(hydroxymethyl)aminomethane deposited on the surface of the microparticles which would have actuated as crystallisation nuclei. We draw the attention to the danger of deriving the size of the microparticles when only SEM technique is used.

Enzymatic activity

The enzymatic activity (EA) for the encapsulated enzyme was calculated and compared with that of the free enzyme in solution by using 4-nitrophenyl disodium phosphate as substrate. The product formation (4-nitrophenol) was monitored at 405 nm by UV spectrometry and the EA was determined by the expression,

$$EA = \frac{\left(\frac{dA}{dt}\right) \cdot V \cdot D}{\xi \cdot d}$$

where (dA/dt) is the slope of the absorption vs. time plot, V is the total volume of the sample (substrate plus enzyme) expressed in mL, ξ is the molar absorption coefficient of the product (18,531 M^{-1} cm^{-1}), d is the optical path (1 cm) and D is the dilution factor of the enzymatic essay. The activity of the free enzyme was $2.5 \cdot 10^4$ µmol min^{-1} whereas that of the encapsulated one was $3.0 \cdot 10^3$ µmol min^{-1}. Encapsulation of the alkaline phosphatase in 5.3% cross-linked polyacrylamide microparticles renders 12% of EA as compared with the free enzyme. We attribute this conspicuous decrease in EA to various factors such as enzyme denaturation during the polymerisation and loss of enzyme, loosely attached to the surface of the microparticles, during the washing with methanol and acetone.

Conclusions

Two steps are employed in the enzyme encapsulation process. The first involves the preparation of a stable W/O concentrated emulsion in which the alkaline phosphatase is included in the aqueous phase together with the monomer, the acrylic acid and the cross-linking agent having as continuous phase the dodecane and the surfactant. In the second step, the system is polymerised at around 25 °C to produce microcapsules containing the enzyme. This encapsulation technique provides microparticles with two size distributions around $d_{av} = 6.2$ µm and $d_{av} = 1.4$ µm. The formation of the gel-like emulsion is a limiting step in this encapsulation technique because small variations in some of the components may avoid the formation of the emulsion.

Acknowledgements The authors acknowledge financial support from DGI (BMF2000-0620 and PGC2000-2246-E) of the Spanish Science and Technology Ministry.

References

1. Stjärnkvist P, Laakso T, Sjöholm I (1989) J Pharm Sci 78:52
2. Deng X, Zhou S, Li X, Zhao J, Yuan M (2001) J Control Release 71:165
3. Oh J, Nam Y, Lee K, Park T (1999) J Control Release 57:269
4. Pizarro C, Fernandez Torroba MA, Benito C, Gonzalez Sainz JM (1997) Biotechnol Bioeng 53:497
5. Gallardo A, Fernández F, Cifuentes A, Diez-Masa JC, Bermejo P, Rebuelta M, Lopez Bravo A, San Roman J (2001) J Control Release 72:1
6. Ortega N, Busto MD, Perez Mateos M (1998) Bioresource Technol 64:105
7. El-Samaligy M, Rohdewald P (1983) Intern J Pharm 13:23
8. Sahoo SK, Tapas KD, Ghost PK, Maitra A (1998) J Colloid Interface Sci 206:361
9. Bum SL, Okano T, Kataoka K (1996) J Pharm Sci 85:85
10. Pekarek KJ, Jacob JS, Mathlowitz E (1994) Nature 367:258
11. Langer R (1998) Nature 392:5
12. Kriwet B, Walter E, Kissel T (1998) J Control Release 56:149
13. Ruckenstein E (1997) Adv Polym Sci 127:1
14. Terreros A (1999) Doctoral thesis, chap. 6. University Complutense of Madrid
15. Pons R, Ravey JC, Sauvage S, Stébe SM, Erra P, Solans C (1993) Colloids Surf A: Physicochem Eng Aspects 76:171
16. Pons P, Carrera I, Erra P, Kunieda H, Solans C (1994) Colloids Surf A: Physicochem Eng Aspects 91:259
17. Ruckenstein E, Park JS (1990) Polymer 31:2397
18. Terreros A, Galera Gomez A, Lopez-Cabarcos E (2000) Progr Colloid Polym Sci 115:50
19. Nagashima S, Ando S, Makino K, Tsukamoto T, Ohshima H (1998) J Colloid Interface Sci 197:377
20. Jingcheng H, Liqiang Z, Ganzuo L, Hanqing W, Zhengwei D (1996) Polymer 37:3117

Progr Colloid Polym Sci (2004) 123: 174–177
DOI 10.1007/b11756

F. Lopez
G. Palazzo
G. Colafemmina
G. Cinelli
L. Ambrosone
A. Ceglie

Enzymatic activity of lipase entrapped in CTAB/water/pentanol/hexane reverse micelles: a functional and microstructural investigation

F. Lopez · G. Cinelli · L. Ambrosone
A. Ceglie (✉)
Consorzio Interuniversitario per lo sviluppo dei Sistemi a Grande Interfase (CSGI), c/o Department of Food Technology, University of Molise, 86100 Campobasso, Italy
e-mail: ceglie@unimol.it
Tel.: +39-0874-404647
Fax: +39-0874-404652

G. Palazzo · G. Colafemmina
Department of Chemistry, University of Bari, Bari, Italy

Abstract The lipase VII (from *Candida rugosa*) activity was studied as a function of the content of pentanol and water in cethyltrimethylammonium bromide (CTAB)/water/pentanol/hexane reverse micelles. The turn-over numbers for the hydrolysis of *p*-nitrophenyl butyrate (*p*-NPB) were determined spectrophotometrically. Reverse micelles size and the partition of *p*-NPB between the aggregates and the oil continuous phase were determined by means of pulsed gradient spin-echo NMR experiments. A strong correlation between enzyme activity and micellar size was found.

Keywords Reverse micelles · CTAB · Pentanol · Lipase

Introduction

There is current interest in the use of water-in-oil microemulsions, as a medium for the study of enzyme properties. The catalytic activity in reverse micelles has been extensively studied for many enzymes [1, 2]. A strong influence of the molar ratio water/surfactant = W_0 on catalytic activity has been reported for a large number of enzymes in reverse micelles, and several models have been proposed in order to explain such a dependence (see [3] for review). In reverse micelles (L_2 phase) usually the aggregate are monodispersed in size, and all the surfactant molecules reside on the water/oil interface. Under this condition the water core radius (R_w) depends on the surfactant polar-head area (a_s) and W_0 according to:

$$R_w = 3v_w W_0 / a_s \tag{1}$$

where, v_w stands for the water molecular volume.

In three-component microemulsion (e.g., AOT/water/oil) a_s depends weakly on W_0. As a consequence Eq. (1) foretells a linear dependence between R_w and W_0. It is, therefore, impossible to decouple changes on the water core size from other properties related to the surfactant hydration (e.g., dielectric constant and ion concentration).

Our previous investigation, on the quaternary system CTAB/water/pentanol/hexane indicate that, the cosurfactant/surfactant molar ratio (hereafter P_0) tunes the CTAB polar head area (a_s), so that increasing the P_0 corresponds to increase the area occupied by the CTAB molecules [4]. Thus, in such a system it is possible to change the micellar size leaving the W_0 unchanged, or to have micellar solutions sharing the same size but having different water-to-surfactant ratios.

In the present study we used as guest enzyme lipase VII from *Candida rugosa*.

Lipases in general catalyse the hydrolysis of fatty acyl esters and are known to act at oil/water interface. This kind of enzymes exhibits an α/β hydrolase fold, and it was found that the active sites, containing the typical Ser-His-Asp (Glu) esterase catalytic triad, are shielded from the solvent by a flexible protein structure element called "lid" [5–9].

Interestingly, while in AOT reverse micelles lipase activity presents a maximum around $W_0 = 5$–11 [10, 11], in CTAB-based reverse micelles it was reported to be scarcely affected by the parameter W_0 [10, 12]. Recent investigations demonstrate that the concentrations of water, alcohol, and Br^- close to the inner interfacial surface are essentially W_0-independent, and indicate that these local concentrations are the key parameters in influencing the lipase activity [12].

Materials and methods

CTAB, *n*-pentanol, *n*-hexane, lipase VII from *Candida rugosa* and *p*-nitrophenyl butyrate (*p*-NPB), were purchased from Sigma. Lipase VII was used without further purification.

Microemulsion

Cetyltrimethylammonium bromide (CTAB) was three times recrystallised from anhydrous ethanol, and stored over dried silica gel under vacuum. *n*-Hexane (spectroscopic grade), and *n*-pentanol have been used without any further purification.

Three parameters are needed in order to define the composition of a four-component system in a single-phase region. Here we use as parameters the molar ratios: water/CTAB (W_0), pentanol/CTAB (P_0) and the (overall) concentration of CTAB.

The basic microemulsions were prepared by weighing in a volumetric flask the amounts of surfactant, alcohol, and buffer (potassium phosphate 50 mM, pH 6.0) needed to obtain the desired P_0 and $W_0 = 5$ (for $W_0 < 5$ the microemulsion does not exist). Hexane was added to obtain a CTAB concentration of 100 mM. To this stock microemulsion at $W_0 = 5$ and at a given P_0, the appropriate amount of lipase dissolved in buffer is added to obtain the final W_0 value. Finally, the reaction was started by adding the substrate (*p*-NPB) as hexane solution. The final concentration of CTAB and lipase were 60 mM and 80 nM, respectively. The ranges of W_0 and P_0 were 6.9–40 and 8.2–20, respectively.

PGSE-NMR

Self-diffusion coefficients measurements have been carried out by the Fourier transform NMR pulsed field gradient spin-echo (PGSE-NMR) method [13] using a BS-587A NMR (Tesla) spectrometer operating at 80 MHz for the proton, equipped with a pulsed field gradient unit (Autodif 504, STELAR s.n.c.). The classical Stejskal-Tanner pulse sequence [14], 90°-τ-180°-τ-echo with two rectangular field gradient pulses of about 0.04 T m^{-1}, separated by a constant interval Δ was used.

The self-diffusion coefficients of water, CTAB, *n*-pentanol, and *p*-NPB were collected.

The results concerning the diffusion of the system's components are not presented here (for the sake of brevity), but they confirm the detailed investigations on the hosting system previously done by us [4, 15]. Accordingly the size of the reverse micelle water core reported in the present paper was evaluated using Eq. (1) and the values of a_s determined at the different P_0 values [4]. The knowledge of the *p*-NPB diffusion coefficient has allowed the evaluation of substrate partition between the micelles and the continuous phase according to the Lindman's law: [16]

$$D_{obs} = P_{mic}D_{mic} + (1 - P_{mic})D_{free} \tag{2}$$

where P_{mic} is the fraction of ester moving with the reverse micelles, $D_{mic} \equiv D_{CTAB}$ [15] and D_{free} is the *p*-NPB diffusion constant in a solution with the same composition of the microemulsion continuous phase. We evaluate D_{free} iteratively from the self-diffusion coefficient of *p*-NPB in hexane and from the viscosity of pentanol/hexane solutions assuming $D\eta$ = constant (for more details on this procedure see Ref. [4]).

Enzyme kinetics

All reactions were carried out at 30 °C, *p*-NPB hydrolysis was determined by recording absorbance change with a Shimadzu spectrophotometer equipped with thermostatted cells. The enzyme concentration and the turnover number (k_{cat}) were calculated using as molecular weight 62,000 Dalton. Enzyme solutions were prepared freshly in 50 mM potassium phosphate pH 6.0. Rates were monitored by means of the absorbance of the *p*-nitrophenol liberated, the monitoring wavelength was selected either at 405 nm ($\varepsilon = 1{,}060\ cm^{-1}\ M^{-1}$) or at the isosbestic point of the nitrophenol/nitrophenolate couple 348 nm ($\varepsilon = 5{,}000\ cm^{-1}\ M^{-1}$). The values of K_{cat} were determined by non-linear regression analysis of the corresponding Michaelis-Menten curves (triplicate experiments) using a Levenberg-Marquardt algorithm (Sigma-plot software).

Results and discussion

Dependence of enzyme activity on W_0

The influence upon enzyme activity of the varying concentrations of water, relative to that of the surfactant for *p*-NPB hydrolysis is shown in Fig. 1. In such a set of experiments we varied the molar ratio water/surfactant at two fixed P_0 values (8.2 and 14). One requires a minimum of 4–5 water molecules per CTAB molecule in order to have a stable transparent microemulsion. Since the enzyme is added to the stock microemulsion (at $W_0 = 5$ see Materials and Methods) as aqueous solution, the lower W_0 value accessible experimentally is about 7. The increase in W_0 parallels an increase in the size of the

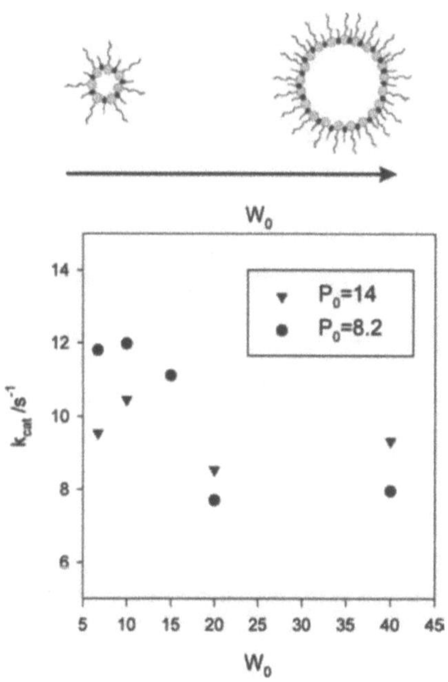

Fig. 1 Effect of molar ratio water/CTAB = W_0 on the turnover number of lipase from *Candida rugosa* solubilised in reverse micelles with two different values of the molar ratio pentanol/CTAB = P_0. The cartoon illustrates the effect of an increase in the W_0 parameter on the micellar size

micellar water core, which at $P_0 = 8.2$ grows from $R_w = 10$ Å at $W_0 = 6.9$ to $R_w = 60$ Å at $W_0 = 40$ (at $P_0 = 14$, $R_w = 6.5$ Å at $W_0 = 6.9$, and $R_w = 39$ Å at $W_0 = 40$). Fig. 1 shows that, for both the P_0 values, variations in the W_0 scarcely affect the enzyme activity (in agreement with previous investigations) [10, 12].

Dependence of enzyme activity on P_0

The influence upon enzyme activity of the varying concentrations of cosurfactant *n*-pentanol, relative to that of the surfactant for *p*-NPB hydrolysis is shown in Fig. 2. We chose to investigate the effect of pentanol on the rate of *p*-NPB hydrolysis at $W_0 = 6.9$ which ensure stable microemulsions at all the P_0 values but is still close to the value ($W_0 = 5$) at which lipase exploit its maximum activity in AOT [11].

In such a set of experiments we tuned the cosurfactant/surfactant ratio leaving unchanged the water/surfactant ratio ($W_0 = 6.9$).

The effective polar head area of CTAB increase with pentanol content and as a consequence the micellar size decreases. Actually, the R_w is equal to 10 Å at $P_0 = 8.2$ and drops to 4.5 Å at $P_0 = 20$. Fig. 2 shows that the increase in the alcohol content induces a decrease in the observed lipase turn-over number. This is a peculiar behaviour, overlooked in previous investigations (that were done at constant P_0).

Basically, there are two explanations for the effect of P_0 on the observed k_{cat}. The first is a true change in the catalytic efficiency of the enzyme. The second is a change in the amount of substrate available to the enzyme. Actually, changing the composition of the interfacial film and of the organic phase could result also in a modification of the substrate distribution between these two sites. If the fraction of *p*-NPB in micelles decreases upon addition of alcohol, we expect a decrease in the rate of its hydrolysis. Fig. 3 demonstrates that this is not the case. Indeed, the fraction of *p*-NPB in micelles increases from 0.25 at $P_0 = 8.2$ to 0.40 at $P_0 = 18$, corresponding to an increase in the ester concentration in micelles that can hardly account for the decrease of the apparent lipase activity.

Several factors could be involved in the dependence of k_{cat} on the pentanol loading. The charge density of the inner micellar surface decreases upon P_0-increase (because the alcohol increases the mean area per CTAB molecule). However, we found that in KBr aqueous solutions the lipase turn-over number *increases* upon KBr dilution (not shown). The P_0 parameter affects also the interfacial film thickness [17] and it is likely to have an effect on the polarity of the environment probed by the enzyme. In the present contribution, however, we are focusing on the influence of the micellar (a parameter of fundamental importance on kinetics of several enzymes).

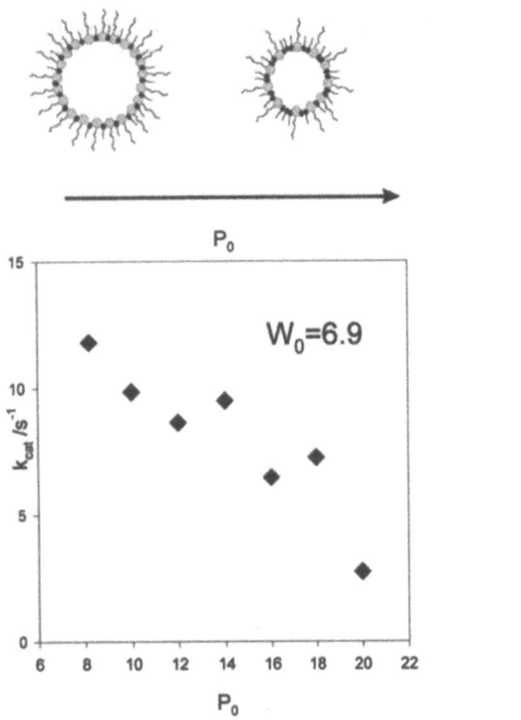

Fig. 2 Effect of molar ratio pentanol/CTAB = P_0 on the turn-over number of lipase from *Candida rugosa* solubilised in reverse micelles at fixed molar ratio water/CTAB = 6.9. The cartoon illustrates the effect of an increase in the P_0 parameter on the micellar size

Enzyme activity as function of the water core size

The reverse micellar size is a function of both the W_0 and P_0 parameters, it is therefore useful to report the turn-over number as a function of water-core diameter.

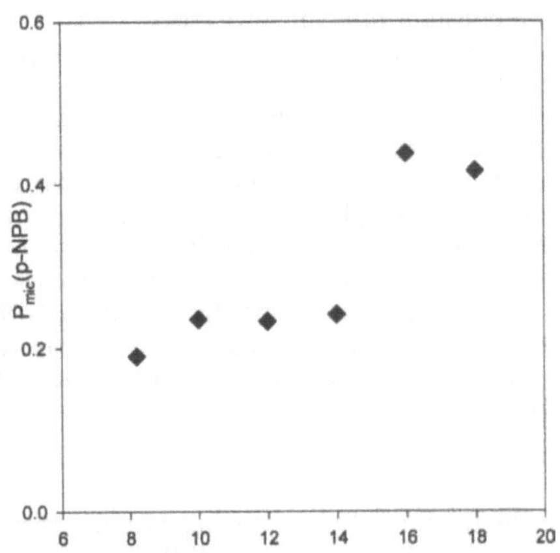

Fig. 3 Fraction of *p*-nitrophenyl butyrate moving with the reverse micelles, P_{mic}(*p*-NPB), as a function of the molar ratio pentanol/CTAB = P_0; $W_0 = 6.9$

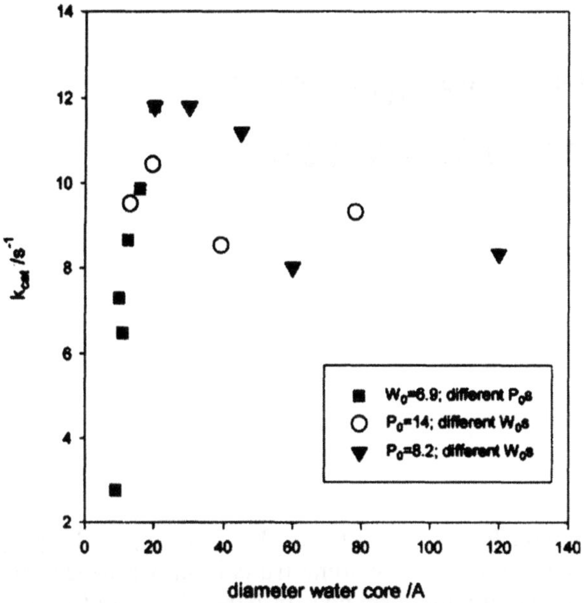

Fig. 4 Dependence of the turn-over number of lipase from *Candida rugosa* on the reverse micelle diameter for microemulsions with different P_0 and W_0 values. Data were from Fig. 1 and Fig. 2

This was done in Fig. 4, and allows the comparison of experiments performed at different W_0 and P_0 values. One can see that micelles with the same size share essentially the same k_{cat} value, although their composition corresponds to different pairs of W_0 and P_0 values. Such a result gives the clear evidence that the micellar size has a strong influence on the enzyme activity. The dependence of k_{cat} on micellar diameter is very sharp for small micelles and levels off for micelles with a water-core diameter larger than 3 nm. Such a critical size is however smaller than the protein dimensions (about 5 nm × 5 nm × 4 nm) in agreement with the scenario of the enzyme solubilised mainly in the interfacial layer but with a (relatively small) portion protruding in the water core.

If this interpretation is correct, the apparent discrepancy between lipase activity in AOT and CTAB based microemulsions disappears. Of course, also factors other than reverse micelles size should affect the enzyme reactivity in the different environments. For example, in the case of air/water monolayer, lipase was found to interact strongly with cationic surfactant, while no interaction was found in the case of anionic surfactant (see [18]). AOT, a double chain anionic surfactant, forms spontaneously reverse micelles without the addition of cosurfactant and of extra-water. CTAB is a single chain cationic surfactant, in order to form water-in-oil microemulsion, it needs the presence of a suitable cosurfactant (pentanol in the present case) and a minimum amount of water ($W_0 \geq 5$). Therefore, in AOT-based systems the W_0 can be continuously tuned from an initial value of $W_0 = 0$. As a consequence, the range of water core diameter, experimentally accessible, encompasses the value of about 20 Å where the lipase exploits its maximum activity. In terms of W_0, assuming for AOT a head group area of about 55 Å^2 this corresponds to the value of $W_0 \approx$ 5–6 previously reported. At variance, in CTAB-based systems, the cut-off for $W_0 < 5$ corresponds to water core diameter larger than 20 Å. Thus, only the region with a weak dependence of enzyme activity on micellar size can be probed. Only at unusually high cosurfactant/surfactant ratios, can one explore a range of sizes that bracket 20 Å.

Acknowledgements The financial support of Italian Ministry for University and Scientific and Technological Research (MURST), grant PRIN/1998 is acknowledged.

References

1. Luisi PL, Giomini M, Pileni MP, Robinson BH (1988) BBA 947:209–246
2. Walde P, Giuliani AM, Boicelli CA, Luisi PL (1990) Chem Phys Lipids 53:265–288
3. Bru R, Sanchez-Ferrer, Garcia-Carmona F (1995) Biochem J 310:721–739
4. Colafemmina G, Palazzo G, Balestrieri E, Giomini M, Giustini M, Ceglie A (1997) Progr Colloid Polym Sci 105:281–289
5. Grochulski P, Bouthhillier F, Kazlauskas RJ, Serreqi AN, Schrag JD, Ziomek E, Cygler M (1994) Biochemistry 33:3494–3500
6. Cygler M, Schrag JD (1999) Biochim Biophys Acta 1441:205–214
7. Brady L, Brzozowsky AM, Derewenda ZS, Dodson E, Dodson G, Tolley S, Turkenburg JP, Christiansen L, Huge-Jensen B, Norskov L, Thim L, Menge U (1990) Nature 343:767–770
8. Brzozowsky AM, Derewenda U, Derewenda ZS, Dodson GG, Lawson DM, Turkenburg JP, Bjorkling F, Huge-Jensen B, Norskov L, Patkar SA, Thim L (1991) Nature 351:491–494
9. Grochulski P, Yuge L, Schrag JD, Bouthhillier F, Smith P, Harrison D, Rubin B, Cygler M (1993) J Biol Chem 268:12843–12847
10. Fletcher PDI, Robinson BH (1985) J Chem Soc Faraday Trans 81:2667–2679
11. Otero C, Rùa ML, Robledo L (1995) FEBS Lett 360:202–206
12. Das PK, Chaudhuri A (2000) Langmuir 16:76–80
13. Stilbs P (1987) Prog NMR Spectrosc 19:1–45
14. Tanner JE, Stejskal EO (1968) J Chem Phys 49:1768
15. Giustini M, Palazzo G, Colafemmina G, Della Monica M, Giomini M, Ceglie A (1996) J Phys Chem 100:3190–3198
16. Nilsson PG, Lindman B (1983) J Phys Chem 87:4756–4761. Nilsson PG, Lindman B (1984) J Phys Chem 88:4764–4769
17. Palazzo G, Lopez F, Giustini M, Colafemmina G, Ceglie A (2003) J Phys Chem B (submitted)
18. Hlomberg K, Nydén M, Lee L-T, Malmsten M, Jha BK (2000) Adv Colloid Interface Sci 88:223–241

Progr Colloid Polym Sci (2004) 123: 178–181
DOI 10.1007/b11757

A.A. Kharlamov
H.D. Burrows

Monitoring of the aroma of fruits at their surface by luminescence

A.A. Kharlamov (✉) · H.D. Burrows
Department of Chemistry,
University of Coimbra,
3004-535 Coimbra, Portugal
e-mail: akh@ci.uc.pt

Abstract Spectrofluorimetry has been used to detect in vivo the photoluminescence of native organic volatiles emitted from the surface of various fruits. Luminescence from above the surface of fruits is observed in the visible spectral region and attributed to species in their aroma. Further, the role of different volatiles in the natural aroma emission has been tested by a laser-induced photoluminescence microscope technique with the vapours of some odours. Monitoring the time evolution of luminescence allows us to study the ripening and ageing of the fruits. The ripening and ageing of some fruits is shown to lead to changes in band shape and to an enhancement of the luminescence. The results demonstrate that luminescence spectroscopy of natural aroma emission is a very promising method for studying biochemical changes in fruits, that may be extended to monitoring of aroma evolution in other living systems.

Keywords Luminescence · Volatile compounds · Fluorimetry of aroma emission · Living systems

Introduction

The improvement of the quality of fruit and vegetables is a topic of widely recognised importance in contemporary research. While it is well known that living systems produce many types of volatiles, there is a lack of information on the formation and emission of compounds responsible for such aromas. The real-time monitoring of the volatiles emitted from fruit and vegetables can give key information to the food industry for determining optimal harvesting and storage conditions to maintain and improve quality. Fluorimetry has been used to detect in vivo the luminescence of native organic volatiles emitted at the surface of fruits [1, 2]. This approach uses the ability of volatile compounds produced by fruit to luminesce in the vicinity of the fruit surface when irradiated by a laser beam. It was suggested that this light emission could be attributed to a gaseous atmosphere of native organic volatiles emitted from the surface of fruits. While the origin of this emission has not yet been confirmed, the spectroscopic capacity of such a method allows for a reliable discrimination between different physiological states of living species, while the optical focusing and scanning of excitation light permits mapping of emitting regions as small as few microns on the plant surface.

Until recently, the analysis of aromas has been based on methods for the extraction, separation and identification of volatile organic compounds by high resolution gas or liquid chromatography hyphenated with mass-spectrometry and/or infra-red spectroscopy. It has been reported that fruits produce complex mixtures of volatiles with more than two hundred at present identified compounds [3–6].

A new powerful method based on proton-transfer-reaction-mass spectrometry has recently been introduced for trace gas detection in real time [7]. This was successfully applied to food control, environmental problems [8, 9], and has been used in real-time monitoring of organic volatiles emitted from fruits during ageing [10]. It has been demonstrated that the ageing processes in fruits are related directly to the emission of dominant volatiles: methanol, ethanol, acetaldehyde, methyl acetate, acetone, acetic acid, ethyl acetate, mixture of ketones and aldehydes, and esters.

We report on in vivo monitoring of the luminescence from fruits in the visible region. Our results show that luminescence spectroscopy of the organic volatiles is a very promising method for monitoring the chemical evolution in living systems.

Experimental

Samples

Fruits (apples) from various origins were purchased locally and were kept at room temperature under normal atmosphere for several weeks (no information is available on when these fruits were picked). A number of fruits were harvested on the day of measurements in central region of Portugal.

Characterisation

Luminescence spectra were measured at room temperature with a Renishaw Image Microscope-2000 model, using an He-Ne laser (Spectra-Physics-127) for excitation. Luminescence measurements were performed in an unpolarised configuration using the back-scattering geometry from an interfacial region located in the confocal microscope focus area of sizes 10–200 microns and power density from 2 W/cm^2 to 8×10^4 W/cm^2. In addition, measurements were made under various conditions of temperature and atmosphere using SPEX DM 3000 F and Jobin Yvon-SPEX Fluorolog-3 spectrofluorimeters with 150 W and 450 W xenon arc lamps, respectively, as excitation sources, and laser diode with excitation wavelength $\lambda_{exc} = 650$ nm and a power of 4 mW at the source.

Results and discussion

Luminescence from fruits

Spectrofluorimetry has been used to detect the luminescence from the surface of fruits. Fig. 1a represents a typical spectrum of an apple measured using excitation by xenon lamp. Significant luminescence from various apples was found in the 600–800 nm region. The spectral bands in this region observed for various excitation wavelengths ($\lambda_{exc} = 400$ nm, 500 nm and 600 nm) were assigned to luminescence. This luminescence from above the surface of fruits was attributed to species in their aroma [1]. The spectra obtained from different fruits have been classified into two sets: 1) very strong contributions at 680–685 nm, and 2) relatively weak broad bands located in the 700–730 nm region.

Luminescence from the gaseous atmosphere near fruits was also detected by the laser-induced photoluminescence microscope technique. The spectra were recorded from the anti-Stokes/Stokes shift of ± 3500 cm^{-1} under excitation at $\lambda_{exc} = 632.8$ nm, the wavelength of an He-Ne laser. Figure 1b shows the spectrum of an apple: the fruit surface was located at a distance of 2 mm from the confocal microscope focus area of size ≈20 microns. A very strong Stokes shifted peak at 685 nm and a relatively weak band or shoulder at 710 nm dominate the spectrum. The appearance of these bands in the Stokes shifted region was attributed to luminescence [1].

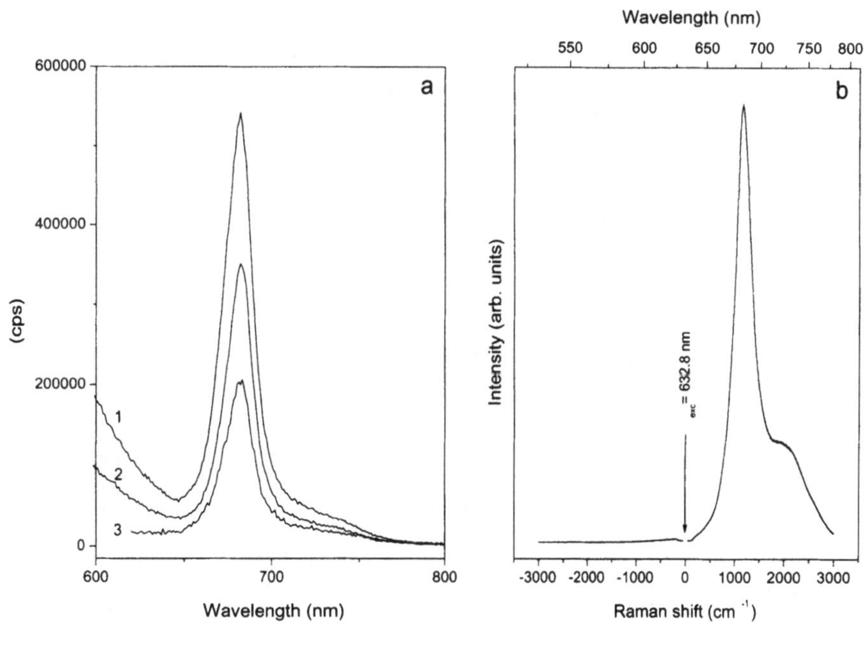

Fig. 1 The spectra from an apple: (**a**) measured by spectrofluorimetry, the spectra-1, -2, and -3 excited, respectively, at $\lambda_{exc} = 400$, 500 and 600 nm wavelengths (xenon arc lamp); (**b**) spectra detected by the laser photoluminescence microscopy over the range from the anti-Stokes to the Stokes shifts detected at the distance of 2 mm between the spectral acquisition area and the surface of the sample, under 4×10^4 W/cm^2 excitation at $\lambda_{exc} = 632.8$ nm wavelength of He-Ne laser

Model experiments

Organic volatiles are seen to luminescence effectively when escaping through the skin of fruit, which makes such a membrane a very interesting system. The role of skin membranes in the natural aroma emission was examined by a laser photoluminescence microscopy for the vapours of aroma compounds (excited at $\lambda_{exc} = 632.8$ nm wavelength of He-Ne laser). In this study, methanol and ethanol were selected for model measurements, as they are the most prominent components in the complex mixture of organic volatiles identified in fruits [5, 10].

As expected, we have not detected any significant photoluminescence from the free vapours of the liquid compounds at room temperature. The successful observation of the spectra was taken in a modified configuration for measurement. The liquid compound was located in a closed glass container with a small aperture (jet) for gas outlet. The volatile molecules in a gas flow from this jet were tested by the laser photoluminescence microscopy. It is thought that there could be some similarity in the molecular motion and rates in this jet configuration and in escaping through the fruit skin, such as through a multijet membrane.

The results of measurements are presented in Fig. 2 for methanol (curves-1) and ethanol (curves-2) in a liquid phase and in the vapours in jet. The spectra for liquid methanol and ethanol (solid lines) demonstrate the normal Raman modes, while the spectra detected in the gas flow from jet (dotted lines) are dominated by strong broad bands at 670–680 nm, which were associated with luminescence.

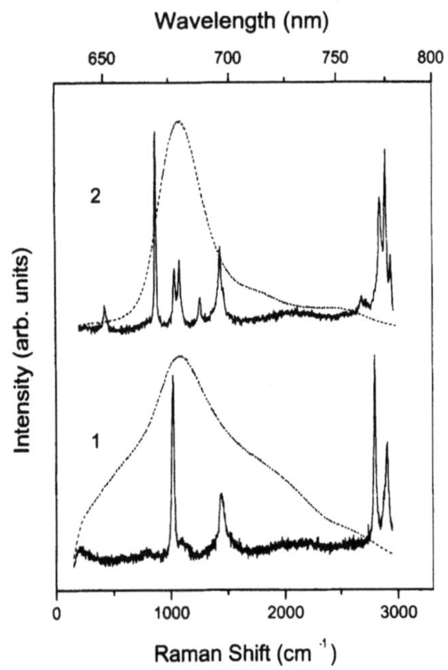

Fig. 2 Spectra of methanol (curves-1) and ethanol (curves-2) detected by the laser photoluminescence microscopy: in a liquid phase-the focal point of microscope was shifted from vapours to liquid (solid lines); and spectra detected in the vapour flow from jet (dotted lines); excitation at $\lambda_{exc} = 632.8$ nm wavelength of He-Ne laser

Ripening and aging of fruits

A ripe fruit smells better than an unripe one, but an overripe one has a misleading smell: it smells more attractive than it really is. To investigate the possibilities of sorting fruits on aroma, we have performed preliminary experiments on the evolution of the luminescence monitored during ripening and ageing of fruits.

A number of apples of different ripeness were harvested in the days of measurements (period of June-August 2001) and these results are presented in Fig. 3a. The samples were examined by spectrofluorimetry using excitation wavelength $\lambda_{exc} = 650$ nm of a laser diode. The ripening of fruits from unripe to overripe was accompanied by modification of spectrum shape and a significant increase in the luminescence contribution in the higher energy region (blue shift).

Luminescence from fruits during ageing was detected by the laser photoluminescence microscopy (excited at

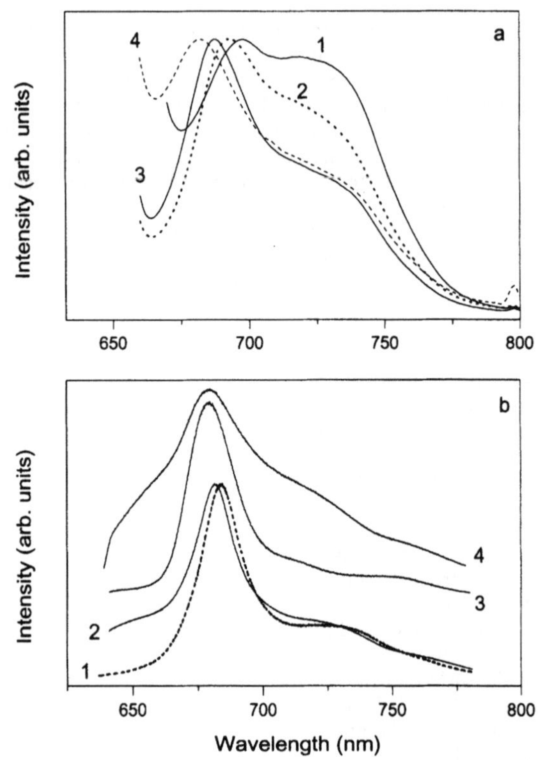

Fig. 3 (**a**) Spectra from apples of different ripeness were examined by spectrofluorimetry using excitation of a laser diode ($\lambda_{exc} = 650$ nm): from unripe (curves-1, -2), ripe (curve-3) to overripe (curve-4). (**b**) The effect of ageing of an apple: initially – spectrum-1, and after a month – spectrum-2, and luminescence spectra of ethanol (curve-3) and methanol (curve-4) in the vapour flow from a jet. Spectra detected by the laser photoluminescence microscopy (excited at $\lambda_{exc} = 632.8$ nm wavelength of He-Ne laser)

$\lambda_{exc} = 632.8$ nm wavelength of He-Ne laser). The effect of one month ageing of an apple is shown in Fig. 3b. The spectra were taken for the same focus for a distance of 2 mm above the fruit skin and under the same excitation level. These results demonstrate that the ageing of apple is also attended by the changes of the luminescence. These spectral transformations can be interpreted in terms of changes in the general abundance of the emitted volatiles and can be related to biochemical changes in fruit. Comparing the changes of spectra with the spectra of more important volatile compounds (methanol and ethanol luminescence spectra in Fig. 3b) we can suggest the connection of these changes to evolution of relative concentration of dominant volatiles (such as methanol and ethanol).

Conclusions

The measurements presented indicate the presence of the luminescence from the fruits. Luminescence from above the surface of fruits is observed in the visible spectral region and attributed to species in their aroma. Following this observation of luminescence from fruits, our attention has been devoted to the nature of this emission. It is possible that this photoluminescence has a vibronic origin, and may be associated with the vibrations of fragments and functional groups of volatiles. While the origin of this luminescence has not yet been confirmed, it allows minute amounts of substance to be successfully detected and characterised. The reported results demonstrate that luminescence spectroscopy of organic volatiles is a very promising method for the control and improvement of the quality of fruits that may be extended to other living systems. We therefore feel this technique shows strong potential for both biochemical and biomedical applications.

Acknowledgements The authors wish to thank the FCT (Portugal) for the award of fellowship (BPD/1525/2000) to A.A.K.

References

1. Kharlamov AA, Burrows HD (2001) J Appl Biochem Microbiol 37(2):206
2. Kharlamov AA, Burrows HD (2001) Sensors Actuators B 77:593
3. Mussinan C, Walradt J (1975) J Agric Food Chem 23(1):482
4. Guichard E (1982) Sci Alim 2:173
5. Nikiforov A, Jirovetz L, Woidich A (1994) Food Quality Pref 5:135
6. Robertson GW, Griffiths DW, Woodford JA, Birch AN (1995) Phytochemistry 38:1175
7. Hansel A, Jordan A, Holzinger R, Prazeller P, Vogel W, Lindinger W (1995) Int J Mass Spectrom Ion Processes 149/150:600
8. Lindinger W, Hansel A, Jordan A (1998) Int J Mass Spectrom Ion Processes 173:191
9. Hansel A, Jordan A, Warneke C, Holzinger R, Lindinger W (1998) Rapid Commun Mass Spectrom 12:871
10. Boschetti A, Biasioli F, Opbergen M van, Warneke C, Jordan A, Holzinger R, Prazeller P, Karl T, Hansel A, Lindinger W, Iannotta S (1999) Postharvest Biol Techol 17:143

Progr Colloid Polym Sci (2004) 123: 182–187
DOI 10.1007/b11759

Laurence S. Romsted
Jianbing Zhang

Determining antioxidant distributions in model food emulsions: development of a new kinetic method based on the pseudophase model in micelles and opaque emulsions

Presented at the European Colloid and Interface Society, XV Conference, 16 to 21 September 2001, Coimbra, Portugal

L.S. Romsted (✉) · J. Zhang
Department of Chemistry and Chemical Biology, Wright and Rieman Laboratories, Rutgers, The State University of New Jersey, New Brunswick, New Jersey 08903, USA
e-mail: romsted@rutchem.rutgers.edu
Tel.: +1-732-4453639;
Fax: +1-732-4455312

Abstract The mechanisms of lipid peroxidation and the distributions of antioxidants in emulsions are not fully understood, in a large part because conventional methods for monitoring reactant distributions in solutions of homogeneous surfactant aggregates cannot be used in opaque emulsions. Here we show that a kinetic method based on monitoring the rate of product formation by HPLC for reaction of an arenediazonium ion probe and the antioxidant TBHQ in aqueous micellar solutions of a non-ionic surfactant, $C_{12}E_6$, gives the same values for the distribution constant of TBHQ as more established spectral shift and kinetic methods. The HPLC method also works in octane/$C_{12}E_6/H_2O$ emulsions. To our knowledge, this is the first application of a pseudophase model to a real biphasic system. Future work will focus on estimating distribution constants of a variety of antioxidants in emulsions and attempting to establish a scale of antioxidant efficiency from their rate constants in the surfactant film of the emulsions.

Keywords Non-ionic micelles · Non-ionic emulsions · Antioxidant distribution · Kinetics · Pseudophase model

Introduction

Antioxidants are commonly used in emulsified foods to slow the peroxidation and degradation of unsaturated lipids and the accompanying bad smells and tastes [1–4]. However, a scientific basis for selecting the most efficient antioxidant for a particular food, ideally based on a scale of antioxidant efficiency, has proved refractory for several reasons. The mechanism of peroxidation, a multi stepped, free radical, process in homogeneous solution is still not fully understood [2]. In addition, the distribution of reactants, unsaturated triglycerides, molecular oxygen, and antioxidants within these metastable, biphasic, sometimes viscous mixtures of oil, water and food emulsifier (called surfactant here because we use highly purified materials) and other additives such as flavorings, salts and buffers in food emulsions is difficult to determine. Indeed, some components such as the triglyceride oils and surfactants are themselves mixtures of variable antioxidant efficiency and distribution within the various regions of emulsions have not been quantified.

This paper describes progress on the development of a kinetic method for determining antioxidant distributions in model systems of emulsified foods such as micelles, microemulsions and opaque macro emulsions (hereafter called emulsions), based on the rates of reaction of an arenediazonium ion, 4-hexadecyl-2,6-dimethylbenzenediazonium ion, $16\text{-}ArN_2^+$, with antioxidants as a function of surfactant concentration. In the absence of antioxidants, product yields from the spontaneous heterolytic dediazoniation of $16\text{-}ArN_2^+$ provide new information on the balance of forces controlling the structures and stabilities of surfactant aggregates from changes in interfacial concentrations of weakly basic nucleophiles such as water, halide ions, and alcohols [5]. Antioxidants such as *t*-butylhydro-

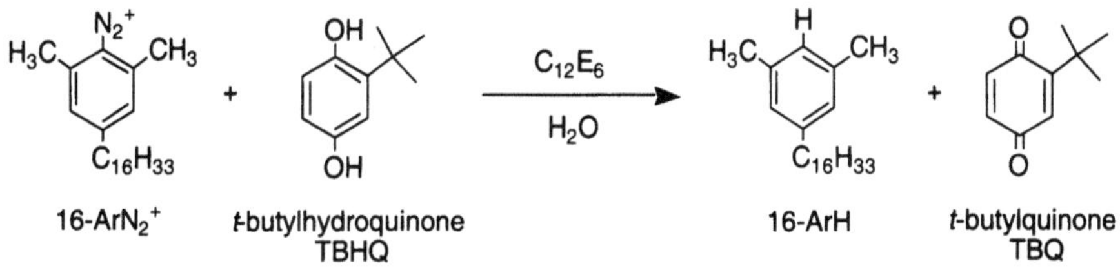

Scheme 1

quinone, TBHQ, which was used in this work, reduce 16-ArN_2^+ to the arene, 16-ArH and *t*-butylquinone, Scheme 1, at a much faster rate than that of heterolytic dediazoniation [6].

Because this reaction occurs only in the surfactant film of micelles, microemulsions and emulsions [5] and not in the aqueous and oil regions, the results are easier to interpret than the antioxidant effects on the reaction with unsaturated triglycerides with O_2 which may occur in any region of an emulsion. This paper shows that our kinetic method gives reproducible values for the distribution constant TBHQ in food model systems and that rate constants can be measured for reactions in opaque emulsions. The long-term goals are to develop a kinetic method for determining distribution constants of antioxidants in emulsions and a practical scale of antioxidant efficiency.

Fig. 1 shows a small section of a droplet in either a water-in-oil or oil-in-water liquid food emulsion, such as mayonnaise or salad dressings and illustrates the probable distributions of unsaturated triglyceride oil, surfactant film and aqueous regions of a water-in-oil or oil-in-water emulsion. The surfactant, e.g., a phospholipid in a food emulsion, coats the boundary between the oil and water regions. Antioxidants, which are often aromatic phenols and are both water and oil soluble and distribute between the oil, surfactant film and water regions of the emulsion. Antioxidant distributions depend on their hydrophilic-lipophilic balance or HLB. Currently there is no established method for determining antioxidant distributions in emulsions or even homogenous model systems such as microemulsions. Results of several groups have been reviewed [4, 7] and they show that determining antioxidant distributions by physical methods is difficult. Because emulsions are generally opaque and antioxidant concentrations are often low compared to other components, spectrophotometric methods such as UV/Visible, fluorescence and NMR cannot be used to estimate antioxidant distributions.

Fig. 2 shows the three regions of a food model system composed of oil, surfactant and water. The model system may be a macro or microemulsion or a micellar solution, i.e., a monophasic or biphasic system. This approach requires assuming that the intermolecular forces controlling distributions of antioxidants are the same in the model systems as in a natural food emulsion. To determine the distribution constants of the antioxidant between the oil and surfactant film, K_o, and between the water and surfactant film, K_w, we apply the pseudophase model for chemical reactivity in solutions of organized amphiphiles [8–10]. In all pseudophase models, the distributions of reactants are assumed to be in dynamic equilibrium because molecular diffusivities between

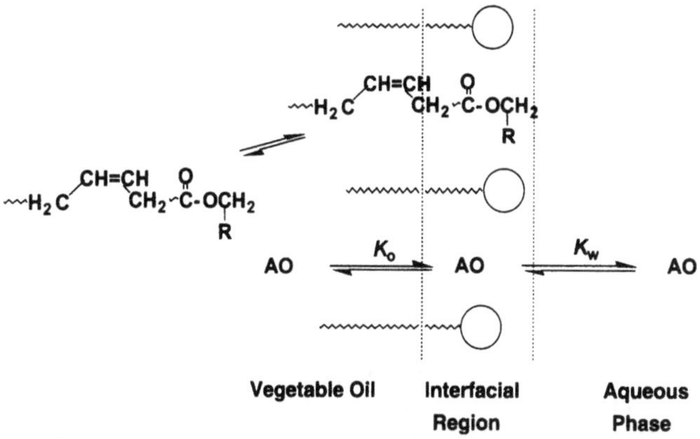

Fig. 1 Cartoon of a small section of the interfacial region of a hypothetical food emulsion showing the location of the surfactant, O, the distributions of unsaturated triglyceride and antioxidant, AO, and its distribution constants, K_o and K_w. Water molecules (not shown) hydrate the surfactant head groups and O_2 diffuses throughout the system

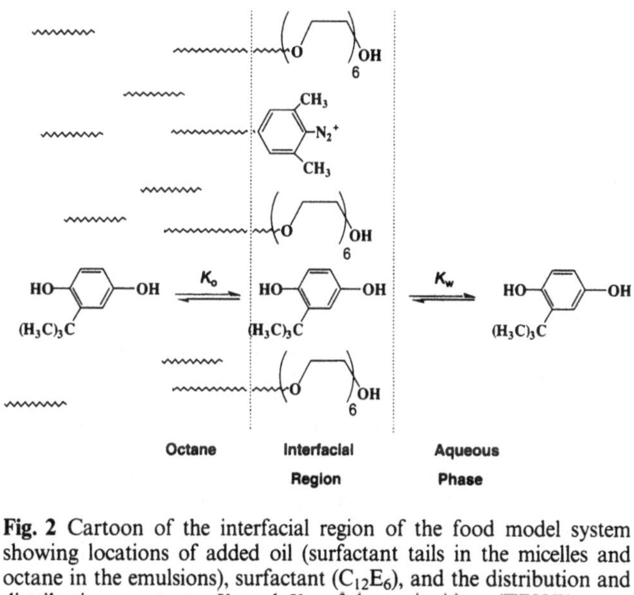

Fig. 2 Cartoon of the interfacial region of the food model system showing locations of added oil (surfactant tails in the micelles and octane in the emulsions), surfactant ($C_{12}E_6$), and the distribution and distribution constants, K_o and K_w, of the antioxidant (TBHQ)

regions are orders of magnitude faster than the rate of reaction and the observed rate is the sum of the rates in the oil, water and surfactant film regions. However, because 16-ArN_2^+ is located only in the surfactant film [5], Fig. 2, the observed rate of reaction depends only on the rate constant for reaction of 16-ArN_2^+ with the antioxidant (not shown) and its concentration in the surfactant film. The distribution and rate constants are obtained from observed rate constant-surfactant concentration profiles by fitting the data using equations based on the pseudophase models. Mathematical treatments for pseudophase models are fully developed [9, 10] and only the final forms of relevant equations are given below.

Experimental

Materials

HPLC grade MeOH, CH_3CN and *i*-PrOH, *t*-butylhydroquinone (TBHQ, 97%), succinic acid (99+%), 5,5-dimethyl-1-pyrroline *N*-oxide (DMPO, 97%), octane (99%) and inorganic reagents were purchased from Aldrich. Hexaethylene glycol mono dodecyl ether ($C_{12}E_6$, 98%) was recrystallized three times from MeOH. The probe and its reaction products needed for HPLC analysis: 4-*n*-hexadecyl-2,6-dimethylbenzenediazonium tetrafluoroborate salt, 16-ArN_2BF_4 4-*n*-hexadecyl-2,6-dimethylbenzene, 16-ArH, 4-*n*-hexadecyl-2,6-dimethylphenol, and 16-ArOH were prepared earlier in our lab [16]. All aqueous solutions were prepared by using distilled water, which was passed over activated carbon and de-ionizing resin and then redistilled. Stock solutions of 0.05 M succinate buffer were prepared by dissolving weighed amounts of succinic acid in a volumetric flask and titrating the solutions to the desired pH by additions of small amounts of 10 M NaOH.

Methods

General

UV spectra were obtained on either a thermostatted Perkin-Elmer UV-VIS 559A or a Perkin-Elmer Lambda 40 UV-VIS spectrophotometer. Data accumulation and analyses were done on attached PC computers using Perkin-Elmer software. Product yields were determined on computer controlled Perkin-Elmer HPLCs equipped a diode array detector and an autosampler fitted with a Ranin C-18 reserve phase column (5 mm particle size, 4.6 mm ID × 25 cm) and a 200 mL sample loop and PE Nelson Turbochrom Version 4.1 software. Solution pH was measured with an Orion combination pH microelectrode, Model 8103 with a Corning Model 100 pH meter. All solutions were prepared with calibrated volumetric flasks, pipettes and syringes.

Determination of distribution constants *Spectra shift method.* 0 to 0.01 M $C_{12}E_6$ solutions were prepared in volumetric flasks from a stock solution of 0.5 M $C_{12}E_6$ in CH_3CN diluted to the mark with 0.05 M succinate buffer, pH = 4.1. These solutions were equilibrated in the spectrophotometer at 30 °C and an aliquot of freshly prepared stock solution of 0.08 M TBHQ in CH_3CN was added to give final [TBHQ] of either 3.0 or 6.0×10^{-4} M and the absorbance, A, at 300 nm was recorded. The value of *K* was calculated from the slope of linear relationship of $(A-A_w)/[C_{12}E_6]$ versus A (see Fig. 3).

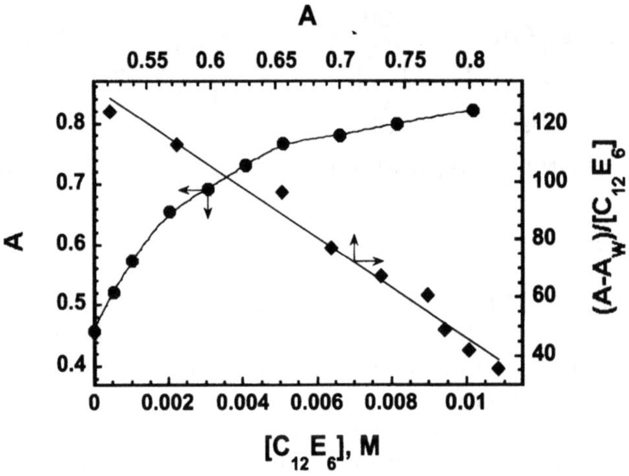

Fig. 3 Plot of absorbance, A, of 3.0×10^{-4} M TBHQ versus $[C_{12}E_6]$ at 300 nm 0.05 M succinate buffer, pH 4.06 at 30 °C. *K* was obtained from the slope of the linear plot of $(A-A_W)/[C_{12}E_6]$ versus A ($y = 286.4 - 301.3x$, $R = 0.993$)

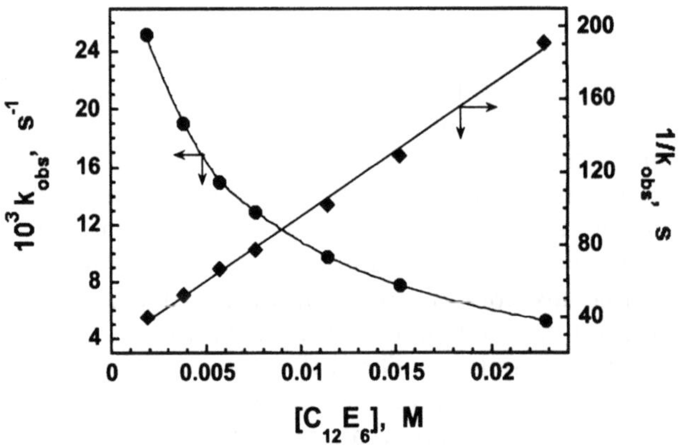

Fig. 4 Plot of k_{obs} versus $[C_{12}E_6]$ for the reaction of TBHQ with 16-ArN_2^+ containing 6.7×10^{-5} M 16-ArN_2^+, 6.3×10^{-4} M TBHQ and 9.6×10^{-3} M DMPO. Reactions were monitored at $\lambda = 275.5$ nm in 0.05 M succinate buffer, pH 4.23 at 30 °C. The value of *K* was obtained from the ratio of the slope to the intercept from linear double reciprocal plot of $1/k_{obs}$ versus $[C_{12}E_6]$ by using Eq. (2) ($y = 24.4 + 7151.7x$, $R = 0.997$ and $K = 7151.7/24.4 = 293$ M^{-1})

Determining k_{obs} by kinetics with UV detection. To succinate buffer at pH 4.2 or 4.6 in the cuvette was added an aliquot of freshly prepared 1.0 M $C_{12}E_6$ in CH_3CN to give final $[C_{12}E_6]$ ranging from 0.001 to 0.05 M was equilibrated in the spectrophotometer at 30 °C. Aliquots of freshly prepared ice cold stock solutions of 0.08 M TBHQ, 0.9 M DMPO and finally, 0.01 M 16-ArN_2BF_4 in CH_3CN were added to the cuvette sequentially via syringe to initiate reaction. The final molarities of TBHQ, DMPO and 16-ArN_2BF_4 were 6.0×10^{-4}, 9.6×10^{-3} and 6.5×10^{-5} M, respectively. The reaction was followed by the decrease in A at 275.5 nm. Values of observed rate constants, k_{obs}, of reaction between 16-ArN_2BF_4 and TBHQ were calculated from the slopes of $\ln[(A_o-A_\infty)/(A-A_\infty)]$ versus time plots. The value of *K* for TBHQ in $C_{12}E_6$ was calculated from the ratio of slope and intercept of the plot of $1/k_{obs}$ versus $[C_{12}E_6]$ (see Fig. 4).

Determining k_{obs} by the HPLC method. Aliquots of freshly prepared 1.0 M stock solution of $C_{12}E_6$ in CH_3CN were added to

20.0 mL of aqueous 0.003 M HCl, pH = 2.6, to give final molarities that ranged from 0.001 to 0.05 M. A volume of CH_3CN was added to keep its total volume constant at 5%. The solution was stirred continuously and the flask was equilibrated at 30 °C for about 15 min. Reaction was initiated by addition of freshly prepared ice cold stock solutions of 0.20 M TBHQ, 1.3 M DMPO and 0.025 M 16-ArN_2BF_4 in CH_3CN sequentially to give final molarities of, respectively, 9.0×10^{-4}, 6.0×10^{-3} and 1.2×10^{-4} M. Aliquots (0.5 mL) were taken periodically and mixed with 0.5 mL of 2.0 M aqueous HCl to quench the reaction between TBHQ and 16-ArN_2^+. The solutions stood until dediazoniation of 16-ArN_2^+ was complete, at least 48 h at room temperature. Each sample was analyzed by HPLC to obtain the peak area, PA, of the reduced products, 16-ArH. The mobile phase for HPLC product separation was 25%/75% *i*-PrQH/MeOH with a flow rate of 1.0 mL per min and the retention time for 16-ArH is typically ca. 12–13 min. The absorbance was measured at 220 nm and the PA values used in the calculations are averages of triplicate injections. Values of the observed rate constants, k_{obs}, of reaction between 16-ArN_2BF_4 and TBHQ were calculated from the slopes of $\ln[PA_\infty/(PA_\infty-PA)]$ versus time plots. The distribution constant of TBHQ was calculated from the slope and intercept of the plot of $1/k_{obs}$.

$C_{12}E_6$-water emulsions by the HPLC method. 10.0 mL of 0.003 M aqueous HCl, pH = 2.6, and 10.0 mL of octane containing the appropriate amount of $C_{12}E_6$ were added to a thermostated vessel fitted with a magnetic stir bar that kept the emulsions mixed throughout the reaction. The final volume fractions of $C_{12}E_6$ in the mixture of octane and water varied from 0.01 to 0.04 and the volume ratio of the aqueous to oil phase was 4 : 1. The reaction mixture was equilibrated at 25 °C for about 15 min. Freshly prepared ice cold stock solutions of 0.20 M TBHQ, 1.3 M DMPO and 0.025 M 16-ArN_2BF_4 in CH_3CN were added sequentially to initiate reaction and their final molarities were, respectively, 9.0×10^{-4}, 6.0×10^{-3} and 1.2×10^{-4} M. 0.05 mL aliquots were withdrawn periodically and mixed with 0.5 mL 2.0 M HCl solutions to quench the bimolecular reaction between TBHQ and 16-ArN_2BF_4. The reaction mixtures stood at room temperature about 72 h until sufficient octane vaporized for the emulsions to become a single phase and the remaining 16-ArN_2^+ decomposed. Each sample was analyzed by HPLC and values of k_{obs}, $R \geq 0.994$ were obtained by following the protocol described above for homogeneous solutions.

Results

In homogeneous solution, the distributions of neutral organic molecules between surfactant aggregates and the surrounding bulk phase, typically water, are traditionally defined as distribution (or binding) constants [11], Eq. (1).

$$K = \frac{[AO_I]}{[AO_w][D_n]} \tag{1}$$

where square brackets here and throughout the text indicate stoichiometric concentration in moles per liter of solution volume, $[AO_I]$ is the concentration of the antioxidant in the interfacial region of the aggregates, $[AO_w]$ is the concentration of the antioxidant in the surrounding aqueous pseudophase, and in these experiments, $[D_n]$ is the concentration of micellized hexaethylene glycol mono dodecyl ether, $C_{12}E_6$, which is assumed to be the same as the stoichiometric concentration because the cmc of $C_{12}E_6$, 6.8×10^{-5} M at 25 °C [12], is much smaller than the $C_{12}E_6$ concentrations.

To demonstrate the viability of a kinetic method for estimating antioxidant distributions in opaque emulsions from product yields measured by HPLC, we compared the HPLC method with values for the distribution constant of TBHQ in homogeneous solutions of micelles obtained by the established methods of spectral shift [11] and kinetics by spectrophotometry [13]. Table 1 summarizes all estimates of K from the different methods discussed below.

Spectral shift

Fig. 3 illustrates one determination of K by the spectral shift method showing the effect added $C_{19}E_6$ on the absorbance of TBHQ at $\lambda = 300$ nm and the linear plot of the data based on the standard equation for estimating the binding constant (see caption). The distribution constant between the micellar surface and water was estimated from the slope as $K = 301$ M^{-1} with a correlation coefficient, R = 0.993.

Kinetics with UV detection

Fig. 4 shows the results of one first order condition for the reaction of 16-ArN_2^+ with excess TBHQ. The reaction mixture contains 5,5-dimethyl-1-pyrroline *N*-oxide, DMPO, a radical quencher and active O_2

Table 1 Binding constants, K, of TBHQ in micellar solutions of $C_{12}E_6$ at 30 °C in 0.05 M succinate buffer except where noted

Method	10^4[TBHQ] M	10^5[16-ArN_2^+] M	10^3[DMPO] M	pH	Wavelength nm	K M-1
UVS-1[a]	5.88	0	0	4.05	302	304
UVS-2[a]	2.80	0	0	4.06	300	301
UVK-1[b]	5.80	6.44	6.42	4.23	320	306
UVK-2[b]	5.79	6.56	6.42	4.23	320	305
UVK-3[b]	5.48	6.38	6.42	4.65	320	287
UVK-4[b]	6.14	6.66	9.59	4.65	275.5	298
UVK-5[b]	6.35	6.66	9.59	4.23	275.5	293
HPLC[c]	8.81	11.63	6.14	2.58		329
				Average Value:		303 ± 1.5%

[a]Spectra shift method
[b]Chemical reaction method, monitored by UV spectrometry
[c]Chemical reaction method, monitored by HPLC of reaction mixture samples

scavenger [14, 15]. TBHQ reacts with dissolved O_2 at a somewhat slower rate than the reaction of TBHQ with 16-ArN_2^+ (experiments not shown) and control experiments showed that added DMPO slows, but does not stop, this background reaction. Repetitive scan experiments of absorbance versus ■ for reaction of TBHQ with O_2 has two isobestic points at 275.5 and 310.7 nm (results not shown). When the reaction of 16-ArN_2^+ with TBHQ is followed at either of these wavelengths, good first order plots (R≥0.999) are obtained.

Because 16-ArN_2^+ is located only in the surfactant film, the observed rate constant, k_{obs}, for reaction of TBHQ with 16-ArN_2^+ versus $C_{12}E_6$ is given by:

$$k_{obs} = \frac{(k_2^M/V)K}{1 + K[C_{12}E_6]} \quad (2)$$

where k_2^M is the second order rate constant for reaction in the micelles and V is the molar volume for the reaction (see ref.[9] for details). A double reciprocal plot of Eq. (2), Fig. 4, gives a value of K = 293 M^{-1} (Table 1).

Kinetics with HPLC detection

Fig. 5 shows the change in k_{obs} for the reaction of 16-ArN_2^+ with TBHQ with added $C_{12}E_6$ and K=328 M^{-1} (Table 1). The natural logarithm, ln, of 16-ArH peak area, PA, versus time plots used to calculate k_{obs} are linear for at least 3 half lives, R ≥ 0.999 for about 25 data points. At the three highest [$C_{12}E_6$], the reaction of 16-ArN_2^+ with TBHQ shows apparent autocatalysis, i.e., the natural log plots of PA versus time are initially linear, but after a certain time deviate strongly upward. The calculated values of k_{obs} based on the initial linear portion of the plot (not shown) are sensible when compared with those obtained at lower [$C_{12}E_6$].

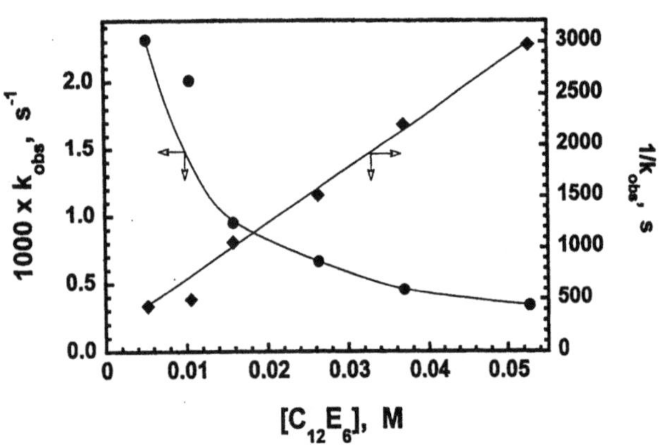

Fig. 5 Plot of k_{obs} versus [$C_{12}E_6$] for the reaction of TBHQ with 16-ArN_2^+ containing 1.15 × 10^{-4} M 16-ArN_2^+, 8.81 × 10^{-4} M TBHQ and 6.14 × 10^{-3} M DMPO in 0.003 M HCl, pH 2.58, at 30 °C. The value of K was obtained from the ratio of the slope to the intercept from linear double reciprocal plot of k_{obs} versus [$C_{12}E_6$] by using Eq. (2) (y = 163.2 ± 53649.1x, R = 0.997 and K = 56349.1/163.2 = 329 M^{-1}). Point at 0.01054 M $C_{12}E_6$ was not included in the estimate of K

Kinetics experiments in emulsions

Fig. 6 shows the rate of formation of 16-ArH measured by HPLC as a function of time for reaction of 16-ArN_2^+ with TBHQ in fluid, stirred, emulsions of 1 : 4 volume ratio of octane to water with 1.52% volume fraction of $C_{12}E_6$. The ln plot is linear for 3 half-lives. Table 2 lists k_{obs} values versus $C_{12}E_6$ volume fraction in 4 : 1 volume ratio of water to octane emulsions. As with determinations of k_{obs} in micelles (see above), the ln plots were non-first order at the three highest $C_{12}E_6$ volume fractions and k_{obs} was calculated from the initial linear sections of the plots. These results show that rate constants can be determined in emulsions by the HPLC method and that

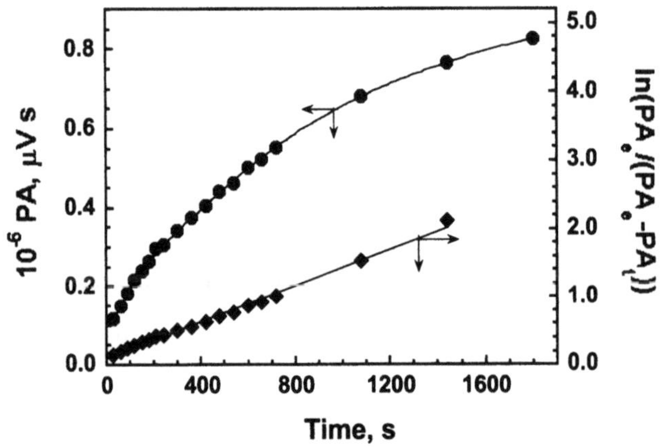

Fig. 6 Change in HPLC peak area versus time for the reaction of 16-ArN_2^+ with TBHQ in an octane/$C_{12}E_6$/H_2O emulsion composed of 4.0 mL octane, 16.0 mL aqueous 0.003 M HCl (pH = 2.6), 1.25% volume fraction $C_{12}E_6$, 1.51 × 10^{-4} M 16-ArN_2^+, 9.11 × 10^{-4} M TBHQ and 4.98 × 10^{-3} M DMPO at 25 °C with continuous stirring. The ln plot versus time is linear for three half-lives (y = 0.0964 + 0.00158x, R = 0.994) and PA = 0.869

Table 2 Values of k_{obs} in $C_{12}E_6$ emulsions composed of 4:1 volume ratio of 0.003 M HCl, pH 2.6, in H_2O to octane and 1.51 × 10^{-4} M 16-ArN_2^+, 9.11 × 10^{-4} M TBHQ and 4.98 × 10^{-3} M DMPO at 25 °C

$C_{12}E_6$ vol, fraction (%)	$10^3 k_{obs}$ (s^{-1})	$C_{12}E_6$ vol, fraction (%)	$10^3 k_{obs}$ (s^{-1})
1.03	2.91	2.23	1.30
1.27	1.89	2.48	0.957
1.52	1.58	2.95	0.839[a]
1.73	1.40	2.42	0.699[a]
1.97	1.28	3.92	0.716[a]

[a]Rate constant obtained from the initial slope

k_{obs} decreases with increasing $[C_{12}E_6]$, consistent with the assumptions used in the pseudophase model.

Application of the pseudophase model to emulsions is more complex than to micelles because the antioxidant is distributed between the oil, surfactant film and aqueous regions, Figs. 1 and 2, and two distribution constants, K_o and K_w, are required to describe the fraction of the antioxidant in each region. Equations based on the pseudophase model have been derived to describe the distribution of molecules in microemulsions [10] and this treatment will be adapted to emulsions, which have the same regions as microemulsions. Obtaining values of K_o and K_w requires solving two equations in two unknowns and the next step is to determine a second set of k_{obs} values at a second octane to water volume ratio. The values of K_o and K_w in emulsions should to be similar to those in aqueous micelles and reverse micelles, respectively, because the solubility properties of TBHQ (and other antioxidants) in the bulk phases and the surfactant films should be similar in both systems. We have already used the spectral shift method to estimate the distribution constant of TBHQ in a reverse micelle of octane/$C_{12}E_6$/water and $K = 198\ M^{-1}$ (results not shown), which will be compared with K_o obtained in emulsions.

Conclusions

Table 1 summarizes the experimental conditions and K values for TBHQ spectral shift, five by UV kinetics and one by HPLC. The results show that $K = 303 \pm 1.5\%$, that the HPLC method works as well as the other methods and that K is insensitive to changes in reactant concentration and pH. The average value of K is similar to that obtained for other aromatic compounds with substituted benzene rings, i.e., on the order of 10^2, although direct comparisons cannot be made because TBHQ has not been measured previously in nonionic micelles [8]. The fact that the change in k_{obs} in emulsions is similar to those in micelles supports our assumption that pseudophase models will work in true biphasic systems. The results also show that mixing with a magnetic stirrer provides sufficient agitation to ensure that component transport across phase boundaries is fast compared to the rate of reaction of 16-ArN_2^+ with TBHQ, at least in relatively fluid emulsions.

Our kinetic approach for estimating distribution constants is, in principle, quite general and the results may lead to a scale of antioxidant efficiency because preliminary experiments show that a variety of antioxidants react with 16-ArN_2^+ including α-tocopherol, ascorbic acid, BHT, BHA, and propyl gallate at measurable rates. The rate equation expression for use in emulsions must contain a term for the rate constant in the surfactant film, just as Eq. (2) does for reactions within micelles. Comparison of rate constants for a variety antioxidants may lead to a practical scale of antioxidant efficiency. We are also exploring the possibility that the concentration of 16-ArN_2^+ can be monitored electrochemically in emulsions, which would permit continuous monitoring of its reaction with TBHQ and other antioxidants.

Acknowledgements This is publication No. D10535-1-02 from the Center for Advanced Food Technology (CAFT). This research was supported in part by the State of New Jersey, and the New Jersey Agricultural Experiment Station. The Center for Advanced Food Technology is a New Jersey Commission on Science and Technology Center. We are also grateful to the National Science Foundation, grant number CHE-9985774, for partial support of this research and to Dr. Donald Hamm of Best Foods for extremely fruitful initial discussions.

References

1. Porter WL (1980) In: Simic MG, Karel M (eds), Autooxidation in food and biological systems. Plenum Press, New York, p 295
2. Coupland JN, Mcclements DJ (1996) Trends Food Sci Technol 7:83
3. Baskin SI, Salem H (1997) Oxidants, antioxidants, and free radicals. Taylor & Francis, Washington, DC
4. Frankel EN, Meyer AS (2000) J Sci Food Agric 80:1925
5. Romsted LS (2001) In: Texter J (ed), Reactions and synthesis in surfactant systems. Marcel Dekker, New York, p 265
6. Brown KC, Doyle MP (1988) J Org Chem 53:3255
7. McClements DJ, Decker EA (2000) J Food Sci 65:1270
8. Savelli G, Germani R, Brinchi L (2001) In: Texter J (ed), Reactions and synthesis in surfactant systems, vol 100. Marcel Dekker, New York, p 175
9. Bunton CA, Romsted LS (1999) In: Kumar P, Mittal KL (eds), Handbook of microemulsion science and technology. Marcel Dekker, New York, p 457
10. Da Rocha Pereira R, Zanette D, Nome F (1990) J Phys Chem 94:356
11. Sepulveda L, Lissi E, Quina F (1986) Adv Colloid Interface Sci 25:1
12. Meguro K, Ueno M, Esumi K (1987) Nonionic surfactants. In: Schick MJ (ed), Physical chemistry, vol 23. Marcel Dekker, New York. p 109
13. Mann MAB, Nome F, Zanette D, Zucco C, Romsted LS (1995) J Phys Chem 99:10879
14. Reszka K, Bilski P, Chignell CF (1992) Free Rad Res Commun 17:377
15. Reszka KJ, Chignell CF (1993) J Am Chem Soc 115:7752
16. Chaudhuri A, Loughlin JA, Romsted LS and Yao J (1993) J Am Chem Soc 115:8351

Progr Colloid Polym Sci (2004) 123: 188–193
DOI 10.1007/b11760

H.A. Wege
J.A. Holgado-Terriza
M.Á. Cabrerizo-Vílchez

Development of a pressure-controlled pendant-drop surface balance

Study of protein adsorption kinetics at the solution-air interface

H.A. Wege · M.Á. Cabrerizo-Vílchez (✉)
Biocolloid and Fluid Physics Group,
Department of Applied Physics,
Universidad de Granada, Spain
e-mail: mcabre@ugr.es
Fax: +34-958-243214

J.A. Holgado-Terriza
Software Engineering Department,
Universidad de Granada, Spain

Abstract A fully automated, pressure-controlled pendant-drop surface balance has been developed. It allows non-invasive surfactant adsorption kinetics studies at liquid-fluid interfaces at constant interfacial pressure π, permitting us to control the height of the energy barrier to adsorption, and to select the different regimes and – in the case of proteins – steps of the adsorption process (adsorption subsurface-surface – molecular rearrangement). In our device a surfactant solution droplet is formed at the tip of a capillary by means of a micro-injector. Drop profiles, extracted from digital drop micrographs, are fitted to the equation of capillarity, yielding π, the drop volume V and the interfacial area A. π is varied by changing V (and hence A). Control is based on a modulated fuzzy-logic PID algorithm able to maintain constant π (or A) over a wide range of experimental conditions. Once generated, the drop is kept at constant A, allowing the surfactant to adsorb freely until the desired π is attained. The adsorption kinetics at constant π are then studied by monitoring the variation of A in time, i.e., determining the relative area dilatation necessary at each instant to compensate the pressure increase due to incorporation of new surfactant molecules into the film or – in the case of proteins – conformational rearrangement of the adsorbed molecules. A study of the adsorption kinetics of β-casein and myoglobin at the air-water interface is presented.

Introduction

Surfactant adsorption and especially protein adsorption at interfaces plays a fundamental role in many technological and biological systems. As the interfacial tension γ is a function of the adsorption density Γ, it is possible to study these processes with interfacial tension techniques. Notable pioneer work on this field was done by Ward and Tordai [1, 2] for simple surfactants, and for the case of protein adsorption by MacRitchie and Alexander [3, 4] and by Graham and Phillips [5].

Among the interfacial tension techniques, ADSA (Axisymmetric Drop Shape Analysis, [6]) is one of the most precise and versatile ones. It fits experimental drop profiles, obtained from digital drop micrographs, to the Laplace equation of capillarity, and provides the interfacial tension γ and area A as outputs. It is non-invasive, i.e., the measuring device is not in direct contact with solvent or adsorbate and does not interact with them. Therefore, it is especially useful for studies in which adsorption from subphase onto the measuring device could take place: it overcomes a known deficiency of the conventional surface tension measurement methodologies like the Wilhelmy technique, where surfactants and particularly proteins may adsorb onto the plate, altering the contact angle and therefore the measured force, and thus complicating the interpretation of the result. And as surface balances with a fixed barrier measure directly the net horizontal force acting on the barrier, which is proportional to the surface tension difference between the surface layer-covered and the free part of the surface, they are not applicable to adsorption-from-bulk studies without the use of rather complicated modifications of the device. The major limitation of the pendant drop methodology was, until now, the impossibility to perform kinetic studies at constant interfacial pressure or area. In this work an upgrade of an ADSA-based surface balance is used: since the interfacial pressure $\pi = \gamma_o - \gamma$ (γ_o and γ are the tensions of pure solvent and surface layer-covered drop surfaces) is a fundamental parameter in adsorption kinetics, control over it opens a new and wide experimental field; working at constant π enables us to control the height of the energy barrier to adsorption, and to select the different regimes and – in the case of macromolecules – steps of the adsorption process, as will be shown.

Many authors agree in distinguishing the following steps in protein adsorption [3–5, 7, 8]:

(1) Transport from subphase to subsurface.
(2) Adsorption from subsurface to the interface.
(3) Conformational rearrangement of the adsorbed molecules.

Step (1) may include stirring or convective transport, but just below the interface there will always be an unstirred diffusion layer. In step (2) the molecules have to overcome the penetration barrier ($\pi\Delta A$), which increases with pressure. The rate-limiting step may be (1) or (2), depending on bulk concentration c_o and on the characteristics of the protein such as bulk diffusivity D, effective surface hydrophobicity and structural hardness. The former case implies local equilibrium between subsurface and interface, and is called Diffusion Limited Adsorption (DLA). The latter implies diffusive equilibrium in the subphase ($c(r) = cte$), and is called Penetration Limited Adsorption (PLA). Step (3) may be irreversible and distinguishes macromolecule from simple surfactant adsorption. Irreversibility implies that $\gamma(\Gamma)$ is not a single-valued state function, and that the states prior to the "final equilibrium" have to be studied in kinetic terms. These usually are carried out in constant A Langmuir troughs and yield dynamic surface tension ($\gamma(t)$) curves like the idealised one shown in the upper part of Fig. 1. Four phenomenological regions can be distinguished: the lag period (a) with no γ decrease, the rapid γ decrease region (b), the slow γ decrease meso-equilibrium region, and the final equilibrium (d). (a) is commonly observed only at low c_o and/or for slowly adsorbing surfactants. For globular proteins (a) might last from milliseconds to hours, and extends up to $\Gamma = 50\%$ monolayer coverage [4, 5, 7], as indicated in the lower part of Fig. 1. (b) extends from half to full surface coverage, and (c) may be related with the protein unfolding [7].

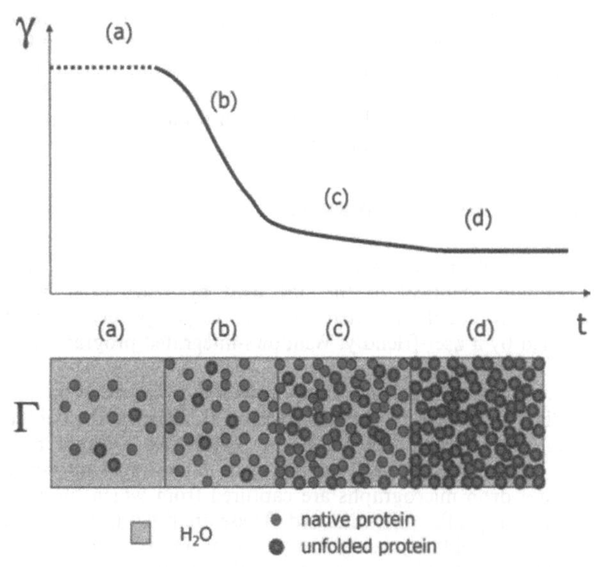

Fig. 1 Idealised diagram of $\gamma(t)$-surface coverage relationship.

However, experimental analysis of the relation between the model steps (1)–(3) and the phenomenological regions (a)–(d) is non-trivial [10, 11]. Therefore, it is interesting to provoke stationary states with a controlled π that correspond to certain (a)–(d) region, in order to relate the resulting surface dilatation with diffusion or the height of the energy barrier to adsorption/unfolding. With an adequate selection of the protein type, $\pi_{control}$ and c_o one can select the regime: low $\pi_{control}$ and low c_o implies DLA, high $\pi_{control}$ and high c_o PLA. Therefore it was necessary to develop and implement a pressure control mechanism to the methodology. With pressure-controlled Langmuir troughs only desorption can be studied at constant π, but due to the possible irreversibility of (3) (surface denaturation) the obtained information is different: the activity of the protein in its native state cannot be studied this way.

Materials and methods

Materials

Lyophilised, ess. salt-free bovine β-casein (90+% by electrophoresis) and lyophilised, crystallised, ess. salt-free equine myoglobin (90–95%), purchased from Sigma Chemical Co., were dissolved in phosphate buffered saline (PBS): 13.0 mM KH_2PO_4 (purissimum, Merck), 54.0 mM Na_2HPO_4 (p.a., Merck), 100 mM NaCl (p.a., Merck), pH 7.4, and used the same day. Millipore MILLI-Q+ water (5.54 μS) was used for buffer preparation and any other purpose. The chemicals were used without further purification. All experiments were performed at T = 23 °C. All solid surfaces in contact with the drop, the bulk phases or the spreading solution are made of glass, stainless steel, Teflon or are Teflon-coated. β-Casein is the main component of the milk protein, whose most characteristic property is to form by autoassembly micelles of always spherical shape but variable size [12]. It is a 24.1 kDa random-coil protein with $D_{bulk}(T = 20\ °C) = 6.05 \times 10^{-11}\ m^2/s$ [13], and is known to adsorb reversibly at the air-water interface in DLA regime [14]. Myoglobin, the protein that stores the oxygen in the muscle tissue, is an 18.8 kDa globular protein with low effective surface hydrophobicity [7], monomeric in solution, and candidate to adsorb in PLA regime at suitable $\pi_{control}$ and c_o.

Set-up and measurement procedures

In Fig. 2, a schematic representation of the set-up is shown, which is essentially identical to the one used by Wege et al. and is described in detail elsewhere [16]. It is completely computer controlled by a user-friendly, Windows-integrated program (Dinaten).

In our device a surfactant solution droplet is formed at the tip of a capillary by means of a micro-injector. (If desired, the capillary tip can be immersed into another – immiscible – liquid phase, permitting this way to perform studies at liquid-liquid interfaces.)

Digital drop micrographs are captured from which the experimental drop profiles are extracted. These are fitted to the Laplace equation of capillarity using Axisymmetric Drop Shape Analysis (ADSA), and yield as outputs γ the drop volume V and the interfacial area A. π is varied by changing V (and hence A) with the micro-injector.

The interfacial control is proportional-integral-derivative, composed of two independent fuzzy PD (proportional-derivative) and PI (proportional-integral) controllers, which commute their response depending on the actual state of the system: the derivative PD is used to obtain a rapid response when the signal is far from the reference value (decreasing the settle and rise time), and the integral PI is used to stabilise the system in the reference value once it is close to it (reducing the steady state error). The internal structure of the controller permits to control the interfacial pressure as well as the area, being possible to combine pressure and area control in one experimental protocol. Since the response of the measurement variables surface pressure and area depends on the relative change of the control variable drop volume rather than on the absolute one, the drop volume has been included as modulating input variable.

Once generated, the drop is kept at constant A, allowing the surfactant to adsorb freely until the desired π is attained. Normally this adsorption is from the drop subphase, being also possible to investigate adsorption from a surrounding liquid phase.

The adsorption kinetics at constant π are then studied monitoring the variation of A in time, i.e., determining the relative area dilatation necessary at each instant to compensate the pressure increase due to incorporation of new surfactant molecules into the film and – in the case of macromolecules – conformational rearrangement of the adsorbed molecules.

Results and discussion

The results of such an experiment (drop formation at $t = 0$) with β-casein 0.1 g/L are shown in Fig. 3, where surface tension (-■-), interfacial area (-●-), total drop volume (-▲-) and control response V_{out} (-▼-) are plotted as a function of time. As the protein adsorption leads to a surface tension decrease (and this would lead to an area increase at constant volume), the area controller must reduce the drop volume slightly to maintain A at $(25.99 \pm 0.02)\ mm^2$, while the pressure controller must increase drop volume to compensate the pressure increase, keeping π at $(20.04 \pm 0.1)\ mJ/m^2$. A set of experiments with β-casein solutions at different concentrations was performed, applying different lateral pressures and following the above protocol. Given that in this kind of experiments the interesting magnitude is the surface dilatation rate, i.e., the relative area change rate $\theta = dA/A\ dt = d(\ln A)/dt$ rather than the absolute one, in the figures Fig. 4 and Fig. 5, the natural logarithm of the area (measured in mm) was plotted instead of the area. This way the dilatation rate θ can be easily obtained by taking the derivative of the $\ln A(t)$ plot. And since the time necessary to reach the control pressure by adsorption at constant area depends on the magnitude of

Fig. 2 Scheme of the set-up

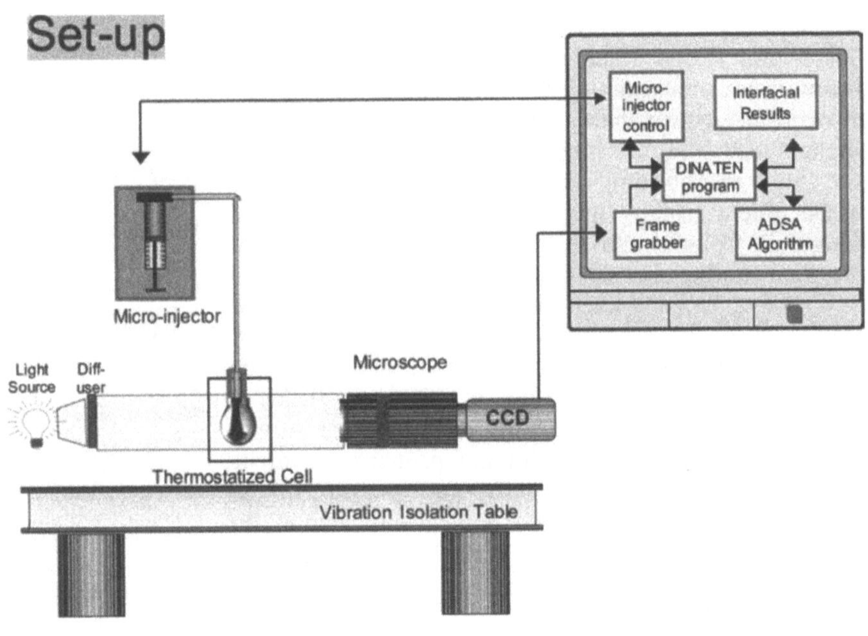

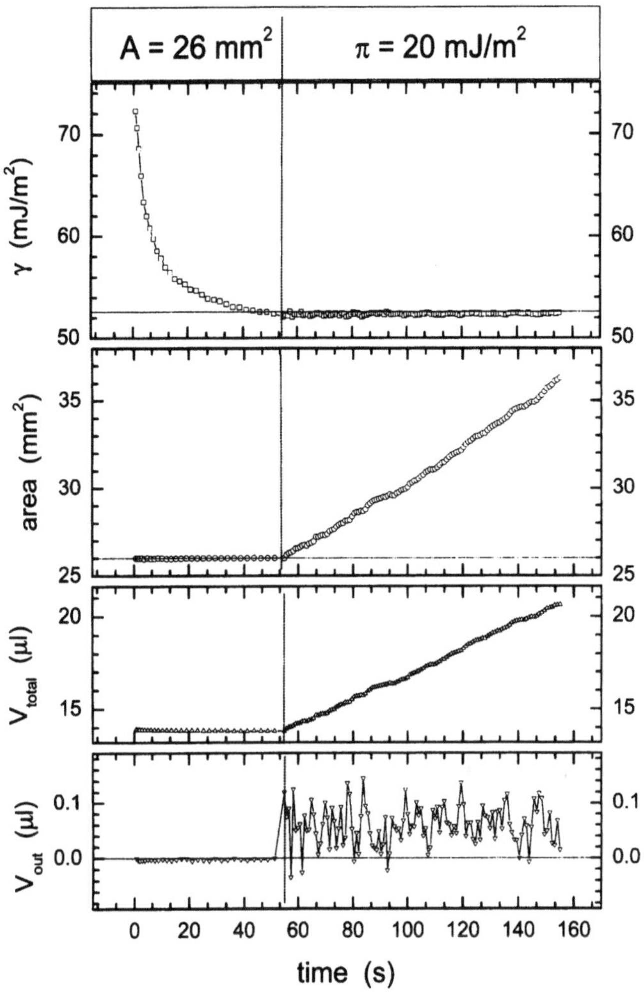

Fig. 3 Typical result adsorption kinetics measurement: a β-casein solution drop (0.1 mg/mL) is generated at $t=0$ and first maintained at constant area (26 mm^2) while the film pressure is less than 20 mJ/m^2, where pressure control sets on, maintaining π constant at this value. Surface tension (-■-), interfacial area (-●-), total drop volume (-▲-) and control response V_{out} (-▼-) are plotted as a function of time

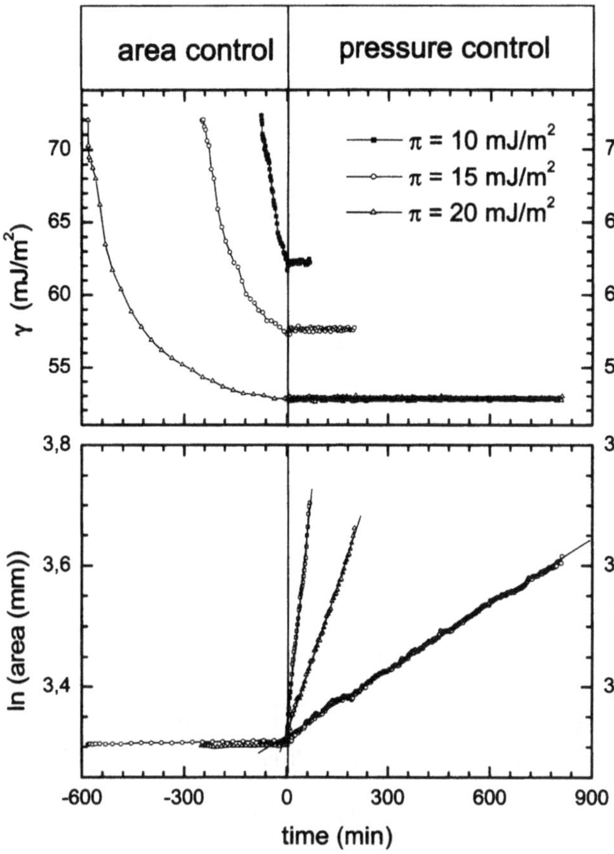

Fig. 4 Effect of the magnitude of the applied $\pi_{control}$ on dilatation for β-casein solutions with a fixed concentration of $[cas]=0.03$ g/L. γ and ln A are plotted as a function of time, with $\pi_{control}=10$ (-■-), 15 (-●-) and 20 (-▲-) mJ/m^2

$\pi_{control}$ and on the concentration, we shifted the time axis in a way that time origin is always at the onset of pressure control. In Fig. 4, the effect of the magnitude of the applied $\pi_{control}$ on dilatation is shown for a fixed concentration of $[cas]=0.03$ g/l. γ and ln A are plotted as a function of time, for $\pi_{control}=10$ (-■-), 15 (-●-) and 20 (-▲-) mJ/m^2. In Fig. 5, the effect of the bulk concentration on dilatation is shown a fixed pressure 15 mJ/m^2: $\gamma(t)$ and ln $A(t)$ curves are shown for $[cas]=0.1$ (-■-), 0.03 (-●-) and 0.01 (-▲-) g/L. In all cases pressure is maintained at $\pi_{control}$ $\pm$ 0.3 mJ/m^2.

The most important feature is the great linearity of the ln $A(t)$ plots: linear fits usually yield regression coefficients r 0.99. Thus, for a given concentration, protein adsorption at constant pressure shows up to occur at constant surface dilatation rates. This is consistent with theoretical considerations, and above all a very important experimental result: we measured dilatation rates over a wide range of pressures and protein concentrations, and except in the limit of bulk depletion (when a significant part of the total amount of surfactant concentrates at the interface, bulk concentration decreases) this was always the case, even at the lowest pressures we could measure. That means that we are able to provoke stationary states in different regimes and steps of protein adsorption, i.e. in the DLA as well as in the PLA regime, and for the penetration as well as for the rearrangement step, depending on c_o and $\pi_{control}$. Each of these stationary states is characterised by a directly measurable and constant dilatation rate. This fact allows direct representation of θ, obtained from the slopes of the linear fits of ln $A(t)$, as a function of bulk concentration and $\pi_{control}$, as shown for β-casein in Fig. 6 and for myoglobin in Fig. 7, where $\theta(\pi_{control})$ is plotted for $[protein]=0.1$ (-■-), 0.03 (-●-) and 0.01 (-▲-) g/L, respectively. The θ values are mean values from 3–10 experiments, depending on scatter. The overall measurable θ range spread over almost 4 decades from 1.5×10^{-5} s^{-1} (myoglobin) to 0.065 s^{-1}(β-casein), between the following limits: i) at low c_o and high π surface

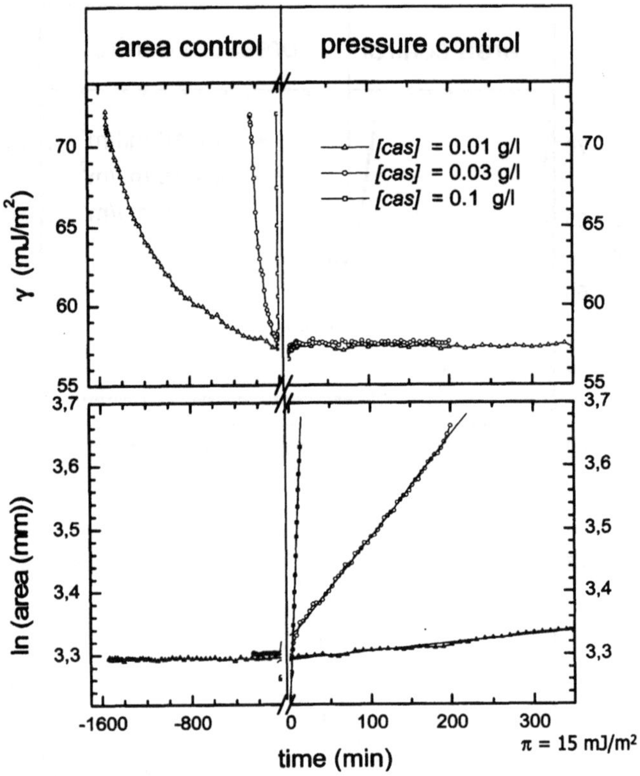

Fig. 5 Effect of the bulk concentration on dilatation for β-casein solutions with a fixed control pressure of $\pi_{\mathrm{control}} = 15$ mJ/m^2, where $\gamma(t)$ and ln $A(t)$ curves are shown for $[cas]=0.1$ (-■-), 0.03 (-●-) and 0.01 (-▲-) g/L

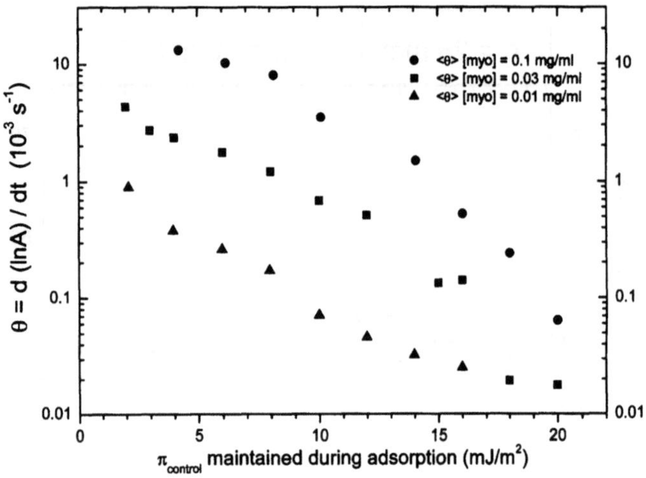

Fig. 6 Dilatation rates θ of β-casein solutions, obtained from the slopes of the linear fits of ln $A(t)$, as a function of π_{control}, for different concentrations: 0.1 (-■-), 0.03 (-●-) and 0.01 (-▲-) mg/mL

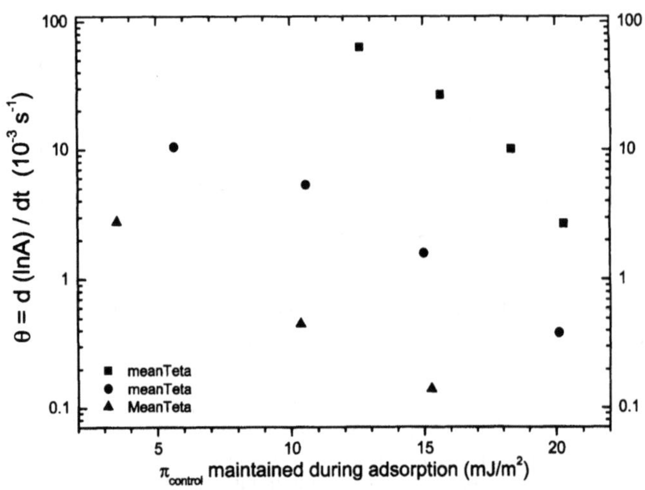

Fig. 7 Same as Fig. 6, for myoglobin

depletion occurs, and the large duration of the experiments favours scatter due to film ageing and contamination; ii) at high c_o and low π, large θ implies short drop lifetimes (detachment), and the pressure control gets in trouble. Note that for each protein, θ varies over almost 3 decades in the measured range of π. β-casein exhibited a significantly greater interface affinity than myoglobin: at the same c_o and π, θ_{cas} was between 5× greater (low π) and 30× greater (high π) than θ_{myo}.

Summary and conclusions

A fully automated, pendant-drop surface balance has been developed, allowing studies of surfactant adsorption kinetics at liquid-fluid interfaces at constant interfacial pressure or area. It offers a wide range of advantages over conventional Langmuir troughs: beside a more stringent control of the environmental conditions and therefore more uniform temperature, pressure and concentration along the interface, small amounts of material needed, it is non-invasive, i.e., it allows adsorption studies without the risk of uncontrolled adsorption onto the measuring device as the case with Wilhelmy plates. The adsorption kinetics at constant π is studied determining the relative area dilatation necessary at each instant to compensate the adsorption-induced pressure increase.

A study of the steady-state adsorption kinetics of β-casein and myoglobin at the air-water interface was performed: protein adsorption at constant pressure was found to occur at constant surface dilatation rates θ (Π, c_o), over a wide range of pressures and protein concentrations. Therefore, the pressure control enables to provoke stationary states at different regimes (DLA-PLA) and – in the case of proteins – steps of adsorption, each being characterised by a directly measurable constant dilatation rate. The overall measurable dilatation rate range spread over almost 4 decades, providing a direct measure of the rate of incorporation and/or unfolding of surfactant molecules.

Acknowledgements This work has been financially supported by MAT2001-2843-CO2-01.

References

1. Ward A, Tordai L (1946) J Chem Phys 14:453
2. Ward A, Tordai A (1952) Rec Trav Chim Pays-Bas 71:572
3. MacRitchi R, Alexander AE (1963) J Colloid Interface Sci 18:458, 464
4. MacRitchie F (1981) J Colloid Interface Sci 79:461
5. Graham DE, Phillips MC (1979) J Colloid Interface Sci 70:403, 415
6. Rotenberg Y, Boruvka L, Neumann AW (1983) J Colloid Interface Sci 93:169
7. Tripp BC, Magda JJ, Andrade JD (1995) J Colloid Interface Sci 173:16
8. Hunter JR, Kilpatrick PK, Carbonell RG (1990) J Colloid Interface Sci 137:462
9. Song KB, Damodaran S (1991) Langmuir 7:2737
10. Chang CH, Franses EI (1995) Colloids Surf A (100):1
11. Joos P, Rillaerts E (1982) J Colloid Interface Sci 88:1
12. BV. Gerritsen B (2001) Protein Spotlight 16 http://www.expasy.org/spotlight/articles/sptlt016.pdf
13. Payens TA, Brinkhuis JA, Van Markwijk BW (1969) Biochim Biophys Acta 175 :434
14. Guzman RZ, Carbonell RG, Kilpatrick PK (1986) J Colloid Interface Sci 114:536
15. Damodaran S, Song KB (1991) Interactions in food proteins. In: Parris N, Burdford R (eds), ACS symposium series, vol 454, chap 8. Am Chem Soc, Washington DC
16. Wege HA, Holgado-Terriza JA, Cabrerizo-Vílchez MA (2002) J Colloid Interface Sci 249:263

Progr Colloid Polym Sci (2004) 123: 194–198
DOI 10.1007/b11624

Y. Dziechciarek
J.J.G. van Soest
A.P. Philipse

Rheology of starch-based colloidal microgels

Y. Dziechciarek · J.J.G. van Soest (✉)
ATO B.V., P.O. Box 17,
6700 AA Wageningen, The Netherlands
e-mail: j.j.g.vansoest@ato.wag-ur.nl
Tel.: +31-317-475029
Fax: +31-317-475347

Y. Dziechciarek · A.P. Philipse
Van't Hoff Laboratory for Physical and Colloidal Chemistry, Debye Institute,
Padualaan 8, 3524 CH Utrecht,
The Netherlands

Abstract Steady state rheology experiments were performed on aqueous suspensions of starch-based particles in dilute and concentrated regimes to quantify the swelling degree of the particles and to determine the zero and infinite shear viscosities, respectively. Results showed that the starch-based particles behaved like microgels (with volumes swelling at least by factor 15), and like polyelectrolytes, since in de-ionised suspensions, the reduced viscosity decreased with increasing particle concentration. Analogous results are obtained for reference charged rigid silica spheres, which approach the hard sphere limit for increasing ionic strengths. Besides changes in shape of the microgels, the expansion of the electrical double layer upon decreasing the particle concentration plays an substantial role in the anomalous viscosity behaviour. The starch-based colloidal microgels exhibit shear-thinning behaviour in aqueous suspensions. In concentrated regimes, the dependence of the zero and infinite shear viscosities on the volume fraction appears to be comparable to that of reference microgels.

Keywords Starch-based colloidal microgels · Viscosity · Dilute regime · Concentrated regime

Introduction

For safety and environmental purposes, there is a strong desire to reduce the use of organic solvents in products like coatings, and to think of aqueous alternatives satisfying similar physical properties requested in the present product formulations (e.g., good flow, and film properties, such as gloss, flexibility). Due to their unique rheological and colloidal properties of suspensions, submicron particles may be used in coatings as rheology modifiers suspended in aqueous solution. In this perspective, aqueous suspensions of lattices based on cross-linked starch have been recently developed and roughly characterised [1, 2, 3]. Nevertheless, to make an actual wide-scale industrial use of these novel particles, more knowledge on the physical properties of the starch-based particles in aqueous suspensions, like the diffusion of the particles or the rheology behaviour of the suspensions, is essential. In this work, steady state rheology experiments were performed on suspensions of starch-based particles in dilute regime for varying ionic strengths, to determine the intrinsic viscosity, and to quantify the swelling degree. To get more insight into the collective behaviour of the starch-based particles, viscosity experiments were carried out on concentrated suspensions. The obtained results for the starch-based particles were subsequently compared with the rheological behaviour of well-characterised colloid systems. The reference colloidal systems consisted of charge stabilised rigid silica spheres [4], the rheological behaviour of which was also investigated in this work, and polymethylacrylate microgels [5].

Materials and methods

Materials

Paselli SA-2, an enzymatically modified potato starch, was supplied by Avebe (moisture content ~10% w/w). Analytical grade epichlorohydrin (ECH), trisodium trimetaphosphate (TSTP), and cyclohexane were supplied by Merck. Sorbitan monooleate (Span 80, Boom) was used as emulsifier and acetic acid (99%, Boom) as emulsion breaking agent. The charge stabilised rigid silica spheres (laboratory code DB373) [4] were de-ionised using mixed bed resins (AG sol-X (8) D) supplied by Biorad.

Methods

Starch-based microgels were synthesised by emulsion cross-linking of the starch chains. Two cross-linking agents were used: TSTP, where phosphate diester links are formed, and ECH, introducing glycerol-like links. The preparation procedures of the suspensions of the starch microgels and of the charged stabilised silica spheres have been extensively described elsewhere [3, 4]. Details on the calculation of the Debye length (κ^{-1}), on the determination of the particle size (a), on the particle density and the electrophoretic mobility were given in a previous paper [3]. The intrinsic viscosity of starch-based microgels and charged stabilised silica spheres was determined on a series of dilute suspensions of particles (pH = 5.5) at 20.0 °C by capillary viscosimetry (Ubbelhode Rheometer). The dilution series were performed at constant electrolyte concentration (c_s), where NaCl was used for starch-based microgels and $LiNO_3$ for charged stabilised silica spheres. To investigate any influence of the electrical double layer on the viscosity behaviour of the particles, the particle dilution series were labelled in terms of ratio of the particle radius and the Debye length (κa). Steady state rheological experiments on concentrated suspensions of starch-based particles, with pH = 5.5 and C ~ 200 μS/cm, were performed with a Stress Controlled Rheometrics rheometer at 20.00 °C, by using parallel plate geometry (40 mm diameter). The experiments were carried out at a shear rate ranging from 0.01 up to 1,500 s^{-1}. The flow curves were fitted according to the Carreau model in order to obtain the zero shear (η_0) and infinite shear (η_∞) viscosities [6]. For all the fits, the regression coefficient was larger than 0.9.

Results and discussion

Dilute suspensions

The radius (a), the zeta-potential, the density, the intrinsic viscosity and the swelling degree of starch-based microgels cross-linked either with TSTP or with ECH are presented in Table 1. The data for the reference charged stabilised silica spheres in ethanol are reported in Table 1, as well. The radius of the starch-based particles is in the sub-micrometer range, and the particles are negatively charged, irrespective of the type of cross-linker used. The density of starch-based particles is comparable with the density of the reference silica spheres.

The determination of the intrinsic viscosity of the starch-based particles and of the charged stabilised silica particles is presented in Fig. 1, where the reduced viscosity (η_{r-1}/c_p) is plotted versus the particle concentration (c_p). The Debye length was considered to be negligible compared to the radius of the colloids ($\kappa a > 15$). In Fig. 1, the increase of the reduced viscosity could be linearly fitted, with a regression coefficient larger than 0.95. By extrapolating the linear regression

Table 1 Particle radius (a) combined with zeta-potential, mass particle density (ρ_{part}), intrinsic viscosity ([η]), and swelling degree (V_s/V_0) of salt-free starch-based and silica particles at 20.00 °C

Sample	a (nm)	Zeta-potential (mV) (±2.5 mV)	ρ_{part} (g/ml)	[η] (ml/g) (κa > 15)	V_s/V_0 (= [η] × ρ/2.5)
ECH-Starch	128	−45	1.70	23.1	15.7
TSTP-Starch	304	−50	1.64	26.1	17.1
TPM-Silica	145	−40	1.60	1.6	1

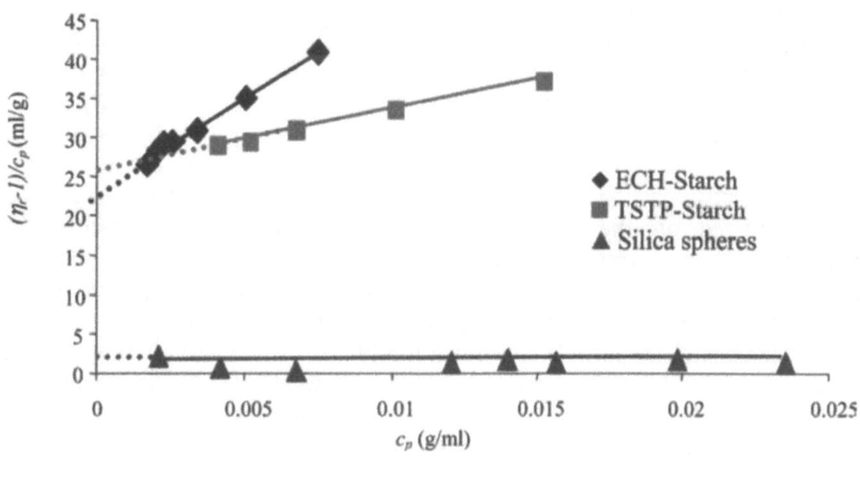

Fig. 1 Reduced viscosity of aqueous suspensions of υ ECH and ν TSTP-starch particles and of δ charge stabilised silica spheres in ethanol as a function of the particle concentration (c_p) for κa > 15, at 20 °C. The lines represent linear regressions of the data. The dotted lines are extrapolations to zero particle concentration

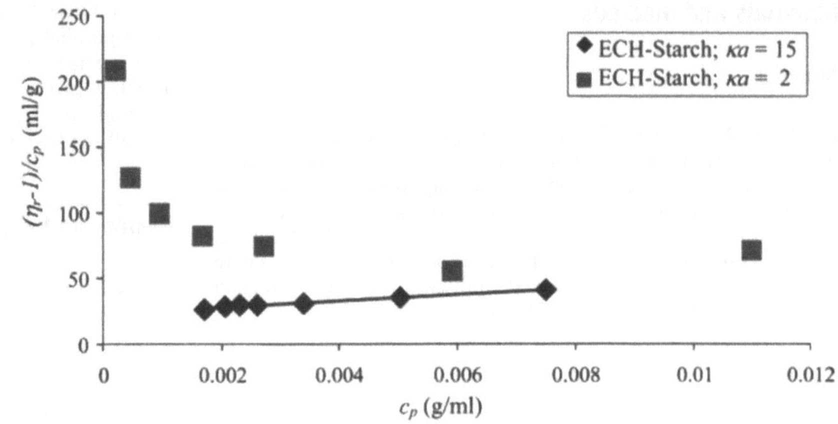

Fig. 2 Illustrative behaviour of the reduced viscosity of ECH-starch particles as a function of the particle concentration (c_p), for $\upsilon\kappa a = 15$ ($a = 128$ nm, $c_s = 1.25$ mM), and $\nu\kappa a = 2$ ($a = 128$ nm, $c_s = 0.02$ mM)

to the zero particle concentration, the intrinsic viscosity ([η]) could be determined. The values of the intrinsic viscosity found for the starch-based particles (cross-linked with ECH or TSTP), and reported in Table 1, were much larger than the value obtained for the charge stabilised silica particles. The swelling degrees (V_s/V_0) for the starch-based particles presented in Table 1, and expressed as $V_s/V_0 = [\eta] \cdot \rho/2.5$, were about 15 times larger than for the reference charge stabilised silica spheres, which behaved like hard spheres. The large swelling degree showed that the starch-based particles behaved like swollen microgels in aqueous suspension [5].

To get more insight into the influence of the salt concentration on the viscosity of aqueous suspensions of starch-based particles, the reduced viscosity of ECH-starch-based microgels in salt-free suspensions ($\kappa a \sim 2$) is plotted versus the particle concentration in Fig. 2. The results are compared with the values obtained for NaCl-aqueous suspensions of ECH-starch-based microgels with a $\kappa a \sim 15$. For $\kappa a \sim 2$, it appeared that the reduced viscosity was decreasing with increasing particle concentration. At sufficiently high salt concentration (i.e., high κa), this non-linear viscosity behaviour was not observed. The decrease of the reduced viscosity with increasing the particle concentration at low salt concentrations is a well-known phenomenon for charged polymers (or polyelectrolytes). It is usually attributed to changes in the shape of the polymer: in salt-free suspensions and for decreasing polymer concentrations, the structure of the polymer is expanding due to electrostatic repulsions, resulting in an increase of the reduced viscosity [7, 8]. Recently, similar observations were reported on the reduced viscosity of charged microgels with varying swelling degree [9]. The non-linear viscosity behaviour was attributed to the expansion of the electrical double layer (increase in κ^{-1}) upon decreasing the particle concentration [9].

To verify whether the expansion of the electrical double layer has any effect on the non-linear viscosity behaviour of charged microgels, extensive experiments on

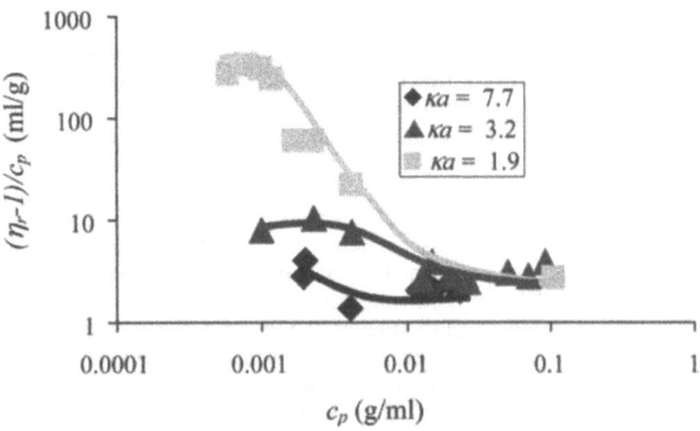

Fig. 3 Dependence of the reduced viscosity on the particle concentration (c_p) of charged stabilised rigid silica spheres dispersed in ethanol for $\upsilon\kappa a = 7.7$($a = 145$ nm, $c_s = 0.25$ mM), ▲ $\kappa a = 3.2$ ($a = 145$ nm, $c_s = 0.044$ mM), and $\nu\kappa a = 1.9$ ($a = 145$ nm, $c_s = 0.015$ mM). The lines guide the eye

suspensions of charge stabilised, rigid silica spheres dispersed in ethanol at low particle concentrations and for increasing ionic strengths were performed. The silica spheres are indeed monodisperse in size, rigid and well-characterised [4], and present convenient reference particles for the more soft starch-based microgels. The results are presented in Fig. 3, where the reduced viscosity of the silica particles is plotted versus the particle concentration and for increasing salt concentrations (increasing κa). At a low salt concentration ($\kappa a = 1.9$), the behaviour of the reduced viscosity was similar to the behaviour of the starch-based microgels. For series of suspensions with larger salt concentrations, the non-linear viscosity behaviour disappeared gradually and, at $\kappa a \sim 7.7$, became negligible. The rigidity of the silica spheres excludes any contribution of the particle swelling to the reduced viscosity. Hence, the ionic strength (including the salt concentration and the number of counter ions produced by the particle itself) will only affect the Debye length (κ^{-1}). An increase of the ionic strength will decrease the Debye length, resulting in a smaller particle radius, and consequently in a smaller reduced viscosity. The results show that, at low ionic strength, besides the expansion of charged microgels due to electrostatic repulsions, the expansion of the electrical double layer contributes strongly to the increase of the reduced viscosity on decreasing the particle concentration.

Concentrated suspensions

The rheology of suspensions of concentrated starch-based microgels was studied for particles cross-linked with ECH and with TSTP, respectively. Illustrative flow curves for both types of starch-based microgels are presented in Fig. 4. All the curves presented a shear thinning behaviour (decrease of the viscosity with increasing shear rate) and were fitted with the Carreau Model in order to determine the zero and infinite shear viscosities (η_0 and η_∞ respectively).

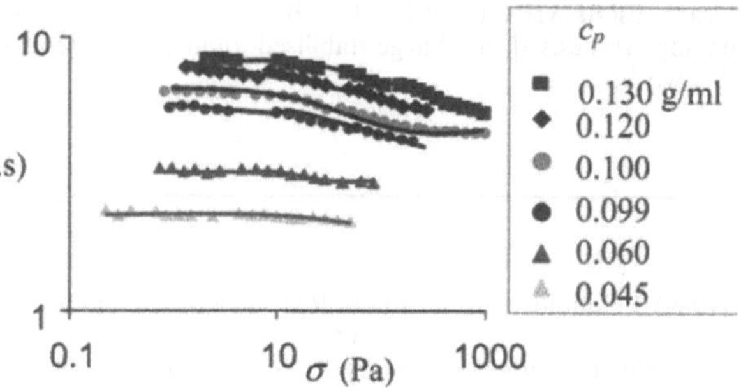

Fig. 4 Illustrative flow curve of starch-based microgels (in this case ECH-starch-based microgels) for varying particle concentration. The curves were fitted according to the Carreau model, with regression coefficients >0.9

The zero and infinite shear viscosities for TSTP and ECH-starch-based microgels are subsequently plotted as a function of the particle concentration in Fig. 5. The results were in accordance with the swelling degrees found for the particles in the viscosity experiments performed in dilute suspensions. The most swollen particles (TSTP-starch-based microgels) had larger viscosities than the less swollen particles (ECH-starch-based microgels).

To compare the obtained results with the viscosities of well-characterised lattices (5), the results are also given in function of the effective volume fraction (ϕ_{eff}), which is defined as $\phi_{eff} = c_p \cdot [\eta]/2.5$ [5] and are illustrated in Fig. 6. The values of the starch-based microgels were reasonably well in accordance with the results obtained for the reference polymethylacrylate microgels [5]. Deviations from the master curve for TSTP-microgels at high particle concentrations are likely due to compression of the microgels, as already observed in the literature [5]. So far, the viscosity of suspensions of starch-based microgels appeared to be comparable with the behaviour of suspensions of well-characterised lattices as polymethylacrylate microgels. Nevertheless, the relationship between the cross-linking efficiency and the swelling degree of the starch-microgels, and hence on the viscosity of the particles needs to be investigated further.

Conclusions

We conclude that the starch-based particles behave like microgels in aqueous suspensions, as the particles still

Fig. 5 Determination of η_0 and η_∞ of starch-based particles differing in $[\eta]$, and in cross-linker. All the experiments were performed in NaCl aqueous suspensions, where $\kappa a > 15$. The lines guide the eye

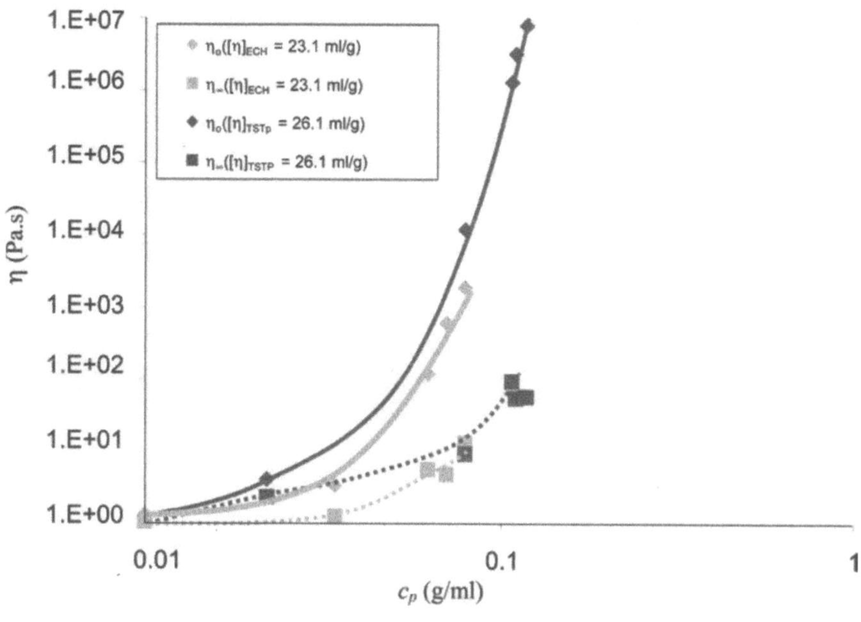

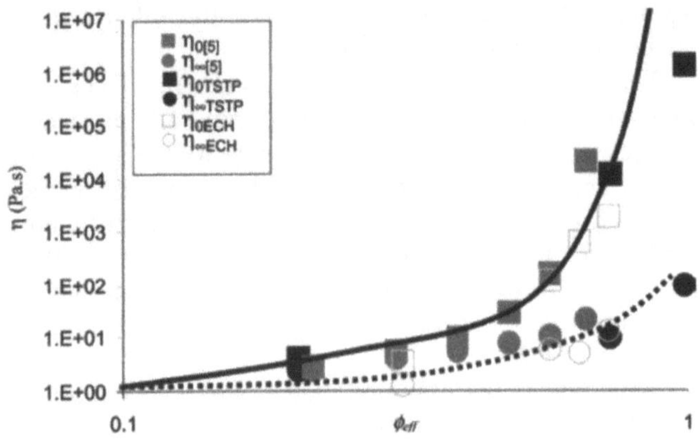

Fig. 6 Comparing plot of the η_0 and η_∞ versus the effective volume (ϕ_{eff}) of starch-based microgels and of reference polymethylmethacrylate microgels [5]. The lines guide the eye

swell in contrast with rigid spheres. The starch-based microgels behave like charged polymers in de-ionised aqueous suspensions, as the reduced viscosity is decreasing with increasing particle concentration. Experiments on charge stabilised, rigid silica spheres in de-ionised ethanol confirm that the increase of the reduced viscosity at low ionic strengths is due to, besides changes in the shape of the starch microgels, the expansion of the electrical double layer upon lowering the particle concentration.

The starch-based microgels in concentrated aqueous suspensions exhibit shear thinning with increasing shear rate. The viscosity behaviour of the starch-based colloidal microgels can easily be compared with the results obtained for petrochemical-based microgels. Future work will be focused on additional information like establishing the relationships between the cross-linking efficiency and the swelling degree of the starch microgels, and the dynamic rheological behaviour of the microgels in suspension. These data are still required for an appropriate use of the starch microgels in industrial applications, such as coatings and paints.

Acknowledgements This work is supported by the Marie Curie Grant ERBFAIRCT985014. Dr. B.J. de Gans is thanked for having synthesised the charge stabilised, rigid silica spheres used in this work.

References

1. Hamdi G, Ponchel G, Duchêne D (2001) J Microencapsulation 18(3):373
2. Soest JJG van, Schijndel RJG van, Gotlieb KFP, Stappers FJM (2000) PCT WO 40617
3. Dziechciarek Y, Soest JJG van, Philipse AP (2002) J Colloid Interface Sci 246:48
4. Philipse AP, Vrij A (1989) J Colloid Interface Sci 28:121
5. Wolfe MS, Scopazzi C (1989) J Colloid Interface Sci 133(1):265
6. Makosko CW (1994) Rheology, principles, measurements and applications. VCH, New York
7. Füoss RM, Strauss UP (1948) J Polym Sci 3:602
8. Flory PJ (1953) Principles of polymer science. Cornell University Press, Ithaca New York
9. Antonietti M, Briel A, Förster S (1996) J Chem Phys 105:7795

Progr Colloid Polym Sci (2004) 123: 199–202
DOI 10.1007/b11625

M. Zoumpanioti
E. Karavas
C. Skopelitis
H. Stamatis
A. Xenakis

Lecithin organogels as model carriers of pharmaceuticals

M. Zoumpanioti · A. Xenakis (✉)
Industrial Enzymology Unit, Institute of Biological Research & Biotechnology, National Hellenic Research Foundation, 48, Vas. Constantinou Ave., 11635, Athens, Greece
e-mail address: arisx@eie.gr
Tel.: +30210-7273762
Fax: +30210-7273758

H. Stamatis
Biological Applications & Technology Department, Univ. of Ioannina, Ioannina, Greece

E. Karavas · C. Skopelitis
Pharmathen Pharmaceuticals, Pallini, Greece

Abstract Lecithin microemulsion based organogels (MGs) formulated with cellulose derivatives such as hydroxypropylmethyl cellulose (HPMC) and hydroxypropyl cellulose (HPC) were studied as matrices for the transdermal transport of drugs. Lecithin, as well as the cellulose derivatives used, are pharmaceutical-grade compounds. The general properties of these MGs containing drugs such as their stability in various organic solvents and ability to incorporate sizeable amounts of model drugs were investigated. Drugs such as fluconazol and diclofenac diethylammonium were dissolved in the hydrophilic compound of the gel. The passage of the encapsulated drugs from the organogel through a membrane was studied with Franz diffusion cells and the amount of the drug was determined spectroscopically. It was found that large amounts of the drugs could be transported from the organogels through membranes to an acceptor solution depending on the nature of the organogel used and the solvent of the microemulsion. Our results show that MGs formulated with cellulose derivatives can be used as good matrices for transdermal transport of drugs.

Keywords Microemulsion · Polymer gel · Organogels · Transdermal drug delivery

Introduction

Lecithin microemulsions are obtained by adding the appropriate amount of water to a solution of lecithin and alcohol in a non-polar organic solvent [1–3]. These microemulsions are not very viscous and can be used in the preparation of polymeric organogels (MGs) consisting of biocompatible natural polymers and water or ethanol [4–6]. These gels are interesting because of their ability to host various guest molecules including drugs [7–9]. This ability makes the stable MGs suitable for pharmaceutical formulations.

Topical administration of therapeutic agents is currently regarded as an important alternative to the classical oral and intravenous administration [10]. Transdermal transport of drugs (i.e., the transport of pharmacologically active compounds through the skin into the blood vessels) offers many advantages over oral and injection delivery. The advantages of the transdermal route have been reported in literature [11, 12]. The absorption of topical drug products involves two consecutive processes: the release of drug from the topical preparation (formulation) and absorption into and through the skin [13]. The topical bio-availability of a drug therefore depends, at least in part, on the rate of release from its formulated product. Formulation of the drug as a gel rather than a solution facilitates drug handling. Also, lecithin and the natural polymers that have been used are all carriers compatible with the skin that interact with it and allow various molecules to pass into the skin [14]. As well as this, components used in the vehicle provide a way to control the transport rate.

Absorption studies are routinely conducted by in-vitro methods with a skin membrane mounted in a diffusion cell. A few investigations have been conducted with synthetic membranes instead of skin in diffusion cells for drug release rate determination. A suitable synthetic membrane must be without significant diffusion barrier effects on the transport into the receptor fluid [15].

The present work takes advantages of two basic properties of lecithin-based organogels, biocompatibility and ability to solubilise drugs, and explores the possibility of using these drug-containing, biocompatible lecithin-based gels as a matrix for transdermal transport. The drugs which have been used are fluconazol and diclofenac diethylammonium.

Experimental

Materials

Fluconazol and diclofenac diethylammonium were supplied by Pharmathen Pharmaceutical Ltd. Hydroxypropyl cellulose (HPC) was supplied by Aldrich Chemical Company, Inc. and hydroxypropylmethyl cellulose (HPMC) was supplied by Sigma. Lecithin, containing 40% L-α-phosphatidylcholine, was supplied by Sigma as well. Millipore purified water was used. The membrane used in Franz diffusion cells was the Visking dialysis tubing 36/32 and Serva supplied it.

Preparation of microemulsions and microemulsions based organogels (MG's).

The lecithin microemulsions were prepared as follows. 3.5% w/w lecithin was dissolved in isooctane. In that solution 5% v/v 1-propanol or ethanol was added. Afterwards was added the appropriate amount of water.

The drug-containing gels were prepared by introducing appropriate amounts of lecithin microemulsion to a second solution of polymer (HPC or HPMC) in water or ethanol containing the pharmaceutical substance. In a typical experiment, 0.5–2 ml of lecithin microemulsion was gelled with 0.5–1 g of HPMC or HPC and 2 ml of water or ethanol containing fluconazol or diclofenac diethylammonium at room temperature. The mixtures were vigorously shaken and stirred until homogeneous (about 5–10 min).

In-vitro drug transfer studies

The passage of the encapsulated drugs from the organogel through a membrane was studied with Franz diffusion cells [16, 17]. A physiological temperature (32 °C) was maintained by connecting the cell drive console to a constant-temperature recycling water bath. The cell orifice diameter was 10 mm and the cell volume was 5 mL. The receptor fluid was 0.05 M isotonic phosphate buffer pH 7.4 and ethanol (2:1 v/v) and was constantly stirred with a small magnetic stirring bar. Ethanol was used in the acceptor solution in order to increase the solubility of drugs.

The membrane was incubated for 24 h in the mixture of buffer and ethanol that was used as the reception fluid. About 1 g of gel was applied above the membrane using a spatula and air bubbles were eliminated from the formulation during the spreading process.

The amount of the drugs in the acceptor solution was determined by UV (HITACHI U-2000) measurements at 280 nm for fluconazol and 260 nm for diclofenac diethylammonium.

Results and discussion

Formation of gels containing pharmaceuticals

Two natural polymers such as the linear hydroxypropyl cellulose (HPC) and the branched hydroxypropylmethyl cellulose (HPMC) have been used for the formation of stable organogels. The aim of our work was to find gels that are compatible with the skin and suitable for transdermal delivery of pharmaceuticals. For this purpose, lecithin w/o microemulsions were gelled to form pseudo-solid gels after mixing with solutions of natural polymers in water or ethanol containing the drug. The drug-containing gel was applied above the synthetic membrane on Franz diffusion cells (Fig. 1).

It has been observed that the exact composition of the gels affects their smoothness. The HPC gels were smoother than the HPMC gels. The higher the ratio of the HPMC was, the harder the gel would be. Furthermore, the gels containing ethanol were smoother than those containing water. There have been made experiments using HPC, two different kinds of HPMC (3,500–5,600 cp and 100 cp) and also a mixture of HPC and HPMC in various weight fractions. All of them have also been used in gel formation with water and with ethanol and in some cases with a mixture of them. However, some of these gels were not smooth enough to be used as matrices for transdermal drug delivery. The most appropriate gels for that purpose were used further in Franz diffusion cells.

Studies of drug transfer yield

In these experiments, the transfer yield of the drug through a membrane was studied using the Franz diffusion cells.

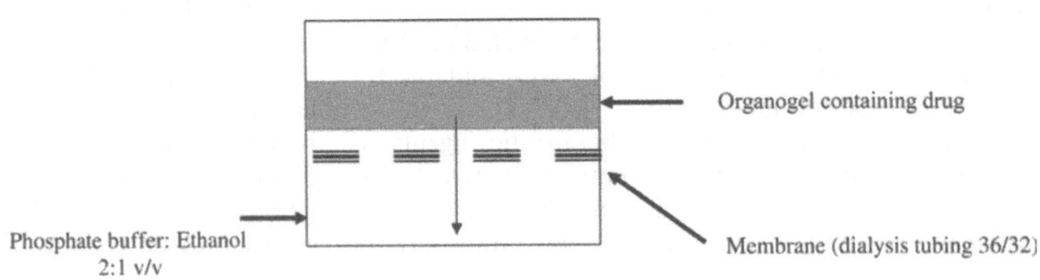

Fig. 1 Schematic representation of the Franz cell

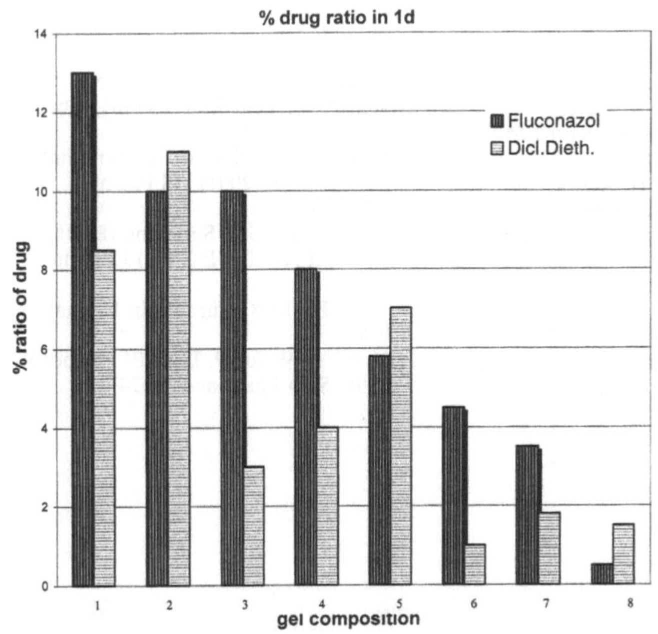

Fig. 2 Drug transfer through a model membrane in a Franz cell after 24 H. Gel compositions: 1) 1 g HPC/2 mL EtOH/no μE. 2) 1 g HPC/2 mL EtOH/0.5 mL 1-propanol in IPP μE. 3) 1 g HPC/2 mL EtOH/0.5 mL EtOH μE. 4) 1 g $HPMC_{100}$/2 mL EtOH/2.5 mL EtOH μE. 5) 1 g $HPMC_{3500}$/2 mL H_2O/1 mL EtOH μE/EtOH as a solvent. 6) 1 g $HPMC_{3500}$/1 mL H_2O/2 mL EtOH μE. 7) 0.75 g $HPMC_{3500}$/0.25 g HPC/2 mL EtOH/1.5 mL EtOH μE. 8) 0.75 g $HPMC_{100}$/0.25 g HPC/2 mL EtOH/1.5 mL EtOH μE

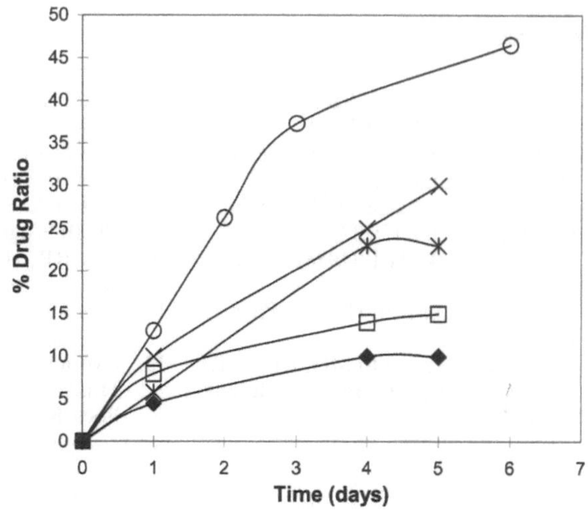

Fig. 3 Effect of gel composition on the transfer yield of fluconazol. Gel composition: ♦ $HPMC_{3500}$/EtOH and EtOH in isooctane μE, □ $HPMC_{100}$/EtOH and EtOH in isooctane μE, X HPC/H_2O and EtOH/and EtOH in isooctane μE, * HPC/EtOH and 1-propanol in IPP μE, ○ HPC/EtOH without μE

In Fig. 2, the transfer yield of the drugs in 24 h, is presented. As we can observe, the gels that have been formed using water appear to have a higher rate than those with ethanol. However, the amount of drug encapsulated in these gels is lower due to the low solubility of the drugs (1.8 mg/1 mL of water for the fluconazol and 3.5 mg/1 mL of water for the diclofenac diethylammonium). In the case of gels formed with ethanol, the amount of the encapsulated drug is higher due to their high solubility (50 mg of fluconazol/1 mL ethanol and 250 mg of diclofenac diethylammonium/1 mL ethanol). However, more days are needed in order to retrieve that amount.

In Fig. 3 is presented the transfer yield of the drugs through time for several gels that have been formed with ethanol. As we can observe, the gels of HPMC present a lower delivering rate independently on the kind of the HPMC. The HPC gels using the same microemulsion and the same solvent show a higher rate than the HPMC gels. A high rate we can also obtain using a different solvent in the microemulsion synthesis (iso-propyl palmitate instead of isooctane). It has also been observed that using the gel formation without the addition of the microemulsion we can get even higher yields. In that case, however, the amount of ethanol in the gel is very high. In addition the use of microemulsions could be more adequate to a gel for a controlled drug release.

Conclusions

The preparation of stable microemulsion-based organogels can be obtained. The organogels, which have been made using cellulose derivatives, can retain their stability for long periods of time and therefore they can be used for encapsulation and transdermal delivery of pharmaceuticals such as fluconazol and diclofenac diethylammonium. The drug amount that passes through the membrane and generally the kinetics of drug delivery via the membrane in a Franz diffusion cell depends on the gel composition.

Acknowledgements The Greek General Secretariat of Research and Technology and Pharmathen Ltd. financed this work, within the frame of the program ΠΑΒΕ 97-BE-93.

References

1. Shinoda K, Araki M, Sadaghiani D (1991) J Phys Chem 95:989
2. Kahlweit M, Busse G, Faulhaber B (1995) Langmuir 11:1576
3. Avramiotis S, Cazianis CT, Xenakis A (1999) Langmuir 15:2379
4. Quellet C, Eicke H-F (1986) Chimia 40:233
5. Haering G, Luisi PL (1986) J Phys Chem 90:5892
6. Rees GD, Nascimento MG, Jenta TRJ, Robinson BH (1991) Biochim Biophys Acta 1073:493
7. Xenakis A, Stamatis H (1999) Progr Colloid Polym Sci 112:132
8. Stamatis H, Xenakis A (1999) J Mol Catal B Enzymatic 6:399
9. Pastou A, Stamatis H, Xenakis A (2000) Progr Colloid Polym Sci 115:192
10. Guy RR, Hadgraft J (1985) J Pharm Sci 74:1016
11. Chien YW (1987) Drugs Pharm Sci 31:1–22
12. Kydonieus AF, Berner B (1987) In: Transdermal delivery of drugs. CRC, Boca Raton, FL, vol 1, pp 101–116
13. Shima K, Matsuka C, Hirose M, Noguchi T, Yamahira Y (1981) Chem Pharm Bull 29:2338
14. Protopappa E, Xenakis A, Avramiotis S, Sekeris CE (2001) US Patent 6,203,791
15. Wu ST, Shiu GK, Simmons JE, Bronaugh RL, Skelly JP (1992) J Pharm Sc 81:1153
16. Franz TJ (1978) Curr Probl Dermatol 7:58
17. Dreher F, Walde P, Luisi PL, Elsner P (1996) Skin Pharmacol 9:124

Progr Colloid Polym Sci (2004) 123: 203–209
DOI 10.1007/b11627

R. Rosmaninho
H. Visser
L. Melo

Influence of the surface tension components of stainless steel on fouling caused by calcium phosphate

R. Rosmaninho · H. Visser · L. Melo (✉)
University of Porto,
Faculty of Engineering,
Department of Chemical Engineering,
LEPAE, Rua Dr. Roberto Frias,
4200-465 Porto, Portugal
e-mail: lmelo@fe.up.pt
Tel.: +351-22-5081458
Fax: +351-22-5081449

Abstract In order to lengthen the processing time of dairy plants, research has been undertaken to study the deposition of calcium phosphate, one of the main components of milk fouling deposits. It is accepted that the surface tension of the solid material in contact with the liquid affects the fouling behaviour and, therefore samples with different surface tension characteristics were tested. These surfaces were non-modified stainless steel 316L food grade (2B and 2R finishes) and surfaces modified by ion implantation (SiF_3^+ and MoS_2), DLC sputtering and coating (Ni-P-PTFE and SiO_x). To understand the fouling process of calcium phosphate, its deposition on the different surfaces was followed by using a rotating disk apparatus and by characterising the deposits by contact angle measurements, image analysis and weighing. From these experiments it could be concluded that:

- In terms of the phenomenology of the build-up of calcium phosphate deposits, surface tension components do not seem to affect the sequence of steps involved in the deposition process. Three phases can easily be distinguished in the process: a) induction period, correspondent to the period of time passed until the start of particle formation in the bulk; b) adhesion and formation of the first layers of crystals and c) growth and compaction of the crystals.
- From a quantitative standpoint, the deposition process seems to depend on the γ^- component of the surface tension of the fouling support, which represents the electron donor part of the acid-base polar component of the solid surface tension. The initial deposition rate showed to be higher for lower values of the γ^- component while the amount of deposit obtained at the end of the deposition process increased for higher values of the γ^- component.

Keywords Fouling · Calcium

Introduction

In order to lengthen the processing time of dairy plants, research has been undertaken to study the fouling behaviour of milk onto different types of stainless steel, the standard material of construction in the dairy industry.

Calcium phosphate mechanism of fouling

Fouling of processing equipment used for heating dairy fluids at high temperatures is mainly caused by calcium phosphate and follows a four-steps mechanism [1, 2]:

1) Formation of calcium phosphate particles in the bulk solution upon heating. Calcium and phosphate ions that are normally present in ionic or sub-colloidal form, start forming a colloidal calcium phosphate complex. As postulated [3] the formation of these calcium phosphate complexes starts with the formation of amorphous calcium phosphate (ACP), which is subsequently transferred via octa-calcium phosphate (OCP) to hydroxyapatite (HAP). This complex formation is detected by an increase in solution turbidity, which seems to indicate that particle formation occurs in the bulk liquid.
2) Transport of these foulant particles to the heated surface.
3) Adhesion of the particles on the heated surface.
4) Deposition of further fouling material on top of the particles already adhered to the surface forming a calcium phosphate open network where small protein aggregates can be entrapped.

The controlling step is often the initial adhesion to the surface, which is dependent on the surface characteristics of the metal support, the surface characteristics of the particles and the physico-chemical interaction forces between them.

Physico-chemical interaction forces

The physico-chemical interaction forces that play a role in the adhesion of the calcium phosphate onto a metal heating surface are, according to van Oss [4], the Lifshitz-Van der Waals interaction forces (LW), the electrostatic double layer interaction forces (EL), the Lewis acid/base interaction forces (AB) and the Brownian motion (Br).

Lifshitz-Van der Waals interaction forces are apolar interaction forces that are established between apolar molecules at the surface of the particles. The free energy of these apolar interactions is directly related with the surface tension. The electrostatic double layer interaction forces have normally a repulsive character since they are related with the interactions between surface charge of the interacting species and in aqueous medium most of the surfaces acquire a negative charge by ion adsorption or ionisation of superficial groups. The Lewis acid/base interaction forces have a polar character and can be either attractive (hydrophobic interactions) or repulsive (hydration pressure). In aqueous medium, the polar interactions are mainly due to the interactions between electron acceptor species (Lewis acid) and electron donor species (Lewis base). The last type of interacting forces, the Brownian motion interactions, are repulsive forces related with the motion of suspended particles at temperatures higher than 0 K.

Material and methods

Calcium phosphate

Calcium phosphate deposition was performed using SMUF (Simulated Milk Ultra Filtrate), which is a water solution that simulates the mineral composition of milk. It was prepared according to Jenness and Koops [5], pH value adjusted to 6.8 and kept overnight at 5 °C before use.

Stainless steel samples

To study the influence of the surface tension of the fouling support, different surfaces with a wide range of surface tensions, were used. The different types of stainless steel samples were provided by a major European steel company and modified at Stuttgart University:

1) Non-modified stainless steel: 316 L samples with a 2R (bright annealed) and a 2B (pickling) finish. Samples provided were protected with a polymeric film.
2) Modified stainless steel (on 316 2R): four types of surface modification were performed on 316 L 2R samples: a) Ion implantation, SiF_3^+ ions with energy of 200 keV were implanted in the surface by ion bombardment with an implantation dose of 5×10^{16} ions/cm^2 in a depth of approximately 0.2 μm; b) DLC sputtering, plating of different layers with a "diamond like carbon" structure; c) Ni-P-PTFE coating, auto catalytic fixing of PTFE particles in a Ni-P matrix with a proportion of about 30% PTFE and d) SiO_x coating: by Plasma CVD (Plasma Chemical Vapour Deposition), 1 μm coating thickness.
3) Modified stainless steel (on 316 2B): four types of surface modification were performed on 316 L 2B samples: a) Ion implantation, SiF_3^+ ions with energy of 200 keV were implanted in the surface by ion bombardment with an implantation dose of 5×10^{16} ions/cm^2 in a depth of approximately 0.2 μm; b) Ion implantation, MoS_2 ions were implanted in the surface by ion bombardment; c) DLC sputtering, plating of different layers with a "diamond like carbon" structure and d) SiO_x coating, by Plasma CVD (Plasma Chemical Vapour Deposition), 1 μm coating thickness.

Experimental equipment – rotating disk apparatus

A rotating disk apparatus is used for studying the deposition of calcium phosphate and whey protein particles onto a steel surface under controlled hydrodynamics (Reynolds number between 4.9×10^3 and 3.6×10^4). For this purpose a stainless steel disk is attached to the copper bottom plate of a hollow heated cone containing silicone oil. The cone can freely rotate in a solution contained in a cylindrical thermostatted vessel whose temperature can be adjusted.

When operating in the laminar flow region (Reynolds number smaller than 5×10^4), the rate of deposition of very small particles from a colloidal dispersion onto the surface of the rotating disk is controlled by molecular diffusion. As the thickness of the diffusion boundary layer is constant over the entire surface of the rotating disk, regardless of the distance from the axis of rotation the conditions for mass transport to any point on the surface of the disk are identical and hence a homogeneous deposit can be obtained [6].

Experimental conditions

SMUF solutions were introduced in the thermostatted vessel of the rotating disk apparatus and its temperature was adjusted to 44 °C (temperature at which solution starts to became turbid indicating calcium phosphate precipitation).

24 h after cleaning, the different stainless steel samples were attached to the heating cone using double-sided adhesive tape and then introduced in the fouling solution. At that moment the cone is started to rotate with a rotational speed of 150 rpm (laminar regime). It is assumed that there is no temperature driving force between the surface of the stainless steel samples and the bulk once the solution is perfectly mixed. The cone with the attached sample was kept rotating for a given period of time after which it was lifted from the solution while rotating and the sample was detached from the cone, fast dried with acetone and the amount of deposit obtained was weighed. After weighing, samples were reintroduced in solution, kept rotating for another certain period of time and removed again.

Cleaning procedures

Before each deposition experiment and the determination of the surface tension of each sample, all stainless steel samples were cleaned with a commercial detergent (RBS35 from RBS Chemical Products) according to the following procedure: 1) samples were immersed in a 2.0% w/v detergent (RBS35) solution in distilled water at 65 °C for 5 min; 2) rinsed with distilled water at 65 °C for 5 min and 3) rinsed with distilled water at 20 °C.

Contact angle measurements

Contact angle values were measured by the sessile drop method in a contact angle meter (KRÜSS-GmbH, Hamburg) using water, formamide and α-bromonaphthalene (α-BR) as reference liquids. The values for the contact angles were measured automatically using an image analysing system (G2/G40). Measurements were performed 24 h after cleaning.

Surface tension determination

Through contact angle measurements and using the approach of van Oss et al. [4] it is possible to evaluate the surface characteristics of a solid in terms of surface tension (γ_s). This approach considers the total surface tension of a solid or a liquid (γ^{TOT}) as the sum of an apolar Lifshitz-van der Waals component (γ^{LW}) and an acid-base polar component (γ^{AB}):

$$\gamma^{TOT} = \gamma^{LW} + \gamma^{AB} \tag{1}$$

Lifshitz-van der Waals interactions arise due to three distinct interactions, induction (Debye), orientation (Keesom) and dispersion (London) the last one being the most significant term [7].

The acid-base forces are always asymmetric since they comprise the electron donating as well as the electron accepting properties of a surface. Thus the acid base component (γ^{AB}) consists of two non-additive parameters, one for the electron donor (γ^-) and one for the electron acceptor (γ^+) contribution [7].

$$\gamma^{AB} = 2\sqrt{\gamma^+\gamma^-} \tag{2}$$

The total interfacial tension between phases i and j can be expressed as

$$\gamma_{ij}^{TOT} = \gamma_i^{LW} + \gamma_j^{LW} - 2\sqrt{\gamma_i^{LW}\gamma_j^{LW}} + 2\left(\sqrt{\gamma_i^+\gamma_i^-} + \sqrt{\gamma_j^+\gamma_j^-} - \sqrt{\gamma_i^+\gamma_j^-} - \sqrt{\gamma_i^-\gamma_j^+}\right) \tag{3}$$

Contact angle values (θ) can be related to the total interfacial tension using simultaneously three forms of the following equation resulting from Young's equation, one for each reference liquid used

$$(1 + \cos\Theta)\gamma_l^{TOT} = 2\left(\sqrt{\gamma_s^{LW}\gamma_l^{LW}} + \sqrt{\gamma_s^+\gamma_l^-} + \sqrt{\gamma_s^-\gamma_l^+}\right) \tag{4}$$

where subscripts s and l mean solid and liquid, respectively.

Weighing

The amount of deposit formed on each sample was determined by weighing the samples after being removed from solution using an analytical balance (AND GR-200).

Results

Qualitative analysis

Surface characterisation by contact angle analysis. Contact angle measurements were performed on every new stainless steel sample after cleaning as well as on the same samples after being used as fouling supports and cleaned again. For all the samples tested different values were found for the contact angles of unused and used samples. Apparently, after being used for deposition studies, the samples acquire different surface characteristics probably due to ineffective cleaning that can be related to the irregularities of the surface where residual components are being entrapped. This evolution in the contact angle values measured for calcium phosphate deposition is illustrated in Table 1.

Surface tension. The total surface tension and its components for each sample tested were calculated using equation [4] from the individual contact angle data measured (Table 2). Comparing the five different types of steel cleaned with detergent, the following conclusions can be drawn:

- Except for the Ni-P-PTFE surface modified sample as expected, the apolar component (γ^{LW}) has approximately the same value for all samples;
- The same holds for the electron acceptor component (γ^+), which is in all cases close to zero;
- The electron donor component (γ^-) is the one that differentiates the different types of finishes for both non-modified and modified stainless steels.

Superficial morphology. "Microwatcher" from Mitsubishi Kasei Corporation, Japan, is an image acquiring instrument based on dark field illumination that allows immediate surface observations in the ambient at an amplification of up to 1000 times and was used to observe the surface morphology of the different types of samples. All the modified samples analysed showed the same surface morphology as the correspondent non modified samples, except for the Ni-P-PTFE coating which is thicker and so is able to mask all the heterogeneities of the support. This means that the differences in contact angles between modified and non-modified samples are

Table 1 Evolution in the contact angle values measured for different stainless steel samples after calcium phosphate deposition and cleaning (standard deviation in brackets for a number of essays between 6 and 10)

	Water [°]	Formamide [°]	α-Bromonaphthalene [°]
Non-Used			
2B	83 (σ 1)	74 (σ 1)	16 (σ 2)
SiO_x (2B)	15 (σ 3)	12 (σ 1)	14 (σ 2)
Ion (2B)	49 (σ 4)	39 (σ 2)	21 (σ 4)
DLC (2B)	59 (σ 4)	33 (σ 5)	12 (σ 1)
MoS_2 (2B)	64 (σ 3)	50 (σ 7)	18 (σ 2)
After Calcium Deposition			
2B	71 (σ 5)	62 (σ 7)	21 (σ 0)
SiO_x (2B)	21 (σ 1)	13 (σ 3)	10 (σ 2)
Ion (2B)	35 (σ 26)	25 (σ 17)	14 (σ 8)
DLC (2B)	36 (σ 7)	16 (σ 2)	9 (σ 0)
MoS_2 (2B)	22 (σ 0)	11 (σ 0)	11 (σ 0)
Non-Used			
2R	25 (σ 5)	21 (σ 5)	14 (σ 4)
SiO_x (2R)	37 (σ 3)	25 (σ 8)	22 (σ 3)
Ion (2R)	31 (σ 8)	25 (σ 6)	14 (σ 6)
DLC (2R)	67 (σ 5)	50 (σ 3)	9 (σ 3)
Ni-P-PTFE (2R)	114 (σ 1)	92 (σ 2)	61 (σ 1)
After Calcium Deposition			
2R	71 (σ 5)	62 (σ 7)	21 (σ 0)
SiO_x (2R)	24 (σ 0)	15 (σ 4)	19 (σ 1)
Ion (2R)	9 (σ 1)	9 (σ 3)	7 (σ 1)
DLC (2R)	65 (σ 3)	53 (σ 7)	6 (σ 1)
Ni-P-PTFE (2R)	106 (σ 1)	88 (σ 2)	70 (σ 1)

Table 2 Evolution in the contact angle values measured for different stainless steel samples after calcium phosphate deposition and cleaning (standard deviation in brackets for a number of essays between 6 and 10)

	γ_s^{lw}(mJ/m^2)	γ_s^+ (mJ/m^2)	γ_s^-(mJ/m^2)	γ_s(mJ/m^2)
Non-Used				
2B	43 (σ 0.5)	0,0 (σ 0.0)	7 (σ 1.1)	43 (σ 0.5)
SiO_x (2B)	43 (σ 0.3)	0,8 (σ 0.1)	52 (σ 1.8)	56 (σ 0.1)
Ion (2B)	42 (σ 1.2)	0,3 (σ 0.1)	30 (σ 4.2)	47 (σ 0.9)
DLC (2B)	43 (σ 0.2)	1,1 (σ 0.5)	14 (σ 3.7)	51 (σ 2.1)
MoS_2 (2B)	42 (σ 0.6)	0,1 (σ 0.2)	18 (σ 2.2)	44 (σ 2.7)
After Calcium Deposition				
2B	42 (σ 0.0)	0,0 (σ 0.0)	18 (σ 1.9)	42 (σ 0.0)
SiO_x (2B)	44 (σ 0.2)	0,7 (σ 0.1)	48 (σ 1.8)	56 (σ 0.9)
Ion (2B)	43 (σ 1.4)	0,6 (σ 0.1)	37 (σ 19.9)	52 (σ 4.8)
DLC (2B)	44 (σ 0.0)	1,2 (σ 0.2)	34 (σ 6.6)	56 (σ 0.3)
MoS_2 (2B)	44 (σ 0.0)	0,9 (σ 0.0)	47 (σ 0.3)	56 (σ 0.0)
Non-Used				
2R	43 (σ 0.9)	0,6 (σ 0.2)	48 (σ 5.4)	53 (σ 2.2)
Ion (2R)	43 (σ 1.0)	0,4 (σ 0.2)	41 (σ 10.4)	51 (σ 3.5)
DLC (2R)	44 (σ 0.4)	0,0 (σ 0.0)	13 (σ 5.1)	45 (σ 1.1)
Ni-P-PTFE (2R)	25 (σ 0.4)	0,0 (σ 0.0)	0 (σ 0.0)	25 (σ 0.4)
SiO_x (2R)	41 (σ 1.0)	0,9 (σ 0.5)	37 (σ 1.7)	52 (σ 3.9)
After Calcium Deposition				
2R	42 (σ 0.9)	0.4 (σ 0.2)	33 (σ 5.4)	50 (σ 2.2)
SiO_x (2R)	42 (σ 0.1)	1.0 (σ 0.2)	46 (σ 0.9)	55 (σ 1.1)
Ion (2R)	42 (σ 1.0)	0.6 (σ 0.2)	42 (σ 10.2)	52 (σ 3.5)
DLC (2R)	43 (σ 0.4)	0.5 (σ 0.0)	17 (σ 5.1)	47 (σ 1.1)
Ni-P-PTFE (2R)	20 (σ 0.4)	0.0 (σ 0.0)	1 (σ 0.0)	20 (σ 0.4)

not due to morphology or roughness but are related to the sample's surface composition.

Calcium phosphate deposition sequence

For all the samples indicated, the various steps in fouling of calcium phosphate (SMUF solution) were followed in time using the Microwatcher optical microscope. The same sequence and timing was observed for all samples. These sequential steps can be summarised as follows:

1) Formation of a homogeneous layer on practically the whole sample. This phase became eyesight visible after about 30 min of rotation of the sample in the heated SMUF solution.
2) Formation of small crystal-like structures spread over the entire surface in a random way. This second phase could be noticed after about 60 min of rotation
3) Crystal like structures grown together forming a compact structure which tends to cover the entire sample surface in a more or less homogeneous way. This happened after about 120 min.

For the 316L 2R samples and its deposition, the various schematised steps are visualised in Figs. 1a–d.

Deposit characterisation by contact angle analysis. Calcium phosphate deposits were prepared on all samples mentioned and their surface characteristics were determined using contact angle measurements. It was found that, independent of the surface used as support, the deposit formed had roughly the same surface tension because similar values for the contact angle were found, as shown in Fig. 2.

Deposit characterisation by X-ray microanalysis. To determine the deposit's composition on each one of the stages previously presented, the various deposits obtained were analysed using X-ray microanalysis. For every stage and for every sample used as support, the deposit was found to contain mainly calcium, phosphorus and oxygen.

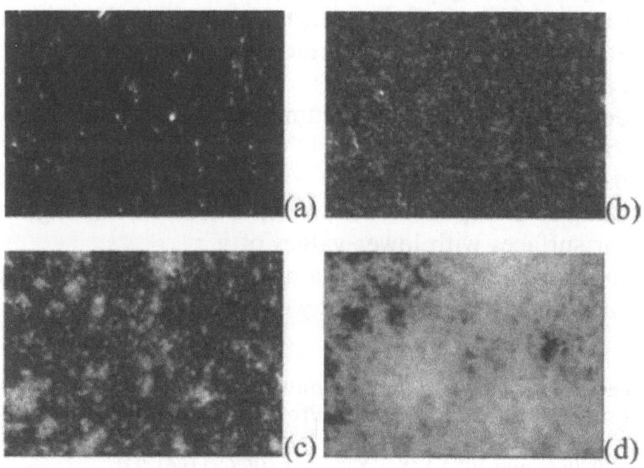

Fig. 1 Calcium phosphate deposition process (1000 ×): (**a**) clean sample; (**b**) homogeneous layer; (**c**) first crystal-like structures; (**d**) compact crystal like structure.

Quantitative analysis

The type of deposit was the same for all the surfaces, which makes a quantitative comparison possible and allows the determination of the influence of the γ^- component of the surface tension of the support on the amount of deposit formed. It should be emphasised that the values considered for the surface tension of the different stainless steels are the ones obtained for used samples since they acquire new surface characteristics after being used.

Crystal deposition curve. For every sample, the amount of calcium phosphate deposited was weighed after regular periods of time. In all cases, the deposition curve displayed a logarithmic behaviour similar to the one presented in Fig. 3 for a 2R non-modified sample. Three different phases can be distinguished in the curve, which are: a) the induction period of 20 minutes, correspondent

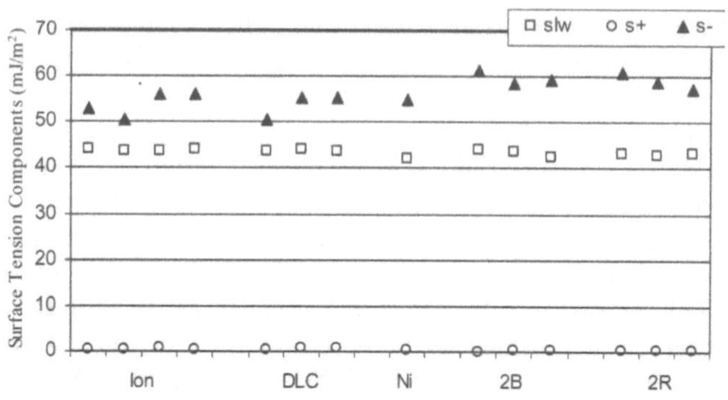

Fig. 2 Surface tension characterisation of calcium phosphate deposits onto different types of stainless steel.

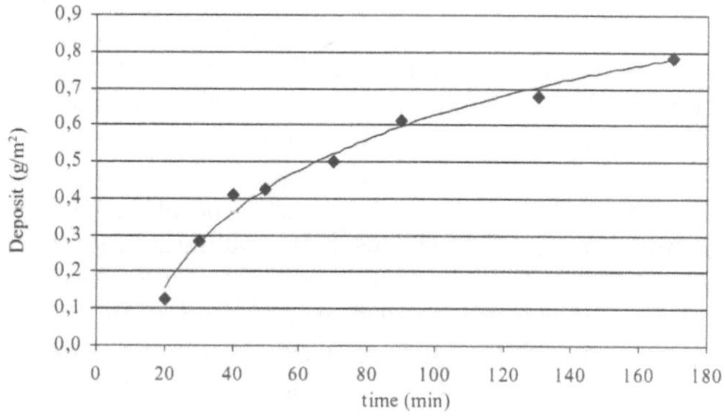

Fig. 3 Calcium phosphate crystal deposition curve of a 2R non-modified sample.

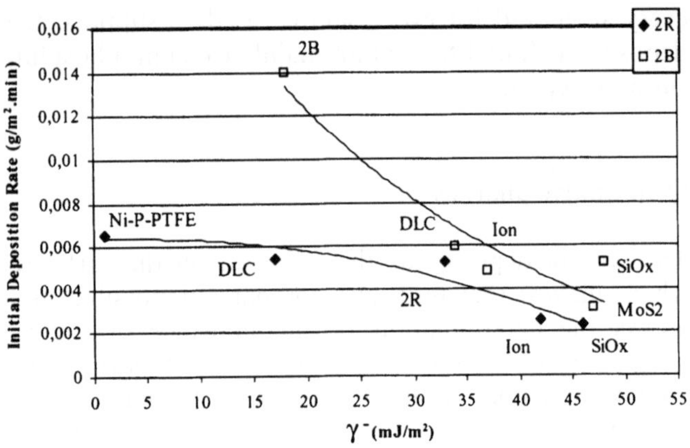

Fig. 4 Initial deposition rate dependence on the γ^- component of the surface tension of the support.

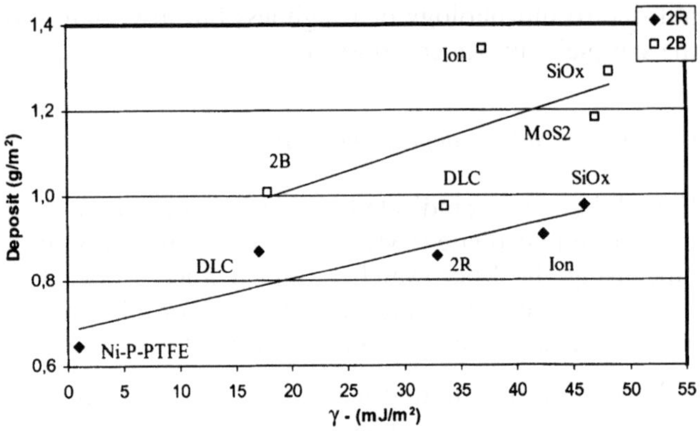

Fig. 5 Calcium phosphate deposition (during 120 min) as a function of γ^-.

to the period of time passed until the start of particle formation in the bulk, after which the deposit started to be weighable; b) the adhesion and formation of the first layers of crystals and c) the growth and compaction of the crystals. These three phases are in accordance with the steps defined by visual observation (see Calcium Phosphate Deposition Sequence).

Initial deposition rate. The particles formed in the bulk are transported to the surface and adhere on it depending on their affinity for the surface. Being so, the affinity of the surface towards calcium phosphate may be measured by the initial evolution of the deposit weighed, the initial deposition rate (IDR). IDR was considered to be the slope of the curve in the second phase previously mentioned.

For all the samples, the initial deposition rate was shown to be dependent on the type of support used. Taking into account that the main characterising parameter of the surface is the γ^- component of the surface tension and that no other obstacle to deposition was found, the relation between the initial deposition rate and this parameter was searched. The initial deposition rate was then plotted against the correspondent γ^- component (Fig. 4) and for both types of stainless steel, it tended to decrease with the increase of the γ^- component. This means that, initially, the deposit is formed much faster for lower values of γ^- than for the higher values. In the case of the 2R samples, the IDR decreased three fold between the highest and the lowest γ^- value while for the 2B samples the decrease was of about fivefold.

Total amount of deposit. In order to know the final amount of deposit formed on each type of surface, all samples were forced to be inside the fouling solution for a fixed period of time and weighed afterwards. Also the amount of deposit formed was related with the γ^- component of the surface used as fouling support (Fig. 5). Once more it was possible to find a dependency on the γ^- component of the surface tension of the support and for both types of stainless steel the amount of deposit obtained increased with the increase of γ^-.

Based on these two last observations it is possible to assume that for lower values of γ^- the deposit is formed much faster but with a more lose structure more sensitive to shear forces, that more easily remove the outer parts of the deposit formed resulting in a smaller amount of deposition at the end of the experiment. On the other side for the higher values of γ^-, the deposit is being formed much slower but with a more consistent structure explaining the fact that at the end of the experiment the amount of material deposit is higher.

Conclusion

The following main conclusions can be drawn from the present work:

1) surface tension of the samples used as fouling supports changes after the deposition experiments;
2) sequential steps of the calcium phosphate deposition are independent of the surface tension of the support used;
3) deposition process of calcium phosphate depended on the surface characteristics of the fouling support namely its γ^- component;
4) calcium phosphate's initial deposition rate is higher for surfaces with lower values of γ^-;
5) the amount of deposit obtained is higher for supports with higher values for the γ^- components.

Acknowledgements The authors gratefully acknowledge the financial support of the MODSTEEL Project (The European Commission, DG Research, Growth Programme, Contract No. G5RD-CT-1999-00066) and PRAXIS XXI grant (BCC/11961/97).

References

1. Visser J (1999) Reducing fouling of heat exchangers in the dairy industry by process optimization. In: Wilson DI, Fryer PJ, Hasting APM (eds), Proceedings of fouling and cleaning in food processing '98. publ. EU
2. Jeurnink ThJM, Walstra P, de Kruif CG (1996) Netherlands Milk Dairy J 50:407–426
3. Visser J (1997) In: Visser HJ (ed), Fouling of heat treatment equipment. Bulletin of the IDF, 328. IDF, Brussels
4. van Oss CJ (1994) Interfacial forces in aqueous media. Marcel Dekker, New York
5. Jenness R, Koops J (1962) Netherlands Milk Dairy J 16:152–164
6. Hull M, Kitchener JA (1969) Trans. Faraday Soc 65:3093
7. Wu W, Giese RF Jr, van Oss CJ (1994), Linkage between ζ-potential and electron donicity of charged polar surfaces. Colloids Surf A 84:241–252

Progr Colloid Polym Sci (2004) 123: 210–216
DOI 10.1007/b11628

L. Pérez
M.R. Infante
M. Angelet
P. Clapés
A. Pinazo

Glycerolipid arginine-based surfactants: synthesis and surface active properties

L. Pérez · M.R. Infante · M. Angelet
P. Clapés · A. Pinazo (✉)
Institut d' Investigacions Químiques
i Ambientals de Barcelona,
CSIC Jordi Girona 18-26,
08034 Barcelona, Spain
e-mail: apgste@cid.csic.es
Tel.: +34-3-4006164
Fax: +34-3-2045904

Abstract The research on biomimetic surfactants is currently increasing given the biocompatibility of the natural systems. Amino acid-glycerolipid conjugates combine in one molecule the chemical, physicochemical and biological properties of the diacyl glycerides and those of the polar amino acid components. A novel class of environmentally friendly surfactants have been synthesised and their surface active properties in aqueous solution evaluated. Symmetrical 1,2-diacylgycerol-3-arginine ester homologues with 8, 10, 12, and 14 carbon atoms in the alkyl chains were prepared by chemical methodologies at lab scale. All final compounds were purified up to 99% by HPLC and characterised by standard spectroscopic procedures. Basic physicochemical properties such as adsorption and self-aggregation in water/surfactant binary systems are reported.

Keywords Amino acid · Glycerolipid · Surfactant · Biocide

Introduction

In the last decades novel high-quality components of consumer friendly market products require surfactants with multifunctional capabilities such as biodegradability, mildness, low potential toxicity, water solubility, wide hydrophilic-lipophilic balance (HLB), formation of lamellar phases and vesicles for using in drug-delivery systems and antimicrobial activity.

Environmentally friendly surfactants are surface active substances with hydrophilic and hydrophobic building blocks taken from natural raw materials. Given their natural and simple structure they show low toxicity and quick biodegradation. These properties have resulted in a fast growth of the commercial interest on these surfactants for the application in consumer friendly market products [1, 2].

An approach to developing new environmentally friendly surfactants with such properties can be accomplished by combining in one molecule plant glycerolipids with natural occurring amino acids or sugars. They would combine in one molecule, the chemical, physicochemical and biological properties of the saturated diacyl glycerides and those of the polar components derivatives [3–6].

In this paper a novel family of natural cationic surfactants, 1-*O*-(L-arginyl)-2,3-*O*-di-acyl-*sn*-glycerol dichlorohydrate, XXR (Fig. 1), is presented. Structurally, the new surfactants are homologues of lecithines with the same type of chemical bonds. The new surfactants will combine in one molecule, the chemical, physicochemical and biological properties of the saturated diacyl glycerides as well as those of the amino acid arginine polar group. The chemical synthesis, preliminary physicochemical characterisation and antimicrobial activity are described.

Experimental

Synthesis

All solvents were reagent grade and were used without further purification. Anhydrous glycerol and the acyl chlorides with 8, 10,12 and 14 carbon atoms were from Fluka.

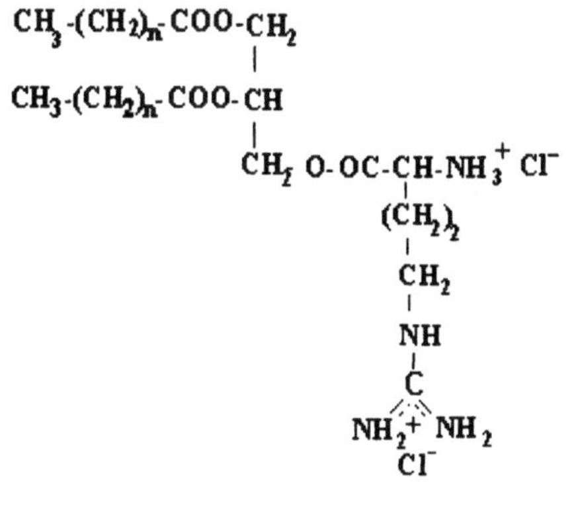

XXR (X = 8,10,12 and 14)

Fig. 1 Chemical structure of 1-*O*-(L-arginyl)-2,3-*O*-di-acyl-*sn*-glycerol dichlorohydrate surfactants

The progress of the reactions was monitored by HPLC, model Merck-Hitachi D-2500 using UV-VIS detector L-4250 at 215 nm. The formation of the diacyl compounds in the reaction mixture as well as the purity of the final products were checked using a Lichrospher 100 CN (propylcyano) 5-μm, 250 × 4 mm column. A gradient elution profile was employed from the initial composition of A/B 75/25 (by volume), changing over 24 min to a final composition of 5/95 where A is 0.1% (vol/vol) of TFA in H_2O and B is 0.085% of TFA in H_2O/CH_3CN 1:4. The flow-rate through the columns was 1.0 mL min^{-1}.

The structures of the pure new compounds were checked by ^1H- and ^{13}C-NMR spectroscopy, which were recorded with a Gemini 300 MHz spectrometer. Chemical shifts are reported in parts per million (δ, in ppm) downfield from tetramethylsilane (TMS). Mass spectroscopy (MS) with fast atom bombardment (FAB) or electro spray techniques was also conducted with a VG-QUATTRO from Fisons Instruments. Elemental analysis of the final compounds was also performed.

Conductivity

Conductivity was measured using a Crisson 525 platinised parallel plates with a constant of 0.998 cm^{-1} working at 1 kHz. The cell constant was calibrated periodically with sodium chloride solutions. Measurements were made at increasing concentrations to minimize errors from possible contamination from electrode.

Surface tension

Equilibrium surface tension measurements were made by the Wilhelmy plate technique with a Krüss K12 tensiometer. The instrument was calibrated against ultra pure distilled water (Milli-Q-4) before measurements were made.

Fluorescence

The fluorescence spectrum of micelle-bound pyrene is sensitive to the polarity of the microenvironment at the site of solubilisation of the fluorophore [7]. The critical micelle concentration (cmc) of the surfactant solutions were determined from the plots of the intensity ratio I_I/I_{III} versus concentration. I_I and I_{III} were, respectively, the first and third vibronic peaks in the fluorescence emission spectrum of pyrene solubilised at 9×10^{-6} M in surfactant solutions. The fluorescence spectra were recorded using a Shimatzu RF 540 spectrofluorometer at the excitation wavelength 335 nm and a bandwidth of 1.5 nm.

Method for qualitative phase behaviour

Qualitative phase behaviour of binary water/88R, water/1010R, water/1212R and water/1414R systems as a function of temperature was studied by optical microscopy. Optical examinations were performed according to the "flooding" or "penetration" method of Lawrence [8]. A polarising microscope equipped with a hot stage was used. In a flooding experiment, water was allowed to diffuse an anhydrous surfactant placed between a slide and a cover slip. After a short time, gradients in composition were produced, and the different mesophases developed as separate around crystalline surfactant.

Antimicrobial activity

The antimicrobial activities were determined *in vitro* on the basis of the minimum inhibitory concentration (MIC) values [9]. The MIC values are defined as the lowest concentration of antimicrobial agent, which inhibits the development of visible growth after 24 h of incubation at 37 °C. The microorganisms used were Gram-negative and Gram-positive bacteria and one yeast.

Results and discussion

Synthesis

The synthetic pathway for the preparation of these compounds, codified as XXR (being X the alkyl chain length), is outlined in Scheme 1.

The strategy of synthesis consisted of three steps. The first step dealt with the condensation of the N-Cbz-arginine to one hydroxy group of glycerol using boron trifluoride etherate as catalyst and the glycerol as solvent medium in order to prepare the 1-*O*,*N*-Cbz-arginyl-*rac*-glycerol (OORZ). The second step was the preparation of the 2,3,*O*-diacyl, 1-*O*,*N*-Cbz-arginyl-*rac*-glycerol (XXRZ) by the esterification of the two remaining free hydroxy groups of the OORZ compound with the corresponding acyl chloride in pyridine at room temperature. The third was the hydrogenation of the XXRZ compounds using Pd on activated charcoal as catalyst. To avoid the hydrolysis of the ester linkages, the reaction was carried out controlling the pH between 4 and 7 by the addition of a solution of the calculated amount of HCl in methanol to obtain the products in hydrochloride form.

The purity of the new compounds was checked by HPLC (Table 1). Satisfactory elemental analyses were obtained from these materials giving FABMS and NMR spectra, which were consistent with the desired compounds (Tables 1 and 2).

Critical micellar concentration

To check the behaviour of the synthesised compounds in solution as surfactants their aggregation properties were studied.

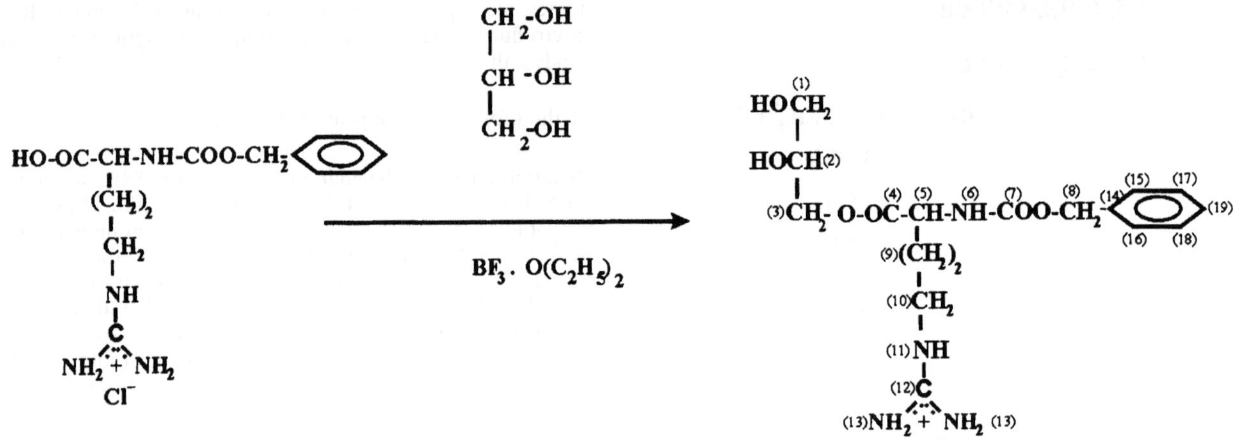

N-Cbz-arginine chlrohydrate

OORZ

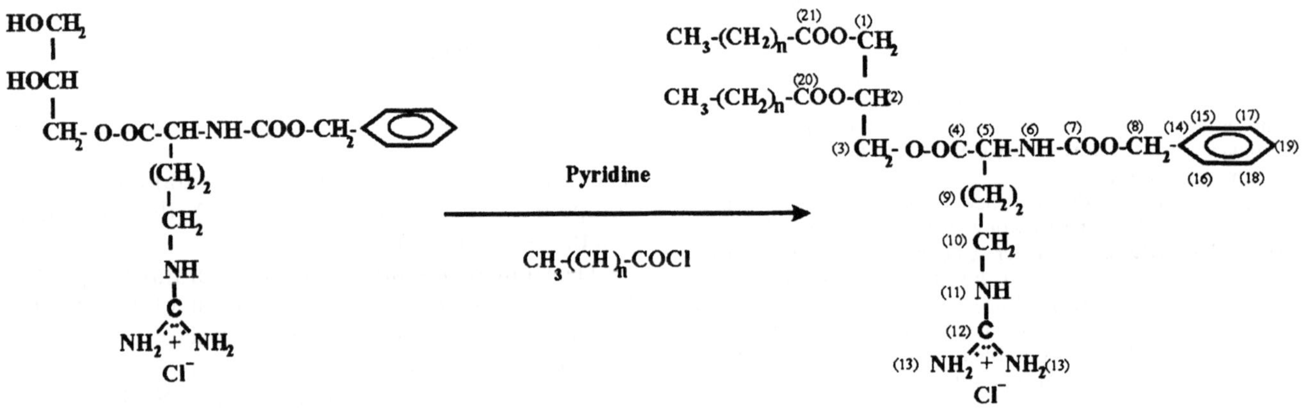

OORZ

XXRZ

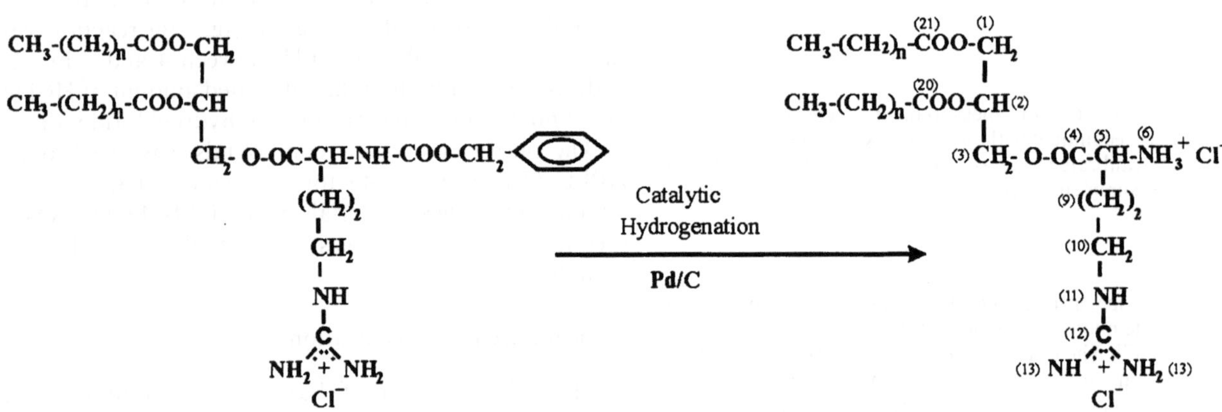

XXRZ

XXR

Table 1 Analytical data of 1,2-diacyl-3-arginyl-glycerols

Compound (XXR)	Molecular formula molecular weight	Yield (%)	HPLC retention time	Elemental analysis calculated/found			
				%C	%H	%N	%Cl
88R	$C_{25}H_{52}N_4O_6Cl_2$	60	12.77	48.46[a]	8.88[a]	9.05[a]	11.47[a]
	574			48.27	8.70	9.31	11.66
1010R	$C_{29}H_{58}N_4O_6C_{12}$	66	16.0	52.33[b]	9.32[b]	8.42[b]	10.67[b]
	629			52.80	9.48	8.72	10.51
1212R	$C_{33}H_{66}N_4O_6C_{12}$	72	18.0	56.33[c]	9.67[c]	7.96[c]	10.09[c]
	685			55.96	9.72	8.32	10.43
1414R	$C_{37}H_{74}N_4O_6C_{12}$	86	19.3	57.71[d]	10.03[d]	7.20[d]	9.23[d]
	741			57.41	9.86	6.84	9.23

[a]Calculated with 2.5 mol H_2O
[b]Calculated with 2.0 mol H_2O
[c]Calculated with 1.0 mol H_2O
[d]Calculated with 2.0 mol H_2O

Table 2 Spectral assignments for XXR compounds according to Scheme 1

Compound	m/e (M^+)	^{1}H-NMR (DMSO), δ ppm	^{13}C-NMR (DMSO), δ ppm
88R	501.2	0.90 [m, 6H, (2 CH_3 of the alkyl chain)] 1.31 [s, 16H, 8 CH_2 of the alkyl chain)] 1.59–2.01 [m, 8H, 2 (-CH_2-CH_2-COO-), (*9*)] 2.31–2.38 [2t, 2 (-CH_2-COO-)] 3.25–3.31 [m, (*10*)], 4.12–4.61 [m, 5H, (*1*)(*3*)(*5*)] 5.32–5.36 [m, 1H (*2*)]	14.435 [CH_3-, alkyl chain]s 23.680–41.751 [-CH_2-alkyl chain and (*9*),(*10*)] 53.573 [(*5*)], 63.173 [(*1*)], 65.424 [(*3*)] 70.306 [(*2*)], 158.643 [(*12*)], 170.098 [(*4*)], 174.448 [(*20*)], 174.806 [(*21*)]
1010R	557.4	0.86 [m, 6H, (2 CH_3 of the alkyl chain)] 1.26 [s, 24H, 12 CH_2 of the alkyl chain)] 1.55–2.02 [m, 8H, 2 (-CH_2-CH_2-COO-), (*9*)] 2.27–2.33 [2t, 2 (-CH_2-COO-)] 3.25 [m, (*10*)], 4.02–4.55 [m, 5H, (*1*)(*3*)(*5*)] 5.28–5.37 [m, 1H (*2*)]	14.466 [CH_3-, alkyl chain] 23.735–41.786 [-CH_2-alkyl chain and (*9*), (*10*)] 53.684 [(*5*)], 63.197 [(*1*)], 65.254 [(*3*)] 70.339 [(*2*)], 158.636 [(*12*)], 170.626 [(*4*)], 174.458 [(*20*)], 174.811 [(*21*)]
12,12,R	613.7	0.90 [m, 6H, (2 CH_3 of the alkyl chain)] 1.29 [s, 32H, 16 CH_2 of the alkyl chain)] 1.57–2.03 [m, 8H, 2 (-CH_2-CH_2-COO-), (*9*)] 2.31–2.38 [2t, 2 (-CH_2-COO-)] 3.27 [m, (*10*)], 4.12–4.60 [m, 5H, (*1*)(*3*)(*5*)] 5.31–5.39 [m, 1H (*2*)]	14.484 [CH_3-, alkyl chain] 23.750–41.734 [-CH_2- alkyl chain and (*9*), (*10*)] 53.560 [(*5*)], 63.193 [(*1*)], 65.298 [(*3*)] 70.295 [(*2*)], 158.618 [(*12*)], 170.011 [(*4*)], 174.358 [(*20*)], 174.763 [(*21*)]
14,14,R	669.7	0.90 [m, 6H, (2 CH_3 of the alkyl chain)] 1.29 [s, 40H, 20 CH_2 of the alkyl chain)] 1.57–2.31 [m, 8H, 2 (-CH_2-CH_2-COO-), (*9*)] 2.31–2.37 [2t, 2 (-CH_2-COO-)] 3.27 [m, (*10*)], 4.12–4.60 [m, 5H, (*1*)(*3*)(*5*)] 5.31–5.39 [m, 1H (*2*)]	14.488 [CH_3-, alkyl chain] 23.761–41.741 [-CH_2- alkyl chain and (*9*), (*10*)] 53.571 [(*5*)], 63.215 [(*1*)], 65.393 [(*3*)] 70.306 [(*2*)], 158.638 [(*12*)], 170.065 [(*4*)], 174.439 [(*20*)], 174.785 [(*21*)]

The conductance of aqueous solution of the new compounds at 25 °C was measured for solutions in concentrations ranging from 0.01 mM to 1.5 mM (Fig. 2).

◀

Scheme 1 Synthetic pathway for the preparation of the compounds XXR

The conductivities of the aqueous XXR solutions increased linearly with increasing concentration up to break points of 5 mM for 88R, 1.1 mM for 1010R, 0.3 mM for 1212R (Fig. 2a) and 0.25 mM for 1414R. These points can be either a cmc or a change on the aggregation size. The possible cmc values were also measured by fluorescence and surface tension for the products 88R and 1212R (Fig. 2b). The fluorescence measurements yielded values of 7 mM and 0.3 mM

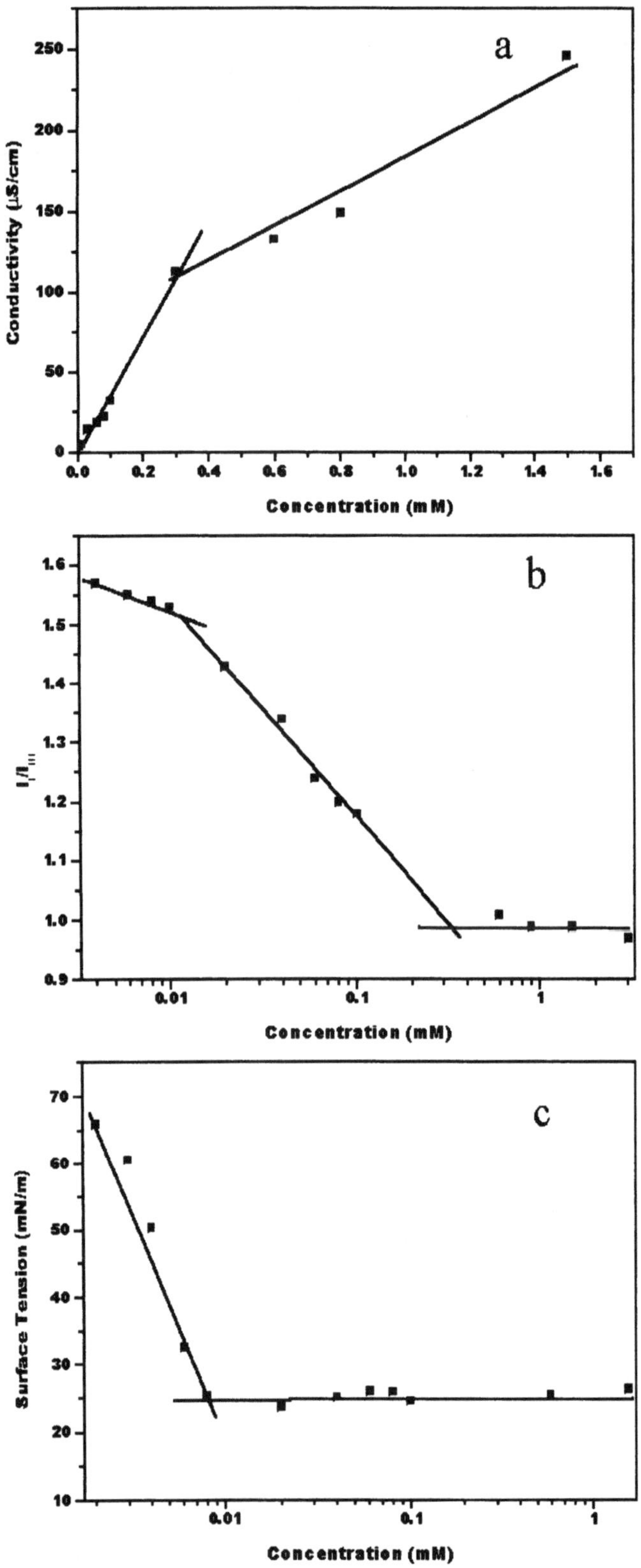

Fig. 2 Critical micellar concentration determination of the 1212R surfactant by conductivity (a), fluorescence (b) and surface tension (c)

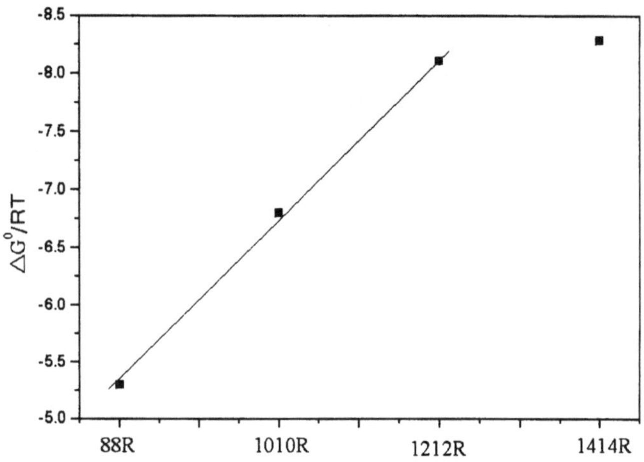

Fig. 3 Variation of $\Delta G^0/RT$ as a function of the surfactant chain for the homologues in the series XXR

respectively, which agrees with the value obtained by conductivity. The surface tension measurements yielded values of 0.07 mM for the compound 88R and 0.008 mM for the compound 1212R (Fig. 2c). The results may suggest that the new surfactants at low concentrations form certain aggregates of size substantial enough to produce fairly constant monomer activity above the break point at the lowest concentration.

The new surfactants differ from short-chain lecithines in terms of aggregation behaviour. Short chain phospholipids aggregate in water forming micelles at concentrations of about 16.5 mM for diC_6-lecithin and 0.24 mM for diC_8-lecithin [10]. Comparing these cmc values with those of the same fatty chain length synthetic homologue, the cmc of the new products are one order of magnitude higher. This increment can be attributed to the high solubility of the new compounds due to the cationic character of the polar group. The solubility in water of XXR compounds depends on the fatty chain length. Compounds 88R and 1010R at 0.1% w/v form an isotropic single phase whereas compounds 1212R and 1414R at 0.1% w/v form a liquid crystalline dispersion.

To evaluate the micellisation undergone by the synthesised surfactants, we applied the mass action micellisation model for monodispersed micelle size. It is the simplest one that can provide a quantitative description of micellisation of real systems. In this model, the standard free energy ΔG^0 for the monomer-micelle association for large association numbers is given by Eq. (1)

$$\Delta G^0 = RT\ \ln(M_1) = RT\ \ln(\mathrm{cmc}) \tag{1}$$

Figure 3 shows the variation of $\Delta G^0/RT$ as a function of the surfactant chain length for the homologues in the series XXR.

For the products 88R, 1010R and 1212R the plots yielded a straight line. The standard free energy of micellisation per mole of -CH_2- group was computed from the slope of the plots yielding a value of −0.70 RT/mol of -CH_2- group (1.7 kJ/mol of CH_2).

In aqueous medium, an increase in one -CH_2- unit in the length of the hydrophobic group resulted in a negative increase of ΔG^0. This variation has been attributed to changes in the degree of non-polarity in the interior of the micelle, changes that, in turn, are induced by the fact that the water penetrates into the micelles and affects the polarity of the hydrophilic head [11, 12].

The variation of ΔG^0 for each extra -CH_2- group in the synthesised cationic compounds was of 1.7 kJ/mol. This value is slightly smaller than expected. For the lecithines, structural homologues, the value is 2.1 kJ/mol of -CH_2- group [13]. We attribute this behaviour to the fact that methylene groups favour micellisation whereas hydration of the cationic polar group during the micellisation process opposes it.

Optical microscopy

The phases of the binary XXR/water system have been determined by visual observation of the samples through crossed polarised microscopy. Qualitative phase behaviour studies applying the flooding method revealed the formation of anisotropic phases in all the binary surfactant systems studied. In Fig. 4 two different anisotropic liquid crystals can be observed.

88R and 1010R compounds formed lamellar and nematic liquid crystals phases at room temperature (Fig. 4a and Fig. 4b). The compounds 1212R and 1414R formed streak liquid crystals at temperatures above 40 °C (Fig. 4c).

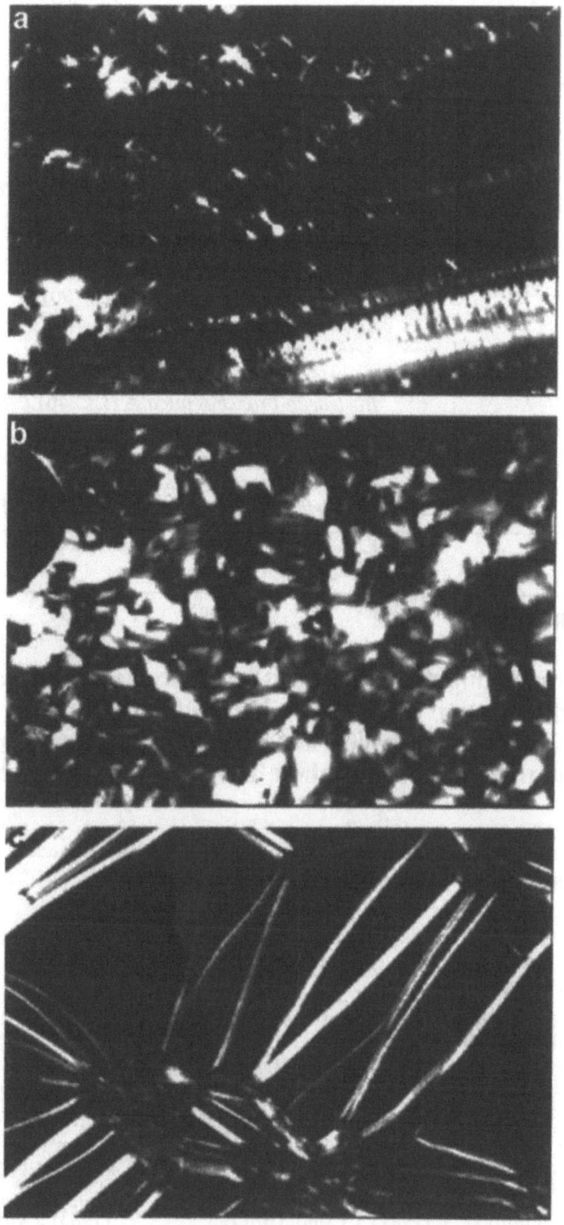

Fig. 4 Crossed polarised microscopy of the binary XXR/water system. (**a**) lamellar liquid crystal, (**b**) nematic liquid crystal and (**c**) streak liquid crystal

Antimicrobial activity

To evaluate the antimicrobial activity of the new compounds the in vivo membrane-disrupting properties were tested using cell bacteria as biological membranes. [9]. The MIC values obtained are given in Table 3. The lower the MIC value, the higher the antimicrobial activity.

The results show that these compounds were active against both Gram-negative and Gram-positive organisms. Given that the MIC values occur at concentrations of surfactants in water below the cmc (Table 3), the species that interact with cells are the surfactant monomers while the aggregates do not have an effect. The antimicrobial efficiency for these surfactants depends on their alkyl chain length. The maximum antimicrobial activity corresponds to the compound with 8 carbon atoms in the alkyl chains then the activity decrease with increasing alkyl chain lengths. A loss of activity occurs for the surfactants with longer alkyl chain lengths because they are less water-soluble.

On the other hand, similar MIC values have been reported for the hexadecyltrimethylammonium bromide (HTAB) [14], a known biocide product. So that, 88R and 1010R compounds can be considered good antimicrobial agents.

Table 3 Minimum inhibitory concentration (μg/mL) of the XXR compounds

	Microorganism	88R	1010R	1212R	1414R
Gram-positives	*Bacillus cereus* var. *mycoide* ATCC 11778	64	16	>256	>256
	Bacillus subtillis ATCC 6633	64	2	>256	>256
	Staphylococcus aereus ATCC 6538	4	16	>256	256
	Staphylococcus epidermidis ATCC 12228	8	>256	>256	16
	Micrococcus luteus ATCC 9341	16	1	4	1
	Candida albicans ATCC 10231	16	32	64	>256
Gram-negatives	*Salmonella typhimurium* ATCC 14028	16	32	>256	>256
	Pseudomonas aeruginosa ATCC 9027	64	>256	>256	>256
	Escherichia coli ATCC 8793	8	64	>256	>256
	Arthrobacter oxidans ATCC 8010	32	16	>256	>256
	Streptoccocus faecalis ATCC 19434	8	4	2	0.5
	Bortedella bronchiseptica ATCC 4617	0.25	0.25	0.25	0.25
	Citrobacter freundii ATCC 22636	32	>256	>256	256
	Alcaligenes faecalis ATCC 8750	8	4	>256	256
	Enterobacter aerogenes CECT 689	32	>256	>256	16
	Klebsiella pneumoniae v. *preumonial* CIP 104216	8	16	>256	4

Conclusions

Novel multifunctional surfactants from diacyl glycerides and the amino acid arginine have been synthesised.

The physicochemical and antimicrobial study reveals that these surfactants show a self-aggregation behaviour similar to lecithines with a relevant antimicrobial activity.

Acknowledgements Grant PPQ2000-1687-CO2-01 from Spain CICYT supported this work.

References

1. Rybinski W von, Rybinski K von (1998) Alkyl polyglucosides. In: Holmberg K, Deker M (eds), Novel surfactants, vol 74. Surfactant Science Series, ch. 2, p 31
2. Infante MR, Perez L, Pinazo A (■) Novel cationic surfactants from arginine. In: Holmberg K, Deker M (eds), Novel surfactants, vol 74. Surfactant Science Series, ch. 3, p 87
3. Okahata Y, Tanamachi S, Nagai M, Kunitake T (1981) J Colloid Int Sci 82:401
4. Infante MR, Garcia Dominguez JJ, Erra P, Julia MR, Prats M (1984) Int J Cosm Sci 6:275
5. Perez L, Torres JL, Manresa A, Solans C, Infante MR (1996) Langmuir 12:5296
6. Infante MR, Seguer J, Pinazo A (1997) J Colloid Surf 49:123
7. Kalyanasundaram K, Thomas JK (1977) J Am Chem Soc 99:2039
8. Rendall K, Tidyy GJT, Trevetham MA (1983) J Chem Soc Faraday Trans I 79:673
9. Jones RN, Barry AL, Gavan TL, Washintong JA II (1985) In: Lennette EH, Ballows A, Hauser WJ (eds), Manual of clinical microbiology. American Society for Microbiology, Washington, DC, p 972
10. Tausk RJM, Karmiggelt J, Oudshoorn C, Overbeek JTG (1974) Biophys Chem 1:175–183
11. Clint JH, Walker T (1975) J Chem Soc Faraday Trans 1 71:946
12. Rosen MJ (1988) Surfactants and interfacial phenomena. Wiley & Sons
13. Hauser H (2000) BBA 1508:164–181
14. Diz M, Manresa A, Pinazo A, Erra P, Infante MR (1994) J Chem Perkin Trans 2 1871

Progr Colloid Polym Sci (2004) 123:217–221
DOI 10.1007/b11761

A. Cuenca

The role of premicellar assemblies and micelles upon the hydrolysis of 2-(2-fluorophenoxy)quinoxaline

A. Cuenca
Departamento de Química,
Universidad Simón Bolívar,
Apartado 89000, Caracas 1080-A,
Venezuela
e-mail: ancuenca@usb.ve
Fax: +58-212-9784410

Abstract In the presence of cationic surfactants ($C_{16}H_{33}NR_3Cl$, R = Me, *n*-Pr, and *n*-Bu), the rate-surfactant concentration profiles for the hydrolysis of 2-(2-fluorophenoxy-quinoxaline) **1** show a double maxima with increasing surfactant concentration and at high substrate concentration. The first rate maximum is ascribed to the formation of catalytically active substrate-premicellar complexes of **1** with the surfactant and the second rate maximum to reaction in micelles. At low substrate concentration, a single maximum is observed. A change of head group in the sequence Me_3N^+, Pr_3N^+, Bu_3N^+ increases the second-order rate constant for the hydrolysis of compound **1**.

Keywords Micelles · Premicelles · Quinoxaline

Introduction

Rates effects of aqueous colloidal self-assemblies such as micelles are generally explained in terms of pseudophase models in which micelles and water are treated as distinct regions [1–3]. However, there is extensive evidence that premicellar assemblies are involved in some reactions [4–10]. The present work involved a search for supramolecular entities that assist bimolecular reactions. The structure of aqueous amphiphilic assemblies is of fundamental importance in understanding the chemical behaviour of solutes reacting in colloidal medium. Different types of aqueous assemblies are obtainable from alkylammonium salts. The microenvironment of aqueous alkylammonium aggregates changes with different combinations of alkyl groups. Ammonium salts that contain a single long-chain alkyl group form globular micelles, but ammonium salts with three octyl chains form tight, small aggregates [8]. Recently, Bunton et al. [10] provided evidence for acceleration by premicelles of the hydrolyses of some dinitroalkoxyphenyl phosphates. In this report, the basic hydrolysis of 2-(2-fluorophenoxy)quinoxaline **1** (Scheme 1) was investigated in the presence of micellised cetyltrialkylammonium chlorides ($C_{16}H_{33}NR_3Cl$: R = Me, CTACl; *n*-Pr, CTPACl; *n*-Bu, CTBACl) and non-micellising tri-*n*-octylmethylammonium mesylate ($C_{26}H_{57}N0_3S$, TOAM). We have examined the alkaline hydrolysis of compound **1** over a wide range of surfactant concentrations, both above and below the critical micelle concentration (cmc). The goal is to provide a deeper understanding of the properties of aqueous amphiphilic aggregates that affect bimolecular reaction rates in colloidal media.

Cationic micelles speed the hydrolyses of some quinoxaline derivatives [9, 11–14]. Quinoxaline derivatives are hydrophobic substrates that bind strongly to micelles and undergo facile nucleophilic heteroaromatic substitution reactions [15].

Materials and methods

Materials

Compound **1** was synthesised from 2-chloroquinoxaline [16] as previously reported [17]. Preparation and purification of the surfactants has been described [18]. There were no minima in plots of surface tension of purified surfactants against log surfactant concentration.

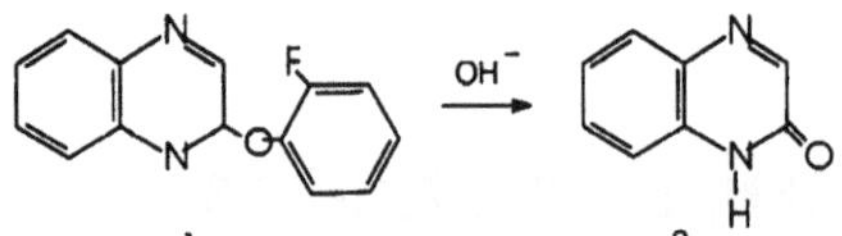

Scheme 1

Kinetics

All reactions were run at 25 °C. The nucleophilic heteroaromatic substitution reaction was followed spectrophotometrically at 362 nm for the appearance of 2-quinoxalone **2** (Scheme 1). The reaction was initiated by the addition of compound **1** (5–10 μ) prepared in acetonitrile to 3 mL of a surfactant solution equilibrated in the cell compartment of a Perkin-Elmer Lambda II spectrophotometer. Solutions were prepared with redistilled carbon dioxide-free water. The first-order rate constants, k_ψ, are in reciprocal seconds. The rate constants are means of three measurements that agreed within 5%. Reactions were followed to infinity (10 half-lives) with correlation coefficients ≥0.999.

Results and discussion

Reaction in water

The alkaline hydrolysis of 2-(2-fluorophenoxy)quinoxaline, **1**, leads to the formation of 2-quinoxalone, **2** (Scheme 1). The second-order rate constant, k_W, for the basic hydrolysis of **1** at 25.0 °C is 1.2×10^{-4} M^{-1} s^{-1}. Repetitive scanning of the spectrum of the reaction mixture showed that no intermediate built up during reaction.

Reaction in Micelles

Figures 1–3 shows the pseudo-first order rate constants, k_ψ, vs. surfactant concentration profiles for reaction

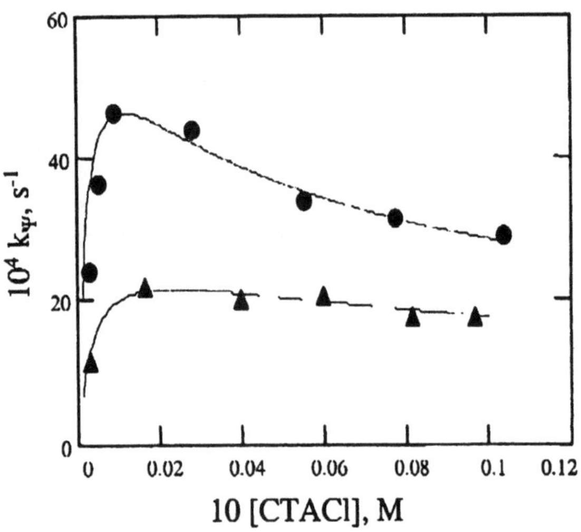

Fig. 1 Basic hydrolysis of compound **1** in CTACl. (●) With 0.01 M NaOH. (▲) With 0.001 M NaOH. [Substrate] = 7.0×10^{-6} M. Curves are calculated

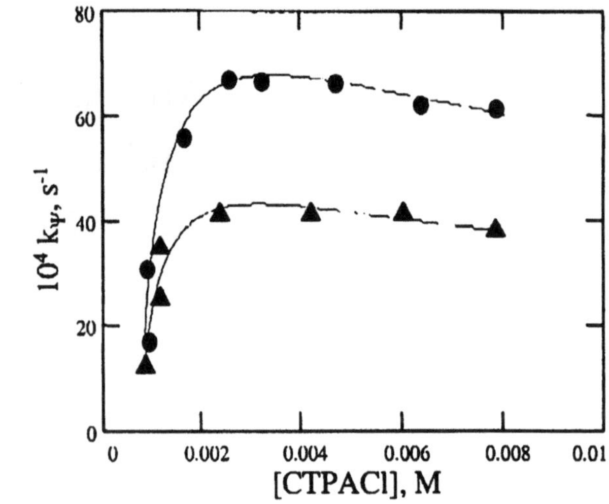

Fig. 2 Basic hydrolysis of compound **1** in CTACl. (●) With 0.01 M NaOH. (▲) With 0.001 M NaOH. [Substrate] = 7.0×10^{-6} M. Curves are calculated

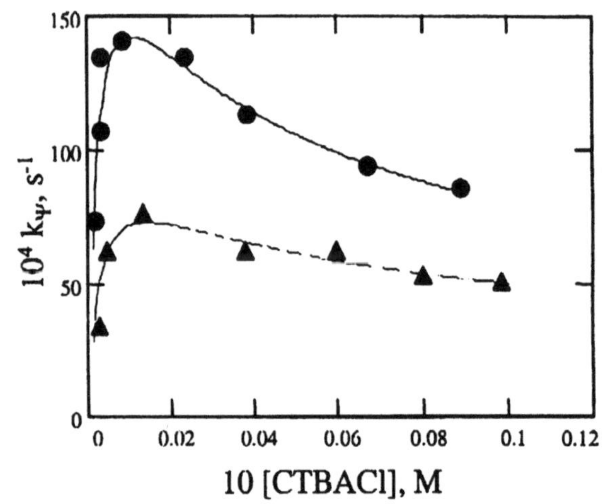

Fig. 3 Basic hydrolysis of compound **1** in CTACl. (●) With 0.01 M NaOH. (▲) With 0.001 M NaOH. [Substrate] = 7.0×10^{-6} M. Curves are calculated

of 2-(2-fluorophenoxy)quinoxaline **1** with OH^- in the presence of cetyltrialkylammonium chlorides, $C_{16}H_{33}NR_3Cl$ (R = Me, CTACl; *n*-Pr, CTPACl; *n*-Bu, CTBACl). At low substrate concentration (7.0×10^{-6} M) the rate-surfactants profiles show a single rate maximum typical of the rate-surfactant concentration profiles for bimolecular reactions [1–7, 10–13, 18]. Below the cmc, the observed rate is the same as that in water. Once micelle formation begins, the rate increases rapidly to a maximum followed by a gradual, but steady decrease in the rate.

At low substrate concentration, micellar effects upon the basic hydrolysis of compound **1** can be treated quantitatively in terms of an equilibrium distribution of substrate (S), between water (W) and micelles (M)

(Scheme 2). In this pseudophase model, K_s, is an association constant with respect to micellised surfactant, D_n, the concentration of which is that of total concentration ($[D_T]$) less that of monomeric surfactant, i.e., $[D_n] = [D_T]$–cmc (the cmc is taken as the concentration of monomeric surfactant at onset of micellization), and k'_W. and k'_M are first-order rate constants in aqueous and micellar pseudophases respectively.

The first-order rate constant for overall reaction, k_ψ, is given by Eq. (1) [1]:

$$k_\Psi = \frac{k'_M K_s [OH^-_M]}{1 + K_S [D_n]} \tag{1}$$

The first-order rate constants can be written as second-order rate constants, k_W and k_M, with the concentration of OH^- in the micellar pseudophase written as a mole fraction, m_M^{OH}:

$$k'_W = k_W [OH^-_W] \tag{2}$$

$$k'_M = k_M m_M^{OH} = k_M [OH_M^{OH}]/[D_n] \tag{3}$$

$[OH^-_W]$ is a molarity in terms of total solution volume.

To fit the kinetic results, the model proposed by Bunton et al. [19] and Bacaloglu et al. [20] and their approach were used. Rate data for reaction of OH^- with compound **1** fit a pseudophase model [Eqs. (1)–(3)], in which the distribution of OH^- and Cl^- between the aqueous and micellar pseudophases is written in terms of the mass-action-like Eqs. (4) and (5) [19, 20]

$$K'_{OH} = [OH^-_M]/\{([OH^-_W][D_n] - [OH^-_M] - [Cl^-_M])\} \tag{4}$$

$$K'_{Cl} = [Cl^-_M]/\{([Cl^-_W][D_n] - [OH^-_M] - [Cl^-_M])\} \tag{5}$$

Rate-surfactant profiles were fitted by using Eqs. (1), (4), (5). Eqs. (4) and (5) were solved simultaneously with an iterative calculation method to give the values of $[OH^-_M]$ and $[X^-_M]$. Curves in Figs. 1–3 were calculated by using the Langmuir isotherms Eqs. (4) and (5). At low substrate concentration, the model adequately fits rate data for reactions of compound **1** as it is shown in Figs. 1–3. Table 1 shows the parameters that best fit the kinetic results for substrate **1** in micelles. K_S and K'_{Cl} were treated as adjustable parameters using cmc values obtained from kinetic data. It is difficult to get good estimates of the binding constant under these conditions

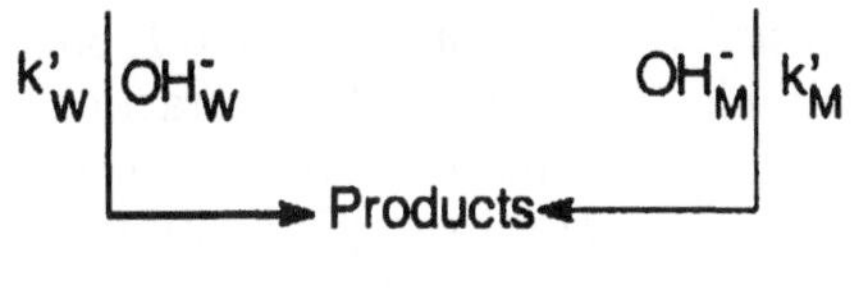

Scheme 2

Table 1 Micellar parameters for the basic hydrolysis of 1 in the presence of cationic surfactants[a]

Surfactant	10^2 $[OH^-_T]$ M	K'_{Cl} M^{-1}	10^3 cmc M	10 k_M s^{-1}	10^2 k_2^m M^{-1} s^{-1}
CTACl	0.1	123	1.0	3.2	4.5
CTACl	1.0	123	1.0	3.2	4.5
CTPACl	0.1	58	0.9	6.7	9.4
CTPACl	1.0	58	0.9	6.7	9.4
CTBACl	0.1	49	0.4	9.1	13.0
CTBACl	1.0	49	0.4	9.1	13.0

[a]At 25.0 °C and with $K_s = 9300$ M^{-1}, $k_W = 1.2 \times 10^{-4}$ M^{-1} s^{-1}, $[1] = 7.0 \times 10^{-6}$ M.

because the fitting is not very sensitive to values of K_s. Values of K'_{Cl} differ slightly from values found in a previous work [9] and are higher from those reported by Bacaloglu et al. [20]. The Langmuir parameters, K'_{OH}, are 55, 25, and 12 for CTA^+, $CTPA^+$, and $CTBA^+$, respectively [21].

The second-order rate constant, k_W, for reaction in the micellar pseudophase has the dimensions of reciprocal time, and cannot be compared directly with second-order rate constant in water, k_w, the units of which are M^{-1} s^{-1}. Second-order rate constants in micellar phase with same dimensions, k_2^m, M^{-1} s^{-1}, are given by Eq. (6) where V_M is the molar volume of the reactive region at the micellar surface, and we take $V_M = 0.14$ L. Estimations of this volume range from 0.14 to 0.37 L [3].

$$k_2^m = k_M V_M \tag{6}$$

Micellar catalysis for the basic hydrolysis of compound **1** is increased by an increase in bulk of head group size (Table 1). A change of head group in the sequence Me_3N^+, Pr_3N^+, Bu_3N^+ increases the second-order rate constant. Polarities of micelle-water interfaces are lower than that of bulk water. Bulky alkyl groups should further decrease the polarity of the micellar surface and increase its hydrophobicity. These observations indicate that reagent's reactivity is increased by a decrease in polarity of the interface and by an increase in hydrophobicity of the head group.

Reaction in submicellar aggregates and micelles

Experiments were performed at approximately 10-fold greater substrate concentration (9.0×10^{-5} M). Figs. 4–6 show the pseudo-first order rate constants, k_ψ, vs. surfactant profiles for reaction of **1** with 0.01 M and 0.001 M OH^- in the presence of CTACl, CTPACl and CTBACl. At high substrate concentration and in dilute aqueous solutions of cationic surfactants, rate

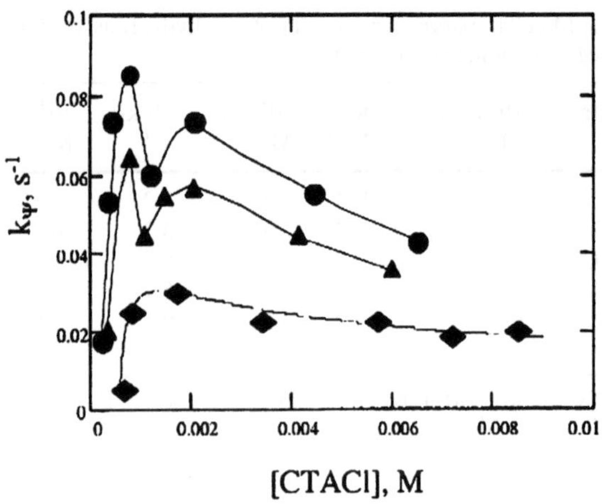

Fig. 4 Basic hydrolysis of compound **1** in CTACl. (●) With 0.01 M NaOH. (▲) With 0.001 M NaOH. (◆) With [NaCl] = 0.02 M. [Substrate] = 9.0×10^{-5} M

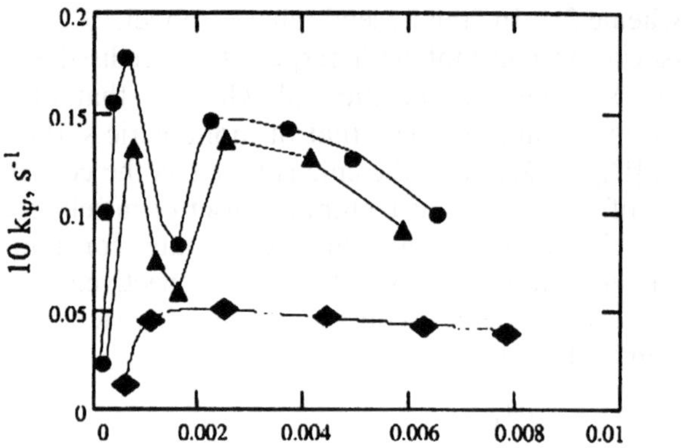

Fig. 6 Basic hydrolysis of compound **1** in CTACl. (●) With 0.01 M NaOH. (▲) With 0.001 M NaOH. (◆) With [NaCl] = 0.02 M. [Substrate] = 9.0×10^{-5} M

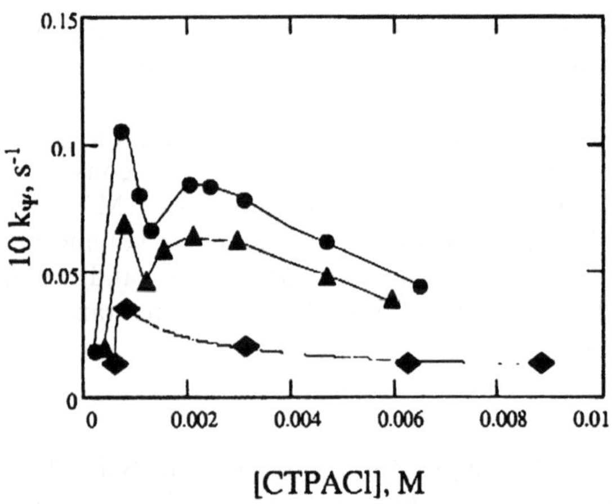

Fig. 5 Basic hydrolysis of compound **1** in CTACl. (●) With 0.01 M NaOH. (▲) With 0.001 M NaOH. (◆) With [NaCl] = 0.02 M. [Substrate] = 9.0×10^{-5} M

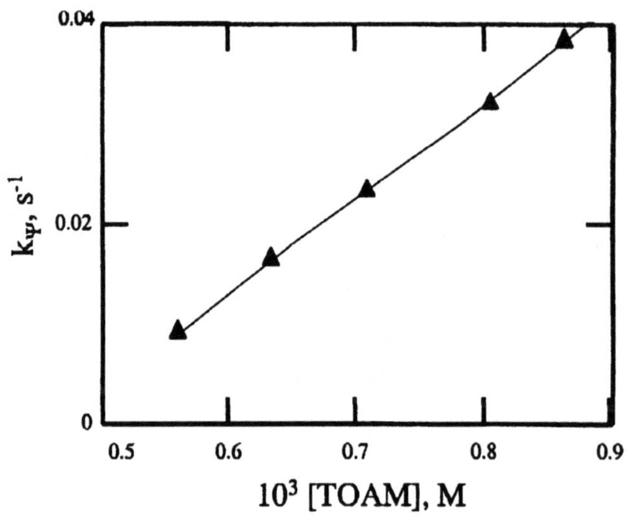

Fig. 7 Basic hydrolysis of compound **1** in TOAM. (▲) With 0.01 M NaOH. NaOH. [Substrate] = 9.0×10^{-5} M

constants for the hydrolysis of 2-(2-fluorophenoxy) quinoxaline **1** go through double rate maxima with increasing concentration of surfactant. The new maxima occur at surfactant concentrations below the cmc. Pseudophase models predict that k_ψ should go through a single maximum with increasing surfactant concentration [1–3]. Submicellar aggregates of the surfactant are probably responsible for the observed rate increase below the surfactant's cmc. In the rate-surfactant profiles the first rate maximum is ascribed to the presence of submicellar aggregates and the second to the presence of micelles. Transient small clusters of the surfactant speed the reaction and they disappear as micelles form at higher surfactant concentrations and take up the substrate.

Added 0.02 M NaCl suppresses the double rate maxima (Figs. 4–6). Addition of salts decrease cmcs [3] converting submicellar aggregates into micelles [7].

To further investigate the catalytic effect of submicellar aggregates on reaction rates, the alkaline hydrolysis of compound **1** was examined in the presence of tri-*n*-octylmethylammonium mesylate, TOAM (Fig. 7). Rate constants for the basic hydrolysis of **1** increase linearly with TOAM concentration. These ions do not micellise but they speed the nucleophilic reaction though the formation of small aggregates, which interact with hydrophobic substrate **1**.

For the basic hydrolysis of compound **1**, polarisability and hydrophobic forces drive the interactions between the quaternary ammonium head group and the heteroaromatic solute. Substrate hydrophobicity seems to be a necessary, but not sufficient, condition for observation of double rate maxima. Nucleophilic heteroaromatic substitution is strongly assisted by electron-withdrawing substituents and occurs readily upon quinoxaline derivatives on account of "aza" activation of the heterocyclic ring [15]. Through association with the heteroaromatic ring, the cationic head group should behave like an electron-withdrawing substituent by polarising the π electron cloud and rendering the C-2 more electrophilic and prone to nucleophilic attack.

Rates are larger in premicelles than in micelles. This result is consistent with a tight interaction in an associative complex between the heteroaromatic compound and the ammonium cationic head group as compared with a relatively loose organisation of substrate in the water-rich interfacial region of the micelle.

Conclusion

For the very hydrophobic and polarisable compound **1** the shape of the rate-surfactant concentration profiles depends on substrate concentration. In very dilute surfactant, the nature of the amphiphilic assemblies is fundamental in identifying the site where reaction is occurring and in understanding their role upon reaction rates. Submicellar assemblies activate the substrate toward nucleophilic substitution but micelles assist reaction by concentrating both reactants in the small volume of the micellar pseudophase.

Acknowledgement Financial support from Decanato de Investigaciones Científicas, Universidad Simón Bolívar (Grant DI-CB-S100117), Caracas, Venezuela, is gratefully acknowledged.

References

1. Menger FM, Portnoy CE (1967) J Am Chem Soc 89:4698
2. Bunton CA, Nome FH, Quina F, Romsted LS (1991) Acc Chem Res 24:357
3. Romsted LS (1984) In: Mittal KL, Lindman B (eds), Surfactants in solution, vol. 2. Plenum Press, New York, p 1015
4. Cerichelli G, Mancini G, Luchetti L, Savelli G, Bunton CA (1994) Langmuir 10:3982
5. Blaskó A, Bunton CA, Hong YS, Mhala MM, Moffatt JR, Wright SJ (1991) Phys Org Chem 4:618
6. Bunton CA, Bacaloglu R (1987) J Colloid Interface Sci 115:288
7. Bacaloglu R, Bunton CA (1992) J. Colloid Interface Sci 153:140
8. Kunitabe T, Okahata Y, Ando R, Shinkai S, Hirakawa S (1980) J Am Chem Soc 102:7877
9. Cuenca A (2000) Langmuir 16:72
10. Brinchi L, Di Profio P, Germani R, Savelli G, Tugliani M, Bunton CA (2000) Langmuir 16:10101
11. Cuenca A (1998) Int J Chem Kin 30:777
12. Cuenca A (1997) Tetrahedron 53:12361
13. Cuenca A, Strubinger A (1996) Tetrahedron 52:11665
14. Flamini V, Linda P, Savelli G (1975) J Chem Soc Perkin Trans 2 421
15. Illuminati G (1964) Adv Heterocycl Chem 3:285
16. Castle RN, Onda M (1961) J Org Chem 26:954
17. Cuenca A, López SE, Garcés I, Aranda A (1999) Synth Commun 29:1393
18. Bacaloglu R, Bunton CA, Ortega F (1989) J Phys Chem 93:1497
19. Bunton CA, Gan LH, Hamed FH, Moffatt JR (1983) J Phys Chem 87:336
20. Bacaloglu R, Bunton CA, Cerichelli G, Ortega F (1990) J Phys Chem 94:5068
21. Bonan C, Germani R, Ponti PP, Savelli G, Cerichelli G, Bacaloglu R, Bunton CA (1990) J Phys Chem 94:5331

Progr Colloid Polym Sci (2004) 123:222–226
DOI 10.1007/b11762

J. Liu
T. Palberg

Crystal growth and crystal morphology of charged colloidal binary mixtures

Abstract We report direct microscopic observations of competitive crystal growth of homogeneously and heterogeneously nucleated crystals and of zig-zag patterned twin-domains in deionised two-component aqueous suspensions of charged polystyrene latex spheres with optical microscopy. The size ratio of these two components is 1:1.47 and the number fraction of PS68 in PS68/PS100 mixtures is 0.2 and 0.5. Growth velocities are measured systematically for increasing particle number densities n. We describe the difference of chemical potential Δμ in our binary system assuming additive pair-interactions with polydispersity term. Crystal growth velocities v_{110} can be fitted to a Wilson-Frenkel law. The limiting growth velocities are 10.04 ± 0.51 $\mu m{\cdot}s^{-1}$, 10.25 ± 0.37 μms^{-1} for number fractions of 0.2 and 0.5, respectively.

Keywords Optical Bragg microscopy · Crystal growth velocity · Wilson-Frenkel law · Diffusion coefficient · Polydispersity

J. Liu · T. Palberg (✉)
Institut für Physik der Universität Mainz,
Staudinger Weg 7,
55099 Mainz, Germany
e-mail: Thomas.Palberg@uni-mainz.de

Introduction

Due to the easy control of the interaction potential of charged spherical colloids, their mixtures may serve as valuable model systems for the investigation of new soft matter materials. Using single-component systems with various repulsive interaction potentials, the validity of the Becker-Döring theory of nucleation [1] and the Wilson-Frenkel theory of growth (WF) [2] have already been tested successfully. A test to above theories of crystal growth for mixed systems is still on demand. Also the nucleation behaviour of colloidal mixtures still needs to be explained.

We here present measurements on a mixture of size ratio 1:1.47 with controlled total particle number density n and number fraction of the smaller component (PS68) p = 0.2 and p = 0.5. Experiments are conducted under deionised conditions and at a particle concentration where samples are completely or partly (fluid-crystal coexistence) solidified at equilibrium. Samples are shear molten and after stop of shear readily solidify via heterogeneous nucleation at the cell wall with subsequent quasi-epitaxial growth or via homogeneous nucleation and radially isotropic growth. The formerly applied shear orients the wall based nuclei and oriented, though twinned crystals result. Their (110) plane is parallel to the cell wall with the <111> direction parallel to the formerly applied flow direction. Growth proceeds inward in the <110> direction. Homogeneously nucleated crystals show no preferred orientation. For the one component system the velocity was found to obey a Wilson-Frenkel (WF) law in both cases:

$$v = v_\infty(1 - \exp(-\Delta\mu/k_BT)) \quad (1)$$

Here v_∞ is the limiting growth velocity. Δμ is the difference of chemical potential between fluid and crystalline phase, k_BT is the thermal energy: $\Delta\mu = B\Pi_f^* = B(\Pi - \Pi_f)/\Pi_f$. Here Π_f^* is the reduced energy density, $\Pi = (1/2)\alpha V(r)$, B is a fitting parameter, α is an effective coordination number, Π_f is Π at freezing, V(r) is the pair interaction energy. In the charged colloidal two-component mixture it is described according to Lindsay and Chaikin [3]:

$$V(r) = \frac{e^2}{4\varepsilon_0\varepsilon_r}\frac{\exp(-\kappa r)}{r}\left[p^2 Z^{*2}_{\sigma,1}\left(\frac{\exp(\kappa a_1)}{1+\kappa a_1}\right)^2 + 2pqZ^*_{\sigma,1}Z^*_{\sigma,2}\frac{\exp(\kappa a_1+\kappa a_2)}{(1+\kappa a_1)(1+\kappa a_2)} + q^2 Z^{*2}_{\sigma,2}\left(\frac{\exp(\kappa a_2)}{1+\kappa a_2}\right)^2\right] \quad (2)$$

where

$$\kappa^2 = \frac{4\pi e^2}{\varepsilon_0\varepsilon_r k_B T}\left[\left(pZ^*_{\sigma,1} + qZ^*_{\sigma,2}\right)n + 2c\right]$$

$q = 1-p$ is the number ratio of the second component, a_1 and a_2 are particle radii of these two components, respectively, and $Z^*_{\sigma,1}$, $Z^*_{\sigma,2}$ is the corresponding conductivity effective charges, c is the salt concentration, which in deionised condition amounts to $c = 0.2$ μM. All limiting growth velocities v_∞ in our measurements were found between 7.2 and 15.9 $\mu m \cdot s^{-1}$, which compares well with previous reports [4].

Crystals were observed by optical Bragg microscopy. Details of this technique have been given recently [5, 6]. In the following, we first give a short outline of the sample conditioning and the optical techniques, then we present our observation of competitive growth between homogeneously and heterogeneously nucleated crystals in binary mixtures. We further measure crystal growth velocities, and describe them via pair-interactions with additive polydispersity term. Then we fit the data to a Wilson-Frenkel law. We also show the microscopic observation of a zig-zag pattern of twin domains in binary mixtures and interpret the microscopic morphology of mixed crystals.

Sample conditioning

Two commercially available species of charged polystyrene spheres in aqueous suspension were investigated (PS68 and PS100). Before preparing mixtures, the particles were carefully characterised by various experiments [7]. The most important results are compiled in Table 1.

For deionising the samples we used an advanced, continuous procedure. Details have recently been given elsewhere [4, 6, 8] . In short, the suspension is peristaltically cycled through a closed tubing system connecting various components. It includes an ion exchange cell filled with mixed bed ion exchange resin (IEX) [Amberlite, Rohm & Haas, F]. This can be bypassed, if desired. A reservoir under inert gas atmosphere allows us to remove or add suspension to adjust particle concentration and sample composition. Further on several measuring cells may be incorporated: a cell for microscopy, a cell for static light scattering etc. The deionisation process is monitored by a conductivity meter [Bridge WTW535, electrode LTA01, WTW, Weilheim, D]. From the conductivity of the suspension, we can also control the particle number density n both for the monodisperse sample [9, 10] and the two-component mixture as: $\sigma = ne\overline{Z^*_\sigma}(\mu_+ + \mu_-) + \sigma_B$ with a mean effective charge of two component mixture $\overline{Z^*_\sigma} = pZ^*_{\sigma,1} + qZ^*_{\sigma,2}$ [11], $\sigma_B \approx 0.06$ μS/cm is the background conductivity stemming from the self-dissociation of water and residual impurities at $T = 297$ k. $\mu_+ = 36.5 \times 10^{-8}\, m^2\, V^{-1}\, s^{-2}$ is the proton mobility and μ_- is that of the particles measured from electrophoresis [9]. Typical values of μ_- are on the order of $(2\text{–}10) \times 10^{-8}\, m^2\, V^{-1}\, s^{-2}$.

Crystal growth and pattern

Measurements of the growth velocity and characterisation of the sample morphology were performed using Bragg-microscopy. The optical cell of rectangular cross-section ($2 \times 10\ mm^2$) is mounted on the stage of an optical microscope [Laborlux 12, Leitz, Wetzlar, D] equipped with a video camera. The cell is illuminated from below using a cold white light source. The latter is adjusted under angles Θ and φ to induce Bragg reflections of the inspected crystals in the direction of the microscope objective, i.e., normal to the cell wall. With properly orientated the white light, we can observe different crystal nucleation behaviour. A competition of growth between homogeneously and heterogeneously nucleated crystals is observed in the binary alloy as depicted in Fig. 1.

Table 1 Pure component properties of polystyrene sample PS68 and PS100. $2a_{nom}$: nominal diameter; Z: effectively transported charge from conductivity; n_m: particle number density at melting point; B: Wilson-Frenkel fitting parameter in $v = v_\infty(1-\exp(-B\Pi_f^*))$; v_∞: Fitted limiting growth velocity at infinite "undercooling"

Sample	Source Batch No:	$2a_{nom}$ [nm]	Z^*_σ	Crystal structure ($n > n_m$)	$n_f/\mu m^{-3}$	n_m [μm^{-3}]	B (k_BT)	v_∞ [$\mu m\ s^{-1}$]
PS68	BASF ZU 2168/7387	68	460 ± 16	bcc	6.0 ± 0.2	6.2 ± 0.2	2.0 ± 0.1	15.9 ± 0.4
PS100	Bangs Lab 3067	100	530 ± 38	bcc	3.8 ± 0.2	4.4 ± 0.2	2.8 ± 0.4	7.2 ± 0.3

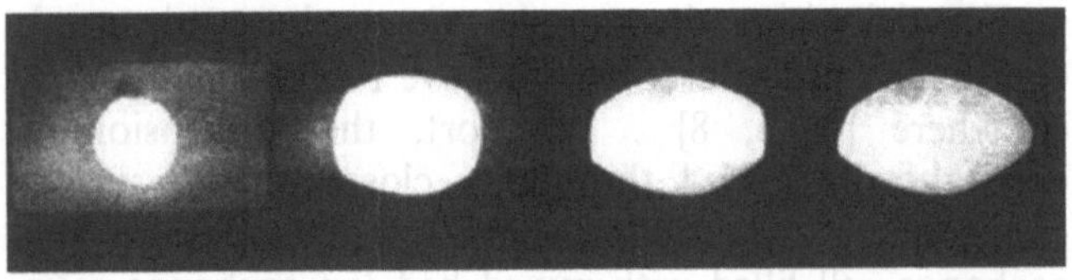

Fig 1 Side view of competitive growth between a homogeneously nucleated crystal (white part) and two heterogeneously nucleated wall crystals (black part). Time is increasing from **a** to **d**, respectively: 37 s; 65 s; 77 s; 98 s. Grey part in the centre of the cell is metastable fluid during the initial nucleation process. The sample corresponds to $p=0.5$ at $n=11.12 \pm 1.50\ \mu m^{-3}$. Due to the competition a lens shape of the central crystal results

The crystal growth as a competitive process is well known for monodisperse samples [12, 13], but it has not yet been demonstrated for mixtures. A heterogeneously nucleated wall crystal grows both from top and bottom (black part), whereas a homogeneously nucleated crystal grows in the central part (white part). A metastable fluid-like region also exists at the initial time (grey colour part in the centre of the cell). At $t=37$ s, we observe the central crystal showing almost perfect round shape. It is only in contact with metastable fluid, the wall crystal is still far away from the central crystal. At $t=65$ s, wall crystals have contacted the centre crystal, thus both are hindered in their growth. With time further increasing, the central crystal can only grow perpendicular to the direction of wall crystal growth, thus a final lens-like crystal is formed enclosed by wall crystals without fluid left. Π_f^* is calculated from the pair-interaction potential with additional polydispersity term [Eq. (2)]. Then a fit to Wilson-Frenkel law is performed. This yields both B and v_{110} for the pure components and the mixtures.

We fit the growth velocities as a function of particle number density n in the style of Wilson-Frenkel (WF) law as $v_{110}=10.04(1-\exp(-0.45(n-2.74))$ for $p=0.2$, (shown in Fig. 2a); $v_{110}=10.25(1-\exp(-0.48(n-4.4))$ for $p=0.5$, (shown in Fig. 2c). Growth velocities as a function of the difference of chemical potential $\Delta\mu$ are fitted with the WF law according to: $v_{110}=10.04(1-\exp(-\Delta\mu/k_BT)$ for $p=0.2$ (shown in Fig. 2b) and $v_{110}=10.25(1-\exp(-\Delta\mu/k_BT)$ for $p=0.5$ (shown in Fig. 2d). For monodisperse samples, we know that the WF law provides very good data fits, especially for samples with a narrow fluid-crystal coexistence region [4]. Here, we demonstrate that the Wilson-Frenkel law can also be applied to binary mixtures. The limiting growth velocities increase from 7.2 $\mu m \cdot s^{-1}$ to 15.9 $\mu m \cdot s^{-1}$ with increasing p from 0 to 1. According to [4], the limiting velocity in the one component case is given as $v_\infty=0.1D_0\, d_{interface}/d_{NN}^2$ with nearest neighbour distance $d_{NN}\approx n^{-1/3}$, interfacial layer thickness $d_{interface}$, and the Stokes-Einstein diffusion coefficient D_0. Assuming formation of bcc crystals with random composition and an unchanged $d_{interface}$ [12], one would expect v_∞ to vary linearly with p due to the linear variation of an average diffusion coefficient $\overline{D_0}=pD_1+qD_2$. In that case, values of 8.9 $\mu m \cdot s^{-1}$ and 11.6 $\mu m \cdot s^{-1}$ are expected for $p=0.2$ and $p=0.5$, respectively. Our data do not

Fig. 2 Crystal growth velocities for colloidal mixtures with number fraction $p=0.2$ (**a, b**) and $p=0.5$ (**c, d**) as function of particle number density n (**a, c**) and difference in chemical potential $\Delta\mu/k_BT$ (**b, d**). Symbols show the experimental data for $p=0.2$ (—○—) at $p=0.5$, (—△—), respectively. Solid lines correspond to Wilson-Frenkel fits

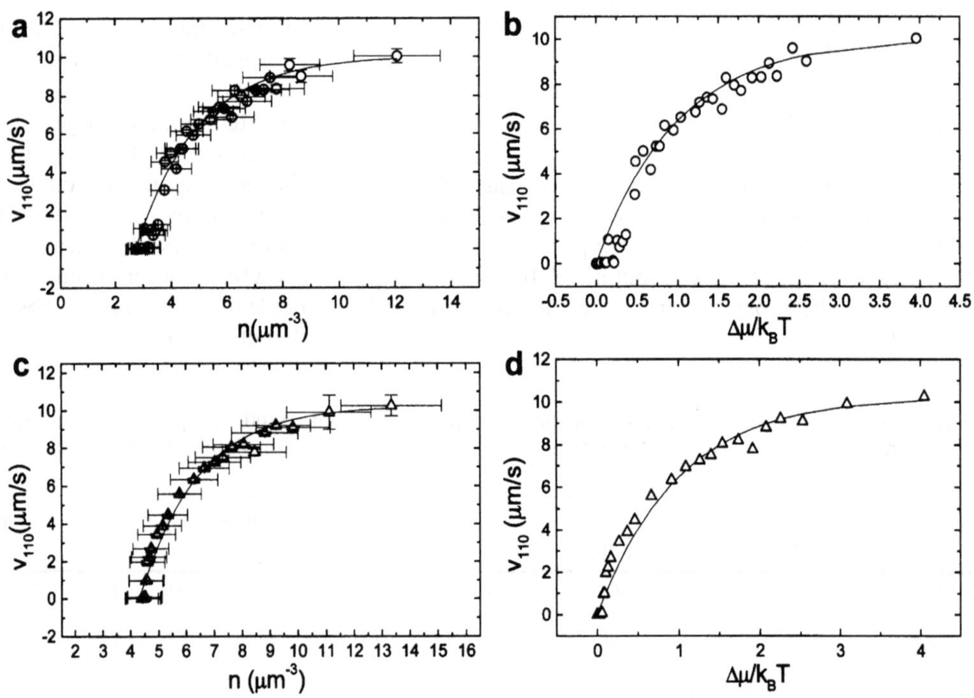

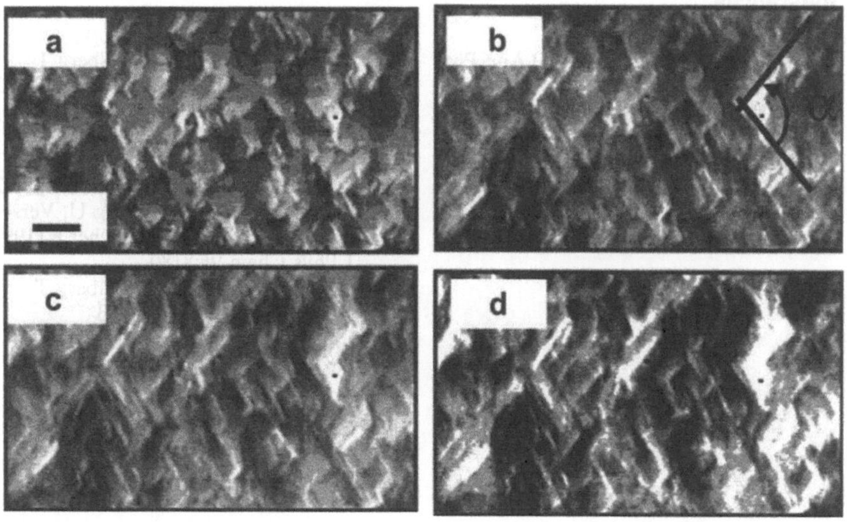

Fig. 3 Top view of a wall crystal. PS68/PS100 binary mixture with p = 0.2, n = 6.74 ± 0.87 μm^{-3}. Images taken at **a**) 9 s; **b**) 20 s; **c**) 30 s; **d**) 69 s. The bar represents 100 μm. A regularly distributed zig-zag pattern is observed to coarsen with time

strictly confirm this hypothesis of random composition. Further experiments on different p are under way.

Finally, we obtained time resolved microscopy images of zig-zag patterns in our two component system. Pictures taken with the cell turned by 90° are shown in Fig. 3. They correspond to a top view of a wall crystal.

At small t, little dark and bright areas are arranged under a mutual angle of $\alpha \approx 98°$. The pattern coarsens in terms of area extension and contrast. Similar, but cloud like pattern were observed before in single component systems and attributed to twin domains. Here we suspect the patterns to be due to twin formation in an alloy structure. In Fig. 3a to Fig. 3d, the zig-zag pattern shows a coarsening process with increasing time. The same Bragg colour for all domains indicates that crystals are orientated in the same direction. We conclude that, like for one component systems, we here observe twin domain patterns. We find such patterns in many particle number densities n both for mixing number ratio p = 0.2 and p = 0.5. Here we show p = 0.2, n = 6.74 ± 0.87 μm^{-3} as an example. We found the one component melting points of PS68 and PS100 as $n_{mPS68} = 6.2 \pm 0.2$ μm^{-3} and $n_{mPS100} = 4.4 \pm 0.2$ μm^{-3}, respectively. In the p = 0.2, n = 6.74 ± 0.87 μm^{-3} mixture, the one component particle number density should be $n_{PS68} = 1.35$ μm^{-3} and $n_{PS100} = 5.39$ μm^{-3}, respectively. So for such a low particle number density n_{PS68} ($<< n_{mPS68}$), PS68 cannot form a crystal lattice by its own. But we do not observe fluid voids in these twin domains. In addition, an even lower melting point $n_{m,p0.2}$ ($\approx 3.03 \pm 0.39$ μm^{-3}) (see Fig. 2a) is found for p = 0.2. A possible explanation for these behaviours could be that compound crystals are formed. Furthermore, the zig-zag pattern, which was only found in the mixture may correspond to a microscopic morphology of this compound crystal.

Conclusion

Bragg microscopy has been shown to serve as a versatile tool to study the morphology and crystal growth kinetics in mixing colloidal suspensions. The limiting growth velocity is found to increase from 7.2 μm·s^{-1} to 15.9 μm·s^{-1} with p increasing from 0 to 1. For all samples both one-component and mixture the Wilson-Frenkel law was confirmed well except samples with a large fluid-crystal coexistence width. We further observed twin domain patterns of morphologies different between pure components and mixtures. Our findings promote more interesting questions: Why do Wilson-Frenkel law fits show a larger deviation for samples with a larger fluid-crystal coexistence region? Is the mixed crystal microscopic structure actually a kind of superlattice or what is the detailed structure of a zig-zag pattern in a two- component mixture? Why does small amount of polydispersity lower the melting point as compared to a one- component system? These questions should be explored in more detail by further research.

Acknowledgements We would like to thank H.J. Schöpe and R. Biehl for helpful discussions, and we would also like to thank Dr. E. Bartsch for careful reading this paper and helpful discussions. Financial support of the DFG (Pa 459/8-1,2 and 459/9-1), the SFB 262 and the Materialwissenschaftliches Zentrum, Mainz (MWFZ) is gratefully acknowledged.

References

1. Becker B, Döring W (1935) Ann Phys 24:719
2. Wilson HA (1900) Philos Mag 50:238; Frenkel JZ (1933) Sowjetunion 1:498
3. Lindsay HM, Chaikin PM (1982) J Chem Phys 76(7):3774
4. Palberg T (1999) J Phys Condens Matter 11:R323
5. Maroufi MR, Stipp A, Palberg T (1998) Prog Colloid Polym Sci 108:83
6. Liu J, Stipp A, Palberg T (2001) Prog Colloid Polym Sci 118:91
7. Wette P, Schöpe HJ, Liu J, Palberg T (2002) Prog Colloid Polym Sci 123:82
8. Palberg T, Härtl W, Wittig U, Versmold H, Würth M, Simnacher E (1992) J Phys Chem 96:8180
9. Hessinger D, Evers M, Palberg T (2000) Phys Rev E 61:5493
10 Liu J, Schöpe HJ, Palberg T (2000) Part Part Syst Charact 17:206
11. Patrick W, Schöpe, HJ, Biehl R, Palberg T (2001) J Chem Phys 114:7556
12. Aastuen DJW, Clark NA, Cotter LK (1986) Phys Rev Lett 57(14):1733
13. Palberg T, Mönch W, Schwarz J, Leiderer P (1995) J Chem Phys 102(12):5082

Progr Colloid Polym Sci (2004) 123:227–230
DOI 10.1007/b11860

D. Pontoni
T. Narayanan
A.R. Rennie

Nucleation and growth kinetics of colloidal silica

D. Pontoni · T. Narayanan (✉)
European Synchrotron Radiation Facility, BP 220, 38043 Grenoble Cedex, France
e-mail: narayan@esrf.fr

A.R. Rennie
Department of Chemistry, King's College London, Strand, London WC2R 2LS, United Kingdom

Abstract A stopped-flow, time-resolved small-angle X-ray scattering study of nucleation and growth of Stöber silica particles is presented. The measured scattered intensity over the scattering wave vector (q) range of, $0.02 \leq q \leq 0.6\ nm^{-1}$ can be described by a polydisperse sphere scattering function during the entire growth process. The smallest observable primary particles are of a few nanometers in radius. In the early stages of growth, the fitted radius increased linearly with time that subsequently crossed over to a smaller exponent between 1/3 and 1/2. In addition, particle number and mass densities remained nearly constant after the initial nucleation stage.

Introduction

The Stöber method [1] is a well known procedure for the synthesis of monodisperse colloidal silica. Although, this method has been widely investigated [1–5], a comprehensive understanding of the mechanism underlying this synthesis is lacking to date. The standard procedure involves the base catalysed hydrolysis and condensation of silicon alkoxides in ethanol solution containing ammonia. In the previous studies [2–5], considerable effort has been made to elucidate the rate-limiting steps, (1) hydrolysis of silicon alkoxide, (2) condensation or formation of primary particles, and (3) aggregation of primary particles to form spherical colloidal particles, involved in the reaction. Based on these studies, there are two leading arguments, (1) the growth rate is determined by the nucleation and aggregation of small primary particles [4], (2) monomers are continuously added on to large particles and the growth rate is limited by the hydrolysis reaction [5]. Static light scattering studies concluded that the growth proceeds through a surface reaction-limited condensation of hydrolysed monomers [2]. However, static light scattering is not sensitive to detect the early structural signal from the primary nuclei. Recent small-angle X-ray scattering (SAXS) studies [6] indicated that initial nuclei are highly ramified fractal structures that subsequently aggregate and condense to form more compact objects.

The availability of high brilliance synchrotron radiation sources permits time-resolved SAXS measurements on dilute dispersions with unprecedented resolution and dynamic range. As a result, it is possible to follow the growth process from the very early stage nucleation to the final colloidal particles in a single experiment. This paper reports a time-resolved SAXS study of the sequence of events from the nucleation of primary particles to the final stable colloidal dispersion.

Materials and methods

The chemicals, TEOS (Fluka), 25% ammonia solution (Prolabo), and absolute alcohol (Prolabo) were used as purchased. The growth process was initiated by mixing two stock solutions of ammonia and TEOS in ethanol in equal volumes. The resulting concentrations of the reacting mixtures were [TEOS] = 0.09, $[NH_3] = 1.45$ and $[H_2O] = 4.15$ moles/liter. These concentrations were chosen as an intermediate between those used in earlier studies [2–4] and to complete the whole growth process within a short period. The stopped-flow apparatus consisted of two pneumatically driven syringes and a mixing chamber which is coupled to a thin walled flow-through capillary (2 mm diameter and wall thickness 10 μm). In order to reduce the parasitic background, the capillary was mounted in vacuum without any windows in the entire flight path of the incident and transmitted beams. The combined mixing and transfer dead times were less than 10 ms.

Small-angle X-ray scattering measurements were performed on the ID2 beamline at the European Synchrotron Radiation Facility, Grenoble, France [7]. The incident X-ray wavelength (λ) was 0.1 nm. In order to cover a wide q-range, $0.02 \leq q \leq 6\ nm^{-1}$, two different sample-to-detector distances were used (1.5 and 10 meters). Here, $q = (4\pi/\lambda)\sin(\theta/2)$, with θ the scattering angle. The 2-dimensional SAXS patterns were recorded with an image intensified CCD camera [7]. The incident and the transmitted fluxes were also registered simultaneously with each SAXS pattern. Typically, a sequence of 120 frames was acquired after each mixing. The dead time between the frames was varied in a geometric progression. The standard data treatment involved different detector corrections for flat field response, spatial distortion, and dark current of the CCD, and normalisation for the incident flux, sample transmission, exposure time and the angular acceptance of the detector pixel elements [7]. The resulting normalised 2-dimensional images were azimuthally averaged to obtain $I(q)$ which essentially refers to the differential scattering cross-section $d\Sigma/d\Omega$ in units of mm^{-1} $sterad^{-1}$.

To complement the time-resolved SAXS measurements, combined SAXS/WAXS (wide-angle X-ray scattering) measurements were also performed with the same reaction protocol and time resolution. Such high sensitive combined SAXS/WAXS measurements can reveal if sub-nanometric primary particles were formed during the nucleation stage. The beam intensity was optimised in order to reduce the beam induced degassing of the dissolved ammonia. During the early stages, the measured $I(q)$ at small q was dominated by these micro bubble scatterings if the full beam intensity (typically 10^{13} photons/s) was used. As a result the beam intensity was reduced by a factor of 20 and the exposure time varied from 100 milliseconds to a few seconds.

Results and discussion

Typical time evolution of SAXS intensity after the mixing is depicted in Fig. 1. Similar features were observed in several data sets acquired under different conditions. The oscillations in the intensity at the later

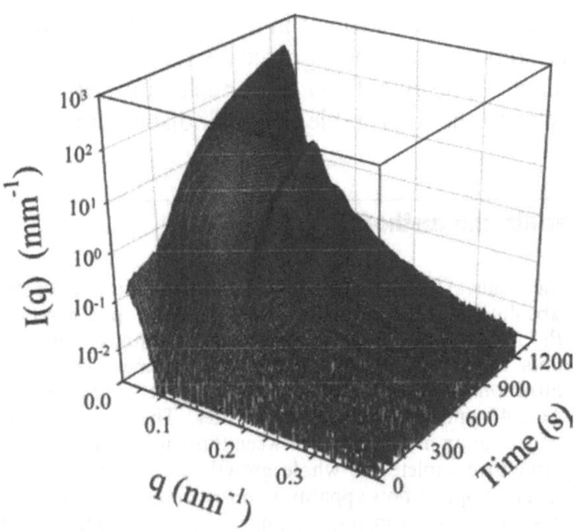

Fig. 1 Background subtracted normalised SAXS intensity during the Stöber synthesis of silica particles. Measurements correspond to a sample to detector distance of 10 m and spanned over 20 minutes

stages of growth process readily indicate the development of form factor of spherical particles. During the initial induction time, the intensity evolved marginally only in the intermediate q-range. The spherical form factor became evident after about 200 seconds of growth. The maxima and the minima progressively shifted to the low-q region signifying the growth of the particles. $I(q)$ of a suspension of uniform non-interacting spherical particles can be described by [8],

$$I(q) = N(\rho_c - \rho_s)^2 V^2 P(q) \tag{1}$$

where N is the particle number density, ρ_c and ρ_s are the average electron densities of the particle and the solvent, respectively, $V = 4/3\pi R^3$ is the volume and P(q) is the form factor of a sphere of radius R which has the following analytical form:

$$P(q) = \left[3(\sin(qR) - qR\cos(qR))/q^3R^3\right]^2 \tag{2}$$

From Eqs. (1) and (2), the intensity at $q=0$, is $I_0 = N(4/3\pi R^3)^2(\rho_c - \rho_s)^2$. If the particle number and mass densities were conserved during the growth process, then I_0/R^6 should remain constant. However, there is a finite distribution of particle size and Eq. (1) has to be weighted over this size distribution. In this case, it is sufficient to use the Schultz distribution function [9, 10] to describe the polydispersity of the particles. Therefore, the measured scattered intensities were fitted to the polydisperse sphere scattering function given by Kotlarchyk and Chen [9] and Aragon and Pecora [10].

The scattered intensity in Fig. 1 at selected time intervals is shown in Fig. 2. The data over the whole q-range can be adequately described by Eq. (1) weighted over the Schultz distribution function as shown by the continuous lines. This implies that particles are dense and have a sharp interface – Porod behaviour [8]. The best fit I_0/R^6 ratios remained nearly a constant ($\sim 4 \times 10^{-15}\ nm^{-7}$) after the initial nucleation stage [11], signifying constant particle number (N) and mass densities $\propto(\rho_c - \rho_m)$ during the growth process. The fitted polydispersity decreased from about 20% to below 10% towards the late stage of growth. The time dependence of the fitted radius is depicted in Fig. 3. The fitted radius increased linearly with time at the early stages of growth followed by a smooth crossover to a smaller exponent between 1/2–1/3. In this time-range, the fitted $I(q)$ at low q's became higher than the experimental data signifying the onset of repulsive interactions between the colloidal particles. The fitted radius for $t < 100$ s deviated from the power law.

The above results demonstrate that $I(q)$ data over the entire Stöber growth process can be described by a model of polydisperse spheres of uniform density. For the reaction conditions used in this study, the induction time associated with the nucleation of particles is fairly short (< 100 seconds). During this time, $I(q)$ evolved margin-

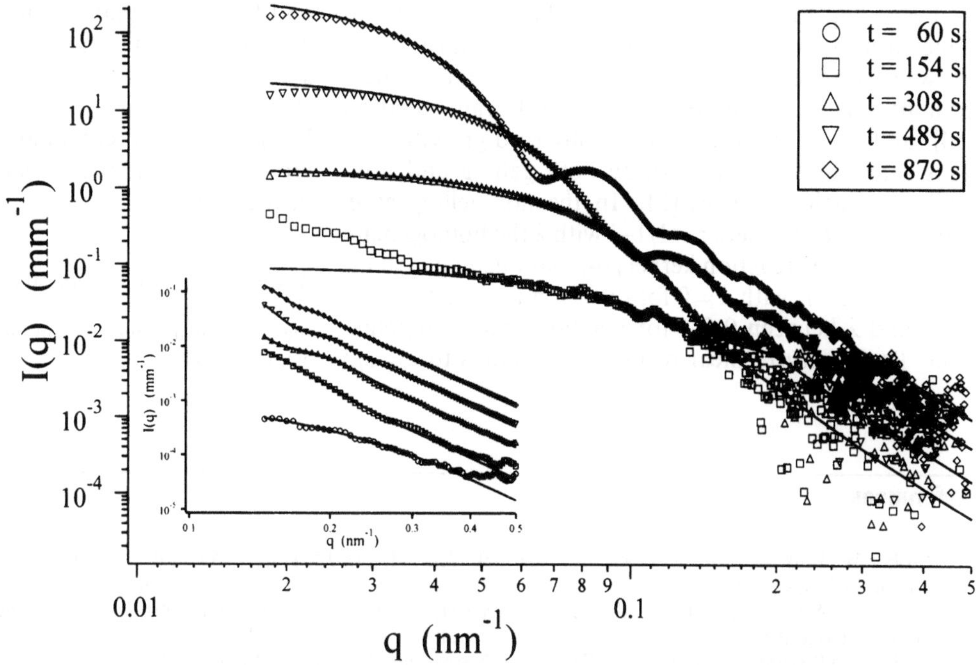

Fig. 2 SAXS intensity at different stages of the growth for the data presented in Fig. 1. The continuous lines are fits to the polydisperse sphere form factor. The earliest analysable data corresponds to nuclei of 3 nm in radius formed within 1 minute after mixing. The inset depicts the data over a small q-window obtained for a sample-to-detector distance of 1.5 m

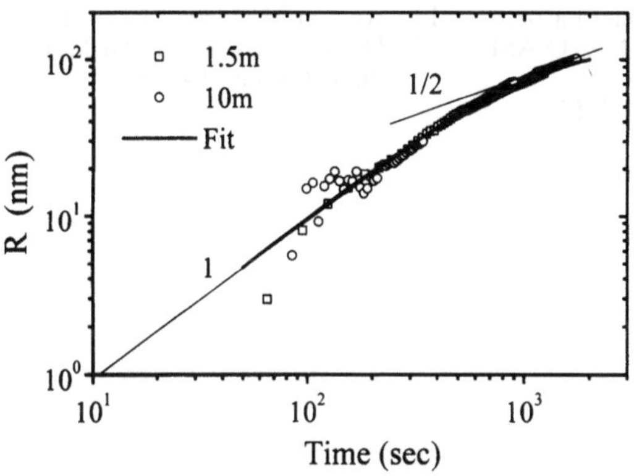

Fig. 3 The time evolution of the deduced radius of the particles during the growth process. The thick continuous line is a fit to the Avrami expression, $R = R(\infty)\{1-\exp[-(t/\tau)^n]\}$, with $R(\infty) = 126$ nm, $\tau = 964$ s and $n = 1.06$. The straight lines with slope 1 and 1/2 are only guides to the eye

ally in the intermediate q-range ($0.1 < q < 0.5$ nm^{-1}) which is attributed to the primary nuclei. At the same time, there was not any observable change in the intensity over the high q range up to 6 nm^{-1} – the maximum q value probed in SAXS and in the WAXS patterns recorded under identical conditions. This suggests that the smallest detectable entities are of nanometric size. Within the statistical uncertainties, the scattering from these primary nuclei is better described by a spherical form factor with q^{-4} behaviour than a fractal structure function as depicted in the inset of Fig. 2. This implies that the initial nuclei (radius~3 nm) have droplet like form. In the subsequent stages of growth, presumably these droplets coalesced to form larger nuclei, thus resulting in dense compact particles.

After the initial nucleation stage, the radius of the particles increased linearly with time. The polydispersity of the system evolved until it reached a steady value of about 10%. In the classical Lifshitz-Slyozov [12] nucleation and growth process, one would expect a power law of either 1/2 or 1/3 during the early stage of nucleation depending on the degree of supersaturation (high and low, respectively). This suggests that the early stage growth of silica particles is different from a purely diffusion controlled process and cannot be readily explained by classical theory of nucleation and growth. Throughout the growth process, the deduced I_0, which is proportional to the square of the molecular mass, remained proportional to R^6 suggesting that the particles are dense (dimensionality, $D = 3$) and the number density of the dominantly growing spheres is constant. However, the possibility of the formation of fresh nuclei cannot be excluded since the measured intensity is completely dominated by the largest, fastest growing particles. In fact, the deduced morphology of the particles suggests that the growth is likely to have proceeded by the addition of the primary nuclei on to the larger particles. The decrease in growth rate at later stages can be attributed to the depletion of the primary nuclei in the reservoir [3].

The time dependence R can alternatively be described by the Avrami expression [13], $R(t) = R(\infty)\{1-\exp[-(t/\tau)^n]\}$, where $R(\infty)$ is the final radius, τ is the growth rate and n is an exponent that depends on the specific nucleation mechanism. The continuous line in Fig. 3 shows that the entire data can be described with $n = 1.06$, which is also in agreement with a previous study [6].

Following the induction period, there was sufficient concentration of primary nuclei and the early stage growth rate is determined by the coagulation rate of these primary nuclei. Therefore, Smoluchowski rate equation [14] can be used to rationalise the observed growth laws. This rate equation can describe both gelling and non-gelling growth processes [14]. In the non-gelling case, typically $R \sim t^{z/D}$, where $z = 1/1-\lambda$ with λ the homogeneity exponent of the reaction kernel [14]. Non-gelling systems pertain to $\lambda < 1$, with $\lambda = 0$ for diffusion limited aggregation, and λ between 0.5–1 for reaction limited aggregation. The observed linear regime corresponds to $\lambda = 2/3$ and the later behaviour refers to λ between 0.33 and 0. This implies that the initial reaction condition corresponds to a reaction limited growth while at the later stage the mechanism cross-over towards a diffusion type behaviour. The colloidal stability is attained in this crossover region and the size distribution approached an invariant form.

Acknowledgements Technical assistance of S. Finet and J. Gorini, and the provision of beam time and the financial support by the European Synchrotron Radiation Facility are gratefully acknowledged.

References

1. Stöber W, Fink A, Bohn EJ (1968) J Colloid Interface Sci 26:62
2. Blaaderen A van, Geest J van, Vrij A (1992) J Colloid Interface Sci 154:481
3. Phillipse AP (1988) Colloid Polym Sci 266:1174; Phillipse AP, Vrij A (1988) J Chem Phys 87:5634
4. Bogush GH, Tracy MA, Zukoski CF (1988) J Non-crystalline Solids 104:95; Bogush GH, Zukoski CF (1991) J Colloid Interface Sci 142:19
5. Matsoukas T, Gulari E (1991) J Colloid Interface Sci 145:557 (and references therein)
6. Boukari H, Lin JS, Harris MT (1997) J Colloid Interface Sci 194:311; Boukari H, Long GG, Harris MT (2000) J Colloid Interface Sci 229:129
7. Narayanan T, Diat O, Boesecke P (2001) Nucl Instr Methods Phys Res A 467:1005
8. Brumberger H (1995) Modern aspects of small-angle scattering. NATO ASI Series. Kluwer, Dordrecht
9. Kotlarchyk M, Chen S-H (1983) J Chem Phys 79:2461
10. Aragon SR, Pecora R (1976) J Chem Phys 64:2395
11. Pontoni D, Narayanan T, Rennie AR (2002) Langmuir 18:56
12. Gunton JD, San Miguel M, Sahni PS (1983) In: Domb C, Lebowitz JL (eds), Phase transitions and critical phenomena, vol 8. Academic Press, London, p 267
13. Avrami M (1939) J Chem Phys 7:1103
14. Dongen PGJ van, Ernst MH (1985) Phys Rev Lett 54:1396

Progr Colloid Polym Sci (2004) 123: 231–235
DOI 10.1007/b11949

J. Klich
M. Paluch

Properties of some mixed adsorption films at the water/air interface

J. Klich · M. Paluch (✉)
Department of Physical Chemistry and Electrochemistry,
Faculty of Chemistry,
Jagiellonian University, Ingardena 3,
30-060 Kraków, Poland
e-mail: paluch@chemia.uj.edu.pl
Tel.: +48-126336377,
Fax: +48-126340515

Abstract The properties of mixed adsorption films of aniline-*p*-toluidine and *p*-fluoroaniline-*p*-toluidine at the water/air interface are presented. These properties were studied by surface tension and surface potential measurements carried out as a function of total concentration of both substances in aqueous solution at the constant molar fraction of one compound (*p*-toluidine or *p*-fluoroaniline). Using the result obtained and applying Gibbs adsorption equation, Motomura's method and Helmholtz formula behaviour of molecules in mixed adsorption films were determined. The weak repulsive interactions between molecules of *p*-toluidine and *p*-fluoroaniline and between molecules of aniline and *p*-toluidine in mixed films were observed.

Keywords Surface tension · Surface potential · Interactions in mixed adsorption films

Introduction

The free water surface has different properties than the interior of the liquid phase due to the fact that intermolecular cohesive forces in the surface phase are uncompensated. As a result of this the surface tension occurs, which can be defined as tangential force, which counteracts the increase of the surface area.

Another property of free water surface is the existence of the electric potential drop. On the free surface, water dipoles passing through the interface in the process of evaporation and condensation become oriented – according to the laws of electrostatics – with that part of the molecule exhibiting greater electric field density pointing towards the medium with greater dielectric permittivity, hence hydrogen atoms point towards the water phase and oxygen atoms towards the air. As a result of water dipole orientation, an electric double layer is formed on the free water surface and an electric potential drop occurs. The free water surface is negatively charged on the air side and positively charged on the side of the bulk phase.

The composition of the water surface layer is changed when molecules of surface active compound are adsorbed. Much more complicated is the situation when there are two or more surface active compounds in solution. Then a mixed adsorption film is generated. Mixtures of surfactants, rather than single surface active compounds, are commonly used in practical applications. This is because the interfacial properties of such mixtures are usually superior to those of system containing only one surfactant [1, 2]. The behaviour of mixed monolayers is mainly dependent on the structure of the surfactants involved. Ideal behaviour is expected for films formed of homologous compounds. Deviations from ideality are observed for monolayers that consist of surfactants with different structures (charges of the head groups, length of hydrocarbon chains, presence of side chains), temperature, pH, the solution ionic strength, and the extent of the surface coverage [3–5].

In this paper we report the results of investigations on the surface properties of aqueous solution of binary mixtures of aniline-*p*-toluidine and *p*-fluoroaniline-*p*-toluidine. These substances have the same hydrophilic part (group $-NH_2$), but different hydrophobic ones (groups C_6H_5-, $CH_3C_6H_4$-, FC_6H_4-). The difference in the polarity of the hydrophobic parts of molecules should give values of an opposite sign for the surface potential of their solution and the change in the surface potential would depend on the composition of the mixed film, the surface orientation angles of adsorbed molecules and interaction between them.

Experimental

Aniline, p.a. (Sigma-Chemie, Germany) and *p*-fluoroaniline, p.a. (Sigma-Chemie, Germany), *p*-toluidine, p.a. (Fluka) potassium chloride, p.a. (POCh, Poland), water (four- times distilled) were used as experimental materials. These substances were not purified further.

The surface potential measurements (ΔV) were based on the flowing jet method, which was described in detail by Kamieński [6] and therefore only a general outline of the method as well as some details directly concerning the apparatus used are given here. The solvent and solution were made to flow, the latter in a column down the centre of a vertical tube, the former along the inside wall of the tube, such that there were constantly renewed surfaces of the solvent and the solution of a large area in fairly close proximity. The vessels with outflowing solutions were connected to Lindemann's quadrant electrometer by means of a 0.1 M. calomel electrode in contact with the central jet, which is connected to the measuring instrument needle. The potential of the flowing system was measured:

Hg, Hg_2Cl_2 |0.1 M KCl solution| air

|0.1 M KCl solution| Hg_2Cl_2, Hg

As long as both solutions were identical, the potential between them cancelled each other out, yielding zero. When, however, a surface active substance was added to one of the solutions, the adsorption of solute dipoles on the free surface changed the potential drop, increasing or decreasing it to some particular extent. The method gives erroneous results for solutions of substances adsorbed slowly to the surface (large size molecules) but it was reliable for solutions of investigated compounds. The above was checked by a radioactive method with a surface of several minutes lifetime. No significant differences in surface potential values provided by the two methods were observed. The entire assembly was placed in Faraday cage. The measurement were performed at room temperature (ca. 293 K), measurements accuracy was ± 5 mV.

The surface tension was measured by tensiometer LAUDA TD1 at constant temperature 293 K. The accuracy of measurements was ± 0.1 mN/m.

Aniline, *p*-fluoroaniline and *p*-toluidine were dissolved in 0.1 M potassium chloride solution in order to reduce the streaming potential, which may occur in the flowing jet method.

Result and discussion

The surface tension and the surface potential were measured as a function of concentration of aniline, *p*-fluoroaniline and *p*-toluidine aqueous solutions and their mixtures at the constant molar fraction of *p*-toluidine and *p*-fluoroaniline depending on the total molarity of aniline-*p*-toluidine and *p*-toluidine-*p*-fluoroaniline, respectively. The molar fraction of *p*-toluidine or *p*-fluoroaniline was 0.2, 0.4, 0.5, 0.6 and 0.8.

Figures 1 to 4 show the results obtained. As we can see all investigated compounds are surface active. They decrease the surface tension (Figs. 1 and 2) of water and change the surface potential (Figs. 3 and 4). The surface potential was changed by *p*-fluoroaniline in the opposite way then aniline and *p*-toluidine. Adsorbed molecules of *p*-fluoroaniline increase of the natural surface potential of water. The increase is to be expected only if we assume that molecules of *p*-fluoroaniline orient themselves at the

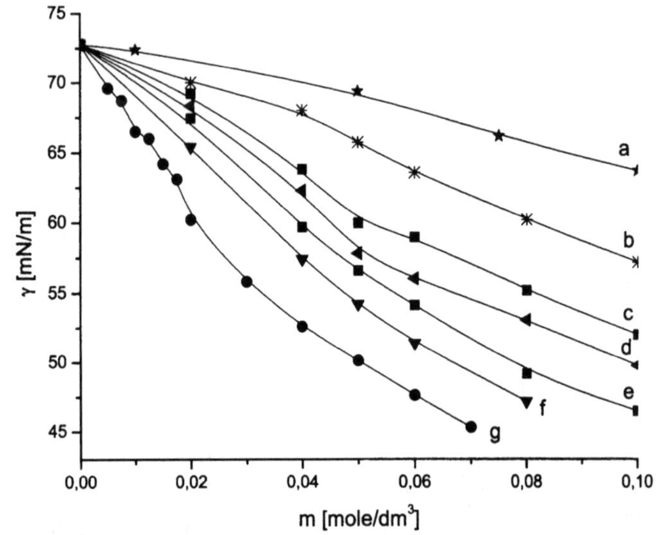

Fig. 1 Surface tension as a function of total molarity concentration of aniline and *p*-toluidine at constant mole fraction of *p*-toluidine in aqueous solution (curves: *a* – pure aniline, *b* – 0.2, *c* – 0.4, *d* – 0.5, *e* – 0.6, *f* – 0.8, *g* – pure *p*-toluidine)

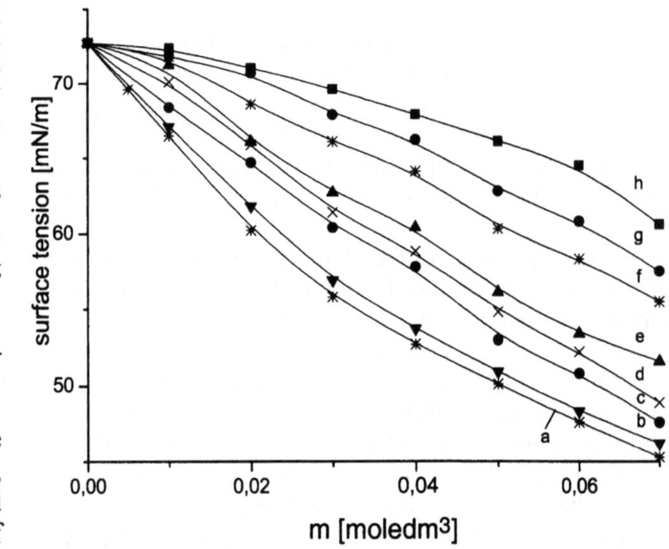

Fig. 2 Surface tension as a function of total molarity concentration of *p*-fluoroaniline and *p*-toluidine at constant mole fraction of *p*-fluoroaniline (curves: *a* – pure *p*-toluidine, *b* – 0.2, *c* – 0.4, *d* – 0.5, *e* – 0.6. *f* – 0.8, *g* – 0.9, *h* – pure *p*-fluoroaniline)

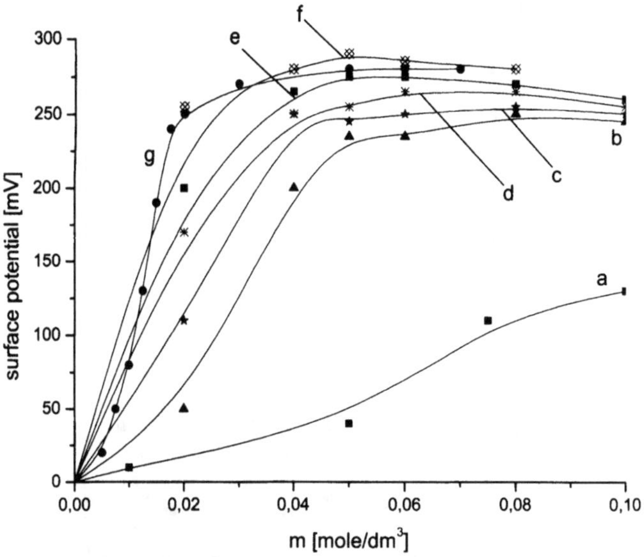

Fig. 3 Surface potential as a function of total molarity concentration of aniline and *p*-toluidine at the constant mole fraction of *p*-toluidine (curves: *a* – pure aniline, *b* – 0.2, *c* – 0.4, *d* – 0.5, *e* – 0.6, *f* – 0.8, *g* – pure *p*-toluidine)

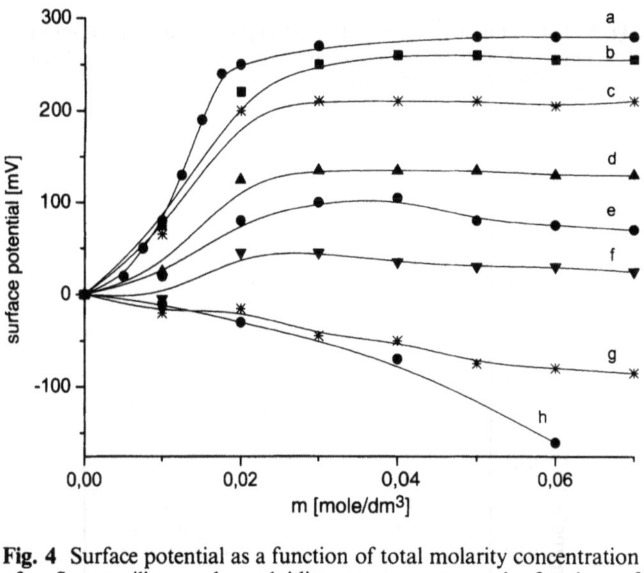

Fig. 4 Surface potential as a function of total molarity concentration of *p*-fluoroaniline and *p*-toluidine at constant mole fraction of *p*-fluoroaniline (curves: *a* – pure *p*-toluidine, *b* – 0.1, *c* – 0.2, *d* – 0.4, *e* – 0.5, *f* – 0.6, *g* – 0.8, *h* – pure *p*-fluoroaniline)

phase boundary with the -NH_2 in the direction of water phase and removing a number of oriented water molecules from the free surface charging the interface their own fields.

Aniline, *p*-toluidine and – *p*-fluoroaniline are weak electrolytes. In aqueous solution their dissociation constant are: 3.8*10^{-10} (pK = 9.42), 4.5*10^{-10} (pK = 9.35), 1.32*10^{-9} (pK = 8.88) for aniline, *p*-toluidine and *p*-fluoroaniline, respectively [7]. From earlier investigations is known that molecules which do not belong to typical surfactants have surface active properties only in undissociated form [8]. Ions do not appreciably change the electric surface potential and surface tension at the free surface of aqueous solutions. The result of measurements of the surface potential and surface tension presented in this paper were carried out for aqueous solutions which pH was between 7.1 to7.3, 7.3 to 8.2, 7.5 to 8.4 for aniline, *p*-toluidine and *p*-fluoroaniline, respectively. It was pH of solutions in the absence of any buffer in the solution and only potassium chloride was present. In this pH values in solution exist 1.26*10^{-3}% for aniline, 1.58*10^{-3}% for *p*-fluoroaniline and 2.51*10^{-3}% for *p*-toluidine molecules in ionised form. Therefore, we do not take it for further consideration.

Addition of *p*-toluidine to the solution of aniline and *p*-fluoroaniline to the solution of *p*-toluidine causes a strong change of the surface tension and surface potential of these solutions (Figs. 1 to 4). The surface tension (Figs. 1 and 2) decreases with increasing total molarity at a given molar fraction of *p*-fluoroaniline and *p*-toluidine and changes regularly with composition from that of pure *p*-fluoroaniline or *p*-toluidine to that of pure *p*-toluidine or aniline solution, respectively.

The composition of mixed adsorption films may be obtained from the Gibbs adsorption equation. For the three-component system (two solutes, 2 and 3, in an aqueous medium, 1) at constant temperature the Gibbs equation has the form presented below (assuming that surface excess of water molecules Γ_1 equals zero):

$$d\gamma = -\Gamma_2 RTd(\ln a_2) - \Gamma_3 RTd(\ln a_3) \quad (1)$$

here γ is the surface tension of solution, Γ_i represents the surface excess and a_i is the activity of the i-th component of the solution. The surface excesses Γ_2 and Γ_3 may be calculated when activity of one component in solution is constant. Hence, the composition of mixed adsorbed film is expressed as:

$$X_2^S = \Gamma_2/(\Gamma_2 + \Gamma_3) \quad (2)$$

$$X_3^S = \Gamma_3/(\Gamma_2 + \Gamma_3) \quad (3)$$

where X_2^S and X_3^S define the molar fraction of surface active compounds 2 and 3 at the aqueous solution/ air interface, respectively.

Motomura and co-workers [9–12] showed that the X_2^S or X_3^S value can be calculated thermodynamically from the experimental results without relying on any specific model of the interface and plays an important role in the examination of the miscibility of the surfactants in the adsorbed film:

$$X_2^S = X_2 - (X_2X_3/m)(\partial m/\partial X_2)_{p,T} \quad (4)$$

$$m = m_2 + m_3 \quad (5)$$

$$X_2 = m_2/m \quad (6)$$

where m, m_2, m_3 and X_2, X_3 are total molarity, molarity of compounds 2 and 3 and molar fraction of compound 2 and 3, respectively.

The total surface density ($\Gamma = \Gamma_2 + \Gamma_3$) can be calculated by applying equation:

$$\Gamma = -(m/RT)(\partial\gamma/\partial m)_{p,T} \quad (7)$$

Using the results presented in Figs. 1 and 2 and Eqs. (2)–(4) and (7) the surface excesses Γ_2 and Γ_3 were determined. The obtained results there are given in Table 1.

The plots of m vs. X_2 and m vs. X_2^S at constant surface tension values appear as analogues of three-dimensional phase diagrams designating the equilibrium compositions of coexisting two phases (Figs. 5 and 6). We can see that molarity vs. surface mole fraction of *p*-toluidine in aniline-*p*-toluidine system and *p*-fluoroaniline in *p*-fluoroaniline-*p*-toluidine system only slightly deviate from the straight line connecting the m values of aniline and *p*-toluidine and of *p*-toluidine and *p*-fluoroaniline. Linearity m vs. X_2 is considered to be a standard for the ideal mixing of surfactants in mixed adsorbed film and is connected with lack of interactions between adsorbed molecules in the monolayers. The results presented in Figs. 5 and 6 indicate the repulsive interactions between molecules in mixed adsorbed film.

The change in the surface potential ΔV can be expressed by the following formula, Eq. (9) [13]

$$\Delta V = \frac{n\bar{\mu}}{\varepsilon_0 \varepsilon} \quad (9)$$

where $\bar{\mu}$ denotes the vertical component of the dipole (μ) of the monolayer-forming molecule, $\bar{\mu} = \mu \cos\theta$, θ is superficial orientation angle of adsorbed molecule, n is the number of molecules on 1 m^2 of surface, and ε is the

Table 1 Surface excesses of *p*-toluidine and *p*-fluoroaniline in mixed adsorption films of aniline-*p*-toluidine and *p*-toluidine-*p*-fluoroaniline, respectively. Data for $\gamma = 65$ mN/m

$X_{p\text{-toluidine}}$	$X_{p\text{-toluidine}}^S$	$\Gamma_{p\text{-toluidine}}*10^{-6}$	$n_{p\text{-toluidine}}*10^{17}$
0.2	0.59	4.70	0.28
0.4	0.85	4.31	0.26
0.6	0.88	3.90	0.23
0.8	0.95	3.21	0.19
$X_{p\text{-fluoroaniline}}$	$X_{p\text{-fluoroaniline}}^S$	$\Gamma_{p\text{-fluoroaniline}}*10^{-6}$	$n_{p\text{-fluoroaniline}}*10^{17}$
0.2	0.017	2.98	0.18
0.4	0.053	3.02	0.18
0.6	0.23	3.40	0.20
0.8	0.48	3.77	0.23

$X_{p\text{-toluidine}}$ – molar fraction of *p*-toluidine in bulk solution
$X_{p\text{-toluidine}}^S$ – molar fraction of *p*-toluidine in surface layer
$\Gamma_{p\text{-toluidine}}$ – surface fraction of p-toluidine
$n_{p\text{-toluidine}}$ – number of *p*-toluidine molecules on 1 m^2
$X_{p\text{-fluoroaniline}}$ – molar fraction of *p*-fluoroaniline in surface layer
$X_{p\text{-fluoroaniline}}^S$ – molar fraction of *p*-fluoroaniline in surface layer
$\Gamma_{p\text{-fluoroaniline}}$ – surface fraction of *p*-fluoroaniline
$n_{p\text{-fluoroaniline}}$ – number of *p*-fluoroaniline molecules on 1 m^2

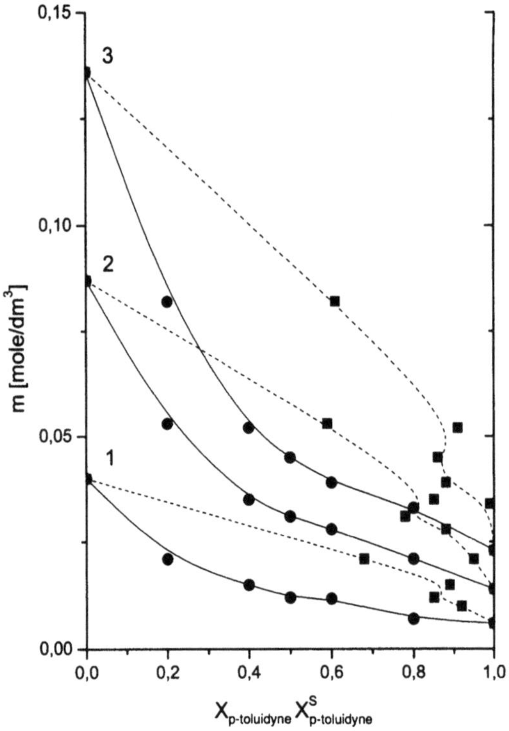

Fig. 5 Total molarity vs. mole fraction of *p*-toluidine in bulk solution (full line) and surface layer (broken line) at fixed surface tension (curves: 1–70 mN/m, 2–65, 3–60)

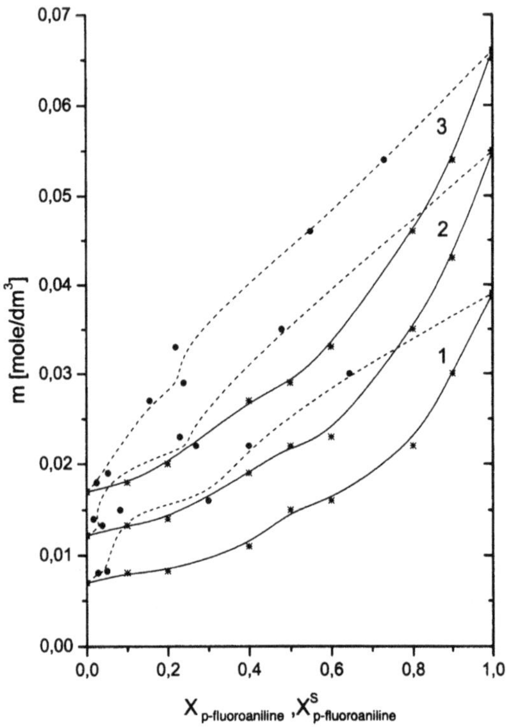

Fig. 6 Total molarity vs. mole fraction of *p*-fluoroaniline in bulk solution (full line) and surface layer (broken line) at fixed surface tension (curves: 1–68 mN/m, 2–65, 3–62)

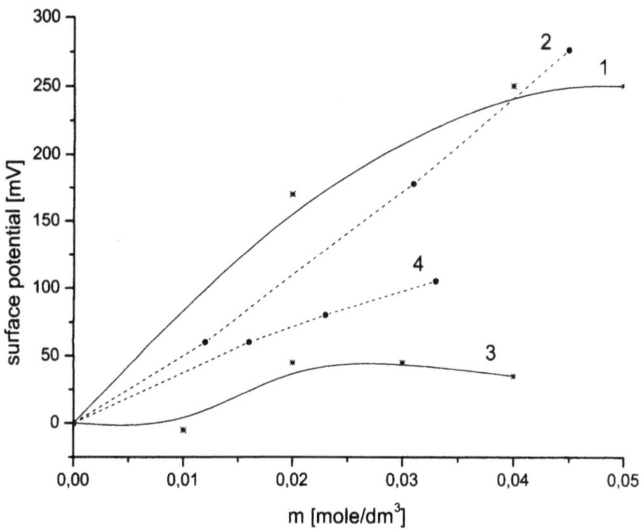

Fig. 7 Experimental (full line) and calculated (broken line) surface potential vs. total molarity concentration of aniline and *p*-toluidine at constant mole fraction of *p*-toluidine X = 0.5 (curves 1–2) and *p*-toluidine and *p*-fluoroaniline at constant mole fraction of *p*-fluoroaniline X = 0.6 (curves 3–4).

dielectric permittivity of the monolayer and ε_0 is the dielectric permittivity of vacuum. The lack of data concerning dielectric permittivity of the interfacial water region is a reason why is in often assumed to be equal $\varepsilon = 1$, hence treating water molecules as isolated entities, although some scientists claim that it is equal to 6 [14].

The Helmholtz formula [Eq. (9)] hold only for the cases when the effective dipole moment has a constant value and does not depend on the number of adsorbed molecules of different kind of substances. For mixed adsorption films the Eq. (9) has the form:

$$\Delta V = \frac{n_2\overline{\mu_2} + n_3\overline{\mu_3}}{\varepsilon_0 \varepsilon} \tag{10}$$

where n_2, n_3 and $\overline{\mu_2}$, $\overline{\mu_3}$ denotes the number of adsorbed molecules on surface area of component 2 and 3 and effective dipole moments, respectively.

We checked validity of Eq. (10) to adsorbed investigated by us films using the data presented in Table 1 and effective dipole moments for aniline ($\bar{\mu} = 1.80 * 10^{-32}$ C m), p-fluoroaniline ($\bar{\mu} = -5.70 * 10^{-32}$ C m) and p-toluidine ($\bar{\mu} = 6.00 * 10^{-32}$ C m) [15]. For the obtained results of surface potential, see Fig. 7. Curves 1–2 correspond to experimented and calculated surface potentials of aniline-p-toluidine (for $X_{p\text{-toluidine}} = 0.5$) and curves 3–4 of *p*-toluidine-*p*-fluoroaniline (for $X_{p\text{-fluoroaniline}} = 0.6$) mixed adsorption films, respectively.

As seen, experimental and calculated values of surface potentials differ significantly. It is probably connected with different of surface orientation angle of adsorbed molecules when the composition of adsorbed film is varied. In this case, the effective dipole moment of adsorbed molecule has not a constant value and in this situation is difficult to conclude something about the interaction between adsorbed molecules.

References

1. Hua XY, Rosen MJ (1982) J Coll Inter Sci 90:212
2. Rosen MJ (1986) ACS Symp Ser 311:144
3. Scamehorn JF (1986) ACS Symp Ser 311:1
4. Gu B, Rosen MJ (1988) J Coll Inter Sci 129:537
5. Góralczyk D (1991) Coll Surf 59:361
6. Kamieński B (1935) Bull Acad Polon Sci Ser A:129, 309, 319
7. Kortüm G, Vogel W, Andrusow K (1961) Dissociation constants of organic acid in aqueous solution. Butterworths, London
8. Kamieński B (1957) Proc 2nd Int Conf of Surface Activity. London, vol III, p 103
9. Motomura K (1978) J Coll Inter Sci 64:348
10. Motomura K, Kanda T, Abe K, Todoraki N, Ikeda N (1992) Coll Surf 67:53
11. Todoraki N, Tanaka F, Ikeda N, Aratono M, Motomura K (1993) Bull Chem Soc Jap 66:351
12. Vileneuve M, Sakamoto H, Minamizawa H, Ikeda N, Motomura K, Aratono M (1997) J Coll Inter Sci 194:301
13. Davies JT, Rideal EK (1963) Interfacial phenomena. Academic Press, New York, London
14. Gileadi E, Kiriva-Eisner E, Penciner E (1975) Interfacial electrochemistry. Addison-Vaseley, London, p 15
15. Paluch M, Klich J (2001) Coll Surf A 191:215

Progr Colloid Polym Sci (2004) 123:236–239
DOI 10.1007/b11950

R. Pieri
G. Carignano
A. Chittofrati
F. D'Aprile
M. Visca

Wetting of low energy surfaces by perfluoropolyether carboxylic salts in aqueous solution

R. Pieri (✉) · G. Carignano
A. Chittofrati · F. D'Aprile · M. Visca
Solvay Solexis, v.le Lombardia 20,
20021 Bollate, Milano, Italy
e-mail: riccardo.pieri@solvay.com
Tel.: +39-02-38356205
Fax: +39-02-38356355

Abstract The study via dynamic contact angle (DCA) of a well-defined mixture of perfluoropolyether (PFPE) carboxylic salts has been carried out to evaluate the wetting features of their aqueous solution against LDPE or PTFE substrates. The preliminary results substantially identify the concentration range to completely wet each substrate. Low interactions between surfactant and polyethylene surface are found, while the surfactant adsorption on the PTFE surface is rather apparent.

Keywords Perfluoropolyether · Contact angle · Surface · Adsorption

Introduction

Fluorinated surfactants meet a widespread interest as wetting agents, since they can reduce water surface tension to 15–20 mN/m, enabling the wetting of low energy surfaces like polyethylene or polytetrafluoroethylene [1, 2], while surfactant adsorption may modify the energy of the substrate.

The aim of this work is a first insight, via dynamic contact angle, of the wetting properties of aqueous solutions of a mixture of carboxylic perfluoropolyether salts, with a particular emphasis on the ammonium salts, on LDPE and PTFE. The results show this surfactant mixture, at appropriate concentration, to wet completely the hydrophobic surfaces. Surfactant adsorption on PTFE could be ascertained.

Experimental

Materials

The surfactant studied is a laboratory sample of a mixture of homologues belonging to the general PFPE class:

$$Cl-(C_3F_6O)_n(C_2F_4O)_mCF_2COO^-X^+$$

where for the individual terms in the mixture, the number of perfluoroisopropoxy units n, varies from 2 to 5 and m may be zero or 1. The counterion is ammonium, sodium or potassium.

The salts have been obtained by neutralisation, with the corresponding hydroxide, of the same acidic surfactant mixture whose gas-chromatographic distribution is reported in Fig. 1.

The features of some of the individual components of the mixture are described elsewhere [3]. Some bulk characteristics of the salts are reported in Table 1.

Water was of Millipore Milli-Q grade.

Low density polyethylene samples (LDPE, 10 mm × 20 mm × 1 mm) were purchased from Exxon Chemical.

Polytetrafluoroethylene films (PTFE, 10 mm × 20 mm × 0.2 mm) were glass fibre fabric by Fother Gill uniformly coated with ALGOFLON® PTFE dispersion (Solvay Solexis), with a mean roughness of 2 μm.

The substrates were thoroughly cleaned by washing cycles with acetone (LDPE) or methanol (PTFE), then dried in oven at 30 °C for 20 hours.

Methods

Equilibrium surface tension has been measured by the Du Nouy method with a tensiometer LAUDA TE1C at different surfactant concentrations; raw force data were corrected with Harkins-Jordan factors [4].

Each value obtained, allowing sufficient time for equilibrium to be attained at each concentration, by the average of 5 measurements at least, with standard deviation reduced upon concentration increase. The experimental error was within 2.0 mN/m at concentration around 0.01 g/L, within 0.8 mN/m around 0.1 g/L and within 0.1 mN/m around 1 g/L.

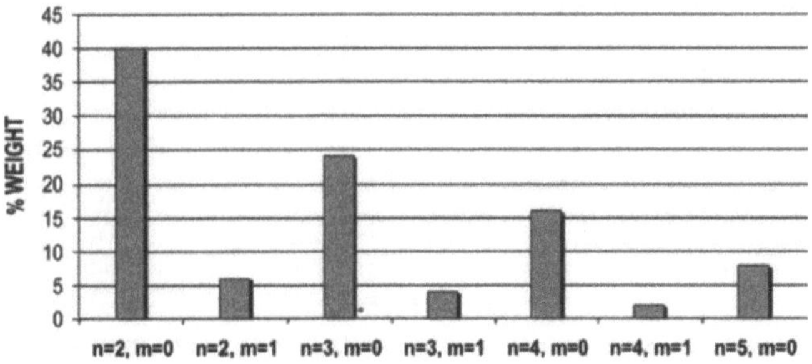

Fig. 1 PFPE surfactant acidic mixture MW distribution

Table 1 PFPE salts bulk characteristics

	NH_4^+	Na^+	K^+
average MW (±5%)	630	622	652
density (g/ml) at 25 °C	1.88	1.72	1.89
T_m (DSC transition), °C	73	46	107
conc. limit (% wt) for phase separation in water, 25 °C	4	32	7

Dynamic contact angle measurements have been performed with KSV Sigma 70 instrument, at 25.0 ± 0.1 °C. Each measurement implied 4 cycles of immersion of the substrate into the surfactant solution. The operating speed range allowed for immersion and emersion of the substrate is in the range 0.5–20 mm/min. One minute stop was allowed between subsequent cycles.

The timeframe selected for the DCA measurements was meant to ensure constant surface tension of the solution, compared to the fast equilibration of the surfactant solution on the time scale allowed by the ring method. More suitable techniques for the study of dynamic surface tension are presently under examination to clarify this point.

To determine the critical surface tension γ_c of the surfaces the Zisman method [2] was chosen. RPE-grade test liquids were used, as purchased, by Aldrich or Carlo Erba; we used *n*-alkanes (C_6-C_{14}) for PTFE and organic polar solvents (with γ in the range 40–50 mN/m) for LDPE.

Results and discussion

Surface tension measurements

The surfactant is a mixture of carboxylic analogues, so ideal mixing might be expected on the basis of previous results on much simpler two-component mixtures [5]; however this issue has yet to be clarified for the present mixture, where the MW of individual components varies in a wide range (from 480 to about 1000).

The results for the three different salts at 25 °C are plotted in Fig. 2, showing little dependence of surface tension on counterion. Ultimate surface tensions of 19–20 mN/m have been measured on these chlorine-containing surfactants, higher than the typical values for surfactants with similar chain-structure but fully fluorinated tail-terminal, for instance, those described in ref.[6]

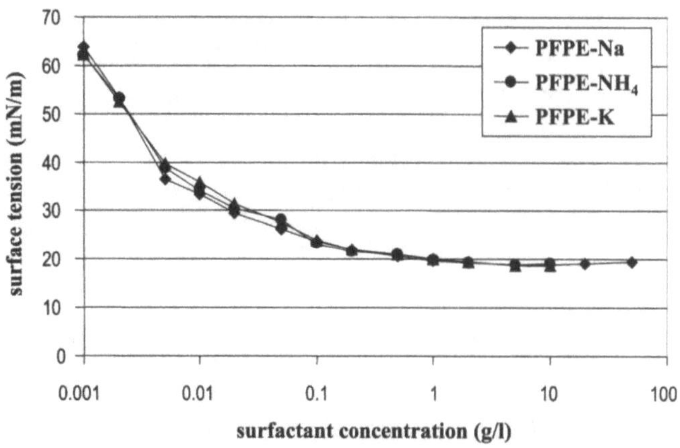

Fig. 2 Equilibrium surface tension measurements of PFPE surfactant salts

where ultimate surface tensions of 15.0 ± 0.3 mN/m were reported. Nevertheless, the present salts enable γ reduction sufficient for most wetting purposes.

Figure 2 shows 0.2–0.4 g/L as the concentration of the present surfactant mixture required to reach constant surface tension.

The surface tension of the ammonium surfactant solutions as a function of time suggested equilibrium to be reached in few minutes in the whole concentration range but the 0.01 g/L solution, where 30 minutes were necessary to reach constant γ value.

The experimental values of critical surface tension of each solid surface were 31.0 mN/m for LDPE, in good agreement with literature data [7], and 20.3 mN/m for the PTFE fabric, against the 18 mN/m reported [7], for a smoother PTFE surface.

Dynamic contact angle measurements

LDPE

Dynamic contact angle measurements on LDPE surfaces were performed with PFPE-NH_4 solutions (concentration range 0.01 to 10 g/L) at 25 °C and two different probe speeds, 6 and 0.5 mm/min. A contact angle of 105° of Milli-Q water at 25 °C on LDPE has been measured. In Fig. 3 the advancing contact angle θ_a (usually employed in discussions of wetting) of the first cycle is plotted vs. surfactant concentration of the solutions at the two different speeds, while in Figs. 4 and 5 the DCA profiles of the most diluted and most concentrated solution at 6 mm/min are reported.

Wetting is achieved at surfactant concentrations ≥1 g/L, with complete spreading on the surface at 10 g/L. No kinetic hysteresis was detected, excluding the surface modification and consequently surfactant adsorption. Lowering the probe speed leads to a spreading improvement.

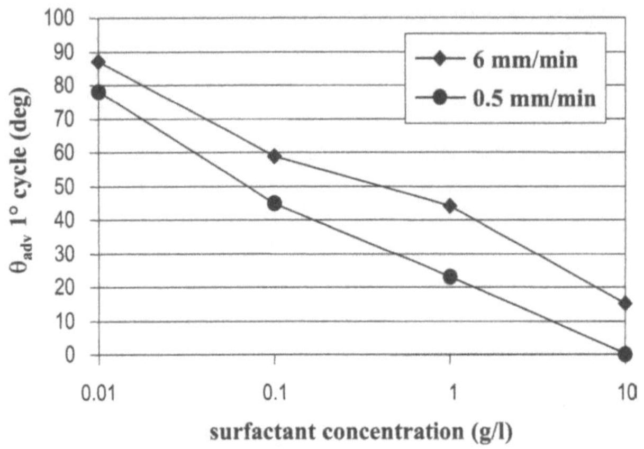

Fig. 3 First cycle advancing contact angle of PFPE-NH_4 solutions on LDPE at 25 °C and two different probe speed as a function of concentration

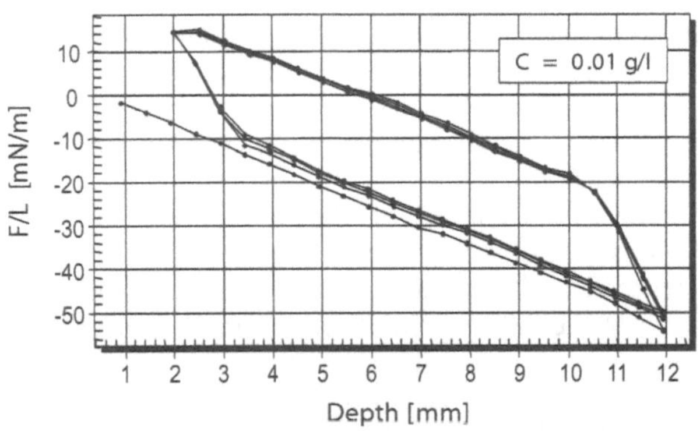

Fig. 4 DCA profile analysis of 0.01 g/L PFPE-NH_4 solutions on LDPE at 25 °C and 6 mm/min probe speed

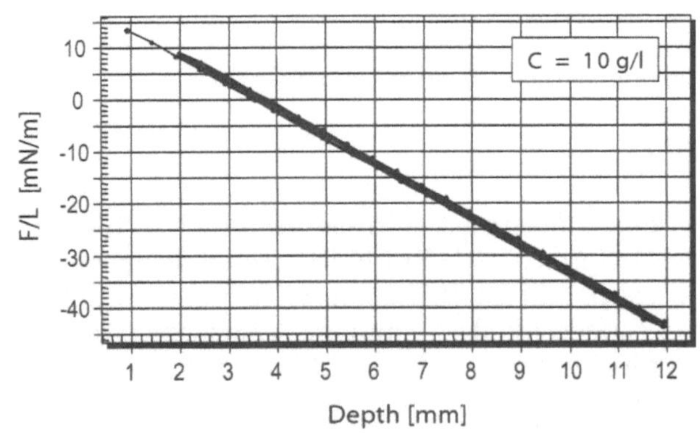

Fig. 5 DCA profile analysis of 10 g/L PFPE-NH_4 solutions on LDPE at 25 °C and 6 mm/min probe speed

Table 2 Effect of temperature on equilibrium surface tension measurements for solution 1 g/L of PFPE surfactant salts

surface tension ± 0.1 (mN/m)	NH_4^+	Na^+	K^+
T = 25 °C	19.6	19.7	20.1
T = 40 °C	18.5	18.8	18.6
T = 60 °C	17.1	17.1	17.3

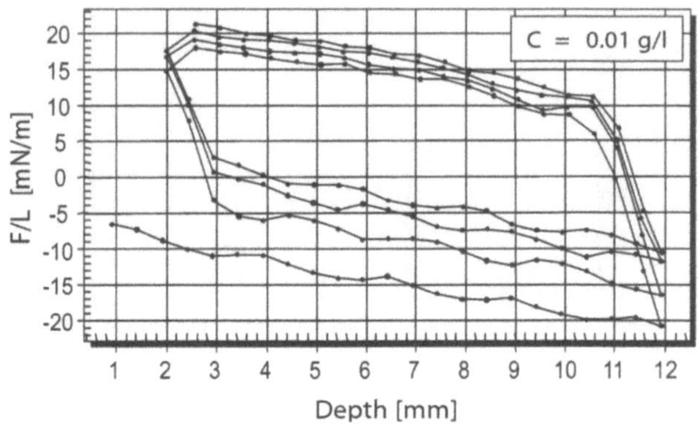

Fig. 6 DCA profile analysis of 0.01 g/L PFPE-NH_4 solutions on PTFE at 25 °C and 6 mm/min probe speed

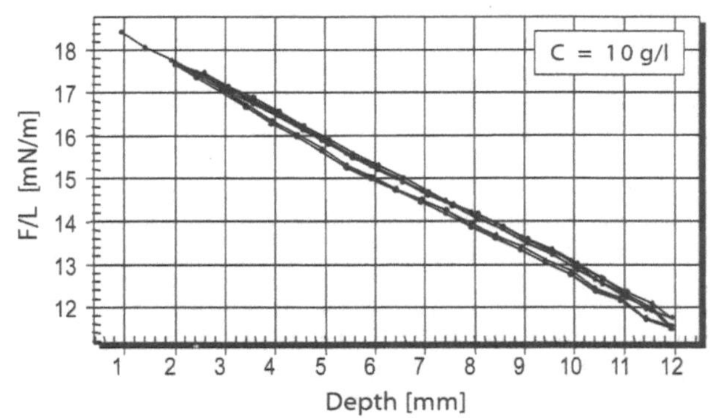

Fig. 7 DCA profile analysis of 10 g/L PFPE-NH_4 solutions on PTFE at 25 °C and 6 mm/min probe speed

Increasing temperature causes a decrease in 1 g/L solution surface tension per each salt, as reported in Table 2.

As a consequence also the contact angle is lowered, with better wetting performance reported for the sodium salt.

PTFE

The study on PTFE has been carried out at 0.5 and 6 mm/min probe speed with the ammonium salt of the PFPE surfactant mixture: Figs. 6 and 7 show the DCA profiles of the most diluted and most concentrated solutions, respectively. At 10 g/L, an almost complete spreading was achieved, while the solution at 0.01 g/L does not reach a surface tension low enough to ensure wetting. The wetting behaviour was not influenced by speed changes. Inhomogeneity of the PTFE surface caused, as for LDPE, a large thermodynamic hysteresis that was reduced with the contact angle itself (lowering the intrinsic contact angle of the different areas of the non-homogeneous surfaces).

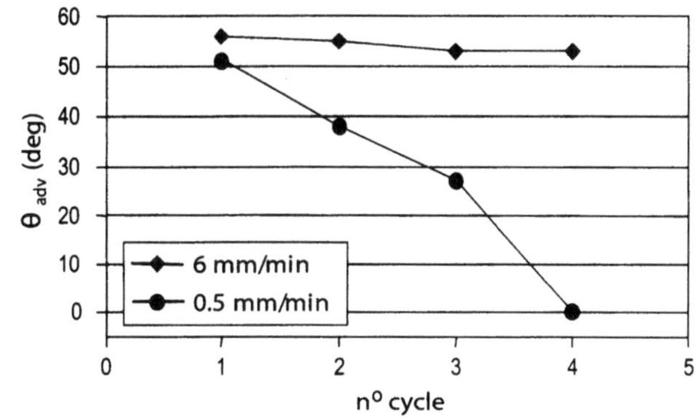

Fig. 8 Advancing contact angles of PFPE-NH_4 0.1 g/L solution on PTFE at 25 °C and two different probe speed as a function of No. of cycles

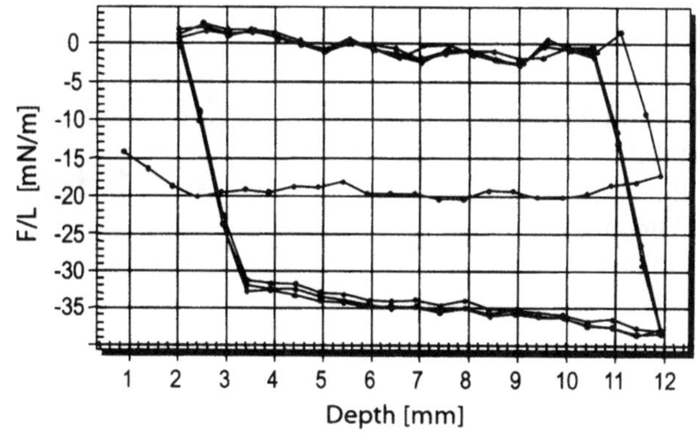

Fig. 9 DCA profile analysis of water on PTFE previously measured vs. PFPE-NH_4 1 g/L solution at 25 °C and 6 mm/min probe speed

A further insight is necessary for the solutions at intermediate concentration, such as 0.1 g/L and 1 g/L, that correspond in Fig. 2 to γ values slightly below and within the plateau, respectively.

In the case of the more diluted solution (Fig. 8), a strong dependence of the contact angle upon the probe speed was noted, with a large kinetic hysteresis at 0.5 mm/min. This behaviour reflects a progressive increase of the hydrophilic characteristic of the substrate, reasonably due to the PFPE surfactant adsorption onto the surface with the polar head towards the solution. A qualitatively similar effect is found with solution at 1 g/L, although with low values of contact angle.

Previous evidence of PFPE surfactant adsorption onto PTFE particles in a latex was reported for the n2-n3 terms [5] confirming the strong dependence of the efficiency of adsorption upon the chain length, while the affinity appeared to be somewhat unchanged. The adsorption features of the mixture studied in the present work are believed to be the consequence of the properties of the partitioning of the single components of the surfactant mixture, like actual concentration of each term and its diffusion in solution (faster for the terms with smaller n value) and PTFE affinity (more favourable to the adsorption of the terms with higher n value); however the present work cannot, and is not meant to, discriminate the contribution of each species.

An indirect evidence of adsorption is shown in Fig. 9, where a PTFE sample, whose contact angle has been previously measured against the PFPE ammonium salt solution at 1 g/L, is reported after subsequent immersion in water. A much lower angle (approx. 105°) was obtained in the first cycle while in the following cycles a contact angle very close to the water/clean PTFE system was measured. This test strongly suggests that adsorption occurs from the surfactant solution but desorption can be easily achieved upon repeated wetting cycles with pure water.

Conclusions

Carboxylic surfactant salts belonging to the PFPE class are shown, at appropriate concentrations, to be able to wet low energy polymeric surfaces, although extremely low surface tension values are not achieved with their Cl-terminal. LDPE or even PTFE can be completely wetted by these PFPE surfactants at fairly low concentrations. In the latter case only, a surfactant adsorption/desorption mechanism, correlated with surface modification, appears to cooperate to the wetting.

References

1. Kissa E (2001) Fluorinated surfactants and repellents In: Surfactant science series, vol. 97. Marcel Dekker, New York
2. Zisman WA (1964) Contact angle wettability and adhesion. In: Advances in chemistry series, vol. 43. American Chemical Society, Washington DC
3. Chittofrati A, Pieri R, D'Aprile F, Lenti D, Maccone P, Visca M (this issue)
4. Harkins WD, Jordan HF (1930) J Am Chem Soc 52:1751
5. Lenti D, D'Aprile F, Chittofrati A, Visca M (1999) Communication at XIII ECIS Conference, Dublin
6. Giannetti E, Chittofrati A, Sanguineti A (1997) Chim Industr 79:22
7. Wu S (1982) In: Polymer and interface adhesion, chap 5. Marcel Dekker, New York, p 170

Progr Colloid Polym Sci (2004) 123: 240–244
DOI 10.1007/b11951

C. Poncet-Legrand
L. Petit
S. Reculusa
C. Mingotaud
E. Duguet
S. Ravaine

Dissymmetrical gold tagging on spherical silica nanoparticles

C. Poncet-Legrand · S. Reculusa
S. Ravaine (✉)
Centre de Recherche Paul Pascal,
CNRS, 115 Av.du Dr Schweitzer,
33600 Pessac, France
e-mail: ravaine@crpp.u-bordeaux.fr
Tel.: +33-5-56845667
Fax: +33-5-56845600

C. Poncet-Legrand · L. Petit · E. Duguet
Institut de Chimie de la Matière Condensée de Bordeaux, CNRS, 87 Av. du Dr Schweitzer, 33608 Pessac Cedex, France

C. Mingotaud
Laboratoire des Interactions Moléculaires, Réactivité Chimique et Photochimique, Université Paul Sabatier, 118 Route de Narbonne, 31062 Toulouse, France

Abstract To tag silica beads with gold nanoclusters in a dissymmetrical way, a dense layer of silica particles functionalised with amino groups was spread along an interface between two media. Gold was introduced through one of these media and then was adsorbed onto the corresponding silica hemisphere. Three interfaces were used: (i) the gas-liquid interface, thanks to the Langmuir technique, (ii) the liquid-solid interface after self-deposition onto large acrylic beads or commercial resins used for solid phase synthesis, or after deposition onto a solid substrate using the Langmuir-Blodgett technique, (iii) gas-solid interface with physical vapour deposition of Au. This communication presents the various synthesis routes, comparing their efficiency based on transmission electron microscopy observations.

Keywords Dissymmetrisation · Silica nanoparticles · Gold nanoclusters · Discriminatory adsorption

Introduction

One of the current challenges in the chemistry of materials is the elaboration of complex nanoscale structures. Our main objective is to synthesise dissymmetrical nanoparticles which can lead to new non-isotropic assemblies and could be useful for various applications.

To reach this goal, we used interfaces as a tool to induce a dissymmetry on an initially symmetrical object adsorbed along this interface. These objects are generally spherical silica particles synthesised using the Stöber process. Recently, we reported the possibility of synthesising silica beads (with an average diameter less than 100 nm) dissymmetrically decorated with gold nanoclusters [1, 2]. The technique consisted in spreading amino-functionalised silica particles along a solution of gold nanoclusters, using the air/water interface as dissymmetrisation tool.

In this paper, the concept of interfacial dissymmetrisation is generalised to other interfaces (liquid/solid, gas/solid) and results obtained with the same submicrometer system (i.e., gold nanoparticles deposited on silica beads) are compared.

Materials and methods

Materials

Tetraethyl orthosilicate (TEOS), 3-aminopropyltrimethoxysilane (γ-APS), trisodium citrate, 1,4-dioxane 99+%, tetrakis(hydroxymethyl)phosphonium chloride, trifluoroacetic acid (TFA) and $NaBH_4$ were supplied by Sigma Aldrich. $HAuCl_4 \cdot 3\,H_2O$ was purchased from Acros Organics, 30% NH_4OH and sodium hydroxide from SDS, absolute ethanol, tetrahydrofurane (THF) and dichloromethane from J.T. Baker. All chemicals were used as received.

Poly(methyl methacrylate) beads "Plexidon F", whose the major population has an average diameter of 60 μm were supplied by Rohm and used as received. Trityl chloride, 2-chlorotrityl chloride and formylpolystyrene resins (mesh size: 100–200) were purchased from Novabiochem.

All reagents were used without further purification. Deionised water (resistivity higher than 18 $M\Omega \cdot cm^{-1}$) was produced with a Millipore system.

Particles synthesis

The synthesis of submicrometer silica beads with amine-functionalised surfaces was previously described by Halas and coworkers [3]. Briefly, it was achieved through the base-catalysed hydrolysis of TEOS, as described by Stöber [4], followed by the grafting of the amine functions on the silica nanospheres using γ-APS.

Colloidal gold particles were prepared as previously described by Meier and Meissner [5] or Duff [6] and resulted respectively in 6–8 or 3–5 nm gold particles.

Dissymmetrisation

Several interfaces were used to induce a dissymmetry, following the scheme on Fig. 1.

Dissymmetry induced by a gas/liquid interface (Fig. 1, Route A)

A dispersion of the amino-functionalised silica particles in chloroform (ca. 2 mg·mL^{-1}) was spread along a solution of gold nanoclusters contained in a KSV 5000 trough. The Langmuir film was then compressed slowly (ca. 1 cm·min^{-1}) until the surface pressure reaches 10 mN·m^{-1}.

Fig. 1 Synthesis procedure of silica beads dissymmetrically gold-decorated

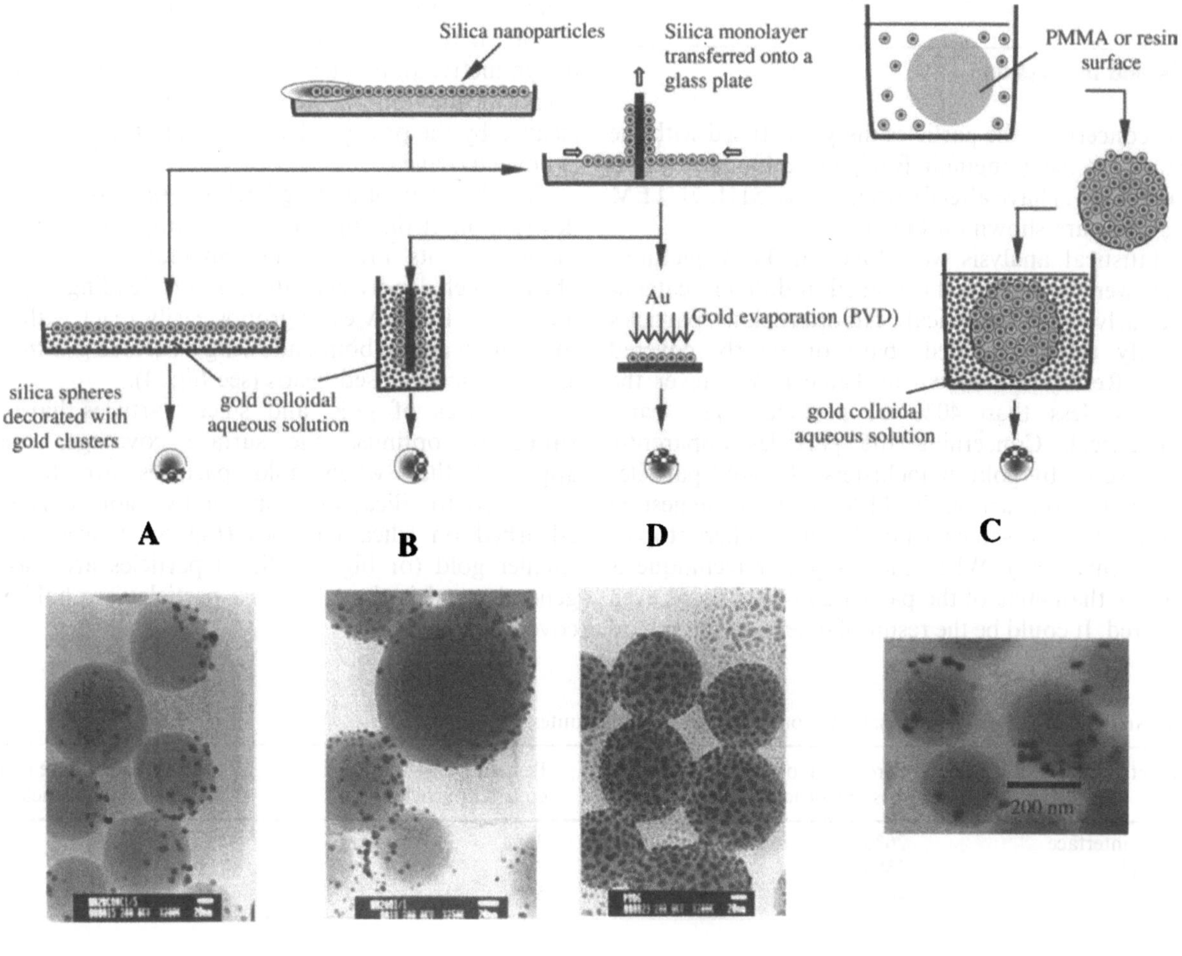

Dissymmetry induced by a liquid/solid interface

Langmuir-Blodgett technique (Fig. 1, Route B). Transfer onto hydrophobic glass substrate (silanated by chloromethylsilane) of the previous Langmuir film was performed using the vertical dipping method (dipping speed ca.1 cm min^{-1}). Immersion of the substrate was performed into the gold nanoparticles suspension during 12 hours. Silica beads were then dislodged from the glass substrate with a spatula in ethanol and redispersed under sonication.

Use of solid synthesis support (Fig. 1, Route C). We used two families of solid phase synthesis support: commercial resins (from Novabiochem) and PMMA beads. The processes followed in both cases are quite similar.

Typically, 300 mg of resin were swollen in 5 mL of dry THF, then 100 mg of amino-functionalised silica dispersed in 5 mL of THF were added and this was stirred overnight at room temperature. Then the resin was filtered, gold was added to the suspension and the resin was finally washed on a filter to remove ungrafted silica and gold nanoparticles. Cleavage of the silica particles from the resin was done as indicated by the supplier: several attempts were done with trifluoroacetic acid (1 to 15%) in dichloromethane at room temperature for 48 h.

Silica particles were also chemically grafted onto commercial PMMA beads. Typically, 2 g of PMMA particles were suspended in 10 mL of ethanol and mixed with 120 mg of silica particles dispersed in 10 mL of a 0.2 M NaCl solution. Ethanol was chosen to allow a good wetting of the PMMA beads and sodium chloride was added to screen electrostatic repulsions between the silica particles and enhance the surface coverage on the PMMA support. The amount of silica added was calculated in order to get a close packing of silica particles on the PMMA beads.

Ungrafted silica was removed by filtration through a PTFE membrane (pore size: 1 μm) and then the silica-coated PMMA beads were immersed in the colloidal gold suspension (diluted in ethanol), filtered and washed to remove the gold excess. A heat treatment at 300 °C overnight led to the depolymerisation of PMMA beads, allowing the recovery of gold decorated silica nanoparticles which were then redispersed in ethanol.

Dissymmetry induced through a gas/solid interface (Fig. 1, Route D)

Glass substrates with a monolayer of silica beads were prepared according to the transfer technique described above. Gold was sputtered from a target source under 0.5 Pa argon pressure and a power density of 1 $W \cdot cm^{-2}$. In such conditions, gold sputtering rate is close to 12 $nm \cdot min^{-1}$.

Transmission electron microscopy (TEM) and scanning electron microscopy (SEM) experiments

In all experiments, the gold-tagged silica beads were recovered onto carbon-coated copper grids, by direct horizontal touching onto the film surface or by evaporating a drop of an alcoholic dispersion of particles. A JEOL 2000 FX operating at an accelerating voltage of 200 kV was used to collect the TEM images.

PMMA beads and resins were observed using a JEOL JSM-840 A scanning electron microscope, with an accelerating voltage of 10 kV. The specimens were gold coated prior to examination.

Results and discussion

Results concerning the particles dissymmetrised with the Langmuir or the Langmuir-Blodgett techniques (A, B and D on Fig. 1) have already been discussed [1, 2]. TEM photographs are shown on Fig. 1.

A statistical analysis was done on TEM pictures. Particles were counted and classified in different categories: clearly dissymmetrical nanoparticles, particles apparently totally covered, bare or poorly covered particles. Results are shown in Table 1. Whatever the method is, less than 40% of particles are clearly dissymmetrical. Concerning the particles apparently totally covered by gold nanoclusters, the gold particles have a similar contrast on the TEM pictures, suggesting that they are on the same part of the silica sphere (top or bottom hemisphere). When the Langmuir technique is used, more than 40% of the particles are poorly or even not covered. It could be the result of defects in the spread silica film (such as localised multilayers), preventing the reaction with gold for every silica nanoparticle.

The techniques derived from the Langmuir-Blodgett films are probably the most efficient (at least one third of the particles are clearly dissymmetrical, up to 100% probably are), but the limitation of these methods are the small quantities produced (a few mg per batch). Therefore we tried to generalise this concept to commercial resins used for the solid phase synthesis of peptides: since the support is micrometric, the developed interface is much more important and the experiments are easier.

These resins are made of polystyrene and bear functional surface groups allowing the grafting of amines or acids. As our silica particles are amino functionalised, we used trityl resins and formylpolystyrene (Fig. 2). The first step of the synthesis (attachment of silica on the resins) is clearly achieved whatever the resin may be, as shown by SEM photos (Fig. 3b). After grafting the silica on the surface, the resins were exposed to a colloidal solution of gold. The last step of the reaction was the cleavage of the amino groups from the resins. When single molecules are attached, this is usually done in a solution of TFA in dichloromethane (typically 1 to 5%). As the first attempts were unsuccessful, we increased the TFA concentration up to 15%, but even at these concentrations no cleavage was observed (Fig. 3c). One of the major difficulties is also that if a chemical dissymmetrisation is achieved (i.e., if we graft a specific chemical function on the surface), this one may be also cleaved by the process. The use of photocleavable resins is now envisaged.

Another way of getting back the silica particles is the destruction of the support. Poly(methyl methacrylate) is known for its low ceiling temperature (200–250 °C) above which depolymerisation occurs, leading to gaseous monomer. PMMA ester groups easily react with amine to form an amide bond, allowing a surface grafting with amino-functionalised beads (see Fig. 4).

The sizes of gold and silica particles have been varied to optimise the surface coverage, since it appeared that when gold particles are too large compared to silica, only one or two gold clusters are adsorbed on silica particles (Fig. 5). However, when smaller gold (or bigger silica) particles are used, the general trend is that the silica particles are half or less covered (Fig. 1C).

Table 1 Analysis of the dissymmetrisations done with the different routes

Dissymmetrisation tool	Percentage of clearly dissymmetric particles	Percentage of totally covered particles	Percentage of bare or poorly covered nanoparticles
Gas/liquid interface	26%	32%	42%
Liquid/solid	25%	75%	–
Gas/solid	40%	55%	5%

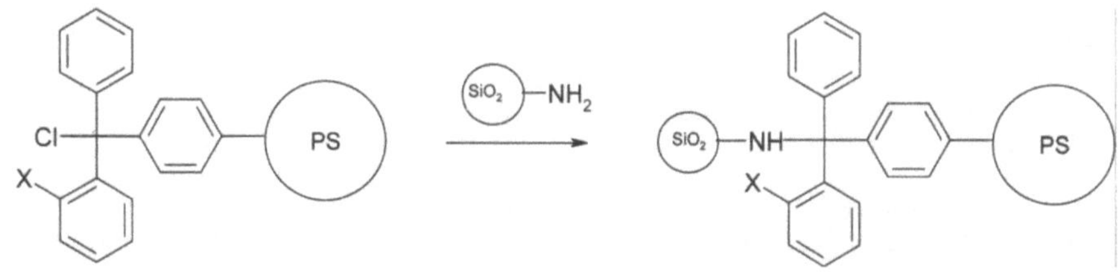

Fig. 2 Typical chemical route of an amine grafting on a commercial resin

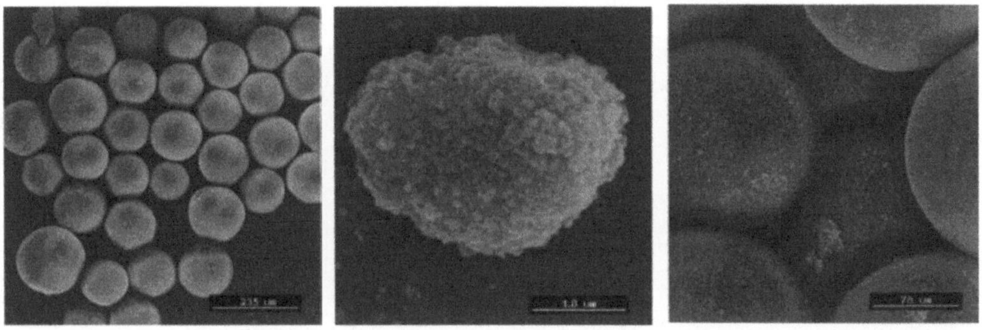

Fig. 3 SEM pictures of a trityl resin before the attachment of the silica particles (**a**), after (**b**) and after the cleavage reaction (**c**)

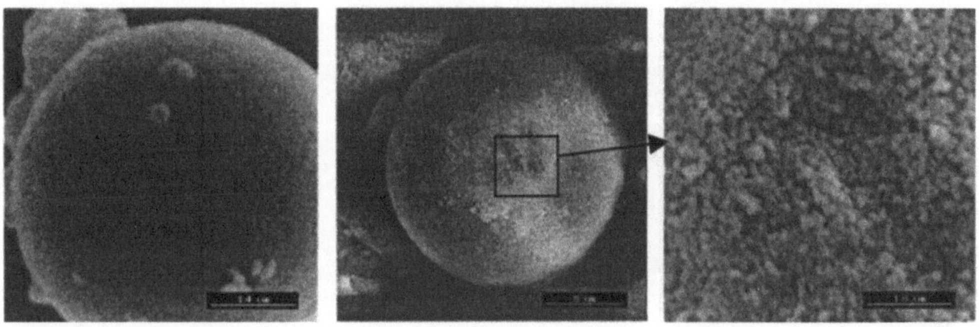

Fig. 4 SEM pictures of a PMMA bead before the attachment of the silica particles (**a**), and after (**b, c**).

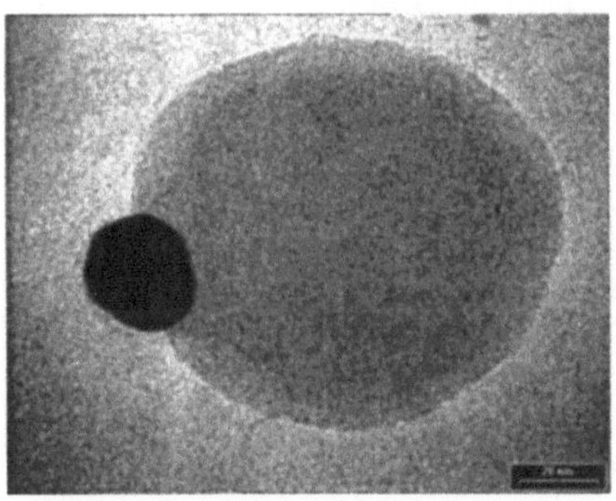

Fig. 5 Silica particle covered with gold after calcination of the PMMA support (scale bar: 20 nm)

Conclusion

Dissymmetrical nanoparticles have been obtained using various interfaces. The techniques derived from the Langmuir films are probably the most efficient (at least one third of the particles are clearly dissymmetrical, 60% probably are), but the limitation of these methods are the small quantities produced (a few mg per batch). To bypass this difficulty, some experiments with commercial resins usually used for solid phase synthesis were performed. The dissymmetrisation step has been successful, but cleavage of the silica particles from the liquid/solid interface is a further difficulty.

The use of PMMA beads as solid phase support has appeared as a promising option because PMMA can be thermally removed. Experiments are currently in progress for optimising the silica grafting yield and the size ratio between silica and gold nanoclusters.

References

1. Petit L, Sellier E, Duguet E, Ravaine S, Mingotaud C (2000) J Mater Chem 10:253
2. Petit L, Manaud JP, Mingotaud C, Ravaine S, Duguet E (2003) Mater Lett (in press)
3. Wescott SL, Oldenburg SJ, Randall Lee T, Halas NJ (1998) Langmuir 14:5396
4. Stöber W, Fink A (1968) J Colloid Interface Sci 26:62
5. Meier A, Meissner D (1996) In: Fendler JH, Dekany I (eds), Nanoparticles in solids and solutions. Kluwer, Dordrecht, p 421
6. Duff DG, Baiker A (1993) Langmuir 9:2301

Progr Colloid Polym Sci (2004) 123:245–250
DOI 10.1007/b11952

B. Gzyl
M. Paluch

Langmuir monolayers of lipids at the water/air interface

B. Gzyl (✉) · M. Paluch
Jagiellonian University,
Faculty of Chemistry,
Department of Physical Chemistry
and Electrochemistry, ul. Ingardena 3,
30-060 Kraków, Poland
e-mails: gzyl@chemia.uj.edu.pl;
paluch@chemia.uj.edu.pl

Abstract In order to establish the mutual miscibility of D,L-dipalmitoylphosphatidylcholine (DPPC) and 4-methylumbelliferyl palmitate in a two-dimensional phase monolayer behaviour have been investigated. For this purpose, the isotherms of surface pressure versus molecular area were measured for monolayers. A classical thermodynamic analysis was performed on these isotherms which involved calculating excess free energies of mixing to determine the miscibility properties. The work shows surface compatibility of DPPC and 4-methylumbelliferyl palmitate with the occurrence of attractive interactions between the two components.

Keywords Monolayers · Lipid membrane · Dipalmitoylophosphatidylcholine · 4-Methylumbelliferyl palmitate

Introduction

Many amphiphilic substances give oriented systems such as monolayers. The study of spreading monolayers of amphiphilic substances is very important because it informs about the surface orientation of the molecules at the interfaces, the mobility of the hydrophobic chains when the area available for a molecule is reduced and about the compatibility of different amphiphilic molecules [1]. It also enables us to predict properties of more complex aggregates, for instance, biological membranes, by the study of molecular interactions in such simple systems [2]. It is well known that biological membranes are made up of bilayer of lipid compounds in which the other components such as proteins and enzymes are immersed or bound to the two interfaces. To understand the role of lipids (which behave as amphiphiles) in various systems at a more detailed molecular level it is necessary to conduct investigations by using a simple model system. Two monolayers make up the bilayer. Therefore phospholipid monolayers are often used as simplified model systems of biomembranes [3–5].

Many studies on the mutual miscibility of lipids at the air/water interface have been reported [6–9]. In this paper we present the study of the behaviour of two lipids with different polar groups and the same hydrophobic chains, and also their respective mixtures at the air/water interface under diverse temperature conditions. The substances examined in this study were D,L-dipalmitoylphosphatidylcholine (DPPC) and 4-methylumbelliferyl palmitate (Fig. 1). The aim of this study was to investigate the DPPC and 4-methylumbelliferyl palmitate binary mixtures as a function of temperature, surface pressure and mixture composition. For this purpose the isotherms of surface pressure versus molecular area were recorded using a Langmuir film balance. Then, a classical surface chemistry thermodynamic analysis was performed on these isotherms, which involved calculating excess free energies of mixing.

Experimental

D,L-Dipalmitoylphosphatidylcholine (DPPC) and 4-methylumbelliferyl palmitate were obtained from Aldrich and Biochemika Fluka, respectively. The spreading solvent was chloroform (distilled before using) from POCh. Water was four times distilled.

The methods of surface pressure (π) measurements versus molecular area (A) were chosen for the investigations. The π-A

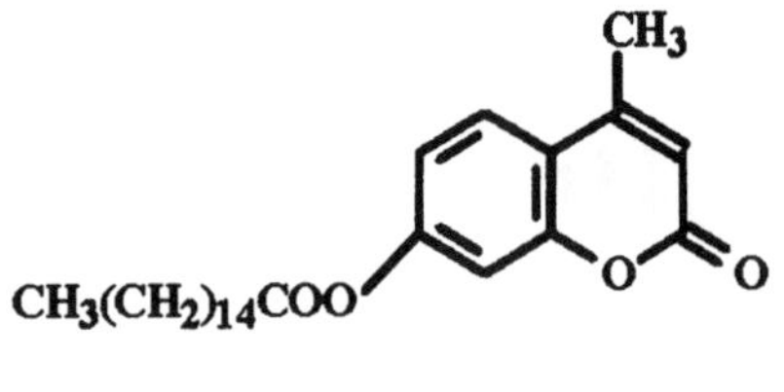

4-Methylumbelliferyl palmitate

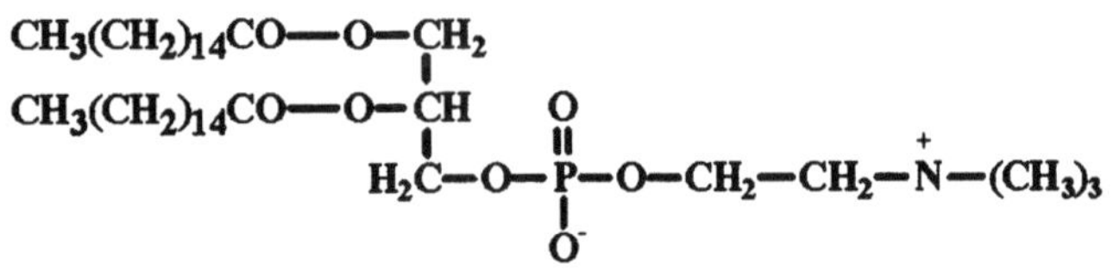

D,L-dipalmitoyl phosphatidylcholine

Fig. 1 Chemical formulas of D,L-dipalmitoylphosphatidylcholine and 4-methylumbelliferyl palmitate

isotherms were measured during continuous compression of the monolayer, using a computerised Langmuir trough (KSV 1000, Finland). The accuracy of measurements was ± 0.01 mN/m for the surface pressure and ± 0.01 A^2/molecule for the area. The experiments were performed at 20, 25, 30 and 35 °C. Subphase temperature was controlled by water circulation from a U3 thermostat. All measurements were performed with a subphase of pure water. Prior to each experiment, the surface was cleaned repeatedly by sweeping the barriers slowly between the open and closed positions and aspirating the surface until the value of surface pressure remained about zero on the reducing surface area. The addition of the spreading solvent alone to a clean surface produced no change to its surface tension.

Langmuir monolayers were prepared by spreading a precisely measured volume of the solutions of investigated lipids dissolved in chloroform. The film spreading was carried out with the help of a microsyringe (Hamilton). The drops of spreading solution were placed carefully on different regions of the entire available surface area. As a standard procedure, monolayers were rested 20 min before compression to allow sufficient solvent evaporation and to reach equilibrium with respect to the initial surface pressure. A symmetric compression was achieved with two moving barriers at a constant rate of 6 mm min^{-1}. Below this compression rate no difference in the π-A isotherms was observed. Each isotherm was measured at least twice. The compression procedure was strictly identical for the two components and for all the mixtures.

Results

In Fig. 2, the surface pressure (π)-molecular area (A) isotherms of system DPPC/4-methylumbelliferyl palmitate at 20, 25, 30 and 35 °C are shown. As can be seen from this figure, π-A isotherms for pure 4-methylumbelliferyl palmitate and DPPC monolayers are completely

Fig. 2 π-A isotherms of DPPC/4-methylumbelliferyl palmitate system on water at 20, 25, 30 and 35 °C

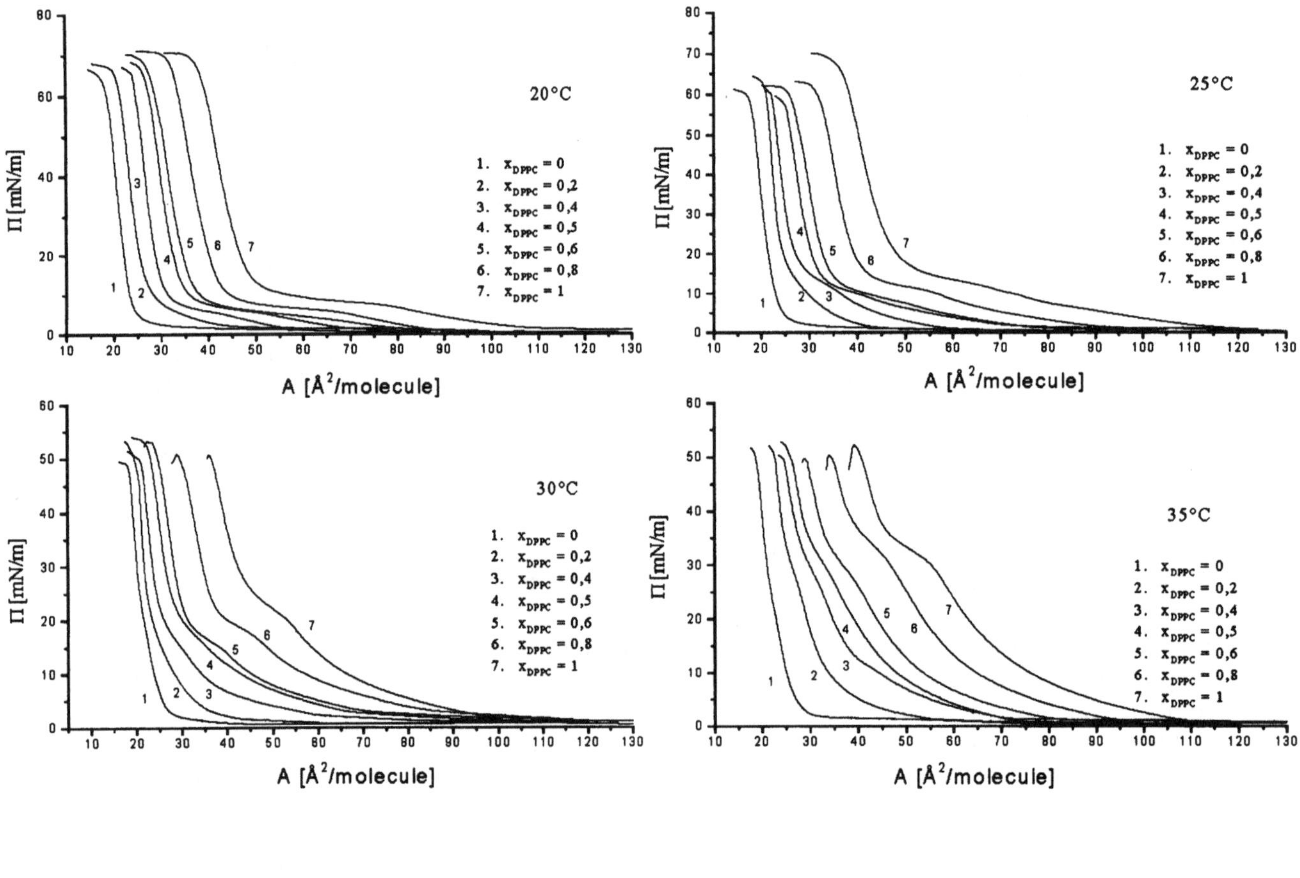

Table 1 Maximum compressibility modulus C_{smax}^{-1} [mN/m] for DPPC and 4-Methylumbelliferyl palmitate, at 20, 25, 30 and 35 °C

	20 °C	25 °C	30 °C	35 °C
DPPC	293.6	264.0	183.5	160.2
4-methylumbelliferyl palmitate	271.1	253.4	258.4	224.0
DPPC + 4-methylumbelliferyl palmitate ($x_{DPPC} = 0.2$)	282.5	387.2	372.9	226.0

different (see curves 1 and 7, respectively). The large differences are very important for the investigation of the mixed monolayers because they allow us to determine the contribution of each component to the mixed monolayer characteristics.

For the DPPC monolayer (Fig. 2), upon compression, the surface pressure increases until a plateau region is reached. The plateau on π-A isotherm of DPPC monolayer is due to the transition from liquid-expanded (LE) to liquid-condensed (LC) state. The surface pressure at which the plateau occurs is temperature dependent: it increases with the temperature increase. The monolayer can be compressed until all the molecules are oriented with their tails towards the air. At this point, where the molecules are in a close packed arrangement (LC phase), further compression leads to a steep increase in surface pressure. Eventually, the collapse pressure π_{coll} is reached. The collapse process consists of the separation of substance from the monolayer and simultaneous formation of bulk phases. Therefore, the collapse pressure can be defined as the maximum to which a monolayer can be compressed without detectable expulsion of molecules from the Langmuir film. It depends upon the details of the experimental procedure used (for instance, the rate at which the film is compressed and procedures to which the film has been subjected). The collapse pressure, determined from the change in the slope of the π/A isotherm, $\pi_{coll} = 59.31$ mN/m (at 20 °C). When the temperature rises π_{coll} decreases to values 48.70 mN/m (at 30 °C) and 49.45 mN/m (at 35 °C), therefore the stability of the monolayer decreases. With increasing temperature the isotherms become progressively more expanded as a consequence of the greater mobility assumed by the hydrocarbon chains. The observation is confirmed by the values of the maximum surface compressibility modulus $C_{s\,max}^{-1}$ (Table 1) which were calculated according to the equation:

$$C_S^{-1} = -A\left(\frac{\partial \pi}{\partial A}\right)_T \quad (1)$$

The value of the limiting area A_{lim} (48.87 A^2/molecule at 20 °C) evaluated from the intersection of the tangent line to the isotherm in the most sloped region (in the LC phase of the film) with the x-axis is higher than for the close packed double chain ($\sim 2 \times 20\ A^2 = 40\ A^2$), assuming vertical orientation of alkyl chains. This indicates

Fig. 3 Surface areas as a function of molar fractions of DPPC, at $\pi = 5$, 20 and 40 mN/m

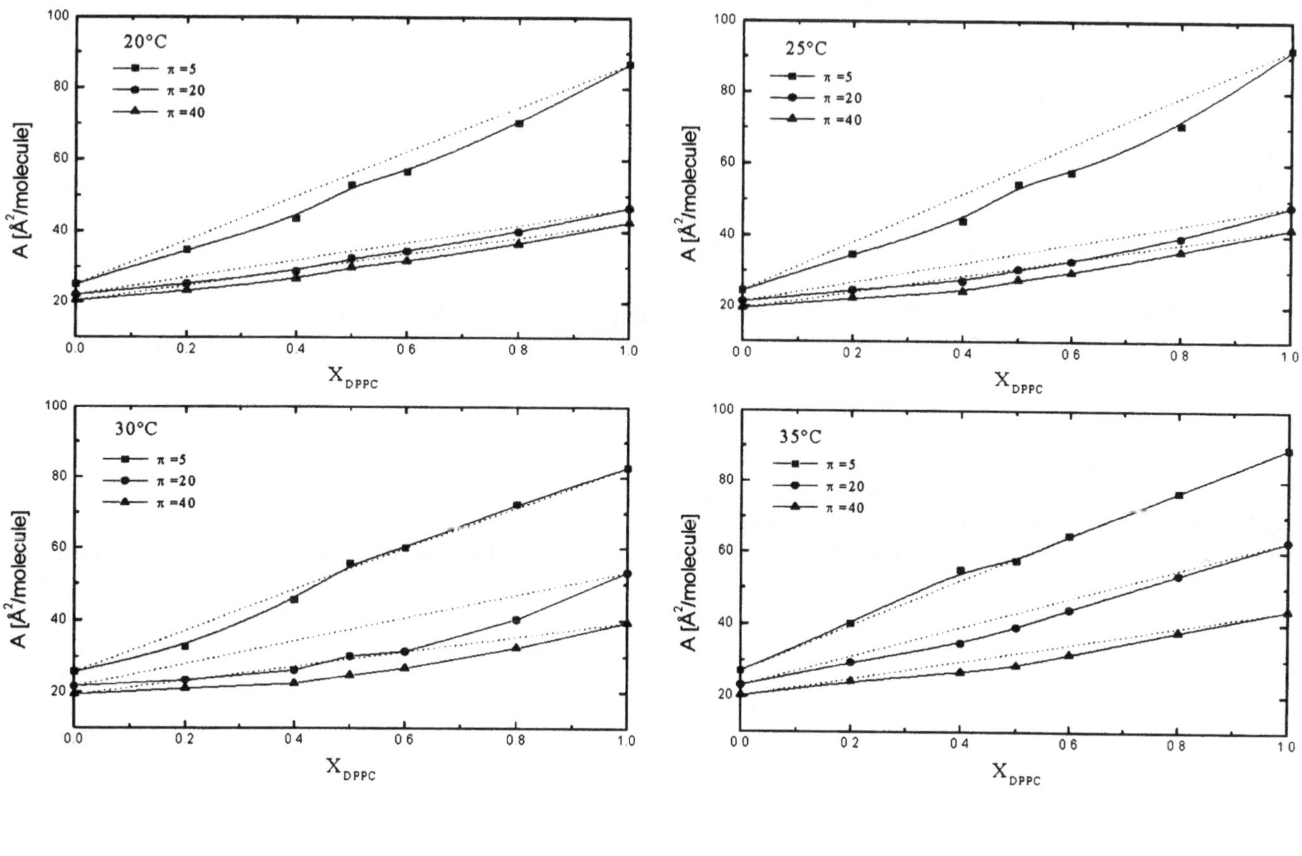

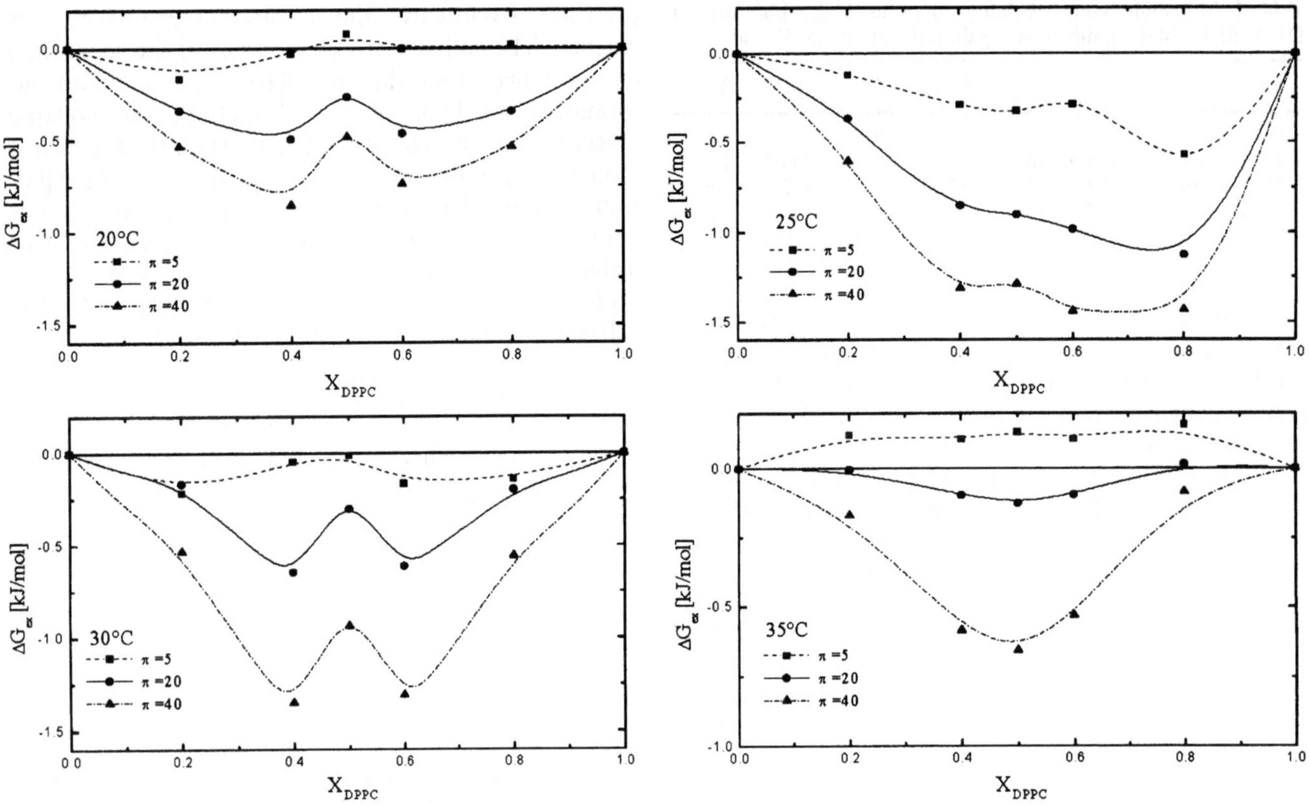

Fig. 4 Excess free energy of mixing of the DPPC/4-methylumbelliferyl palmitate system as a function of molar fractions of DPPC, at $\pi = 5$, 20 and 40 mN/m

inclination of the hydrocarbon chains of the DPPC molecules to the surface.

4-Methylumbelliferyl palmitate gives condensed monolayers at the water/air interface. Compression of that film at small surface areas leads to the very steep increase of surface pressure until 55.75 mN/m (at 20 °C). Above that point, the rise of surface pressure is slower and collapse process begins to start. The area per molecule extrapolated to zero surface pressure is 23.75 A^2/molecule. It seems that the change of the temperature has a very small influence on compressibility of the monolayer in the region of high surface pressures: the maximum compressibility moduli (Table 1) are in the range: 271.1–224.0 mN/m (20–35 °C).

The π-A isotherms of mixed 4-methylumbelliferyl palmitate-DPPC monolayers are included in between the isotherms of the pure components (Fig. 2) in a regular sequence of mole fraction of one component in the binary mixture. The addition of 4-methylumbelliferyl palmitate shifts the isotherm of pure DPPC (curve 7) to progressively lower mean molecular areas due to the lower area occupied by the single-chain 4-methylumbelliferyl palmitate. Interactions between 4-methylumbelliferyl palmitate and DPPC molecules produce changes in the molecular packing inside phospholipid monolayers. The most striking feature of the isotherms of mixed monolayers is decrease of the surface pressure at which the LC state appears by rising the proportion of 4-methylumbelliferyl palmitate in the mixture. 4-Methylumbelliferyl palmitate causes the condensation of the DPPC monolayers: mixtures with the greatest content of that component showing very high maximum compressibility moduli, even higher than $C_{s\,max}^{-1}$ values for the pure monolayers. However, this condensation effect refers to high surface pressures. In the lower part of the isotherms of the mixed monolayers the influence of the liquid-expanded state of the DPPC monolayer is evident.

The interphasal miscibility of the two components in the bi-dimensional state at the water-air interface is deduced as a result of behaviour of molecular areas as a function of mixture composition at constant surface pressure. In fact, if the two components are immiscible or if they behave like an ideal mixture, the following relationship is valid [10]:

$$A_{12} = x_1A_1 + x_2A_2 \qquad (2)$$

where A_{12} is the molecular area in the mixed monolayer at the fixed surface pressure π, A_1 and A_2 are the molecular areas in the pure component monolayers at the same π, and x_1 and x_2 are the molar fractions of the pure components in the mixture, such that $x_1 + x_2 = 1$. Positive or negative deviations from Eq. (2) indicate the presence of repulsive or attractive interactions between the two components in the mixed monolayers.

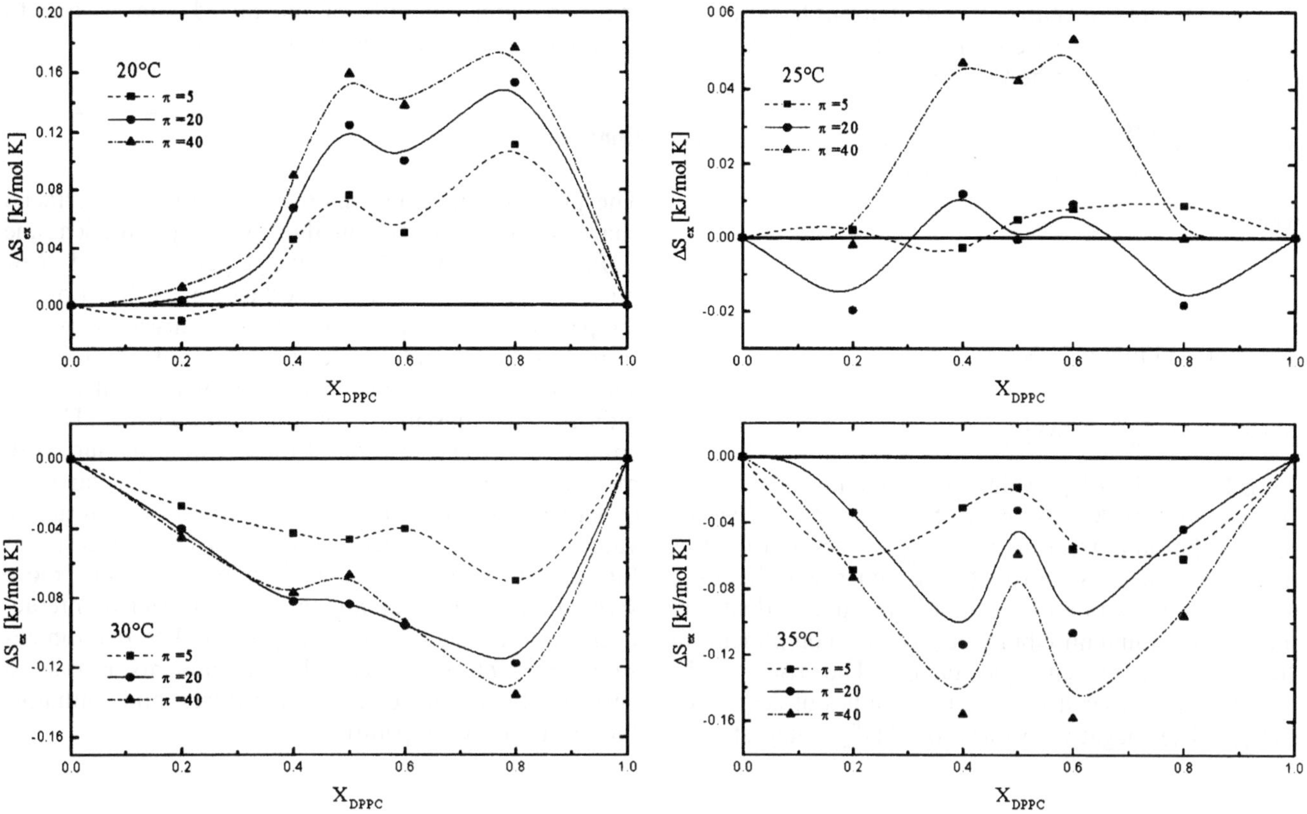

Fig. 5 Excess entropy of mixing of the DPPC/4-methylumbelliferyl palmitate system as a function of molar fractions of DPPC, at $\pi = 5$, 20 and 40 mN/m

The values of mean molecular area as a function of the mixed monolayer composition at different surface pressures (5, 20 and 40 mN/m) are given in Fig. 3. The calculated straight lines in that figure are expected in an ideal mixture ...[Eq. (2)]. For all four temperatures studied the deviations from additivity are mainly negative. These results show that a contraction of molecular area takes place and they constituted one of the first indications that the components are miscible with prevalent attractive interactions. In the temperature range 20–30 °C a local maximum for the equimolar mixture of DPPC and 4-methylumbelliferyl palmitate is present. It suggests that the molecular organisation in that mixed monolayers is less favourable for the bidimensional miscibility of DPPC and 4-methylumbelliferyl palmitate.

A further proof of the interphasal miscibility comes from the surface phase rule developed by Crisp [10], stating that when the film components are immiscible the collapse pressure π_{coll} does not vary with the mixture composition. A true mixture will give only a single collapse. Therefore, the collapse pressure variation with the mixture composition is additional evidence for the DPPC-4-methylumbelliferyl palmitate interactions in the mixed monolayers (Fig. 2).

In Fig. 4, the excess free energy of mixing ΔG^E_{mix} as a function of the mole fraction of DPPC is presented. The excess free energy of mixing was determined by integrating the π-A isotherms up to π lower than the collapse pressure, following the Goodrich method [11]:

$$\Delta G^E_{mix} = N_A \int_{\pi_1}^{\pi_2} (A_{12} - x_1 A_1 - x_2 A_2) d\pi \qquad (3)$$

where π_1 and π_2 are two fixed surface pressures and N_A is the Avogadro number.

The behaviour of ΔG^E_{mix} is analogous to that observed for the mean molecular area (Fig. 3). In fact, the negative values of ΔG^E_{mix} are observed. For the lowest surface pressure ($\pi = 5$ mN/m) the ΔG^E_{mix} is close to zero. However, the more the monolayer is compressed the greater negative values are observed, regardless of mole fraction. This suggests that mixtures are thermodynamically more stable than the pure components and attractive interactions between the components prevail. They exhibit a marked temperature dependence – initially become more negative with rising temperature. However further rise of the temperature disturbs the interactions between the hydrophobic tails. The presence of a local maximum of ΔG^E_{mix} at a molar fraction of DPPC equal to 0.5 offers a proof of the lower thermodynamic stability of the equimolar mixture. This is in agreement with previously observed behaviour of mean molecular areas (Fig. 3).

The enthalpic and entropic contributions to the excess free energy of mixing were calculated according to the Bacon and Barnes method [12], i.e.,

$$\Delta S^{E}_{mix} = -\left(\frac{d\Delta G^{E}_{mix}}{dT}\right)_{\pi} - \Delta A^{E}_{mix}\left(\frac{d\gamma}{dT}\right) \quad (4)$$

where:

$$\Delta A^{E}_{mix} = A_{12} - x_1A_1 - x_2A_2$$

$$\frac{d\gamma}{dT} = -0.154 \text{ mN m}^{-1} \text{ K}^{-1}$$

$$\Delta H^{E}_{mix} = \Delta G^{E}_{mix} + T\Delta S^{E}_{mix} \quad (5)$$

In Fig. 5, the ΔS^{E}_{mix} values as a function of the mole fraction of DPPC are presented. The ΔH^{E}_{mix} values present the same trend as the ΔS^{E}_{mix} values and for this reason are not shown. From the dependence of the ΔS^{E}_{mix} with the composition of the monolayer, it is evident that the bidimensional miscibility between the components at 20 °C is due to entropic factor. The rise of the temperature causes decrease of the values of ΔS^{E}_{mix} and ΔH^{E}_{mix}. The negative values of ΔH^{E}_{mix} suggest that thermodynamic stability of the mixed films is due to enthalpic factor at higher temperatures.

Conclusion

The present work concerns the study of the interfacial characteristic of D,LD,L-dipalmitoylphosphatidylcholine (DPPC) and 4-methylumbelliferyl palmitate and their mixtures at the air/water interface, as a function of temperature. The rise of temperature causes the expansion of isotherms and the decrease of monolayer stability. The deviations from additivity rule for molecular areas indicates the miscibility of the components. This is confirmed by the variation of the collapse pressure with mixture composition. The contraction of molecular areas (negative deviations from additivity rule) and negative values of excess free energy of mixing ΔG^{E}_{mix} suggest that the mixed monolayers are thermodynamically more stable than the pure components and attractive interactions between the components prevail. The two-dimensional miscibility between the components is due to entropic factor at lower temperatures and enthalpic factor at higher temperatures.

References

1. Chattoraj DK, Birdi KS (1984) Adsorption and the Gibbs Surface Excess, chap. 6. Plenum Press, New York London
2. Petty MC (1996) Langmuir–Blodgett films, an introduction. Cambridge University Press, Chap. 4
3. Vila N, Puggelli M, Gabrielli G (1996) Colloids Surf 119:95
4. Berthelot L, Rosilio V, Costa ML, Chierici S, Albrecht G, Boullanger P, Baszkin A (1998) Colloids Surf B 11:239
5. Barnoski Serfis A, Katzenberger R, Williams K, Tran N (1999) J Colloid Interface Sci 215:356
6. Gehlert U, Vollhardt D (1994) Progr Colloid Polym Sci 97:302
7. Saulnier P, Foussard F, Boury F, Proust JE (1999) J Colloid Interface Sci 218:40
8. Lawrie GA, Barnes GT, Gentle IR (1999) Colloids Surf 155:69
9. Gzyl B, Paluch M (2001) Progr Colloid Polym Sci 118:22
10. Gaines GL (1966) Insoluble monolayers at liquid-gas interfaces. Wiley, New York
11. Goodrich FC (1957) Proc 2nd Int Congress Surf Activity, vol 1, p 85
12. Bacon KJ, Barnes GT (1978) J Colloid Interface Sci 67(1)

Progr Colloid Polym Sci (2004) 123: 251–254
DOI 10.1007/b11953

A. Fernández-Nieves
A. Fernández-Barbero
F.J. de las Nieves

Static light scattering from fractal aggregates of microgel particles

A. Fernández-Nieves ·
A. Fernández-Barbero ·
F.J. de las Nieves (✉)
Group of Complex Fluids Physics,
Department of Applied Physics,
University of Almería,
04120 Almería, Spain
e-mail: fjnieves@ual.es
Tel.: +34-950-015434;
Fax: +34-950-015434

Abstract In this work, we present static light scattering results from clusters formed from mesoscopic gel particles. The resultant $I(q)$ curves show two power law behaviours within the experimentally accessible q window. At low q, the scattered intensity is sensitive to the aggregate structure factor, allowing cluster fractal dimensions to be determined. The particles form factor becomes noticeable at higher scattering wave vectors and the intensity is dominated by the scattering from the interfaces between the colloidal particles and the surrounding fluid. The exponents characterising the $I(q)$ behaviour in this region are similar to those obtained for thermo-sensitive microgels.

Keywords Microgel · Aggregation · Fractal · Scattering

Microgels form an interesting subset of polymer gels since they have properties in common with macrogels, as well as features typical of colloidal systems. The aggregation of such particles is an example of this, since the swelling affects both the particle and by inference the nature of the growth process.

In this work, we study light scattered by clusters formed from mesoscopic *soft* particles. The clusters were grown by aggregation under high salt conditions (far above the critical coagulation concentration, *ccc*). Addition of salt above the *ccc* reduces the polymer solvency leading to particle deswelling. The size reduction is detected in the $I(q)$ curves by the presence of a high q region that seems to be controlled by the particle form factor. As the microgel deswells, this region displaces to higher q values, as expected. In addition, the low q intensity dependence allows the cluster fractal dimension to be obtained. The results show that the aggregates present a more compact structure than expected for *DLCA*, which can be attributed to the *soft* character of these colloids, that are able to swell or deswell depending on the environmental conditions.

In a static light scattering experiment, the measured mean scattered intensity may be expressed in the following factorised form:

$$I(q) \sim P(q)\, S(q) \qquad (1)$$

where $P(q)$ and $S(q)$ are the form and structure factors, respectively, for the scattering wave vector $q = 4\pi/\lambda \sin(\theta/2)$, with λ the light wavelength in the solvent and θ the scattering angle. λ and θ set the characteristic observation length scale by tuning the value of q. The form factor contains information concerning the interference of light coming from different volume elements within a particle, whilst the structure factor is determined by the relative positions of all particles.

In the limit $qR > 1$, with R the aggregate size, the structure factor depends on q through the power law relation:

$$S(q) \sim q^{-d_f} \qquad (2)$$

where d_f is the cluster fractal dimension

This asymptotic behaviour is valid for spatial length scales smaller than the aggregate size. For sufficiently small q, the particle form factor q-dependence can be neglected and $I(q) \sim S(q)$. In spite of this, measurements of scattered intensity as a function of q will permit the cluster fractal dimension to be obtained. For scattering wave vectors corresponding to length scales of the order of the particle size d, only the Porod region of $P(q)$ should contribute significantly to the scattering intensity and $I(q) \sim q^{-4}$. For higher q, as achieved with small angle neutron scattering, the intensity becomes dependent on the particle inner structure [1].

Microgel particles will be employed as *soft* spheres. The synthesis of the system is described elsewhere [2]. The spherical and monodisperse particles are based on poly(2-vinylpyridine) (2VP), cross-linked with divinylbenzene (0.25% by weight). The initiator used in the synthesis was 2,2′-azobis(2-amidinopropane) dihydrochloride (V50, Wako). Two types of chemical groups are able to confer charge to the colloidal particles: (i) amidinium groups arising from the initiator, located essentially at the periphery of the particles, and (ii) the constituent monomer 2VP, located within the particles. In this work, deswollen particles are employed with charge only at their surface [3].

Experiments were performed using a slightly modified Malvern Instruments 4700 System (UK) working with a 632.8 nm wavelength He-Ne laser. The q range extends from 0.0023 nm^{-1} to 0.026 nm^{-1}, corresponding to scattering angles between 10° and 150°, respectively. On the other hand, the particle mean hydrodynamic diameter was measured using dynamic light scattering. Intensity autocorrelation functions were determined at a scattering angle of 90°, using a commercial equipment (Zetaxnaster-S, Malvern Instruments). Data analysis based on cumulant analysis was performed using home-made computer programs. All measurements were carried out under very dilute conditions, with the particle volume fraction equal to $\sim 10^{-5}$.

The temperature was equal to (25.0 ± 0.1) °C for all experiments and the ionic concentration was adjusted with NaCl.

The hydrodynamic diameter was measured as a function of ionic concentration in order to establish the *ccc* for a particular surface charge state; specifically, the one corresponding to pH~5.8 (free pH). The results are presented in Fig. 1. Two well defined regions can be distinguished. In the first one, the effect of the ionic concentration translates in a particle diameter reduction. The addition of salt modifies the polymer-solvent interaction, favouring the polymer-polymer contacts in relation to the polymer-solvent ones [4], thus increasing the Flory χ parameter of the polymer gel particles. In other words, the solvent becomes *poorer* with the addition of salt and the particles deswell. In the second region (above [NaCl]~30 mM), the mean hydrodynamic diameter increases due to destabilisation of the colloidal dispersion. The electrostatic interactions responsible for the colloidal stability are screened out, giving rise to

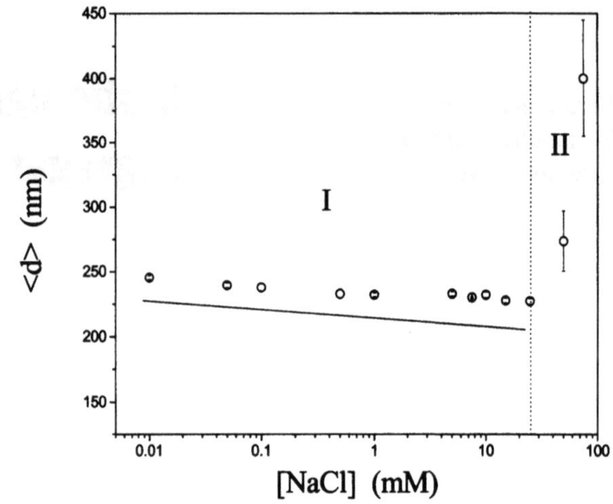

Fig. 1 Hydrodynamic diameter versus salt concentration for the surface charge state corresponding to *pH* 5.8. The particle deswelling due to an increase in the χ parameter can be observed (region I. The line is a guide to the eye) as well as the destabilisation of the system caused by the screening of the electrostatic interaction between particles (region II)

aggregation. All aggregation experiments will be performed far above the encountered *ccc*, thus assuring that the particles approach upon aggregation is diffusive since no bug range electrostatic repulsion is present.

The $I(q)$ curves are shown in Fig. 2 and Table 1 summarises the main features. As is clearly seen, (i) the $I(q)$ curves show two well distinguished power laws, being q_0 the scattering wave vector separating the two regions. This break shifts to lower q values, as the salt concentration is decreased. At [NaCl] = *3* M and *4* M, the break point disappears. (ii) In the low q region, $I(q) \sim S(q)$, allowing d_f to be determined with the aid of Eq. (2). The fractal dimensions obtained in such a way increase from ~1.80 to ~2.45, as the ionic concentration decreases (Table 1). These values agree with small angle light scattering estimates [5]. (iii) The power law dependence between I and q for scattering wave vectors above q_o can be characterised by an exponent $3 \lesssim \Gamma < 4$.

All characteristics of the $I(q)$ curves are consequence of the colloidal and gel-like properties of the experimental system under consideration, since the addition of salt reverts in changes on the particles (deswelling) and also on the interaction between them (as noticed by variations in d_f). The change in the $I(q)$ power law relation is caused by the particle form factor, that becomes apparent for $q > q_o$. The inverse of q_o thus establishes the spatial dimensions below which the form factor becomes noticeable and must then be related to the particle dimensions. The displacement of q_o to higher scattering wave vectors as the salt concentration increases and its eventual disappearance for [NaCl] ≥ 3 M, point in this direction, since the increase of salt causes the deswelling of the microgel particles. In addition, the fact that Γ is almost salt independent supports that in this region, the

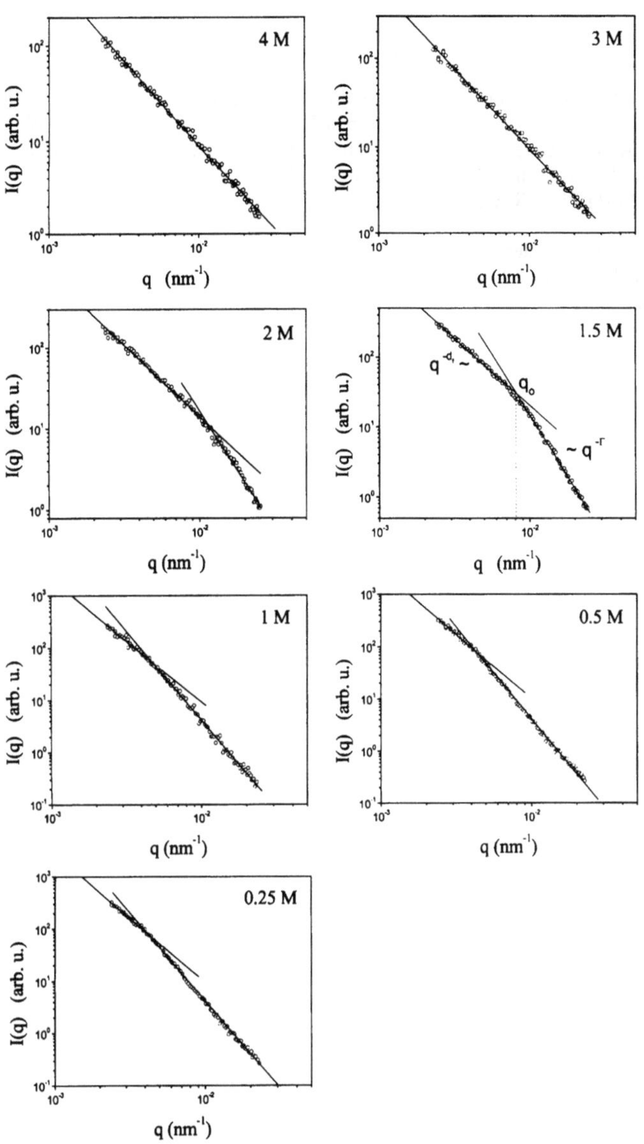

Fig. 2 Intensity versus scattering vector, at different salt concentrations. The break separating the two scattering regions moves towards lower q as the salt concentration decreases, corresponding to an increase in particle size

intensity q dependence is due to the effect of single particles, irrespective of the aggregation process. For particles with a smooth surface, the scattered intensity should follow Porod's law. However, Γ is not equal to 4, indicating that the microgel surface is not completely smooth and thus emphasising the fact that the particles are soft. Small angle neutron scattering of thermosensible microgels show that Γ is below 4 for swollen particles and approaches 4 near their maximum deswollen degree, in agreement with our observations [1].

The rough microgel particle surface controls the aggregation mechanism, as noticed by the change in d_f. If this surface effect was not present, we expected that once the *ccc* had been surpassed, any further increase of salt should leave the aggregation mechanism unchanged with

Table 1 Cluster fractal dimension, exponent Γ and break (q_o), for different salt concentrations

[NaCl] (M)	q_o (nm^{-1})	d_f	Γ
4	–	1.793 ± 0.012	–
3	–	1.837 ± 0.014	–
2	~0.0119	1.779 ± 0.019	2.96 ± 0.24
1.5	~0.0073	1.930 ± 0.023	3.49 ± 0.10
1	~0.0051	2.35 ± 0.07	3.43 ± 0.08
0.5	~0.0049	2.45 ± 0.06	3.444 ± 0.03
0.25	~0.0045	2.45 ± 0.05	3.380 ± 0.03

a constant fractal dimension of 1.7–1.8. The structure of the clusters formed by the aggregation of microgel particles changes, indicating that the aggregation mechanism is being altered. The explanation for this unexpected behaviour has its origin in the *soft* character of the microgel surface, which permits interpenetration [6]. Recall that water is a *poor* solvent for the microgel under the actual experimental conditions. This effect induces a short-range attraction between microgel particles that is counterbalanced by a steric repulsion. The result is an energy minimum, whose depth increases as the polymer concentration in the interpenetration region increases. Thus as the salt concentration is raised, the particles deswell and the energy minimum becomes deeper.

The interpretation of the structure results can then be understood in terms of a reversible growth model, firstly introduced by Shih, Aksay and Kikuchi [7]. It is basically a simulation model built by modifying the classical cluster–cluster aggregation model with a finite interparticle attraction energy, V_{min}. The aggregation reversibility is attained by introducing a probability for unbinding to occur: $P \sim \exp[-nV_{min}/(kT)]$, where n is the number of closest neighbours of every single particle. Reducing V_{min} results in increasing the probability for unbinding. In the limit $V_{min} \to \infty$, $P \to 0$ and the predictions of the cluster-cluster aggregation model are recovered. Due to the finiteness of the attraction minimum, the cluster structure may be more compact in comparison with that arising from the cluster–cluster aggregation model, since the binding forces within the cluster are weaker. The deeper V_{min} is, the smaller is the aggregate fractal dimension. The decrease in the fractal dimension with the increase in salt concentration (see Table 1) could be understood as an aggregation process taking place in a finite energy minimum, which is increasing in depth as the ionic concentration is raised. As the minimum becomes deeper, the probability for unbinding decreases and the cluster structure becomes more branched. As the energy minimum depth is reduced, the number of particle neighbours n has to increase in order to reduce the probability for unbinding thus, yielding higher fractal dimensions.

Acknowledgements Financial support by Ministerio de Ciencia of Tecnologia under project *MAT-2001-2767*.

References

1. Fernández-Barbero A, Fernández-Nieves A, Grillo I, López-Cabarcos E (2002) Phys Rev E 66:051803
2. Loxley A, Vincent B (1997) Colloid Polym Sci 275:1108
3. Fernández-Nieves A, Fernández-Barbero A, Vincent B, Nieves FJ de las (2000) Macromolecules 33:2114
4. Zhu PW, Napper DH (1995) Colloid Surf A 98:106
5. Fernández-Nieves A, Fernández-Barbero A, Duijneveldt JS van, Vincent B, Nieves FJ de las (2001) Phys Rev E 64:051603
6. Fernández-Nieves A, Fernández-Barbero A, Vincent B, Nieves FJ de las (2001) Langmuir 17:1841
7. Shih WY, Aksay JA, Kikuchi R (1987) Phys Rev A 36:5015; Lin J, Shih WY, Sarikaya M, Aksay JA (1990) Phys Rev A 41:3206

Progr Colloid Polym Sci (2004) 123: 255–259
DOI 10.1007/b11954

J.J. Valle-Delgado
J.A. Molina-Bolívar
F. Galisteo-González
M.J. Gálvez-Ruiz

Stabilisation of an amphoteric latex by hydration forces

J.J. Valle-Delgado · J.A. Molina-Bolívar
F. Galisteo-González (✉)
M.J. Gálvez-Ruiz
Grupo de Física de Fluidos y Biocoloides,
Departamento de Física Aplicada,
Facultad de Ciencias, Universidad de Granada, 18071 Granada, Spain
e-mail: galisteo@ugr.es
Tel.: +0034-958-243212
Fax: +0034-958-243214

J.A. Molina-Bolívar
Departamento de Física Aplicada II,
Escuela Universitaria Politécnica,
Universidad de Málaga,
29013 Málaga, Spain

Abstract The colloidal stability of an amphoteric polystyrene latex has been studied using a low-angle light scattering technique (nephelometry). Measurements were carried out at different pH values and NaCl concentrations. The results show a behaviour in agreement with DLVO theory when the pH of the medium is below the isoelectric point of the latex. However, at a pH above the isoelectric point, the latex is completely stable even at high electrolyte concentrations. Electrophoretic mobility measurements indicate that the stability at basic pH values is not due to electrostatic repulsion. This stability could be explained by means of repulsive interactions due to the structure of water molecules (that accompany hydrated cations) around the hydrophilic latex surface. These interactions are the so-called hydration forces, which have been observed in some other systems like proteins, lipids, DNA, etc.

Keywords Amphoteric latex · Stability ratio · Hydration forces · Electrophoretic mobility

Introduction

The DLVO theory (developed by Derjaguin and Landau [1] and by Verwey and Overbeek [2]) has been and is still very useful to explain the stability of colloidal dispersions [3, 4]. This theory considers that the total interaction potential (V_T) between two colloidal particles is the sum of a repulsive electrostatic interaction energy (V_E) and an attractive London-van der Waals (dispersion) energy (V_A).

V_T as a function of the distance between the particles presents a minimum (the primary minimum) close to a maximum at small separations. The maximum represents the energy barrier (V_{max}) which opposes coagulation. If particles approach each other with sufficient kinetic energies to overcome V_{max}, coagulation will occur in the primary minimum and the suspension will be unstable.

V_{max} depends on the electrolyte concentration in the medium. An increase in the electrolyte concentration causes a decrease in V_{max} because the ions diminish the repulsive electrostatic interactions between the particles. The electrolyte concentration at which $V_{max} = 0$ is the so-called critical coagulation concentration (ccc). Thus, the colloidal dispersion aggregates at electrolyte concentrations high enough.

However, the classical DLVO theory is not able to explain the stability of very hydrophilic systems, where repulsive forces arise at very short particle (or surface) separations. These forces are called hydration forces (if the medium is aqueous). There are a lot of examples of hydration forces in the literature [5–17].

The origin and nature of the hydration forces have long been controversial, especially in the colloidal and biological literature. A well-known interpretation of this force is that a polar surface induces an ordering of the solvent, which exponentially decays away from the surface. An overlap of the ordered-solvent layers near the two mutually approaching surfaces gives rise to a repulsive force.

The DLVO theory is a continuum approach that necessarily breaks down at short distances since it does not take into account either the discreteness of the solvent or that of the interacting surfaces.

Experimental

All chemicals used were of analytical grade quality. Water was double distilled by reverse osmosis, followed by percolation through charcoal and a mixed bed of ion-exchange resins (Milli-Q System). In aggregation experiments, pH was controlled using different buffers (acetate at pH 4–5, phosphate at pH 6–7, borate 8–10, constant ionic strength 2 mM); 1 mM citric acid was used to keep the samples at pH 3.

The latex was supplied by Ikerlat Polymers (Guipúzcoa, Spain). The diameter of the latex is 210 ± 10 nm (measured by TEM). Its maximum positive charge (due to amine groups) and its maximum negative charge (due to carboxyl groups) obtained by means of potentiometric titrations are 50 ± 10 $\mu C/cm^2$ and −110 ± 15 $\mu C/cm^2$, respectively.

Particle aggregation studies were carried out using a low angle light scattering technique (nephelometry) for the measurement of aggregation rates. Scattered light intensity was followed at 20° during 120 s at different concentrations of NaCl and several pH values. The light source is an He-Ne laser (633 nm, 10 mW). The scattering cell shape is rectangular, with 2-mm path length. Equal quantities (1 mL) of salt and latex solutions were mixed and introduced in the cell by an automatic mixing device. A concentration of 7×10^9 particles per mL was used to minimise multiple scattering effects.

A Zeta-PALS instrument (Brookhaven) was used to carry out the electrophoretic mobility measurements of the latex as a function of pH at different NaCl and $NaNO_3$ concentrations. This instrument is based on the principles of phase analysis light scattering (PALS) [18]. A latex particle concentration of 0.01 mg/mL was chosen.

Results and discussion

Colloidal stability

The stability ratio (W) is a criterion for the stability of a colloidal system:

$$W = \frac{K_r}{K} \tag{1}$$

in which K_r is the rate constant that describes rapid aggregation, and K is a rate constant of a particular aggregation. Thus, the inverse of the stability ratio provides a measure of the effectiveness of particle collisions leading to aggregation.

Theoretically, W depends on the total interaction energy and it can be obtained through the following equation [19, 20]:

$$W = \frac{\int_0^\infty \frac{\beta(H)}{(H+2a)^2} \exp\left(\frac{V_T(H)}{K_B T}\right) dH}{\int_0^\infty \frac{\beta(H)}{(H+2a)^2} \exp\left(\frac{V_A(H)}{K_B T}\right) dH} \tag{2}$$

where a is the particle radius, H is the distance between the surfaces of the particles and $\beta(H)$ is the hydrodynamic correction factor given by [21]:

$$\beta(H) = \frac{6H^2 + 13Ha + 2a^2}{6H^2 + 4Ha} \tag{3}$$

According to the DLVO theory:

$$V_T = V_E + V_A \tag{4}$$

The attractive London-van der Waals energy V_A is expressed as [22]:

$$V_A = -\frac{A}{6}\left[\frac{2a^2}{H(4a+H)} + \frac{2a^2}{(2a+H)^2} + \ln\frac{H(4a+H)}{(2a+H)^2}\right] \tag{5}$$

where A is the Hamaker constant of the studied system.

In the Linear Superposition Approximation (LSA) [23], the expression for the repulsive electrostatic energy V_E is [2, 22]:

$$V_E = 2\pi\varepsilon_o\varepsilon_r a\psi_o^2 \exp(-\kappa H) \tag{6}$$

where ε_o is the dielectric constant of vacuum, ε_r is the relative dielectric constant of the medium, ψ_o is the surface potential of the particles and κ is the inverse of the Debye length. κ is given by the following equation:

$$\kappa = \sqrt{\frac{\sum_i n_i e^2 z_i^2}{\varepsilon_o \varepsilon_r K_B T}} \tag{7}$$

where e is the electron charge, K_B is the Boltzmann constant, T is the absolute temperature, n_i is the number concentration in the bulk of ions of type i, and z_i is the valence of such ions.

The aggregation rate constants of our colloidal particles have been obtained using the low angle light scattering technique developed by Lips et al. [24, 25]. According to this theory, scattered light intensity increases linearly with time at low scattering angles (in the initial stages of the coagulation):

$$\frac{I_t(\theta)}{I_o(\theta)} = 1 + 2Kn_s t \tag{8}$$

where $I_o(\theta)$ is the initial intensity of light scattered at angle θ; $I_t(\theta)$ is the intensity of light scattered at angle θ in an instant t; n_s is the initial number of spherical particles and K is the aggregation rate constant (between monomers to give rise to dimers). K can be obtained from the slope of the previous linear relation.

The dependence of log W on log NaCl concentration at three different pH values smaller than the i.e.p. (≈6, see later) are shown in Fig. 1. The stability factor decreases gradually with increasing salt concentration until a certain value is reached (the critical coagulation concentration, ccc) and the curve remains parallel to the concentration axis. Theoretical curves of the stability factor have been computed by numerical integration using Eqs. (2), (3), (5) and (6), with A and ψ_o as fitting parameters (see Table 1). We can observe that the latex is

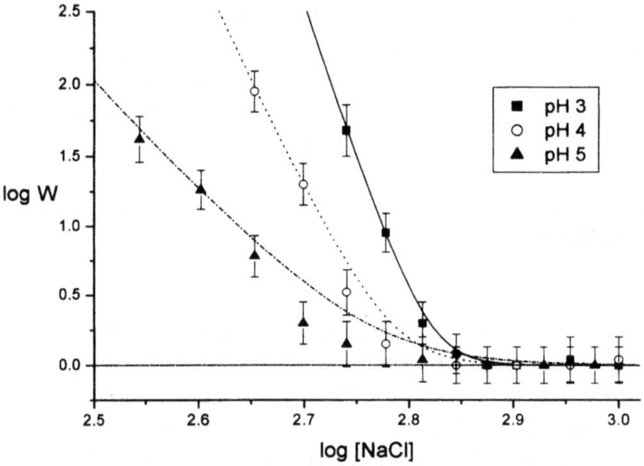

Fig. 1 *log W* versus *log [NaCl]* (mM) for amphoteric latex at different pH values

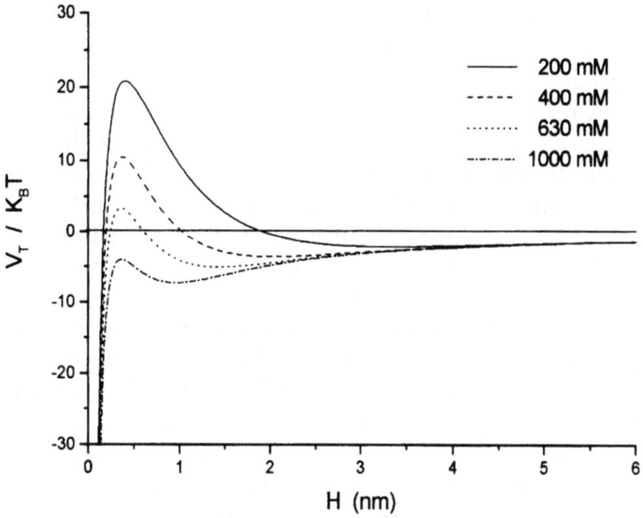

Fig. 2 Interaction potential energy curves at different NaCl concentrations at pH 4

Table 1 Surface potential (ψ_o) and Hamaker constant (A) calculated using DLVO theory at different pH values

	ψ_o [mV]	A [10^{-20} J]
pH 5	20.0	0.27
pH 4	27.6	0.54
pH 3	32.5	0.74

more stable as pH decreases. This is expected since the charge of the latex (and therefore ψ_o) grows up as pH separates from the i.e.p. However, a strange dependence of A on pH has been observed. Similar behaviours have been obtained by other authors [20, 26]. Some authors state that A is not a real constant, but it depends on ionic strength and separation distance [27].

Using the surface potential and Hamaker constant values obtained (Table 1) and Eqs. (4), (5) and (6) we can compute interaction energy curves for different pH values. As an example, Fig. 2 shows these interaction energy curves at pH 4 and at different NaCl concentrations. At this pH, the experimental ccc was 630 ± 50 mM NaCl. It can be observed that the increase in electrolyte concentration provokes a decrease in the height of the energy barrier, and so does stability. This finally disappears, or is lower than 3 K_BT, when the electrolyte concentration is similar to the experimental ccc. Within the experimental error, we can say that the DLVO theory explain satisfactorily the latex stability at this pH. Similar results have been obtained at pH 3 and 5.

The latex was very unstable at pH 6. It aggregates quickly in the low ionic strength buffer (without adding salt). However, it was completely stable at any NaCl concentration when the pH was above the i.e.p. (pH 7, 8, 9, 10). This phenomenon is, of course, quite contrary to the DLVO theory, since addition of electrolyte is generally expected to cause aggregation. These results

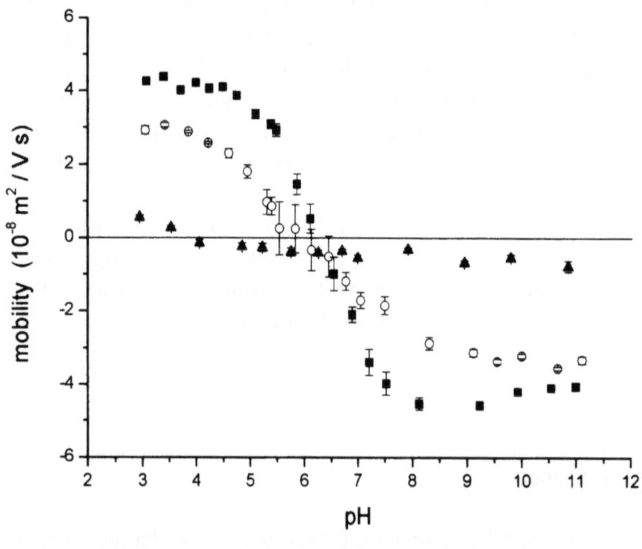

Fig. 3 Electrophoretic mobility of the amphoteric latex as a function of pH at different electrolyte concentrations: (■) 2 mM NaCl; (○) 20 mM NaCl; (▲) 500 mM $NaNO_3$.

are in agreement with those obtained by Healy et al. [13], who studied the stability of an amphoteric latex using KNO_3 and $LiNO_3$ as electrolytes. These authors concluded that this stability may be due to the energy required to dehydrate the adsorbed counterions and to allow a close approach of the surfaces.

Electrophoretic mobility

Fig. 3 shows electrophoretic mobility data as a function of the pH at different electrolyte concentrations. As expected for an amphoteric latex, its mobility depends to a large extent on pH, becoming more negative the higher the pH. The i.e.p. is around 6. The higher is the ionic strength the smaller is the mobility.

A reaction between Cl^- anions and palladium [28] of the electrode takes place at NaCl concentrations higher than 20 mM. So we decided to change Cl^- anions for NO_3^- anions. The mobility curves obtained at 2 mM and 20 mM $NaNO_3$ and the corresponding curves with NaCl are identical. Therefore there is no influence of the anion type on the latex mobility.

At 500 mM $NaNO_3$ the mobility is practically zero at any pH. At this concentration the latex aggregates at pH 4 and 5, but it is stable at pH 7, 8 and 9. However the mobility is practically the same at these pH values. This indicates that the stability at pH 7, 8 and 9 is not due to electrostatic repulsion. A steric stabilisation mechanism could also be invoked in order to explain this anomalous stabilisation. However, the different behaviour observed at pH values over or below the latex i.e.p. does not support this assumption, since a steric effect should be, in principle, independent of pH. At the basic pH values where these colloidal particles are completely stable, the net surface charge is negative, and an excess of cations probably exists around the particles. Since cations are highly hydrated in aqueous media, while anions practically are not hydrated [5], the stabilisation mechanism could be related with the hydration of adsorbed cations on the hydrophilic particle surface. An overlap of the hydration layers of two mutually approaching particles should create a repulsive hydration force. This hydration force would reflect the work required to remove this bound water layer around the particle.

It might be very interested to study these interactions in systems with industrial applications. For example, different latex particles are used in film formation on solid substratum (coating processes) from drying of stabilised dispersions in water [29–32].

Conclusions

We have studied the stability of an amphoteric polystyrene latex using a low angle light scattering technique (nephelometry) for the measurement of aggregation rates. The results show a behaviour according to DLVO theory at pH below the i.e.p. (≈6) of the latex. However, the latex is stable at any electrolyte concentration when the pH is larger than the i.e.p.

Electrophoretic mobility measurements point out that the stability of the latex at high electrolyte concentrations and basic pH values can not be explained by electrostatic repulsions.

Taking into account that the surface of our latex is hydrophilic and that Na^+ cations are hydrated, we propose the existence of hydration forces, like other authors [13–17], in order to explain the stability of our latex. These repulsive forces dominate the interaction at short range, when the electrostatic repulsions are negligible.

Acknowledgements This work was supported by the "Comisión Interministerial de Ciencia y Tecnología-CICYT" (projects MAT 99-0662-03-CO2, MAT 2001-1743 and AGL 2001-3843-C02-02) and by a grant of the "Ministerio de Ciencia y Tecnología" to J.J.V.D.

References

1. Derjaguin BV, Landau L (1941) Acta Physicochim USSR 14:633
2. Verwey EJW, Overbeek JTG (1952) Theory of the stability of lyophobic colloids, vol 1 and 2. Elsevier, Amsterdam
3. Adamczyk Z, Weronski P (1999) Adv Colloid Interface Sci 83:137
4. Banash MA, Croll SG (1999) Prog Org Coat 35:37
5. Israelachvili J (1991) Intermolecular and surface forces, 2nd edn. Academic Press, New York
6. Horn RG, Smith DT, Haller W (1989) Chem Phys Lett 162:404
7. Rabinovich YI, Derjaguin BV, Churaev NV (1982) Adv Colloid Interface Sci 16:63
8. Peschel G, Belouschek P, Muller MM, Muller MR, Konig R (1982) Colloid Polym Sci 260:444
9. Pashley RM (1981) J Colloid Interface Sci 80:153
10. Pashley RM (1981) J Colloid Interface Sci 83:531
11. Pashley RM (1982) Adv Colloid Interface Sci 16:57
12. Pashley RM, Israelachvili JN (1984) J Colloid Interface Sci 97:446
13. Healy TW, Homola A, James RO, Hunter RJ (1978) Faraday Discuss Chem Soc 65:156
14. Delgado-Calvo-Flores JM, Peula-García JM, Martínez-García R, Callejas-Fernández J (1997) J Colloid Interface Sci 189:58
15. Molina-Bolívar JA, Galisteo-González F, Hidalgo-Álvarez R (1997) Phys Rev E 55:4522
16. Molina-Bolívar JA, Galisteo-González F, Hidalgo-Álvarez R (1998) J Colloid Interface Sci 206:518
17. Molina-Bolívar JA, Galisteo-González F, Hidalgo-Álvarez R (1996) Colloids Surf B 8:73
18. McNeil-Watson F, Tscharnuter W, Miller J (1998) Colloids Surf A 140:53
19. McGrown DNL, Parfitt GD (1967) J Phys Chem 71:449
20. Puertas AM, de las Nieves FJ (1999) J Colloid Interface Sci 216:221
21. Honig EP, Roebersen GJ, Wiersema PH (1971) J Colloid Interface Sci 36:97
22. Hunter RJ (1987) Foundations of colloid science, vol. 1. Oxford University Press, New York
23. Chan D, Perram JW, White LR (1975) J Chem Soc Faraday I 71:1046
24. Lips A, Smart C, Willis E (1971) Trans Faraday Soc 67:2979
25. Lips A, Willis E (1973) J Chem Soc Faraday Trans 69:1226

26. Molina-Bolívar JA (1999) PhD Thesis. University of Granada
27. Mahanty J, Ninham BW (1976) Dispersion forces. Academic Press, London
28. Huheey JE, Keiter EA, Keiter RL (1997) Química inorgánica: principios de estructura y reactividad, 4th edn. Oxford University Press – Harla México, México
29. Tang JS, Dimonie VL, Daniels ES, Klein A, El-Aasser MS (2000) Macromol Symp 155:139
30. Tang JS, Dimonie VL, Daniels ES, Klein A, El-Aasser MS (2000) J Colloid Interface Sci 77:644
31. Chesne AD, Bojkova A, Gapinski J, Seip D, Fischer P (2000) J Colloid Interface Sci 224:91
32. Buckmann F, Overbeek A, Nabuurs T (2001) Eur Coat J 6:53

Progr Colloid Polym Sci (2004) 123: 260–263
DOI 10.1007/b11955

M. Medebach
T. Palberg

Flashing of colloidal crystals in square wave electric fields

M. Medebach · T. Palberg (✉)
Institut für Physik der Universität Mainz,
Staudinger Weg 7, 55099 Mainz, Germany
e-mail: Thomas.Palberg@uni-mainz.de
Tel.: +49-6131-3923638
Fax: +49-6131-3925156

Abstract We have investigated the optical properties of colloidal crystals grown in confined parallel plate geometry and subjected to an alternating square wave electric field. A transient increase in the Bragg-scattering intensity was observed upon each reversal of the field direction. Careful analysis of the full 2D flow profile showed that this effect could be attributed to a change of the strain direction in the crystals connected with the changed flow direction in the crystals. Friction at the external wall and intrinsic friction in connection with electro-osmotic flow are discussed as underlying reasons of the observed shear flow.

Keywords Colloidal crystals · Confined geometry · Optical properties · Shear flow · Electrophoresis · Electro-osmosis

Introduction

The equilibrium properties of colloidal crystals consisting of either hard spheres in organic solvents or of charged spheres suspended in aqueous electrolyte have gathered keen attention over the past few years [1, 2]. Recently focus has shifted to non-equilibrium properties like phase transition kinetics or behaviour in external fields [3, 4]. In particular the influence of electric fields on the stationary structure as well as on the kinetics of structural rearrangements has been addressed with optical, and preferably microscopic methods [5, 6, 7]. While the electrophoretic motion of single particles has been widely studied and is theoretically well captured [8], the richness of phenomena observed in crystalline systems is just starting to be explored.

In the present paper we report on an optical phenomenon, which can be traced back to the particular mode of motion of a colloidal solid within a homogeneous electric field. To be specific, a colloidal solid is prepared in a closed flow through cell of rectangular cross section. It is subjected to a square wave alternating electric field and polarisation microscopy and laser Doppler velocimetry monitor its response. As the field changes direction the image of the crystal appears very bright for a short period of time. This we term flashing. Further the intensity distribution over the crystal changes. This we term switching. In what follows we carefully characterise this unexpected transient changes of optical properties and explore the underlying reason in some detail. We discuss a simple phenomenological model explaining our observation as due to an alternation of colloid crystal strain, caused by the direction reversal of the electrophoretic and electro-osmotic flow of the particles and the suspending fluid, respectively.

Experimental

We studied monodisperse polymer latex particles PnBAPS68, which were a kind gift of BASF, Ludwigshafen, suspended in salt free water. PnBAPS68 is a *n*-butyl acrylate-styrene copolymer. Before use the particles were carefully characterised by different methods and their properties and solidification kinetics are discussed in detail in [9]. Sample conditioning used a continuous deionisation procedure and in situ determination of the particle number density n by Bragg scattering and conductivity. Details were recently reported elsewhere [10]. All measurements were performed at residual salt concentrations of $c < 2 \times 10^{-7}$ Mol L^{-1} and $10\ \mu m^{-3} < n < 52\ \mu m^{-3}$, i.e. significantly above the first order fluid-solid phase transition.

The sample cell is a quartz optical flow through cell of rectangular cross section ($X \times Y = 10 \times 0.5\ mm^2$) [Rank Bros., Bottisham, Cambridge, UK]. The suspension is shear molten during conditioning and readily solidifies once the flow is stopped and the cell separated from the conditioning circuit. At low $n < 17\ \mu m^{-3}$ wall oriented crystals are obtained, while at larger $n > 25\ \mu m^{-3}$ homogeneous nucleation leads to randomly oriented polycrystalline solids. At intermediate n mixed morphologies result [11]. After complete solidification an alternating square wave electric field is applied in Z-direction. The effective electrode spacing was calibrated from conductivity to be 65 mm. Maximum field strengths amounted to $E \leq |\pm 80|\ V\ cm^{-1}$ and the field switching frequency was $f_{AC} = 0.05$ Hz.

The sample was monitored by polarisation microscopy (PM). For cubic crystals a PM signal is observed under crossed polarisers only, if a crystal Bragg-scatters light with a difference in polarisation between incident and scattered light [12]. Accordingly the composition of transmitted light with respect to colour and polarisation is changed. Without any field applied, we always observed PM intensity of uniform colour for the oriented wall crystals. In the polycrystalline regime most of the randomly oriented crystals showed differently coloured PM intensity. With the field applied, we observed a short transient increase of PM intensity upon each sign reversal. Interestingly this effect was restricted to wall crystals and few polycrystals. Further the intensity distribution within a wall crystal or such an individual polycrystal usually was observed to be altered or reversed upon sign reversal of the field. An example of the latter case is given in Fig. 1, where the central crystal shows both effects. Images were taken each 40 ms. Due to the shortness we termed the intensity effect "flashing", the distribution effect "switching". Note in the fourth image that the intensity distribution remains stable at longer times.

We further performed Laser Doppler Velocimetry (LDV) to measure the particle velocity. In electrophoresis the measured velocity comprises two contributions. The particle electrophoretic velocity is superimposed on the electro-osmotic flow profile of the suspending medium. For closed cells of rectangular cross section both the electro-osmotic and the particle velocity profile are parabolic for non-interacting particles. Defined locations exist with zero electro-osmotic flow and the electrophoretic mobility can be directly measured there. Such points are called stationary level. This may not be the case for solid materials with strongly interacting particles. This fact necessitates measurements of the complete 2D-flow profile.

We used real fringe illumination and reference beam heterodyne detection followed by frequency analysis. Each individual particle velocity contributes to the Doppler frequency $\omega = q \cdot v(X)$. The present set-up integrates over the complete X-direction and thus yields the complete velocity distribution P(v) from an evaluation of

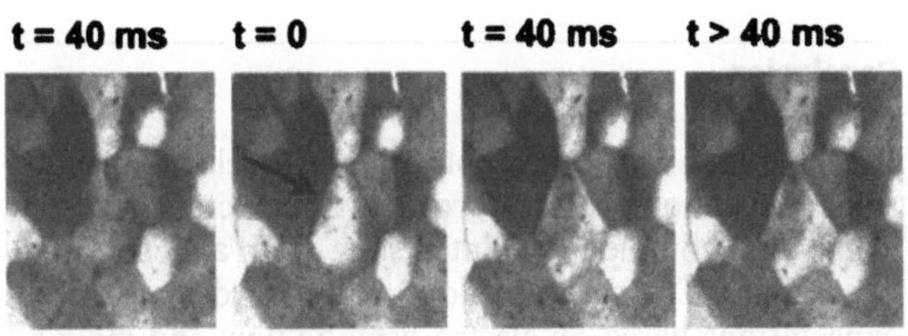

Fig. 1 Flashing and switching effects, which occur at the sign reversal of the field. The first picture is picked up before, the second immediately after, the third another 40 ms later. The last one is taken some 200 ms after the reversal of the field. Note that both flashing and switching occur in this sequence taken on a suspension with $n = 28\ \mu m^{-3}$ and $E = \pm 80$ V/cm

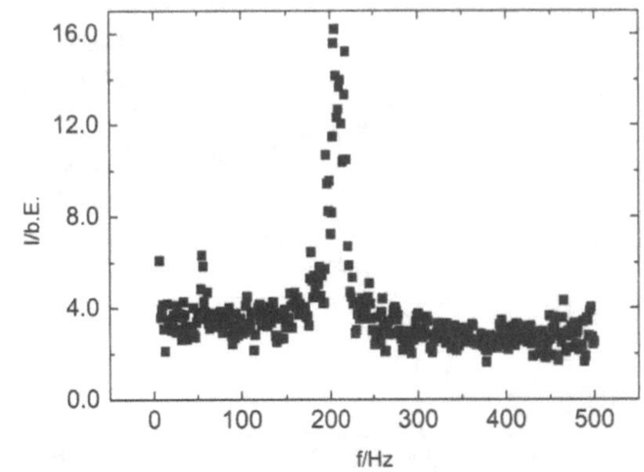

Fig. 2 An example of a spectrum recorded in the crystalline state

the homogeneously diffusion-broadened spectrum [13]. For good statistics spectra were averaged over at least five minutes. An example of a spectrum spectra recorded in the crystalline state is given in Fig. 2. In contrast to the melt or the fluid state the spectra recorded in the crystalline state contain practically no inhomogeneous broadening of the spectral line. We conclude that all particles move with the same velocity, i.e. we observe plug flow in Y-direction.

Results

Our microscopic investigations in the Y-Z-plane revealed a small but significant motion of crystals relative to each other. No relative motion was resolvable in the X-direction. To further explore this point we performed height depended LDV measurements. We first report on the time-averaged results. Measurements were conducted at different cell heights Y ($Y \equiv 0$ in the cell centre). The results for the time-averaged mobilities are shown in Fig. 3 as closed squares. For $Y \leq 3$ mm the mobilities are roughly constant at $\mu = 7 \times 10^{-8}\ m^2\ V^{-1}\ s^{-1}$, they

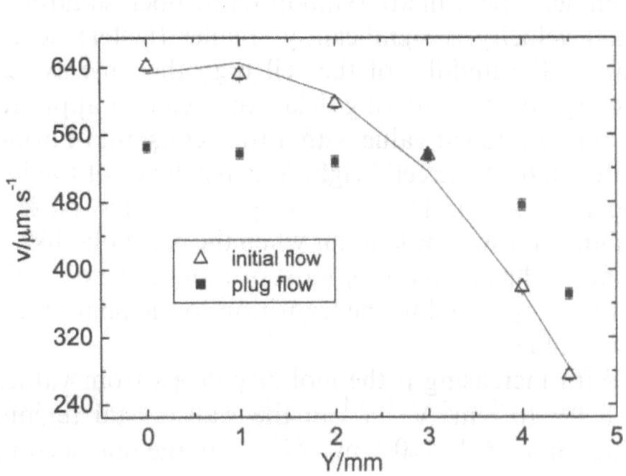

Fig. 3 The height dependent results for the initial flow and the final flow shows that at short times the particles move faster (slower) than at long times for $Y < 3$ mm ($Y > 3$ mm). The line is a fit by a parabolic flow profile. Note that the long time flow profile considerably deviates from parabolic flow. This example is measured at $n = 53\ \mu m^{-3}$ and $E = 79.7\ V\ cm^{-1}$

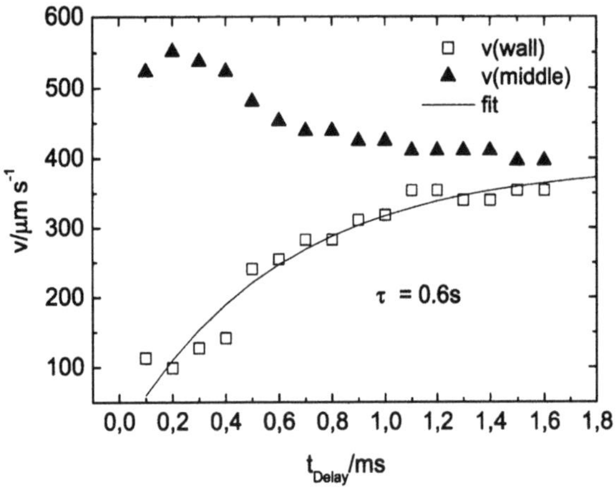

Fig. 4 Time dependence of the particle velocity as measured close to the cell wall, X = 4.5 mm, and in the middle of the cell, X = 0, at $n = 70\ \mu m^{-3}$ and $E = 79.7\ V\ cm^{-1}$. The data sampling over t = 100 ms was triggered with an increasing delay time $t_{Delay} = n \times 0.1$ s after sign reversal of the field. The velocity increases with time and approaches the long time limit v_∞. The fit of $v = v_\infty\ [1 - \exp(-t/\tau)]$ yielded a relaxation time of $\tau = 0.6$ s

drop to significantly lower values at the cell wall. The time averaged flow profile is more plug-like than parabolic, as would be expected (and is indeed observed) for fluid like order. Plug-flow has also been observed before under DC conditions and under steady flow through a tube cell [14, 15]. Also there it was attributed to the coherence of the (poly-)crystalline colloidal solid.

We further performed time resolved measurements. Data sampling over $\Delta t = 100$ ms was triggered with an increasing delay time $t_{DELAY} = n\ 0.1$ s after sign reversal of the field. Again the peak position was evaluated for the average flow velocity. The results are shown in Fig. 4 in dependence on t_{DELAY}. Velocities taken in the middle of the cell Y = 0 are shown as closed triangles, velocities taken near the wall are symbolised by open squares. The initial velocity is significantly smaller (faster) near the wall (in the middle) of the cell than the time averaged velocity. With increasing delay the velocity approaches the time-averaged value with a time constant of approximately 0.6 s. The cell height Y dependence of the initial velocity is shown in Fig. 3 as open triangles. Note that flashing and switching occur when the parabolic-like flow profile of the initial flow is most pronounced, i.e., at times small as compared to the transition to the final plug-like flow profile.

With increasing n the mobility drops from values of $\mu = 8.4 \times 10^{-8}\ m^2\ V^{-1}\ s^{-1}$ in the wall crystal regime to values of $\mu = 6.7 \times 10^{-8}\ m^2\ V^{-1}\ s^{-1}$ in the polycrystalline regime. In contrast to the unexpected increase of mobilities upon the onset of interaction [10], a decrease in the crystalline state may be expected for several reasons. At present, we cannot decide whether this is caused by a drop of the particle potential or by the additional friction of the colloidal crystal at the cell walls.

Discussion

Microscopic observations and LDV revealed the presence of a shearing process in the Y-Z plane, which was absent for the X-Z plane. Further the shear was found more vigorous over the first 1 s. The flashing and switching effects occur in the beginning of shearing. We propose the following scenario: Suppose a field is applied and the particles move towards the anode. Due to friction at the cell walls and/or electro osmotic flow (cf. Fig. 3) a shear field is created straining the crystals. Upon field reversal particles and electro-osmotic flow change direction. Again, particles in the cell centre acquire the larger final velocity, but with reversed sign. The sign of the final strain also reverses. This is supported by the changed long time PM intensity distribution. In the course of velocity reversal the strain in the crystals is changed, too. Then a (presumably unstrained) state may be reached transiently, which fully Bragg-reflects before it is strained further and in the opposite direction. Accordingly the PM intensity will show a maximum on the same time scale as the flow reversal and different intensity distributions before and after field reversal. Using alternating square wave fields with different offsets performed a test of this hypothesis. Fields used above were centred at zero, $E_1 = -E_2$. Flashing and switching occurred at each sign reversal. We then used fields of same frequency but with only positive voltages applied $E_3 = 2E_1$, $E_4 = 0$. Flashing did not appear, but switching was clearly visible. Finally we alternated the field between $E_5 = 3E_1$ and $E_6 = E_1$. Neither flashing nor switching was visible. This shows that the flow and strain *reversal* are necessary conditions for the appearance of flashing. Switching may however also appear upon turning the field on or off. This may correspond to straining under flow and partial elastic relaxation after cessation of flow. Both are states of different strain. A complete relaxation to an unstrained situation may, however not be reached within one field cycle. Therefore no flashing is observed. Frequency dependent measurements could help to further resolve this point.

Conclusion

We have shown that the application of an electric field to a charged colloidal crystal grown in a restricted geometry leads to a complex and interesting flow and straining behaviour. The overall flow patterns deviate from expectations based on electro-osmotic flow and individual particle motion. The presence of plug-flow could be qualitatively attributed to the finite rigidity of the colloidal crystal. A strain is build up over a first short parabolic flow period. As in the case of mechanic shear the elastic forces of the crystals restrict further shearing and initial flow ceases. Where the shear forces surmount the elastic forces, plastic flow results. At long times we therefore observe a plug flow of strained crystals in the central parts of the cell. The phenomenon is similar, but also shows significant differences to the patterns observed under flow through a tube cell [16]. A theoretical description of the observed flow patterns is eagerly awaited.

We further observed the appearance of a transient change of scattering properties upon field reversal. We showed that this was due to a corresponding change of the state of straining in the crystals. The origin of the strain is not yet clear. Possibly both the friction of the crystal at the cell walls and the electro-osmotic flow contribute.

The phenomenology of electric field induced shear flow is thus observed to differ considerably from the case of purely mechanical strain. We hope to have stimulated further investigations in that direction.

Acknowledgements We thank BASF, Ludwigshafen, Germany, for the kind gift of PnBAPS68. Financial Support of the DFG (SFB TR6, Pa459/11-1) and the Materialwissenschaftliches Forschungszentrun, Mainz are gratefully acknowledged.

References

1. Sood AK (1991) Solid State Phys 45:1
2. Bartlett P, Megen W van (1994) In: Mehta A (ed), Granular matter. Springer, New York, p 195
3. Palberg T (1999) J Phys Condens Matter 11:R323
4. Löwen H (2001) J Phys Condens Matter 13:R415
5. Palberg T, Mönch W, Schwarz J, Leiderer P (1995) J Chem Phys 102:5082
6. Dassanayake U, Fraden S, Blaaderen A van (2000) J Chem Phys 112:3851
7. Okubo T, Ishiki H (1999) J Colloid Interface Sci 211:151
8. Palberg T, Evers M, Garbow N, Hessinger D (1999) In: Müller SC, Parisi J, Zimmermann W (eds), Transport versus structure in biophysics and chemistry. Springer, Berlin Heidelberg, p 191
9. Wette P, Schöpe HJ, Liu J, Palberg T (2004) Prog Colloid Polym Sci 123:
10. Evers M, Garbow N, Hessinger D, Palberg T (1998) Phys Rev E 57:6774
11. Liu J, Palberg T (2004) Prog Colloid Polym Sci 123:
12. Gast AP, Monovoukas Y (1991) Nature 351:552
13. Palberg T, Versmold H (1989) J Phys Chem 93:5296
14. Palberg T, Würth, König P, Simnacher E, Leiderer P (1992) Progr Colloid Polym Sci 87:125
15. Palberg T, Streicher K (1994) Nature 367:51
16. Palberg T, Würth M (1996) J Phys I (France) 6:237

Progr Colloid Polym Sci (2004) 123:264–268
DOI 10.1007/b11956

P. Wette
H.-J. Schöpe
J. Liu
T. Palberg

Characterisation of colloidal solids

P. Wette · H.-J. Schöpe · J. Liu
T. Palberg (✉)
Institut für Physik der Universität Mainz,
Staudinger Weg 7, 55099 Mainz, Germany
e-mail: Thomas.Palberg@uni-mainz.de
Fax: +49-6131-3925156

Abstract Charged colloidal particles may form fluid, crystalline or glassy phases even at low packing fraction, if the experimentally adjustable repulsion becomes sufficiently strong. Here we report a comprehensive overview of the characterisation of charged colloidal solids, formed by thoroughly deionised suspensions. A description of the used experimental methods like Bragg microscopy, static and dynamic light scattering, torsional resonance spectroscopy and conductometry is given. For demonstration, these methods are applied to a suspension of polystyrene spherical charged colloids with 68 nm diameter to investigate our system in terms of conductivity, structure, shear modulus, fluid-solid phase transition, crystal growth and nucleation.

Keywords Colloids · Charged spheres · Phase transition · Light scattering · Shear modulus · Crystal growth

Introduction

Colloidal suspensions have acquired a model system status in experimental research on the fundamentals of statistical mechanics. They exhibit many features such as Brownian dynamics or the accessibility on an interparticle scale by light scattering and microscopic techniques [1]. A astounding property of colloidal solids is their extreme softness: Shear moduli are on the order of a few Pa only [2]. Further colloidal systems have also become a valuable model system for the study of kinetics of crystal nucleation and growth. The precise control of interaction parameters and new instrumental developments have allowed for quantitative determination of nucleation rate densities [3, 4]. We here conduct a number of different experiments to comprehensively characterise our investigated colloidal system of polystyrene spheres with 68 nm diameter. Besides the structure analysis we measured the conductivity, the shear modulus and the growth velocity. We further report the fluid solid phase transition and compare it with theoretical models. Finally we have investigated the nucleation kinetics and compare our results with classical nucleation theory.

Experimental

Experiments were performed using polystyrene latex spheres with 68 nm diameter (PS68 2168/7387, BASF Ludwigshafen, Germany). The suspension was prepared from diluted and precleaned stock suspension of approximately 10% packing fraction. The suspension was pumped peristaltically through a tubing system connecting the ion exchange chamber to the measuring cells for conductivity and light scattering experiments, respectively [5].

Conductivity is measured in situ allowing for control of both the particle number density n (in the deionised state) and electrolyte concentration c (for known n). In the full deionised state the conductivity of suspensions is well described by $\sigma = n \cdot e \cdot Z_\sigma^*(\mu_p + \mu_{H^+}) + \sigma_B$ with the elementary charge e, the particle and proton mobilities μ_p and $\mu_H{}^+$, respectively and the background conductivity σ_B (mainly stemming from dissociated water and ionic impurities). The particle mobility μ_p is of the order of $(2–10)\cdot 10^{-8}$ (m^2/Vs) as determined by laser Doppler velocimetry [6]. The conductivity measurements were in the full deionised state were calibrated by static light scattering (SLS).

To determine the structure and the morphology of the colloidal solid SLS was used. Here the scattered light intensity I(q) is detected for different scattering vectors $q = (4\pi \cdot \upsilon/\lambda) \cdot \sin(\theta/2)$, where ν is the refractive index of the suspension medium, λ the vacuum wavelength of the used laser light and θ the scattering angle.

In the powder averaged experiments on polycrystalline samples (Debye-Scherrer technique) sharp diffraction peaks can be observed at the scattering vector $q_{hkl} = (2\pi/g)\sqrt{h^2+k^2+l^2}$ where g is the lattice constant of the crystal and h,k,l are the Miller Indices of the considered diffraction plane. For an ideal crystal the structure factor S(q) takes the form of delta functions. In the case of finite crystal size considerable broadening may be observed.

For torsional resonance spectroscopy (TRS) the polycrystalline sample is put into weak oscillations about its vertical axis which excites the solid in the cell to resonant vibrations if the resonance frequencies (eigenfrequencies) for the given sample geometry are met. This leads to periodically changing lattice constant and a periodic shift of the Bragg peaks result which can be detected via a position sensitive detector. From the resonance frequencies we determined the shear modulus G [7, 8]. The deviation of values for the shear modulus G in close analogy to the shear modulus predictions for atomic solids includes nearest neighbour interactions . The expression for the shear modulus of bcc ordered may be described as [9]:

$$G_{bcc} = f_A \frac{4}{9} n \cdot V(r)|_{r=d} \kappa^2 d^2 \tag{1}$$

where f_A is known factor for orientational averaging in a polycrystalline sample. In most cases a value of $f_A = 0.5$ is encountered. $V_{(r=d)}$ is the pair potential at nearest neighbour distance d:

$$V(r) = \frac{Z^* e^2}{4\pi\varepsilon_0\varepsilon_r} \left(\frac{\exp(\kappa a)}{1+\kappa a}\right)^2 \frac{\exp(-\kappa r)}{r} \tag{2}$$

with the screening parameter calculated via $\kappa^2 = (e^2/\varepsilon_0\varepsilon_r k_B T)(n \cdot Z_G^* + n_{salt})$, where $\varepsilon_0\varepsilon_r$ is the dielectric permittivity of the suspension, k_BT is the thermal energy and n_{salt} the particle number density of added salt. Z_G^* is the effective charge of the particle. In the present experiment the salt concentration is set to zero and the Z_G^* is used as a fit parameter, to get an effective shear modulus charge from the particle number density dependence of the shear modulus.

The SLS and TRS measurements were performed using a recently reported multi-purpose light scattering instrument. In this static structure, shear modulus and dynamic structure factors may be measured quasi simultaneously on the very same (crystalline) sample. The apparatus is equipped with a double goniometer. It uses two counterpropagating illumination detection schemes for static and dynamic light scattering and a third independently adjustable scheme for torsional resonance spectroscopy. The cell for the light scattering experiments is connected with the particle preparation tubing system as mentioned above. Details have recently been given elsewhere [10].

Bragg microscopy is used to detect the light scattering of growing crystals by illuminating the sample with white light and placing a camera under an angle fulfilling the Bragg condition of an individual crystal. For Bragg microscopy a rectangular flow through cell on the stage of a microscope, illuminating the cell from below in a manner that the wall based crystals scatter into the microscope objective. The crystallisation proceeds either via immediate heterogeneous nucleation at the wall, or after same induction time via homogenous nucleation in the bulk. Using a thin optical cell the first mechanism dominates at low packing fractions. Hence twinned single crystals grow from the walls once the shear process by pumping the suspension through the tubing system has been stopped. In case of body centred cubic (bcc)-formed crystals their ⟨111⟩ direction are aligned along the flow direction while their (110) layers are parallel to the cuvette wall. The cold white light source is adjusted to monitor the growth in the [110] direction off the cell wall [11].

Application on colloidal solids of PS68 spheres

The PS68 system was prepared under full deionised conditions. The suspension showed pronounced fluid or at higher n crystalline order. The system was first prepared in the crystalline state. Then the particle number density is lowered by stepwise diluting the suspension. At each given n (determined by SLS measurements) we measured the conductivity to obtain a linear relation. Fig. 1 shows the result. The vertical line indicates the position of the fluid solid phase boundary. From this linear relation we derived the effectively transported charge $Z_\sigma^* = 450 \pm 15$.

Structure was determined by static light scattering (SLS) measurements. The scattering volume corrected scattering pattern I(q) of solid PS68 samples are shown in Fig. 2. From these data the crystal structure was identified to be body centred cubic (bcc). The angular position of the first peak θ_{max} was then used to calculate n according to

$$n_{bcc} = 2\left[\frac{2\upsilon}{\lambda}\frac{1}{\sqrt{h^2+k^2+l^2}}\right]^3 \sin^3\left(\frac{\theta_{hkl}}{2}\right) \tag{3}$$

where ν is the refractive index and λ is the vacuum wavelength. At a particle density $n < 2.4 \cdot 10^{19} (1/m^3)$ statistics are too bad for particle density determination. In this case we used the conductivity to determine n. Thus SLS and conductivity measurements complement each other.

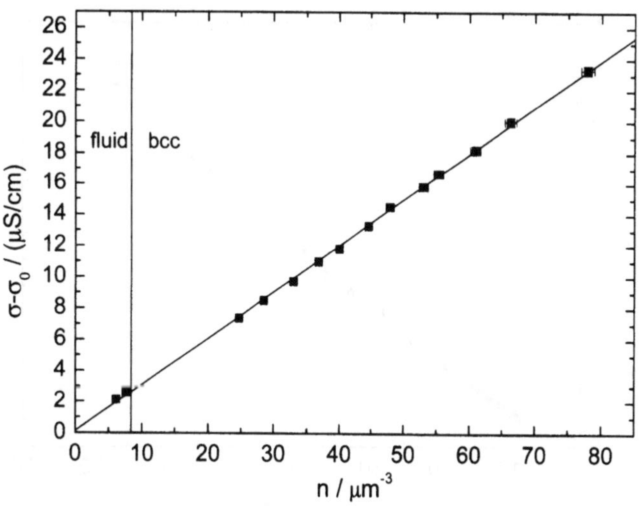

Fig. 1 Conductivity of PS68 as a function of the particle number density n. The vertical line indicates the position of the fluid solid phase boundary. The conductivity is not influenced by the phase transition. The solid line is a linear fit with $Z_\sigma^* = 450 \pm 15$ as fit parameter

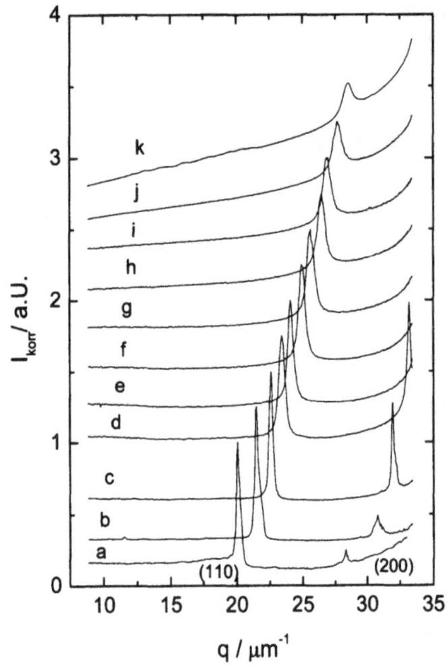

Fig. 2 Representative Debye-Scherrer light scattering patterns of deionised polycrystalline PS68 sample (wavelength $\lambda = 488$ nm). The curves are shifted for clarity and normalised to the first peak height. The particle number density increases from the bottom ($n = 24.7\ \mu m^{-3}$) to the top ($n = 66.3\ \mu m^{-3}$)

In Fig. 3 we plot the shear modulus G versus n. The solid line is the best one parameter fit according to Eq. (1) with Z_G^* as free fit parameter using $f_A = 0.5$. At full deionised conditions we obtain an effective shear modulus charge $Z_G^* = 326 \pm 7$.

Next we compare the phase behaviour of the PS68 sample with the phase diagram of Robbins, Kremer and Grest [12] and Meijer and Frenkel [13]. In these phase diagrams the inverse normalised interaction energy $k_B T / V(\bar{d})$ at averaged inter particle distance $\bar{d} = n^{-1/3}$ is plotted versus a coupling parameter $\lambda = \kappa \bar{d}$. In dependence of the experimental parameters n, c, a and Z^* different states in this diagram can be reached. In Fig. 4 we plotted the phase behaviour of the PS68 system using the interaction potential determined by conductivity measurements, i.e., by using the effectively transported charge Z_σ^*. We did the same for the interaction potential determined by the shear modulus measurements, i.e., by using the effective shear modulus charge Z_G^*. The error bars indicate the position of the determined melting point under full deionised conditions. With the effective transported charge no agreement with the theoretical predictions was found. Whereas using the interaction potential determined by the effective shear modulus charge we found good agreement with the theoretical melting line of Robbins, Kremer and Grest.

A further morphological information ma be extracted from the measured Debye-Scherrer patterns. The width of the principal maximum Δq is correlated with the crystallite size $L = (2\pi \cdot K_{hkl}) / \Delta q$ where K_{hkl} is the Scherrer constant which is of order one [14], and L is the average edge length of the crystals which are assumed to be cube shaped. Fig. 5 shows the result of this evaluation. The crystallite size decreases with increasing n to a minimum value of a few microns only.

In general, the crystallite size is determined by the nucleation rate density J and the crystal growth velocity v. Crystals stop growing once they intersect. Figure 6 shows the result for growth velocity in the [110]-direction as inferred from Bragg microscopy. Here we followed the

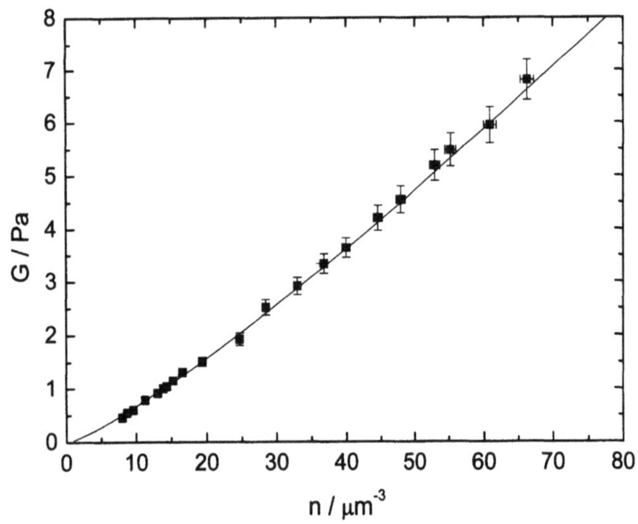

Fig. 3 Particle number density dependence of the shear modulus of the PS68 solid at deionised conditions. The solid line is a fit to Eq. (1) with $Z_G^* = 326 \pm 7$. No phase transition to fcc is observed in the studied range of n

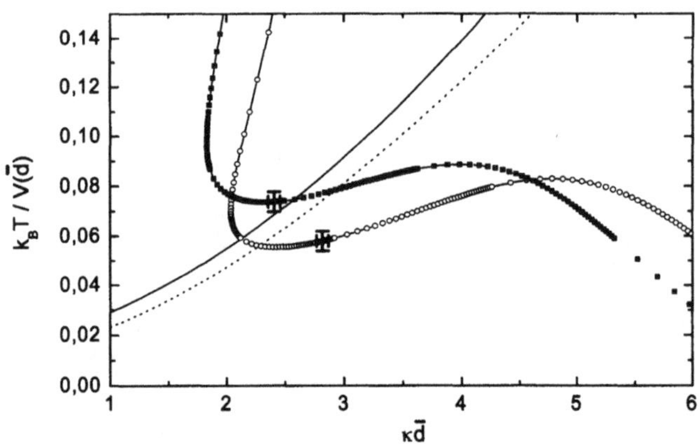

Fig. 4 Comparison of the phase behaviour of PS68 sample with the phase diagrams of Robbins, Kremer, Grest (*solid line*) and of Meijer, Frenkel (*dotted line*). The calculated inverse pair potential for the effectively transported charge (-○-) and for the shear modulus charge (-■-) is plotted versus the coupling parameter $\lambda\bar{d}$. The boarders of the coexistence region of the fluid solid phase transition are indicated by the error bars. A good agreement is found with the theoretical prediction of Robbins, Kremer and Grest, if Z_G^* is used

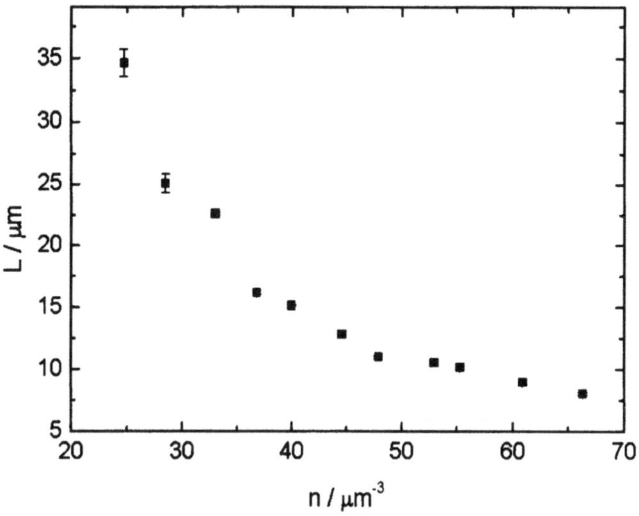

Fig. 5 Average linear dimension L of the cube shaped bcc crystal as inferred from the (110) peak width plotted versus n

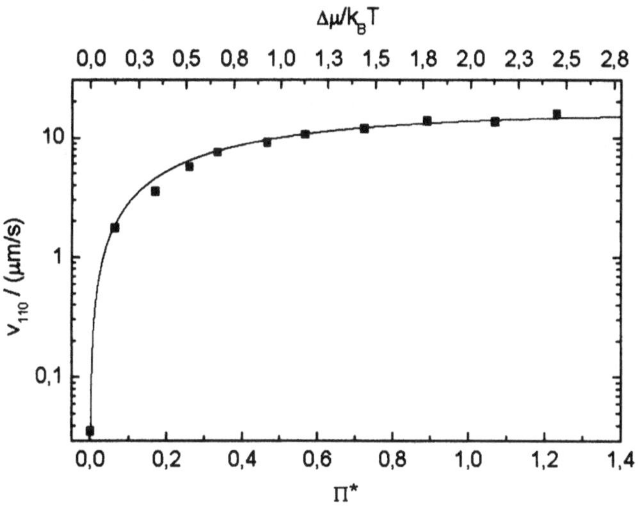

Fig. 6 Growth velocity in [110] direction v_{110} versus the reduced energy density Π^*. The solid line is a fit to a Wilson- Frenkel growth law using Eq. (5). The best fit yields $B' = 1.99 \pm 0.1$ and $v_\infty = (15.9 \pm 0.9)$ μm/s

procedure of Würth et al. [15] and plot the growth velocity against the reduced energy density Π^*:

$$\Pi^* = \frac{\Pi - \Pi_f}{\Pi_f}; \quad \Pi = \tfrac{1}{2}\alpha \cdot n \cdot V(r) \tag{4}$$

where α is the effective coordination number and V(r) the pair potential. The suffix f correspond to the value of freezing. The growth velocity was well described by a Wilson-Frenkel law:

$$v = v_\infty \left[1 - \exp\left(\frac{-\Delta\mu}{k_B T}\right)\right] \tag{5}$$

where v_∞ is the maximal growth velocity, limited by the elementary step of integrating the particles into the crystal surface and Δμ is the difference in chemical potential between the fluid and the crystalline phase. We express Δμ in terms of the reduced energy density $\Delta\mu/k_B T = B'\Pi^*$ with the conversion factor B′. With this expression it is possible to fit the Wilson-Frenkel law to the measured data set with the fit parameters v_∞ and B′. The best fit (fixing $v(\Pi^* = 0) = 0$ μm/s) yields $B' = 1.99 \pm 0.2$ and $v_\infty = (15.9 \pm 0.5)$ μm/s.

Further we calculated the particle density dependence of the nucleation rate density J following Astuen et al. [16] $J = 1.158 \cdot v \cdot n_k^{4/3}$ where $n_k = 1/L^3$ and v is the maximum growth velocity according to the Wilson-Frenkel fit.

The crystallisation kinetics are controlled by the exponential dependence on the ratio between some intrinsic energy scale and the thermal energy $k_B T$. From this observation the classical rate equations result which give the nucleation rate density as $J = J_0 \exp(-\Delta G^*/k_B T)$ where ΔG^* is the height of nucleation barrier and J_0 is a kinetic prefactor. The nucleation barrier is $\Delta G^* = 16\pi\gamma^3/(\Delta\mu \cdot n)^2$. Here γ is the surface tension between the melt and the solid state and Δμ is the chemical potential difference according to the Wilson-Frenkel fit. In Fig. 7 we plotted the calculated nucleation rate densities logarithmically against the particle number density n (*left*) and the same curve versus $(n \cdot \Delta\mu)^{-2}$ (*right*). With this plot the surface tension was determined from the slope at large supersaturation to be $\gamma = (98.5 \pm 8.3)$ nJ m^{-2} . The plot shows that with decreasing particle number density the nucleation rate deviates from the exponential decay. This could be due to the starting of inhomogeneous nucleation at the walls of the sample cuvette or from impurity nucleation at low particle concentration.

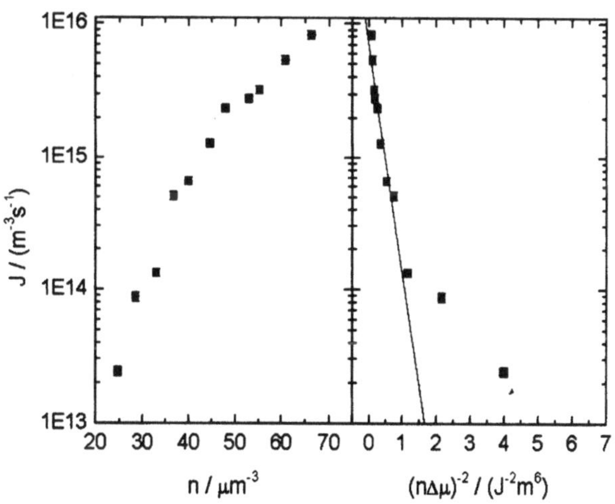

Fig. 7 Nucleation rate densities versus the particle number density (left) and versus $(n \cdot \Delta\mu)^{-2}$ (*right*). Except for points at lowest n we fitted the data linear to determine the surface tension between the melt and the solid: $\gamma = (98.5 \pm 8.3)$ nJ m^{-2}

Conclusion

We characterised systematically the crystalline phase of charged polystyrene spheres with 68 nm diameter by using a combination of static light scattering (SLS), torsional resonance spectroscopy, conductivity and Bragg microscopy measurements under full deionised conditions. The structure determinations shows a crystallisation in the bcc-ordered state over the whole range of the investigated particle number densities. The conductivity were found well compatible to the theoretical predictions and was not influenced by the fluid solid phase transition. The shear modulus in dependence of the particle number density can described well by using a Debye-Hückel potential proposed in the PBC model. Using the interaction potential determined by TRS we are able to describe the experimental phase boundaries in a universal theoretical phase diagram proposed by Robbins, Kremer and Grest for the first time. Thus the phase behaviour and the elastic properties in charged colloidal systems can well described by the same interaction potential available experimentally. Further we have introduced investigations of the basic processes of growth and nucleation of colloidal crystal from the melt. To obtain this we used a combination of static light scattering and Bragg microscopy measurements. Good agreement with Wilson-Frenkel growth and classical nucleation theory at large supersaturation for our system was found.

References

1. Pusey PN, Hansen JP, Levesque D, Zinn-Justin J (1991) Liquids, freezing and glass transition. Ecole dété, Les Houches 1898, 51. Elsevier, Amsterdam, p 763
2. Lindsay HM, Chaikin PM (1982) J Chem Phys 76:3774
3. Palberg T (1999) J Phys Condens Matter 11:R323–R360
4. Aastuen DJW, Clark NA, Cotter LK, Ackerson BJ (1986) Phys Rev Lett 57:1733
5. Wette P, Schöpe HJ, Biehl R, Palberg T (2001) J Chem Phys 114:7556
6. Evers M, Garbow N, Hessinger D, Palberg T (1998) Phys Rev E 57:6776
7. Dubois-Violette E, Pieranski P, Rothen F, Strzelecki I (1980) J Phys 41:369
8. Joanicot M, Jorand M, Pieranski P, Rothen F (1984) J Phys 45:1413
9. Flügge S (1955) Handbuch der Physik, Band VII: Kristallphysik (1). Springer Verlag, Berlin
10. Schöpe HJ, Palberg T (2001) Journal Colloid Interface Sci 234:149–161
11. Maaroufi MR, Stipp A, Palberg T (1998) Progr Colloid Polym Sci 110: 83–88
12. Robbins MO, Kremer K, Grest GS (1988) J Chem Phys 88:3268
13. Meijer EJ, Frenkel D (1991) J Chem Phys 94:2269
14. Hammond C (1998) The basics of crystallography and diffraction. Oxford University Press
15. Würth M, Schwarz J, Culis F, Leiderer P, Palberg T (1995) Phys Rev E 52:6415
16. Astuen DJ, Clark NA, Swindal JC, Muzny CD (1990) Phase Transitions 21:139

Progr Colloid Polym Sci (2004) 123: 269–274
DOI 10.1007/b11957

Md. H. Uddin
Y. Yamashita
H. Furukawa
A. Harashima
H. Kunieda

Phase behavior of poly(oxyethylene)-poly(dimethylsiloxane) surfactant (copolymer) with water or silicone oil

Md. H. Uddin · Y. Yamashita
H. Kunieda (✉)
Graduate School of Environment and Information Sciences,
Yokohama National University,
Tokiwadai 79-7, Hodogaya-Ku,
Yokohama 240-8501, Japan
e-mail: kunieda@ynu.ac.jp
Tel.: +81-45-3394190
Fax: +81-45-3394190

H. Furukawa, A. Harashima
Dow Corning Toray Silicone Co. Ltd.,
Chigusa-Kaigan 2-2, Ichihara 299-0108,
Japan

Abstract Phase diagrams of A-B type silicone surfactant or copolymer, $Me_3SiO\text{-}(Me_2SiO)_{m-2}\text{-}Me_2\text{-}SiCH_2CH_2CH_2\text{-}O\text{-}(CH_2CH_2O)_nH$, ($Si_mC_3EO_n$) in water or octamethylcyclotetrasiloxane ($C_8H_{24}O_4Si_4$ or D_4) were constructed as a function of poly(oxyethylene) (EO) chain length (n) ranging from 3.2–51.6. Two series of surfactants: $Si_{14}C_3EO_n$ and $Si_{25}C_3EO_n$ form reverse micellar solution (O_m), reverse discontinuous cubic, (I_2), reverse hexagonal (H_2) and lamellar (L_α) phases in water with increasing EO chain lengths, n, or its volume ratio to the surfactant, f were investigated. However, the H_2 and I_2 phases are stabilized in a wide range of f in water + $Si_{25}C_3EO_n$ than that in water + $Si_{14}C_3EO_n$. Similar phase sequence, O_m-I_2-H_2-L_α was observed in $Si_mC_3EO_n/D_4$ systems with the increase of the volume in the hydrophilic moiety. In water binary systems, all the homogeneous phases are in equilibrium with excess water, whereas the liquid crystalline phases are systematically changed to the O_m phase via inter-existing phases in D_4 systems because the surfactant layer curvature changes towards more negative due to the increase of effective lipophilic volume by added D_4.

Keywords Silicone surfactant · Phase behavior · Liquid crystal · Silicone oil

Introduction

Poly(oxyethylene)-type non-ionic surfactants (C_mEO_n) including poly(dimethylsiloxane) ($Si_mC_3EO_n$) in water, block copolymers in selective solvents (water and oil), and copolymer melts form various self-organized structures depending on the hydrophilic and lipophilic property of the amphiphiles [1–5]. In all the cases, the volume ratio of the hydrophilic moiety to the amphiphiles (surfactant or copolymer), f, which is also related to the Griffin's HLB number in the surfactant field, can modulate the layer curvature and the resulting microstructures [6]. It is known that amphiphilic molecules self-assemble in solution when a solvent is compatible with one part but incompatible with the other part by enhancing the segregation between these two distinct parts. As the hydrocarbon chain length in C_mEO_n is limited in the range C_8-C_{18}, otherwise the Krafft temperature is increased, C_mEO_n as well as EO_x-PO_y-EO_x block copolymers rarely form aggregate in oil in the absence of water because of the low segregation of EO and hydrocarbon or PO chains. Hence, the dependence of self-organization and phase behavior in oil on the EO-chain length is yet to be well understood. On the other hand, long poly(dimethylsiloxane)-chain surfactant or copolymer can be used because the siloxane chain is in a liquid state due to its flexibility. Since the compatibility between poly(oxyethylene) and poly(dimethylsiloxane) chains is extremely low, A-B type silicone surfactants self-organize in water, as well as in polar and non-polar oil, and even in the absence of solvent, if A or B is not very short [3, 7–9].

In this context, we studied the phase behavior of poly(oxyethylene)-poly(dimethylsiloxane) surfactant ($Si_mC_3EO_n$: $m=14$ or 25, $n=3.2$–51.6) as a function of poly(oxyethylene) length in water (compatible with EO chain) or silicone oil (D_4) (compatible with poly(dimethylsiloxane) chain) and the effects of the solvents quality on the phase behavior are also discussed.

Experimental

Materials

Poly(oxyethylene)-poly(dimethylsiloxane) surfactants or A-B type copolymers with the general formula Me_3SiO-$(Me_2SiO)_{m-2}$-Me_2-$SiCH_2CH_2CH_2$-O-$(CH_2CH_2O)_nH$, abbreviated as $Si_mC_3EO_n$, were obtained from Dow Corning Toray Silicone Co. Ltd., Japan. Me is a methyl group attached to Si, *m* is the total number of silicone, and *n* is the average number of ethylene oxide (EO) units. To construct the phase diagrams as a function of the EO-chain length, intermediate EO chain lengths were obtained by mixing two neighbors with $n=3.2$, 7.8, 12, 15.8, 33.1 or 51.6 for $Si_{14}C_3EO_n$ and with $n=3.2$, 7.8, 12.2, 15.8 or 51.6 for $Si_{25}C_3EO_n$. The purity, and the polydispersity index of poly(oxyethylene) and poly(dimethylsiloxane) chains, of the surfactants were stated in refs.[3, 8]. The main impurity is water-soluble unreacted polyether, $CH_2=CHCH_2$-O-$(CH_2CH_2O)_nH$. Octamethylcyclotetrasiloxane ($C_8H_{24}O_4Si_4$ or D_4) was obtained from Dow Corning Toray Silicone Co. Ltd. Millipore filtered water was used.

Phase diagrams

Various amounts of constituents were weighed and sealed in ampoules. Homogeneity was attained using a vortex mixer and repeated centrifugation through a narrow constriction in the sealed sample tubes. The phase equilibria were detected by visual inspection of the samples in transmitted light and with crossed polarizers for birefringence. The types of liquid crystals were identified by a video-enhanced microscope (Nikon X2F-NTF-21) and small-angle X-ray scattering measurements [3, 10].

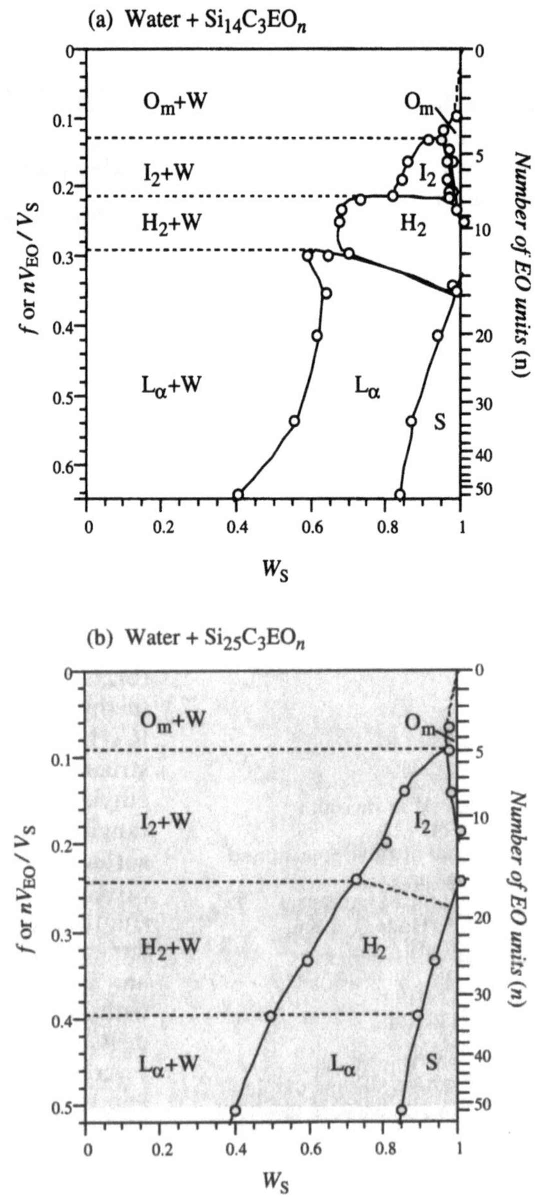

Fig. 1 Phase diagrams of water-$Si_mC_3EO_n$ systems as a function of the EO-chain length at 25°C. *f* is the volume ratio of EO chain to the surfactant. I_2, H_2, L_α and O_m indicate reverse discontinuous cubic, reverse hexagonal, lamellar and surfactant liquid or reverse micellar solution phase, respectively. W indicates excess water phase and S indicates solid present phase. **(a)** $Si_{14}C_3EO_n$, **(b)** $Si_{25}C_3EO_n$

Results

Phase diagrams of water-$Si_mC_3EO_n$ systems as a function of EO-chain length

The isothermal (25 °C) phase diagrams of the water-$Si_{14}C_3EO_n$ and the water-$Si_{25}C_3EO_n$ systems constructed as a function of the volume ratio of the hydrophilic moiety to the surfactant, nV_{EO}/V_S or *f* (vertical axis) and the weight fraction of the surfactant in the system, W_S (horizontal axis) are presented in Fig. 1a and 1b, respectively. The corresponding number of EO units, *n*, is also shown on the right-hand axis, vertically.

The surfactant liquid or reverse micellar solution (O_m), discontinuous reverse cubic (I_2), reverse hexagonal (H_2), and lamellar (L_α) phases are successively formed with the increase of the EO-chain length of the surfactant in the both systems as is shown in Fig. 1. Pure anhydrous surfactants also form liquid crystals above their solid transition temperature. The viscous, transparent and optically isotropic I_2 phase existing between the surfactant liquid (O_m) and the birefringence H_2 phase is considered to consist of a discontinuous-type reverse micellar cubic phase. In our previous study, the structure of the I_2 phase formed in the solvent absent $Si_{25}C_3EO_{15.8}$ system (Fig. 1b) was confirmed as a face-centered cubic belonging to the *Fd3m* space group by a synchrotron radiation SAXS measurement [7]. The H_2 and L_α phases were ascertained based on the characteristics SAXS patterns and microscopic textures.

The very short EO chain surfactants form only O_m phase solubilizing a small amount of water. There is a general tendency of increasing the solubilization of water by the aggregates with the increase of the EO chain length. The liquid crystalline phases are in equilibrium with excess water beyond their solubilization limits. It was difficult to determine the phase boundary between a single liquid crystalline phase and excess water region in Fig. 1 because it is gradually introducing a little turbidity and the intensity of the SAXS peak is considerably low at high water content. Therefore, the boundary was determined by visual inspection. However, we observed a clear single-phase region up to the phase boundary.

In the two-phase regions, the boundaries were determined by direct visual observation of the samples and inspection through crossed polarizers and optical microscope. In the O_m+W region, an isotropic solution (O_m) phase coexists with excess water making the sample a turbid fluid, whereas the mass of an isotropic viscous turbid phase separated with excess water is present in the I_2+W region. The boundary between O_m+W and I_2+W regions is located at $n=4.5 \pm 0.5$ in the $Si_{14}C_3EO_n$ and $n=5.5 \pm 0.5$ in the $Si_{25}C_3EO_n$ system. In the H_2+W region, a dispersion of an optically anisotropic liquid crystal was observed by optical microscope, whereas vesicles were found in the $L_\alpha+W$ region. The boundaries between I_2+W and H_2+W regions are $n=8 \pm 1$, for the $Si_{14}C_3EO_n$ and $n=16 \pm 0.5$ for the $Si_{25}C_3EO_n$ and those between H_2+W and $L_\alpha+W$ regions are $n=12 \pm 0.5$ for the $Si_{14}C_3EO_n$ and $n=34 \pm 1$ for the $Si_{25}C_3EO_n$.

The same type of phase diagrams were constructed for other poly(oxyethylene) dodecyl or oleyl ethers and poly(oxyethylene) trisiloxane surfactants [1, 2, 11]. It is observed that as the hydrophobic-chain length of surfactant increases, a same type of self-organized structure tends to form at longer EO-chain length. In our previous study, it is reported that the f (nV_{EO}/V_S) ranges for each liquid crystal are similar for poly(oxyethylene) dodecyl and oleyl ethers and even for the present $Si_{14}C_3EO_n$ as far as a linear chain surfactant is concerned [3]. From Fig. 1, it is observed that f ranges for H_2 and I_2 phases in $Si_{25}C_3EO_n$ system is wider than that in $Si_{14}C_3EO_n$ system. This will be discussed in the following section.

Phase diagram of $Si_{14}C_3EO_{15.8}$-D_4 system as a function of temperature

It is observed from Fig. 1 that bulk surfactants, namely copolymer melts also form liquid crystalline phases because of the strong segregation between the poly(oxyethylene) and poly(dimethylsiloxane) chains of the surfactant or copolymer. $Si_{14}C_3EO_{15.8}$ ($f=0.35$) forms reverse hexagonal phase (H_2) in the pure melted state. A small amount of water changes the H_2 phase to the L_α phase [3]. Because the hydration of EO chain increases the repulsion between EO chains and hence the surfactant layer curvature changes from negative to zero. On the other hand, D_4 is compatible with poly(dimethylsiloxane) chain and is expected to be solubilized in the corona of reverse cylinder in the H_2 phase changing the layer curvature toward more negative. The phase diagram of the $Si_{14}C_3EO_{15.8}$-D_4 was constructed as a function of temperature and is shown in Fig. 2. The melting temperature of the H_2 phase decreases and H_2-I_2-O_m phase transitions take place upon addition of D_4. The regions of the H_2 and I_2 phases shrink with increasing temperature. The H_2 and I_2 phases changes to the O_m phase with the increase of either D_4 concentration or temperature because in the both cases, the attraction between the reverse cylinders or reverse micelles decreases.

Phase maps of $Si_mC_3EO_n$-D_4 systems as a function of EO-chain length

The phase maps of the $Si_{14}C_3EO_n$-D_4 and the $Si_{25}C_3EO_n$-D_4 systems constructed as a function of f (vertical axis) and W_S (horizontal axis) are presented in Figs. 3a and 3b, respectively. The solid present regions are not shown in the phase diagrams. Since, even a very long poly(dimethylsiloxane) chain is in a liquid state due to its flexibility, the melting temperature of the solid phase of A-B type silicone surfactants corresponds to that of its EO-chain. The melting temperature increases up to 50 °C with increasing EO-chain lengths up to 51.6. As the EO chain is completely insoluble in D_4, the melting temper-

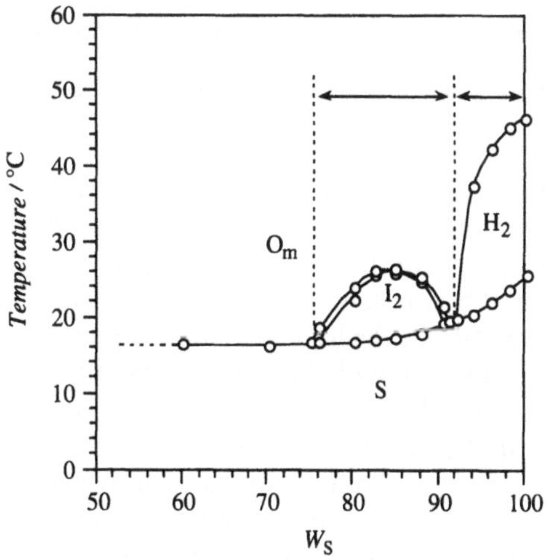

Fig. 2 Phase diagram of $Si_{14}C_3EO_{15.8}$-D_4 system as a function of temperature. Phase notation as in Fig. 1

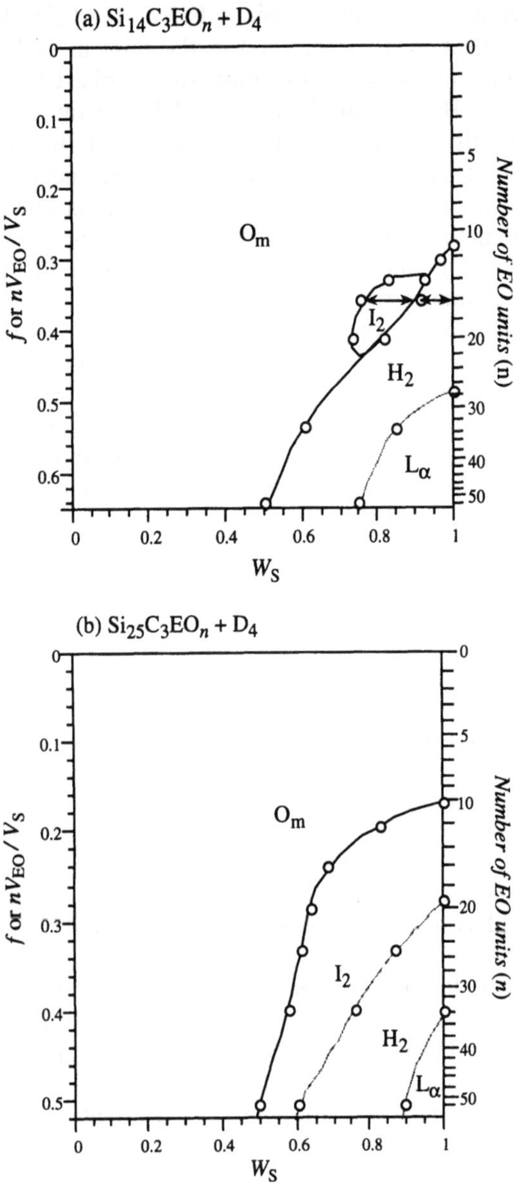

Fig. 3 Phase maps of $Si_mC_3EO_n$-D_4 systems as a function of the EO-chain length at 10–50°C. f is the volume ratio of EO chain to the surfactant. Phase notation as in Fig. 1. (**a**) $Si_{14}C_3EO_n$, (**b**) $Si_{25}C_3EO_n$

ature of the solid present region is not decreased much upon addition of D_4. If we construct the phase diagrams at room temperature, the most of the regions in Fig. 3 will be covered by the solid present region, whereas if we construct it at 50 °C, the liquid crystalline phases will be observed only in the long EO chain regions. For example, $Si_{14}C_3EO_{12}$ form H_2 phase in the temperature range 10–15 °C and $Si_{14}C_3EO_{33.1}$ form L_α in 43–109 °C and H_2 in 41–70 °C [8, 9]. Therefore, we have drawn the phase maps just above the melting temperature of the solid phase ranging from 10–50 °C to show the maximum possible regions of the occurrence of liquid crystals in the D_4 binary systems, e.g., the maximum regions of the occurrence of the I_2 and H_2 phases in Fig. 2 indicated by arrow mark are taken in Fig. 3a.

From Fig. 3, it is observed that the surfactants form liquid crystals even in the absence of solvent when EO-chain length, $n > 10$. Similarly, a short poly(dimethylsiloxane) surfactant $Si_{5.8}C_3EO_{36.6,51.6}$ does not form a liquid crystal in a pure state [3, 12]. In the solvent absent system, any of the blocks of A-B type surfactants or copolymers should be long enough to segregate each other to form ordered structures. The surfactant liquid, O_m is changed to the L_α through the H_2 in the $Si_{14}C_3EO_n$-D_4 and through the I_2 and H_2 in the $Si_{25}C_3EO_n$-D_4 system with increasing n or f. Hence, like water binary systems, the surfactant layer curvature changes from negative to zero with increasing the volume fraction of the hydrophilic moiety in the surfactant. $Si_{14}C_3EO_n$ does not form the I_2 phase in the absence of solvent and the H_2-L_α phase transition occurs at $n = 27 \pm 0.5$. In the $Si_{25}C_3EO_n$-D_4 system, the I_2-H_2 and H_2-L_α phase transitions take place at $n = 19 \pm 0.5$ and $n = 34 \pm 1$, respectively. Increased EO-chain length enhances the segregation between A and B blocks and hence increases the order of the system. Ordered lyotropic liquid crystalline phases are stable over a comparatively wider range of composition at a long EO-chain length. However, L_α-H_2-O_m and L_α-H_2-I_2-O_m phase transitions take place at $f > 0.48$ in the $Si_{14}C_3EO_n$-D_4 and $f > 0.4$ in the $Si_{25}C_3EO_n$-D_4 systems, respectively. The dotted lines are the approximate boundaries between the homogeneous liquid crystalline phases. In our previous study [7, 8], it is observed that D_4 molecules penetrate into the poly(dimethylsiloxane) chains corona of the reverse micelle forming the I_2 phase in the $Si_{25}C_3EO_n$-D_4 system increasing the effective cross-sectional per surfactant molecule. Hence, the surfactant layer curvature changes to more negative and the I_2-O_m phase transition takes place [8]. In the present systems, therefore, it is considered that D_4 molecules penetrate into the poly(dimethylsiloxane) moiety changing the surfactant layer curvature toward more negative and finally, the liquid crystalline phases diluted to the O_m phase. $Si_{14}C_3EO_n$ and $Si_{25}C_3EO_n$ form reverse micelles in D_4 or decane at a very low surfactant concentration (0.5~2 mmol/L) when $n = 7.8$ or more [8].

Macroscopic phase behavior and f

In the case of poly(oxyethylene) type non-ionic surfactants, the macroscopic phase behavior is highly related with the f, which is also co-related to Griffin's HLB number:

$$\text{HLB number} = 20(\rho_{EO}/\rho_S)f$$

where ρ_{EO} and ρ_S are the densities of EO chain and surfactant, respectively. Since both densities are not very

Table 1 Ranges of f, f' and f'' for each liquid crystal in water-$Si_mC_3EO_n$, and $Si_mC_3EO_n$-D_4 systems

Surfactant/ Phase	$Si_{14}C_3EO_n$				$Si_{25}C_3EO_n$			
	f in water	f'	f in D_4	f''	f in water	f'	f in D_4	f''
O_m	0–0.13	0–0.27	0–0.27	0–0.32	0–0.09	0–0.18	0–0.16	0–0.16
I_2	0.13–0.21	0.13–0.36		0.27–0.34	0.09–0.24	0.09–0.44	0.16–0.27	0.16–0.3
H_2	0.21–0.28	0.21–0.52	0.27–0.48	0.27–0.48	0.24–0.39	0.27–0.7	0.27–0.39	0.27–0.45
L_α	0.28–0.64-	0.28–0.85	0.48–0.64-	0.46–0.64	0.39–0.5-	0.39–0.8-	0.4–0.5-	0.4–0.5-

different, the vertical axis, f, in Figs. 1 and 3 are compatible with the Griffin's HLB number on the volume basis. f can be also defined as the effective volume ratio of the hydrophilic moiety in the system: $f' = (\phi_S f + \phi_W)$ for the water-$Si_mC_3EO_n$ binary system and $f'' = \phi_S' f$ for the $Si_mC_3EO_n$-D_4 binary system, where ϕ_W and ϕ_S are the volume fractions of water and the surfactant, respectively in the water binary system, and ϕ_S' is the volume fraction of the surfactant in the oil binary system. The f, f' and f'' ranges for various self-organized structures in water-$Si_mC_3EO_n$ and Si_m-C_3EO_n-D_4 systems are summarized in Table 1. f' and f'' indicate the maximum ranges of the occurrence of the each phase in the water and D_4 systems, respectively.

It is very interesting that the f ranges for each liquid crystal formed in the water-$Si_mC_3EO_n$ systems are similar. However, as mentioned in the previous section f ranges for the H_2 and the I_2 phases in the water-$Si_{25}C_3EO_n$ system is wider than that in the water-$Si_{14}C_3EO_n$ system. The long poly(dimethylsiloxane) chains are very flexible and shrink in a coiled structure. To change the layer curvature from negative to zero or to positive the coiled chain should be stretched and the longer the chain, greater the entropy loss is. Hence, the long poly(dimethylsiloxane) chain surfactant prefer the negative layer curvature, i.e., the hydrophilic (water + EO) core and poly(dimethylsiloxane) corona, and consequently the H_2 and I_2 phases are stabilized in a wide range of f for $Si_{25}C_3EO_n$ compared to $Si_{14}C_3EO_n$. f ranges for each phase in the $Si_mC_3EO_n$-D_4 systems are also comparable. But it is clear that both the $Si_{14}C_3EO_n$ and $Si_{25}C_3EO_n$ form liquid crystals at higher f values in the D_4 systems than that in the aqueous systems. Water enhances the segregation of EO and poly(dimethylsiloxane) more than that is attributed by the addition of D_4 and hence the surfactants form the same aggregate at comparatively shorter EO chain in the water binary systems. From Fig. 1, it is observed that at a constant EO chain length the surfactant generally forms only one type of liquid crystal and the surfactant layer curvature is not changed in the whole range of composition. Hence, the each type of liquid crystal appears in the wide range of the effective hydrophilic volume, f' as compared to f. On the other hand, each phase in the D_4 systems occur at the similar ranges of f and f''. D_4 or cyclic tetra(dimethylsiloxane) molecules are very compatible with poly(dimethylsiloxane) chains and can be easily distributed in the lipophilic moiety resulting the surfactant layer curvature changes toward more negative by expanding the effective interfacial area per molecule. In this case, the amount and proportion of the effective lipophilic moiety can modulate the layer curvature and the resulting morphology. These results attribute a big difference between water binary and oil binary systems. Although water increases the segregation between A and B blocks of the surfactant in the greater extent than D_4, water cannot change the surfactant layer curvature toward positive. Because it is energetically unfavorable to form a positive layer curvature in which the shrunk long poly(dimethylsiloxane) chains should be stretched. Moreover, water has less affinity to EO chain as compared to that in D_4 to poly(dimethylsiloxane) chain. Instead of water, if we select such a solvent, which is structurally similar and has the same affinity to the EO chain as between the D_4 and poly(dimethylsiloxane) chains each type of aggregate will appear in the same range of f and f'. In fact ethylene glycol or low molecular weight poly(ethylene glycol) has greater affinity to EO chain than water and consequently form the aggregates with positive layer curvature even in the case of long poly(dimethylsiloxane) surfactant [9]. Therefore, as far as polyoxyethylene-type nonionic surfactants are concerned, the f or effective f in the selective solvents systems can accurately predict the layer curvature or self-organized structure.

Conclusion

Poly(oxyethylene) poly(dimethylsiloxane) surfactants ($Si_{25}C_3EO_n$ and $Si_{14}C_3EO_n$) form different liquid crystals in water or silicone oil and the layer curvature changes from negative to zero with the increase of f regardless of any solvent. The reverse type liquid crystals (I_2 and H_2) are stabilized in a wide range of f for the long poly(dimethylsiloxane) surfactant as compared to shorter one. The same type of aggregate is formed at lower f value in water system than that in the D_4 system because water enhances the segregation in the greater extent than D_4. At a fix *f* value, water is unable to change the layer curvature, whereas D_4 changes the layer curvature toward more negative because D_4 molecules are very compatible with the poly(dimethylsiloxane) chain. When the solvents are very compatible or structurally similar with one of the blocks of A-B type non-ionic surfactant, the amounts or the proportion of the solvents can predict the layer curvature. Overall, *f* controls the layer curvature and the resulting microstructure in the both solvents systems.

References

1. Huang K-L, Shigeta K, Kunieda H (1998) Progr Colloid Polym Sci 110:171
2. Shigeta K, Suzuki M, Kunieda H (1997) Progr Colloid Polym Sci 106:49
3. Kunieda H, Uddin MH, Horii M, Furukawa H, Harashima A (2001) J Phys Chem B 105:5419
4. Svensson B, Olsson U, Alexandridis P (2000) Langmuir 16:6839
5. Khandpur AK, Forster S, Bates FS, Hamley IW, Ryan AJ, Bras W, Almdal K, Mortensen K (1995) Macromolecules 28:8796
6. Griffin WC (1954) J Soc Cosmet Chem 5:249
7. Uddin MH, Rodriguez C, Watanabe K, Lopez-Quintela A, Kato T, Furukawa H, Harashima A, Kunieda H (2001) Langmuir 17:5169
8. Rodriguez C, Uddin MH, Watanabe K, Furukawa H, Harashima A, Kunieda H (2002) J Phys Chem B 106:22
9. Kunieda H, Uddin MH, Yamashita Y, Furukawa H, Harashima A (2002) J Oleo Sci 51:113
10. Kunieda H, Shigeta K, Ozawa K, Suzuki M (1997) J Phys Chem B 101:7952
11. Kunieda H, Taoka H, Iwanaga T, Harashima A (1998) Langmuir 14:5113
12. Rodriguez C, Uddin MH, Furukawa H, Harashima A, Kunieda H (2001) Progr Colloid Polym Sci 118:53

Progr Colloid Polym Sci (2004) 123: 275–279
DOI 10.1007/b11958

V. Dugas
Y. Chevalier
G. Depret
X. Nesme
É. Souteyrand

The immobilisation of DNA strands on silica surface by means of chemical grafting

V. Dugas · Y. Chevalier (✉)
Laboratoire des Matériaux Organiques à Propriétés Spécifiques,
UMR 5041 CNRS-Université de Savoie,
BP 24, 69390 Vernaison, France
e-mail: Yves.Chevalier@lmops.cnrs.fr
Tel.: +33-4-78022271
Fax: +33-4-78027187

G. Depret · X. Nesme
Laboratoire d'Écologie Microbienne,
UMR 5557 CNRS-Université Lyon 1,
69622 Villeurbanne, France

V. Dugas · É. Souteyrand
Ingénierie et Fonctionnalisation de Surfaces, UMR 5621 CNRS-École Centrale de Lyon, 69131 Écully, France

Abstract The surface chemistry and physicochemical phenomena involved in the chemical grafting process of short DNA single strands on silica or glass slides has been studied in order to be able to prepare reusable DNA-arrays (DNA-chips) with an optimum signal-to-noise ratio. Some crucial steps of the immobilisation of oligonucleotides on silica surfaces by means of their chemical grafting at their 3′ terminus (aminolinker) were studied carefully. A prior grafting of the surface with an organosilane is performed. The surface of the silica should have been completely covered by a dense anionic grafted layer in order to minimise DNA adsorption with respect to chemical grafting and allow an easy desorption of non grafted materials. The efficiency of the covalent immobilisation of DNA strands performed from a very dilute solution in a small volume is drastically increased by means of a slow in situ evaporation of the solvent. The discrimination between covalently bound and adsorbed oligonucleotides requires a correct control of the rinsing processes after the immobilisation reaction. An efficient washing process increases the signal-to-noise ratio. Ready hybridisation of complementary oligonucleotides or large double strands bearing the complementary sequence at their centre (PCR fragments) could be obtained at the surface. A clean and robust immobilisation process allows a clear-cut discrimination between hybridisation and non-specific adsorption and very low levels of background (noise) in the radioactivity measurements. Several cycles of hybridisation and denaturation were carried out. Glass plates functionalised according to the same process might be used for the mass preparation of DNA-arrays.

Keywords DNA · DNA chip · Oligonucleotide · Chemical grafting

Introduction

DNA-arrays (DNA-chips) are miniaturised devices used for a parallel and fast analysis of DNA or RNA samples in quite different biological applications going from the identification of organisms by the recognition of their DNA, the investigations of gene expression and genetic modifications in biological research, to the genome mapping at the highest degree of complexity [1]. DNA-arrays are elaborated by the immobilisation of a set of different DNA strands relevant for the biological problem studied, in a localised manner (as spots) on a solid support (glass plate, polymer membrane). Upon hybridisation with a biological sample, the spots where the hybridisation reaction is detected are those for which the complementary sequence was present in the sample as in the reverse dot-blot technique [2]. Most current devices are prepared by an immobilisation process involving a simple deposition with a spotter. The DNA strands adsorbed from the solution can be successfully hybridised once, but cannot resist repeated hybridisation and denaturation cycles. Those DNA-arrays are disposable, that is, single-use because all adsorbed targets are eliminated during denaturation. Another important point is the necessity of a precise quantification of the detected signals in biological studies where small differences in the DNA sequence are looked for (single nucleotide polymorphism, studies of genetic biodiversity). Chemical grafting allows one to overcome these limitations.

The process allowing a chemical grafting of oligonucleotides on silica involves several steps [3]. Firstly, the silica surface has to be functionalised with chemical groups which are reactive towards the oligonucleotides to be attached. The surface functionalisation step is ensured by means of organosilanes. The next step is the immobilisation reaction itself where the oligonucleotides bind to the functional surface. The oligonucleotides have to be synthesised in such a way that the aminohexyl residue (amino-linker) is attached at the 3′ end of the strands. Because of the expense of the oligonucleotides, very low concentrations are used and the reaction is quite sluggish. The reaction rate which is proportional to the concentration is very slow. Lastly, the non-grafted (adsorbed) oligonucleotides have to be removed from the surface by means of efficient washings. In the case of inefficient washing, the residual adsorbed oligonucleotides may contribute to the hybridisation of complementary strands, but they also slowly leak into the solution during the utilisation. The consequence of an incorrect washing is then a drift of the hybridisation measurements which may reach the complete lost of the signal, and poor reproducibility and repeatability of the measurements. These processes involve chemical and physicochemical phenomena that deserve careful investigations in order to improve the performances of DNA-arrays.

In this paper, we describe first the chemical grafting process studied. Secondly, the most important steps are discussed: 1) the preparation of the silica surface required for attaining to a robust (covalent) immobilisation and a low level of adsorption, 2) the immobilisation process itself where the grafting reaction of oligonucleotides is carried out, 3) the subsequent washings of the surfaces. Thirdly, the biological validation of the grafting quality by hybridisation tests with both oligonucleotides and large PCR-products is presented.

The oligonucleotides used for the immobilisation on the surface were 15-mers purchased from Eurogentec (Seraing, Belgium) having the H_2N-$(CH_2)_6$- amino linker attached at their 3′ terminus. The sequence $^{(5')}$GCTTG CGGGGCGTTC$^{(3')}$-$(CH_2)_6$-NH_2 was a complementary part of the gene 16S rRNA of *Agrobacterium tumefaciens*. The oligonucleotide having the exact complementary sequence was used for hybridisation experiments and the non-complementary sequence used to estimate the non-specific adsorption was $^{(5')}$CAGCAGCCGCGGTAA$^{(3')}$.

Description of the immobilisation process studied

Commercially available oligonucleotides with a functional chain end have either a primary amine or a thiol group at their 3′ or 5′ terminus. The immobilisation reaction necessarily takes place in a medium containing water where the oligonucleotides are soluble; this is a severe limitation as regards to the choice of the grafting reaction. The reactions more often used [4] are that of a thiol with a maleimide or an iodoacetamide, and the reactions of a primary amine with either carboxylic acids in the presence of carbodiimide, an isothiocyanate or an aldehyde followed by a reduction step. The robust reaction chosen for the present study is the direct acylation of an activated acid grafted to the surface with the primary amine attached to the DNA strands (amino-linker) [5]. The functional organosilane bears an acid group in a protected form for that purpose. Thus, the full chemical process shown in Fig. 1 involves the grafting of *t*-butyl dimethylaminodimethylsilylundecanoate on silica, the deprotection of the *t*-butyl ester into a carboxylic acid, the activation of the acid by *N*-hydroxysuccinimide (NHS) and finally the covalent immobilisation of the DNA strands by their amino-linker.

Grafting of the silica plates with a functional organosilane

The first step is the grafting of an α,ω-difunctional molecule bearing a silane group allowing a reaction on the surface of silica and an ester group at the other end of the molecule. Activated acids and their acid precursors are not chemically compatible with silanes. Thus, a silane bearing an ester group was grafted on silica; it was

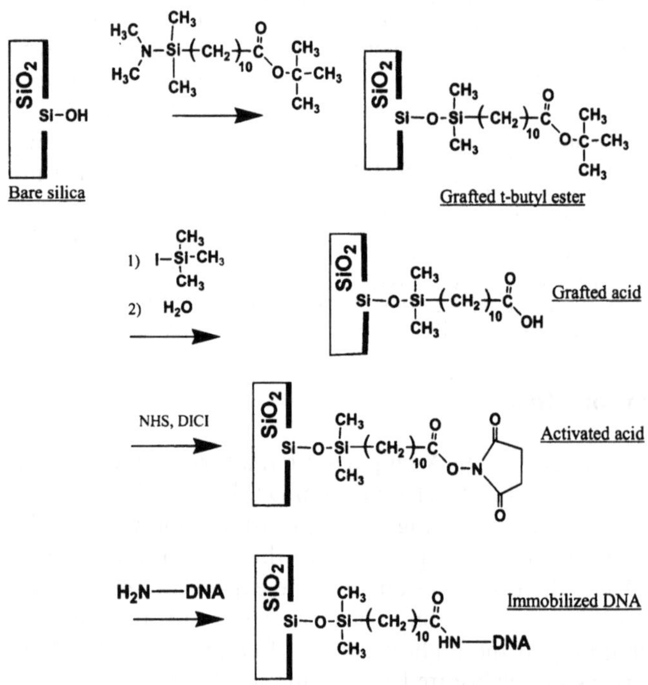

Fig. 1 Full multi-step chemical process involving the grafting of the silane, the functionalisation into an activated acid and the immobilisation of DNA

subsequently converted into the corresponding acid and activated in a third step. The choice of the type of silane is of importance. There are quite a large variety of reactive silanes described in the literature which differ by their type of leaving group [alkoxysilanes, chlorosilanes or (dimethylamino)silanes] [6] and their functionality [7]. Two important requirements have to be fulfilled: the grafted layers have to be monomolecular and dense. The trouble with the use of multifunctional silanes having several leaving groups at the silicon atom is the polycondensation at the surface in the presence of traces of water, which leads to thick and ill-defined grafted layers; the difficulties encountered in the control of moisture lead to irreproducible results. A monofunctional silane avoids this problem. The dimethylamino leaving group was chosen among the most reactive ones because monofunctional silanes are much less reactive than multifunctional ones [8]. A reproducible grafting reaction of *t*-butyl 11-(dimethylaminodimethylsilyl)undecanoate was carried out in two steps called "impregnation" and "condensation" as in previous works [9]. The analysis of the reaction was easy on the surface of fumed silica because of its high specific area (CabOSil, 200 m^2/g), allowing direct elemental chemical analyses and infrared spectroscopy. Reproducible grafting densities of 3.5 $\mu mol/m^2$ were obtained. But the surface of fumed silica is different to that of the thermal silica present on silicon wafers. Thermal silica was grown by a thermal oxidation of silicon ATR internal reflection elements, so that IR spectra could be measured by ATR (Fig. 2). The grafting density was then 1.4 $\mu mol/m^2$ because of the lack of silanol group at the surface of thermal silica.

The deprotection of the *t*-butyl ester leading to the acid, the activation with NHS and the reaction with a model amine (benzylamine) could also be monitored in the same way by IR spectroscopy in ATR mode (Fig. 2).

The cleavage of the *t*-butyl ester was carried out under mild conditions with iodotrimethylsilane [10]. This method allowed an efficient capping of the silica surface because iodotrimethylsilane reacted with the residual silanol groups at the silica surface, yielding trimethylsilyl ether groups. This capping reduced the adsorption and made the desorption of unreacted DNA strands easier since the polar silanol groups were converted into non-polar and chemically inert groups.

Covalent Immobilisation of Oligonucleotides

The reaction used for the immobilisation of oligonucleotides has been successfully tested with a primary amine (benzylamine) 0.1 M in water solvent. But the concentration of oligonucleotides is far below this concentration range, so that the reaction rate is very slow. For the immobilisation reaction on silica plates, a 80 μL drop of a solution in a 9:1 mixture of acetonitrile and water, is

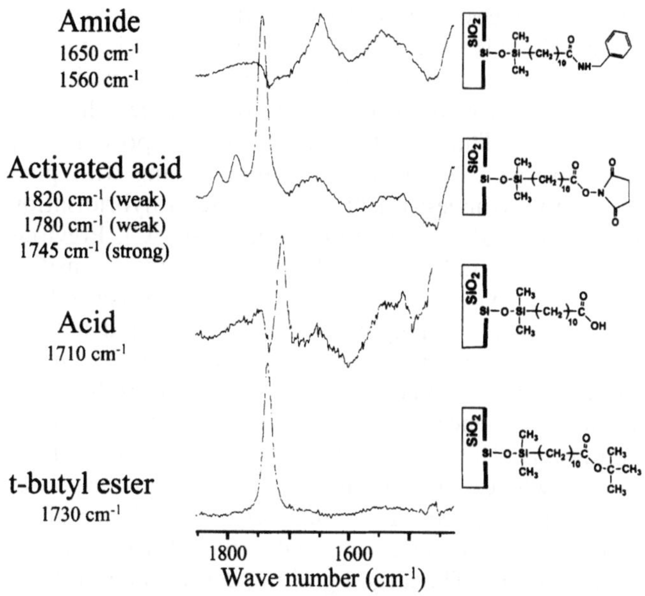

Fig. 2 Infrared spectra of the organic molecules grafted on the thermal silica surface of the ATR crystal recorded in ATR mode. The spectra corresponding to the different steps of the process are shown: the grafting density is 1.4 $\mu mol/m^2$ and each reaction is close to completion

deposited on the plates; the oligonucleotide concentration is about 10 μmol/L. It is not possible to run the reaction for very long times because the hydrolysis of the activated acid by water slowly consumes the reactive species at the surface. The grafting density of oligonucleotides is approximately proportional to the concentration used during the immobilisation reaction [11] and the concentrations required in order to obtain dense layers of DNA strands looking like polymer brushes would be at least of the order of 100 mM.

A decisive improvement consisted in a slow drying of the drop in a vapour-saturated atmosphere. The concentration of oligonucleotides increased as the solvent was evaporating, and the reaction rate increased concomitantly. The reaction rate was maximum in the last moment of the drying process when the solution was still fluid, allowing the molecules to move for the reaction to proceed. The reaction stopped when the viscosity rose too much or when the materials precipitated. The in situ concentration of the oligonucleotides by slow drying allowed reaching grafting densities of the order of 10^{12} strands/cm^2 (~20 $nmol/m^2$) which could easily be detected using radioactive or fluorescent labelling of the oligonucleotides. If the drying was prevented, the grafting densities were below the detection limit of the detection by radioactive labelling (~5×10^8 cm^{-2}).

The importance of efficient washing

Once the immobilisation reaction was completed, chemically grafted and adsorbed oligonucleotides were present at the surfaces. It is difficult to discriminate between them. The adsorbed oligonucleotides should be removable by means of rinsing with water but our experiments and a critical reading of the literature showed that this not the case. Thus, most grafting densities reported in the literature were taken after a simple rinsing with pure water; the strongest washings made use of hybridisation conditions (saline citrate buffer at 50–60 °C). Previous works were indeed intended to elaborate sensitive analytical devices which would be single use (disposable): a large signal in a hybridisation test required a large amount of immobilised materials, whatever the immobilisation mode, and the most severe conditions the plates would experience in this utilisation are hybridisation conditions. These ideas lead to severe limitations of the devices where significant drifts and erroneous responses are often observed. A reusable device would also allow repeating the measurements and varying the experimental conditions. The largest grafting densities reported so far [11, 12] pertain to samples having a cationic reactive layer [deposited poly(lysine), grafted aminopropyltriethoxysilane] where the adsorption of anionic DNA strands is very strong; it is obvious that such layers contain predominantly adsorbed oligonucleotides.

In the present work, successive washings having increasing stringency were applied: water at room temperature, hot water, detergent solution (10% SDS) at 80 °C. The immobilised amount decreased drastically during the first washings with water. The removal of adsorbed DNA strands was made easy because the surface contains mainly residual acidic groups which are anionic. When the immobilised amount remained constant after repeated washing cycles with hot water, an additional more stringent washing with SDS solution could remove again significant amounts of DNA strands. This shows that the desorption of the last adsorbed DNA strands is difficult. Hydrophobic interaction between DNA and the surface may operate since a detergent could overcome them.

The decrease of the immobilised amount along successive washings may prove disastrous as regards the sensitivity of detection in a hybridisation test. This was not the case because the sensitivity of the fluorescence or radioactive labelling methods is more than enough. The immobilised amount and the hybridisation signal indeed decreased drastically during the washings, but the background radioactivity (or fluorescence) decreases to a larger extent, so that the signal-to-noise ratio increased as the washing process proceeded. Referred to as "the noise" in an immobilisation experiment, is the signal detected for the immobilisation of oligonucleotides that do not bear the amino-linker. The noise of a hybridisation experiment is the signal measured for the hybridisation of non-complementary DNA strands. The main origin of the noises is the non-specific adsorption.

Hybridisation tests

The hybridisation tests with ^{32}P labelled DNA strands were performed overnight at 47 °C in conventional hybridisation buffer (SSC buffer, SDS and Denhardt). The hybridisation at the surface was selective since only the complementary DNA strands gave a significant radioactivity, which was easy to quantify with standard equipment (Fig. 3). On the contrary, hybridisation of non-complementary DNA strands gave a very low radioactivity signal. Oligonucleotides without amino-linker have been immobilised according to the same process on some plates. Since they cannot react with the activated acids on the surface, the immobilisation density was very low, and the signal in the hybridisation test was also very low, even when the hybridisation was carried out with complementary DNA strands. This showed that the washing process was efficient. The hybridisation could be repeated after a denaturation at the surface (Fig. 3). The plates could resist several hybridisation and denaturation cycles because of the covalent immobilisation. The hybridisation yield defined as the ratio of the hybridised amount to the immobilised amount was of the order of 20% for oligonucleotides of 15 bases hybridised with their exact complementary oligonucleotides. The hybridisation was also successful with large double strands (up to 1.5 kbase) containing the complementary sequence in a central position. Thus, the double strands were first denaturated at high temperature (10 minutes at 100 °C) and subsequently contacted with the plates in hybridisation conditions. However, the yield of hybridisation to the oligonucleotides attached to the surfaces was much lower, mainly because a part of strands were hybridised again with their complementary inside the solution.

Conclusions

A process for the robust chemical grafting of single strand DNA with respect to the application to DNA-chips is presented. The main concerns are discussed: the choice of the grafting reaction and chemical reagents, the reaction process, the discrimination between chemical grafting and strong adsorption. A good signal-to-noise ratio in the specific detection of the hybridisation of complementary strands at the surface requires a large enough grafted density, but the highest density is not necessary with sensitive detection methods such as radioactivity or fluorescence. A low level of non-specific adsorption (low background noise) and a low drift could be obtained with an adequate washing of the non-grafted oligonucleotides. A good hybridisation yield required a correct control of the accessibility of DNA strands at the surface. Repeated hybridisation – denaturation cycles of

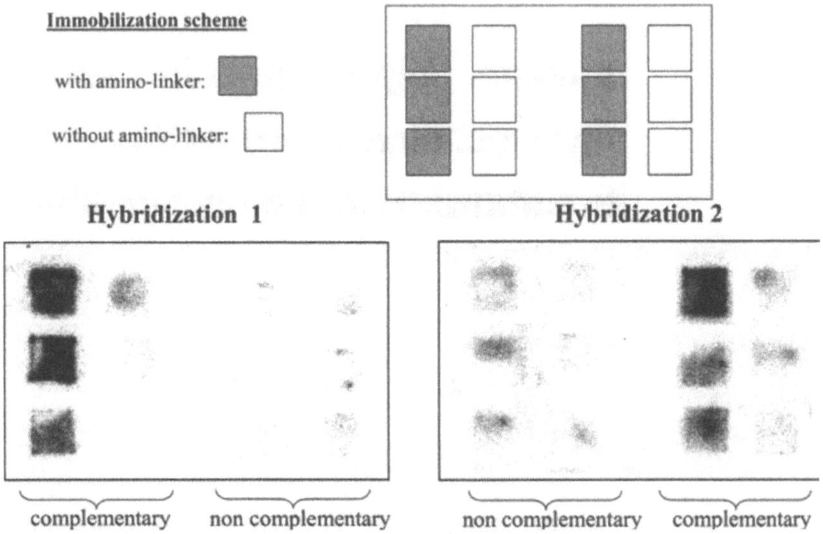

Fig. 3 Two successive hybridisation tests where radioactive labelled (^{32}P) complementary and non-complementary oligonucleotides were hybridised on 1×1 cm^2 silica plates. Oligonucleotides either with or without an amino-linker were immobilised on the different plates. The signal was observed only for plates where the complementary strand was hybridised to an oligonucleotide immobilised by means of an amino-linker. A first hybridisation was performed (left), followed by a denaturation at the surface and a second hybridisation (right) where the plates receiving the complementary and non-complementary DNA strands were inverted

short complementary oligonucleotides were successful. Large double stranded PCR fragments could also be hybridised, but with lower yields. The main improvements of an optimised control of the surface chemistry in this the field of DNA-chips is the reusability and the higher signal-to-noise ratio which allows a better quantification of the hybridisation yield useful in some precise biological studies.

References

1. a) Southern EM (1996) Trends Genet 12:115; b) Chetverin AB, Kramer FR (1994) Bio/Technology 12:1093; c) Souteyrand E, Cloarec J-P, Martin J-R, Cabrera M, Bras M, Chauvet J-P, Dugas V, Bessueille F (2000) Appl Surface Sci 164:246
2. a) Southern EM (1989) PCT application, WO 89/10977; b) Lysov YuP, Khorlin AA, Khrapko KR, Shick VV, Florentiev VL, Mirzabekov AD (1988) Proc USSR Acad Sci 303:1508; c) Khrapko KR, Lysov YuP, Khorlin AA, Ivanov IB, Yershov GM, Vasilenko SK, Florentiev VL, Mirzabekov AD (1991) DNA Sequence. J DNA Sequencing Mapping 1:375
3. Dugas V (2001) PhD thesis, École Centrale de Lyon
4. a) Agrawal S (1994) In: Agrawal S (ed), Protocols for oligonucleotide conjugates. Methods in molecular biology, vol 26. Humana Press, Totowa NJ, p 93; b) Fidanza JA, Ozaki H, McLaughlin LW (1994) In: Agrawal S (ed), Protocols for oligonucleotide conjugates. Methods in molecular biology, vol 26. Humana Press, Totowa, NJ, p 121; c) Yang M, McGovern ME, Thompson M (1997) Anal Chim Acta 346:259
5. a) Staros JV, Wright RW, Swingle DM (1986) Anal Biochem 156:220; b) Sehgal D, Vijay IK (1994) Anal Biochem 218:87; c) Bhatia SK, Shriver-Lake LC, Prior,KJ, Georger JH, Calvert JM, Bredehorst R, Ligler FS (1989) Anal Biochem 178:408
6. a) Lork KD, Unger KK, Kinkel JN (1986) J Chromatogr 352:199; b) Morel D, Serpinet J (1982) J Chromatogr 248:231. c) Szabó K, Ha NL, Schneider P, Zeltner P, Kováts ESZ (1984) Helv Chim Acta 67:2128
7. Plueddemann EP (1982) Silane coupling agents. Plenum Press, New York
8. Evans B, White TE (1968) J Catalysis 11:336
9. a) Elbhiri Z, Chovelon J-M, Jaffrezic-Renault N, Chevalier Y (1999) Sensors Actuators B 58:491; b) Chevalier Y, Elbhiri Z, Chovelon J-M, Jaffrezic-Renault N (2000) Progr Colloid Polym Sci 115:243
10. Ho T-L, Olah GA (1976) Angew Chem Int Ed Engl 15:774
11. Guo Z, Guilfoyle RA, Thiel AJ, Wang R, Smith LM (1994) Nucleic Acids Res 22:5456
12. a) Chrisey LA, Lee GU, O'Ferrall CE (1996) Nucleic Acids Res 24:3031; b) O'Donnell MJ, Tang K, Köster H, Smith CL, Cantor CR (1997) Anal Chem 69:2438. c) Boncheva M, Scheibler L, Lincoln P, Vogel H, Åkerman B (1999) Langmuir 15:4317

Progr Colloid Polym Sci (2004) 123:280–283
DOI 10.1007/b11959

E. Carretti
L. Dei
P. Baglioni

Aqueous polyacrylic acid based gels: physicochemical properties and applications in cultural heritage conservation

E. Carretti · L. Dei (✉) · P. Baglioni
Department of Chemistry and Consortium CSGI, University of Florence,
Via della Lastruccia, 3,
50019 Sesto Fiorentino (FI), Italy
e-mail: dei@csgi.unifi.it
Tel.: +39-55-4573045
Fax: +39-55-4573036

Abstract Solutions of polyacrylic acid form strong gels upon the addition of bases. Two different gels have been studied by differential scanning calorimetry (DSC) and rheometry in order to examine both their viscoelastic behaviour and the fraction of the free solvent in the system. These properties are important from a technological point of view, because of their application in cultural heritage conservation field as cleaning agents. When the continuous phase is constituted of water the study was focused on the influence of pH on the gel properties. Rheological properties have been studied as a function of the NH_3 concentration, the base necessary to stabilise the aqueous gel. The rheological behaviour of the system polyacrylic acid-water-ammonia, has been characterised using rotating and oscillating measurements. Dynamic (oscillatory) shear tests were carried at 25 °C in the frequency range 0.01–100 Hz; rheological parameters (such as loss modulus G″ and storage modulus G′) were also correlated to the pH of the blends. To vary the gel strength, different ammonia concentrations were used, while keeping the temperature and cross-linker concentration constant. Both the loss and storage moduli increased with the NH_3 concentration for this gel system. The viscosity and the elastic modulus G′ increased when the pH is between 3 and 6. When all the acidic groups have been neutralised by a solution of 0.1 M NH_3 G′ reaches a maximum (pH = 6.9); further increases of pH produced a decrease of G′. The excess NH_3 decreases the fraction of bound water within the system lowering the strength of the network.

Keywords Rheology · Polyacrylic acid · Gels · Cultural heritage conservation

Introduction

It is well known that polyacrylic acid can form gels both with water and organic solvents [1–3]. Polyacrylic acid has a structure of polymer chains interconnected by cross-links between the carboxylic groups of the monomers. If a solvent is present, the neutralisation of the acidic groups by a basic substance can cause the breakage of these intramolecular structures [1] leading to the formation of a gel-network. This process is characterised by a drastic increase of the network extension: polyacrylic acids and their derivatives swell up to 1000 times their original mass [2]. Polyacrylic acid based gels are largely used in various technological applications such as drug delivery [3], as medium for the synthesis of nanoparticles by the sol-gel method [4] and so on. These gels are largely used also in the cultural heritage conservation field as cleaning agents, especially for the restoration of easel paintings [5] in the place of pure solvents. In fact the application of pure solvents like alcohols, xylene and chloroderivatives presents two main negative side effects related both to their toxicity, and to the swelling of the organic materials covering the paint layer. Due to this swelling there is an increase of their volume from 5 to 80%, the partial solubilisation of the painting protective layer and, when dried, its detachment from the pigment surface. This causes both an alteration of the optical properties of the work of art's surface and a weakening of the pigment layer [6]. The use of a polymer based gel can minimise the penetration of the solvent into the layers, drastically reducing the interaction between the dispersed system and the work of art [7]. The physico-chemical properties of these systems are connected both to the percent of the polymer solubilised (the solubility is related to their molecular weight, estimated between 700,000 and 3 or 4 billion) and decreases with increasing MW [8] and to the amount of the base, i.e. to the fraction of the neutralised acidic groups.

The aim of the present study was to characterise aqueous gels based on the use of polyacrylic acid in order to understand their potentiality in easel painting conservation.

Rheological and thermal properties were studied to individualise the best parameters to be adjusted in order to achieve highly performing gels for easel paintings cleaning [1–3].

Results and discussion

The behaviour of some rheological parameters measured for aqueous polyacrylic acid based gels was studied as a function of the pH. The polyacrylic acid used as gelator was produced by BFGoodrich (Carbopol Ultrez10®). Polyacrylic acid solutions were prepared by adding H_2O to the corresponding quantity of polymer (1% w/w). The gel network was obtained by adding NH_3 0.1 M. All the samples were maintained under agitation for 2 hours. The mixture obtained was twice heated to 80 °C and cooled to 25 °C and afterwards heated to 40 °C until it became transparent and all the air bubbles disappeared (24–48 hours).

Fig. 1 shows the dependence of the pH of the aqueous gels obtained with this procedure, as a function of NH_3 concentration; all these measurements were carried out using a CRISON pH-meter (mod. Basic 20) with an Ag/AgCl electrode. The curve in Fig. 1 indicates that the increasing of the pH up to an NH_3 fraction of ca. 0.5 produces the progressive neutralisation of all the acidic groups of the polyacrylic acid chains. However, when the total percent of NH_3 in the sample is between 0.5 and 3, pH is always in the range typical of the ammonia buffer, meaning that all the acidic groups of the polyacrylic polymer have been neutralised. The possibility to choose the correct pH is an important characteristic for a cleaning agent to be applied in cultural heritage conservation; in fact, the efficacy in the removal of some

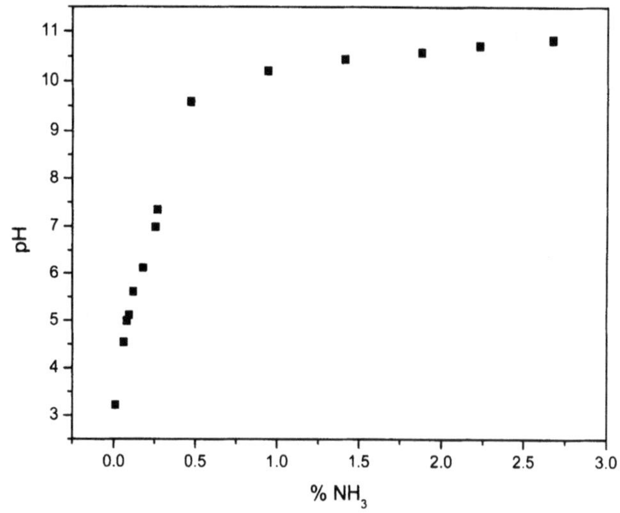

Fig. 1 Behaviour of the pH against the weight fraction of NH_3 for the aqueous polyacrylic acid based gels

hydrophobic (lipid) and hydrophilic (proteins) [9] is strictly related to the pH of the system.

Rheological measurements were performed on a Paar Physica UDS 200 Rheometer. The plate/plate geometry (diameter 2.5 cm), which requires 2 mL of sample, was used. Measurements were performed maintaining the temperature constant (25 °C) during the whole run. The kinetics of the gelation process was studied using dynamic rheological methods. The gelation time has been calculated studying the behaviour of G′ as a function of the time and observing when G′ = G″. The crossover of G′ and G″ has been suggested as a criterion for gelation [10, 11, 12], however it is frequency dependent in most polymer systems and is observable in polymer fluids where significant molecular entanglements are present but no permanent connectivity extends throughout the solution [13]. The graph in Fig. 2 indicates, for a generic system, the behaviour of G′ and G″ as a function of the time. The figure shows that initially the sample is purely viscous, with a high value of the loss modulus, G″, and a very low elastic modulus, G′. After a gelation induction time (the so called gelation time Tg), both G′ and G″ begin to increase. G′ increases more rapidly than G″ and passes beyond it. Above the gelation point the values of G′ and G″ tend towards a plateau value. This behaviour is typical of the aqueous polyacrylic based gels. We report in Fig. 3 the gelation time as a function of pH. A drastic decrease of Tg value for the samples characterised by a pH > 5 is observed. This indicates that an increasing percent of neutralised acidic groups reduces the time needed to form the gel system, because of the increasing number of the groups available to form the network.

In each case the viscoelastic behaviour of the systems was measured as a function of frequency. Fig. 4 shows the spectra obtained from some frequency sweep tests for

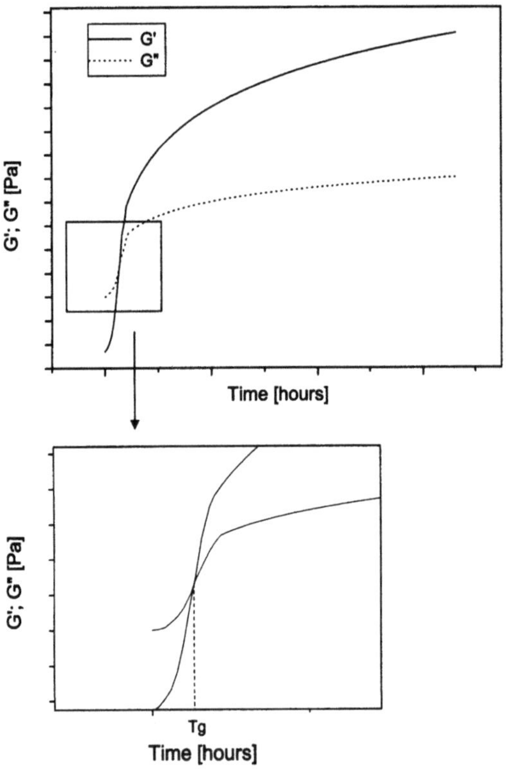

Fig. 2 Elastic modulus G′ (*full line*), loss modulus G″ (*dashed line*) as a function of time, for a generic aqueous polyacrylic acid based gel. The crossover point between G′ and G″ indicates the gelation time T_g

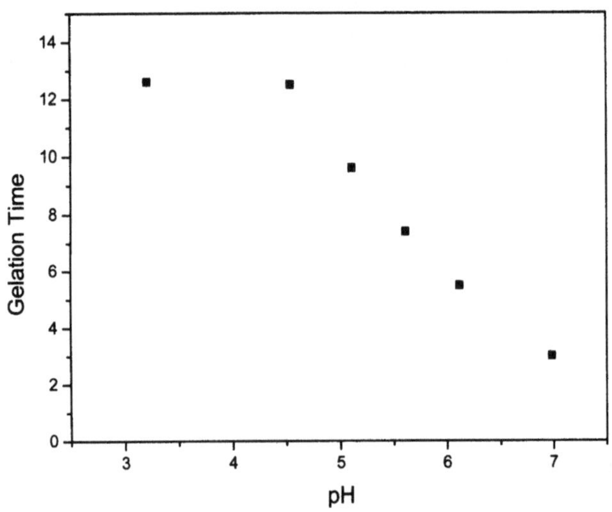

Fig. 3 Behaviour of the gelation time as a function of the pH for aqueous polyacrylic acid based gel

samples at different pH values. As expected for a gel system, at all the pH investigated, both G′ and G″ showed a very small increase with frequency. The measurements were checked to be in the linear viscoelastic range by making amplitude sweep experiments prior to the dynamic frequency sweep experiments. We observed that for all the samples, G′ and G″ have a

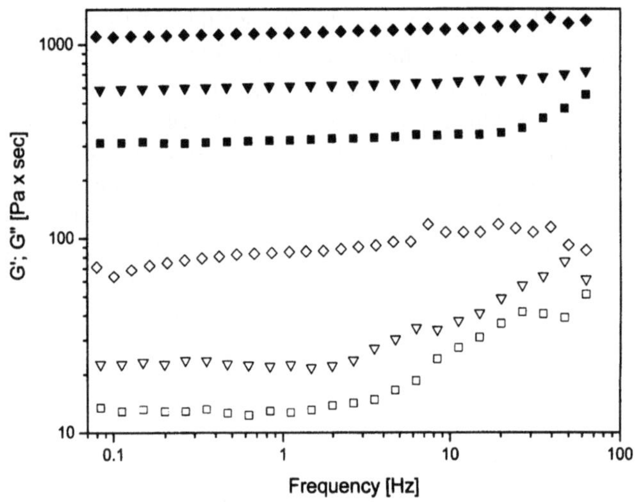

Fig. 4 G′ and G″ parameters as a function of frequency for the studied gels

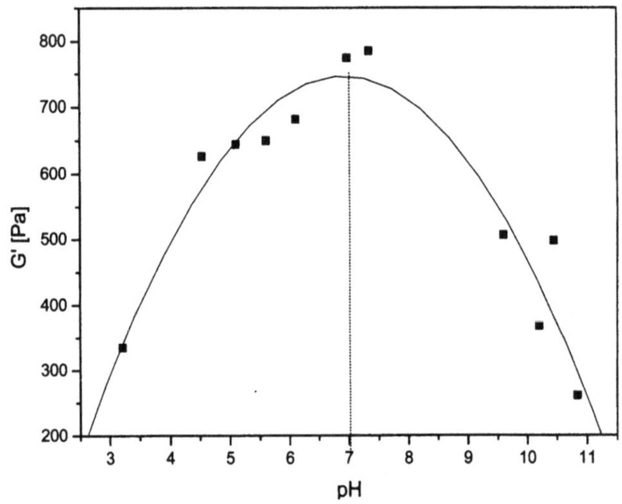

Fig. 5 Behaviour of the G′ parameter as a function of pH

linear behaviour if the displacement is between 0.1 and 10 mrad (frequency test: 1 Hz) so all the frequency sweep spectra were collected with an amplitude of 1 mrad. The spectra obtained at pH 3.2 are typically gel-like for a gelling system where the degree of cross-linking is sufficient to give a continuous network. Both log G′ and log G″ vary linearly with log ω in the frequency range between 0.1 and 1.5, with about the same slope for both moduli G′ and G″ that have very low frequency dependances. It is interesting to note that on increasing the pH to 6.9, the gel-like character becomes more pronounced, with greater separation of G′ and G″ and even less variation with frequency. The elastic modulus, G′, at 1 Hz was plotted as a function of pH to describe the neutralisation process using a rheological approach (Fig. 5) [14]. Though typical of most of polyacrylic acid gel systems, the value of the elastic modulus remained small (less than 1,000 Pa) even in the region of the complete neutralisation of carboxyl groups of the polymer chains. The increasing in the elastic modulus (G′) value confirms the increase of the gel network extension, increasing the pH. The full line in the graph is the polynomial fitting curve where the maximum is the point indicating the complete deprotonation of the acidic groups (see dotted line in Fig. 5). The elastic modulus G′ increased for $3 < pH < 6$. When all the acidic groups have been neutralised, the log G′ value at 1 Hz reaches a maximum at about pH = 6.9. A further increase of pH produced a decrease of G′ and a loss in efficiency of the polyacrylic acid as gelator. This behaviour was attributed to the characteristic of NH_3 to behave as a water structure breaker and is typical of all the polyacrylic acids, neutralised with ammonia and water-soluble amines in water systems. investigated.

The fraction of free water, w_f, and that of freezable bound water, w_{fb}, expressed in % with respect to the total water amount, were estimated by DSC measurements [15], evaluating the shape of the ice melting endothermic peak. We observed that there is an increase of the w_{fb} until the pH value is 6.3. After this point, the freezable bound water fraction remains about constant because of the complete formation of the gel network. The w_f and $w_{fb\ values}$, expressed in g/g dry gel, were estimated by the following equations, respectively [15]:

$$w_f = \Delta H(0°C)/\Delta H_m^{ice}\ (0°C)$$

$$w_{fb} = \Delta H(< 0°C)/\Delta H_m^{ice}\ (0°C)$$

$\Delta H_m{}^{ice}$ (0 °C) is the enthalpy change of the ice melting above 0 °C (that is determining the fraction of the DSC peak area above 0 °C), $\Delta H_m{}^{ice}$ (0 °C) is the reference value for pure ice, and $\Delta H_m{}^{ice}$ (<0 °C) is the enthalpy change of the ice melting below 0 °C determined with the same procedure as for $\Delta H_m{}^{ice}$ (≥0 °C).

Conclusions

The results achieved show that for the aqueous polyacrylic acid based gels the physicochemical properties of the system are strictly related to the degree of neutralisation of the acidic groups of the polymer chains: the higher is the fraction of deprotonated carbonyl groups, the greater is the strength of the gel network. For pH values below 7, the increase of pH produces an increase in the extension of the gel network: the neutralisation of the acidic groups makes possible the formation of oxygen bridges and Van der Waals bonds between different chains. The complete neutralisation of the acidic groups at pH = 6.9. This is indicated by both pH measurements, and the behaviour of G′ versus pH. The dynamic frequency sweep experiments give spectra that are typical of a gelling system where the degree of cross-linking is sufficient to give a continuous network: the response is typically gel-like with a small frequency-dependence in G′ and G″ moduli, increase of the difference between G′ and G″, and a value of the elastic component G′ greater than the loss modulus G″. Adding further ammonia causes a strong alteration of the gel network properties that is probably due to the water structure breaker characteristics of NH_3. From a technological point of view we conclude that the best performance of these aqueous gels as cleaning agents for easel paintings is at pH 7.

Acknowledgments Thanks are due to the Consorzio Interuniversitario per lo Sviluppo dei Sistemi a Grande Interfase, CSGI, for financial support.

References

1. Xue W, Champ S, Huglin M (2001) Polymer 42:3665–3669 (and references therein)
2. Cavasino FP, Hoffmann H, Sbriziolo C, Turco Liveri ML (2001) Colloids Surf A 183–185:689–697
3. Various Authors In: Carbopol high performance polymers for pharmaceuticals. Bulletin 3, BF Goodrich Specialty Chemicals, Brecksville, Ohio, USA
4. Michalski S (1990) In: Cleaning, retouching and coatings. Preprints to the contributions to the Brussels congress, 3–7 September 1990. IIC, London, pp 85–92
5. Kandekar N (2000) Reviews Conserv 1:10–20
6. Srividya B, Cardoza RM, Amin PD (2001) J Control Release 73:205–211
7. Zepeda Martinez G (1997) Uso de geles en la limpieza gradual de peinturas sobre lienzo. Instituto Nacional de Antropologia e Historia, Mexico DF, Mexico, pp 102–107
8. Chen DH, He XR (2001) Mater Res Bull 36:1369–1377
9. Mora P, Mora L, Philippot P (1983) Conservation of wall paintings. Butterworths, London, pp 282–338
10. Djabourov M, Leblond J, Papon P (1988) J Phys (France) 49:333–343
11. Rodd AB, Cooper-White J, Dunstan DE, Boger DV (2001) Polymer 42:185–198
12. Tung C-YM, Dyne PJ (1982) J Appl Polym Sci 27:569
13. Winter HH (1987) Polym Eng Sci 27:1698
14. Various Authors In: Carbopol high performance polymers for pharmaceuticals, Bulletin 11. BF Goodrich Specialty Chemicals, Brecksville, Ohio, USA
15. Shibukawa M, Aoyagi K, Sakamoto R, Oguma K (1999) J Chromatogr A, 832:17–27

Progr Colloid Polym Sci (2004) 123: 284–286

AUTHOR/TITLE INDEX

Progr Colloid Polym Sci (2004) 123 : 287

KEY WORD INDEX

MIX
Papier aus verantwortungsvollen Quellen
Paper from responsible sources
FSC® C105338

If you have any concerns about our products, you can contact us on
ProductSafety@springernature.com

In case Publisher is established outside the EU, the EU authorized representative is:
Springer Nature Customer Service Center GmbH
Europaplatz 3, 69115 Heidelberg, Germany

Printed by Libri Plureos GmbH
in Hamburg, Germany